Cropping Systems Modeling Under Changing Climate

Mukhtar Ahmed • Shakeel Ahmad •
Ghulam Abbas • Sajjad Hussain •
Gerrit Hoogenboom

Cropping Systems Modeling Under Changing Climate

Springer

Mukhtar Ahmed
Department of Agronomy
Pir Mehr Ali Shah Arid Agriculture University
Rawalpindi, Pakistan

Ghulam Abbas
Institute of Agronomy
Bahauddin Zakariya University
Multan, Pakistan

Gerrit Hoogenboom
Global Food Systems Institute &
Department of Agricultural and Biological Engineering
University of Florida
Gainesville, Florida, USA

Shakeel Ahmad
Institute of Agronomy
Bahauddin Zakariya University
Multan, Pakistan

Sajjad Hussain
Department of Horticulture
Bahauddin Zakariya University
Multan, Pakistan

ISBN 978-981-97-0330-2 ISBN 978-981-97-0331-9 (eBook)
https://doi.org/10.1007/978-981-97-0331-9

This Springer imprint is published by the registered company Springer Nature Singapore Pte Ltd.
The registered company address is: 152 Beach Road, #21-01/04 Gateway East, Singapore 189721, Singapore

Preface

Deterioration of land due to intensive agriculture is creating trouble for food security. Hence, appropriate measures should be taken to prevent land from being destroyed. This includes adoption of sustainable cropping systems under changing climate through the aid of process-based models. Land on planet earth is mainly used for agriculture. Extensive usage of land due to modernization in agriculture has resulted in the climate change and threatens biodiversity. Hence, it is necessary to reduce the usage of resource-intensive products and bring sustainability in existing cropping systems. Area under crop cultivation is decreasing day by day mainly due to population pressure as well as because of land degradation and changes in land-use patterns. Furthermore, in future, intensive agricultural practices will be questionable because of diminishing stocks of natural resources (e.g., fossil fuels and nutrients). Similarly, ongoing patterns of environmental changes will seriously hamper agricultural production as the intensity of extreme events across the globe has increased at a rapid pace. Degradation of natural resources, loss of biodiversity, and climate change due to anthropogenic activities are big concerns for future food, fuel, and fiber production. Unsustainable cropping systems in the form of intensive monoculture farming have resulted in the destruction of flora and fauna. Thus, it is necessary to bring sustainability in the agricultural system as agriculture is also a major contributor of greenhouse gases. The option can be regenerative agriculture that is an approach to farming and land management that aims to restore and enhance the health and vitality of ecosystems while also improving agricultural productivity. It is often seen as a response to the environmental and sustainability challenges associated with conventional industrial agriculture. Other terms used for "regenerative" agriculture include sustainable agriculture, green agriculture, alternative agriculture, agroecological farming, biodynamic agriculture, carbon farming, nature-inclusive farming, conservation agriculture, and organic regenerative agriculture. Models can be used to quantify the efficiency of cropping systems as well as to design a sustainable agriculture system. They can also assess agricultural production and environmental risks. Similarly, crop models can help to design adaptation (e.g., agronomic, nature based, technological, and financial) options under future changing climate. This book, ***Cropping Systems Modeling Under Changing Climate***, presents the views of agricultural experts. The 15 chapters—contributed by internationally recognized scientists from Asia and the USA—have been written under the theme of climate change, cropping systems, and modeling. The vast array of subject areas discussed in the book ranges from sustainable agriculture to process-based modeling, from main cropping systems to new proposed cropping systems, from resource-intensive systems to resource conservation system, and from quantification of climate risk to suggestions of adaptation options under changing climates to have sustainable production. As far as possible, the language of the chapters has been kept simple so that educated nonexpert readers may enjoy reading and may benefit from the information provided herein. This book will serve as an educational tool for budding scientists, will provide a comprehensive overview for advanced researchers, and will lay guidelines for important policy decisions.

Rawalpindi, Pakistan Mukhtar Ahmed
Multan, Pakistan Shakeel Ahmad
Multan, Pakistan Ghulam Abbas
Multan, Pakistan Sajjad Hussain
Gainesville, Florida, USA Gerrit Hoogenboom

Contents

About the Authors

Mukhtar Ahmed's research focuses on the application of IoT in agriculture and impact of climate change on crop ecology, crop physiology, cropping system, and rain-fed ecosystem management. He specializes in environmental sciences, precision agriculture, agro-ecosystem modeling, climate change, and climate variability, with extensive teaching and research experience. His work has an impact factor of over 500, with an H-Index of more than 40. He has published 05 books. According to Mendeley Data, his name was listed among top 2% of renowned researchers in the world in 2022. Dr. Ahmed has research experience from Sydney University, Australia, Swedish University of Agricultural Sciences, Sweden, and APCC South Korea where he worked on the application of process-based modeling and remote sensing. Dr. Ahmed was part of the Regional Approaches for Climate Change (REACCH) project in the USA and was part of Washington State University where he developed multi-model ensemble approaches to minimize the uncertainties in modeling. He is a project co-leader in the Model Calibration Group of the Agricultural Model Intercomparison and Improvement Project (AGMIP) focusing on wheat, maize, evapotranspiration, and CO_2.

Sajjad Hussain is working as a Professor of Horticulture in the Faculty of Agricultural Sciences & Technology at Bahauddin Zakariya University, Multan, Punjab, Pakistan. He earned his Ph.D. in 2011 at the University of Corsica Pascale Paoli, France. Apart from teaching various graduate and postgraduate horticultural courses, his primary areas of research are plant physiology under abiotic stresses, management of nutrient uptake of horticultural crops, and plant adaptation under climate uncertainty. The overall aim of his research is to understand the underlining mechanisms for the management of abiotic stresses (salt, drought, temperature, waterlogging, etc.) by combining physiological, eco-physiological, molecular, and biogeochemical approaches. He has been working in the field of horticulture for the last 15 years. He has published over 70 peer-reviewed publications, one edited book, over 40 book chapters, and several conference papers. He has successfully supervised three Ph.D. students and 25 M.Phil. students. He has been involved as principal and coprincipal investigator in various research grants worth around 25 million PKR from different funding agencies. He is also actively working as a reviewer for different impact factor journals.

Shakeel Ahmad is a Professor of Agronomy at the Institute of Agronomy (IOA), Bahauddin Zakariya University Multan, Pakistan. In 2006, he received his Ph.D. from the University of Agriculture, Faisalabad, Pakistan. Later, he completed his postdoctoral research from the University of Georgia, USA. He joined as a Lecturer in the Department of Agronomy, Bahauddin Zakariya University, in October 2002. He was promoted to Professor in 2016. He has devoted himself to teaching and researching in the field of arable crops, especially focused on crop modeling, climate change impact assessment, and adaptation strategies. He has published over 200 articles and 150 book chapters. Furthermore, he has published six books [(1) *Cotton Production and Uses*; (2) *Modern Techniques of Rice Crop Production*; (3) *Citrus Production: Technological Advancements and Adaptation to Climate Change*; (4) *Disaster Risk Reduction in Agriculture*; (5) *Biotechnology and Omics Approaches for Bioenergy Crops*; and (6) *Principles of Field Crop Production*]. His publications received over 7500 citations with an h-index of 50 (according to Scopus). He is an editor and reviewer of many peer-reviewed international journals. He is also an active and life member of professional societies like Pakistan Society of Agronomy and Pakistan Botanical Society. He has attended and presented papers and posters in national and international conferences in different countries.

Ghulam Abbas is working as an Agriculture Officer in the Punjab Agriculture Department. He holds a Ph.D. in Agronomy from Bahauddin Zakariya University. His areas of interest encompass agronomy, climate change and agro-meteorology, crop growth modeling, remote sensing, and geographic information system (GIS). He has expertise in cropping system models "Decision Support System for Agrotechnology Transfer (DSSAT) and The Agricultural Production Systems sIMulator (APSIM)." He has more than 5 years of research experience in different research projects. He has published over 50 articles in national and international peer-reviewed journals and books in his areas of expertise. Additionally, he serves as a reviewer for numerous esteemed national and international journals.

Gerrit Hoogenboom is a Preeminent Scholar in the Global Food Systems Institute and Professor of Agricultural and Biological Engineering at the University of Florida, USA. Prior to his appointment at the University of Florida he was Professor of Biological Systems Engineering and Director of the AgWeatherNet Program at Washington State University, USA, specializing in providing near real-time weather information for decision support by growers and producers. Professor Hoogenboom has over 30 years of experience in the development of crop simulation models and decision support systems. Applications range from freeze forecasting to climate variability and climate change, water resources management, biofuels, economic and environmental sustainability, and food and nutrition security. He currently coordinates the development of the Decision Support System for Agrotechnology Transfer (DSSAT), one of the most widely used crop modeling systems across the world. He frequently organizes and facilitates international training workshops on crop modeling and decision support systems. He has published over 500 scientific papers in refereed journals as well as numerous book chapters and proceedings. In 2022, he was elected as a fellow of the American Association for the Advancement of Science.

Cropping Systems and Application of Models

1

1.1 Cropping Systems

Land on planet earth is mainly used for agriculture. New world map of land-use systems has been provided by Helmholtz Centre for Environmental Research (UFZ) so that appropriate measures can be taken to prevent land from destruction (Fig. 1.1) (Václavík et al. 2013). Different land-use indicators have been used to elaborate each archetype as shown in Fig. 1.2. Extensive usage of land due to modernization in agriculture resulted in the climate change and threatens biodiversity. (Ahmed and Ahmad 2023; Ahmed 2023; Abbas et al. 2023; Liu et al. 2023; Ahmed et al. 2022a; Ahmed et al. 2022b; Nadeem et al. 2022; Ahmed 2020; Khan et al. 2020; Asseng et al. 2019; Ahmed et al. 2017; Jabeen et al. 2017; Aslam et al. 2013; Ahmed et al. 2012). Hence, it is necessary to reduce the usage of resourc e-intensive products and bring sustainability in existing cropping systems.

Cropping system refers to the crops and crop sequences and the management techniques used on particular piece of land over a period of years. Similarly, cropping systems refer to the practices and strategies employed in the cultivation of crops within a specific agricultural system. Furthermore, multiple cropping systems refer to the practice of growing two or more crops on the same piece of land within a single growing season. Multiple cropping, defined as harvesting more than once a year, is a widespread land management strategy in tropical and subtropical agriculture. It is a way of intensifying agricultural production and diversifying the crop mix for economic and environmental benefits (Waha et al. 2020). It involves carefully planning and managing the timing, spacing, and selection of crops to maximize productivity and resource utilization. These systems involve decisions regarding crop selection, planting techniques, intercropping, crop rotation, and management practices. Different cropping systems are designed to optimize productivity, resource-use efficiency, soil health, and sustainability. Here are some commonly practiced cropping systems:

M. Ahmed et al., *Cropping Systems Modeling Under Changing Climate*,
https://doi.org/10.1007/978-981-97-0331-9_1

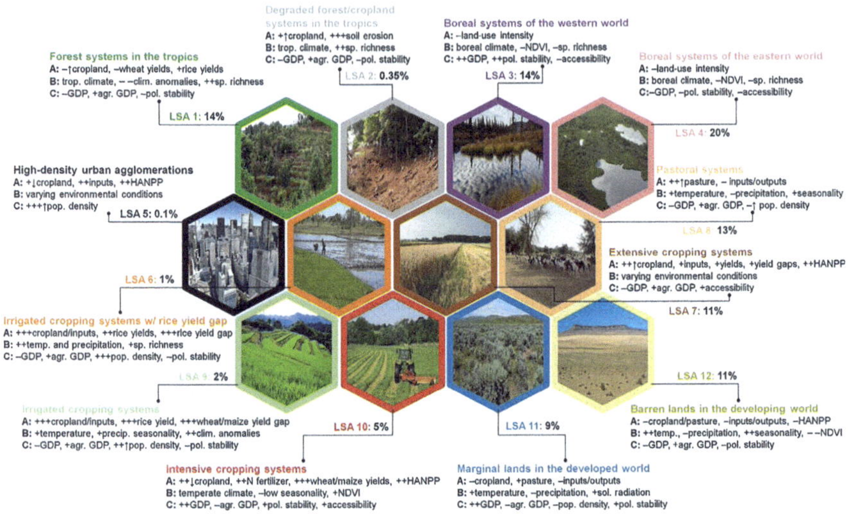

Fig. 1.1 Global map showing land usage. (Source with permission: Václavík et al. 2013; https://www.ufz.de/index.php?en=35349)

1. Monoculture
   Monoculture involves the cultivation of a single crop species on a given piece of land. It is a straightforward system that allows for focused management practices and specialized equipment. However, monoculture can lead to increased pest and disease pressures and nutrient imbalances over time.
2. Double Cropping
   Double cropping involves growing two different crops successively on the same field within a year. After harvesting the first crop, a second crop with a different growth cycle is planted to take advantage of the remaining growing season. This system is common in regions with long growing seasons and sufficient moisture.
3. Crop Rotation
   Crop rotation involves growing different crops in a planned sequence over multiple seasons or years. This system helps to break pest and disease cycles, enhance soil fertility, and reduce weed pressure. Crop rotation can also improve nutrient utilization and reduce the need for chemical inputs.
4. Interplanting
   Interplanting, also known as mixed cropping, involves growing multiple crops together in the same field, either in rows or mixed randomly. The crops are selected based on their compatibility, growth habits, nutrient requirements, and pest interactions. Interplanting can provide benefits such as pest control, efficient use of resources, and increased biodiversity.
5. Strip Cropping
   Strip cropping involves alternating strips of different crops on the same field. It is often practiced on sloping land to reduce soil erosion by breaking up the flow of water. The strips can be planted with different crops or cover crops to provide ground cover and stabilize the soil.

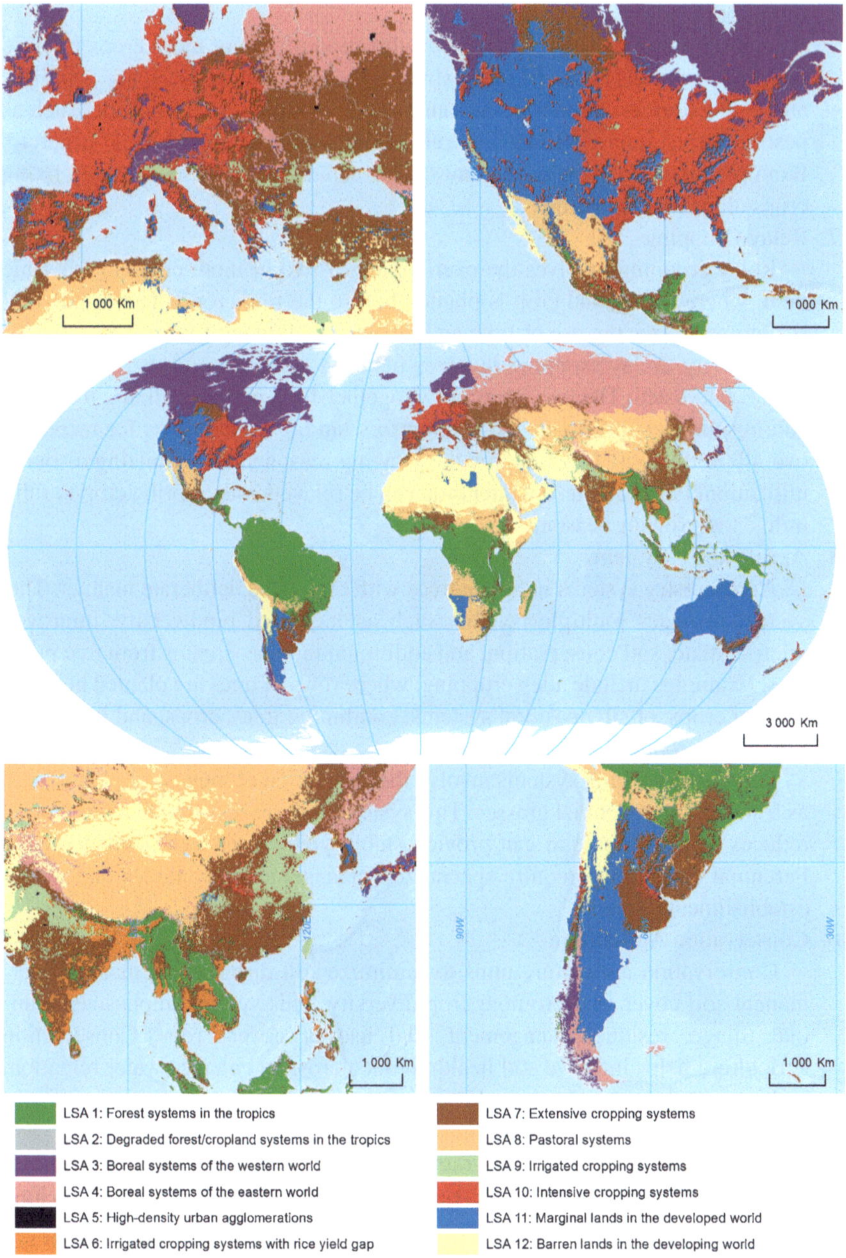

Fig. 1.2 Land system archetype overview showing major land-use intensity indicators (**a**) and environmental (**b**) and socioeconomic factors (**c**) that best characterize each archetype. Here, + and – signs show above and below global average. (Source with permission: Václavík et al. 2013)

6. Intercropping
   Intercropping is the simultaneous cultivation of two or more crops in close proximity on the same field. It maximizes the use of available space, sunlight, and soil resources. Intercropping can provide complementary benefits such as pest control, improved nutrient utilization, and enhanced soil structure. Examples include growing legumes with cereals or planting nitrogen-fixing crops alongside cash crops.
7. Relay Cropping
   Relay cropping involves the overlapping of two or more crops in the same field, where the second crop is planted before the first crop is harvested. This system optimizes the use of time and resources, allowing for increased overall productivity. Relay cropping involves planting a second crop before the first crop is harvested. The two crops grow together for a period, utilizing the available resources simultaneously. It optimizes land use and allows for increased overall productivity by extending the growing season and maximizing resource utilization. For example, a winter crop can be relayed with a spring crop to fully utilize the growing season.
8. Agroforestry Systems
   Agroforestry systems integrate trees with crops in a deliberate manner. This system provides multiple benefits such as increased biodiversity, improved microclimate, soil conservation, and additional income streams from tree products. Examples include alley cropping, where rows of trees are planted between rows of crops, or silvopastoral systems, combining trees, crops, and livestock.
9. Perennial Cropping
   Perennial cropping systems involve the cultivation of long-lived plants, such as fruit trees or perennial grasses. This system requires less frequent replanting, reduces soil erosion, and can provide stable yields over an extended period. Perennial crops often require specialized management techniques and longer establishment times.
10. Conservation Agriculture
   Conservation agriculture aims to minimize soil disturbance, maintain permanent soil cover, and promote crop diversity. This system emphasizes minimal tillage, residue management, and use of cover crops. Conservation agriculture helps improve soil health, reduce erosion, enhance water retention, and increase long-term sustainability.
11. Mixed Farming
   Mixed farming involves integrating crop production with livestock rearing. The crops and livestock are managed together, allowing for nutrient cycling, efficient resource utilization, and increased farm productivity.
12. Regenerative Agriculture
   Regenerative agriculture is an approach to farming and land management that aims to restore and enhance the health and vitality of ecosystems while also improving agricultural productivity. It is often seen as a response to the environmental and sustainability challenges associated with conventional industrial agriculture (Fig. 1.3). Other terms used for "regenerative agriculture" include

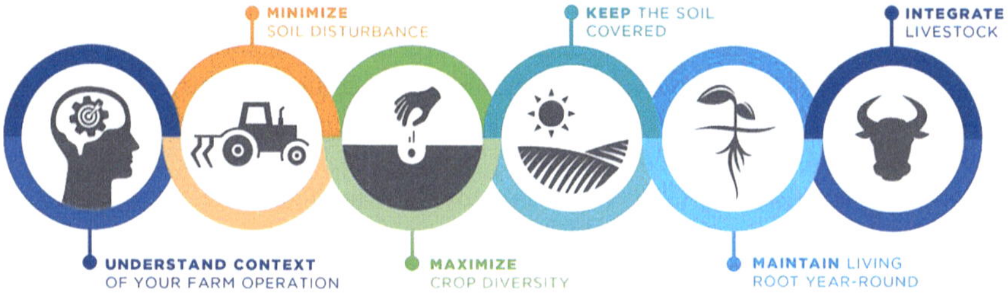

Fig. 1.3 Principles of regenerative agriculture

sustainable agriculture, green agriculture, alternative agriculture, agroecological farming, biodynamic agriculture, carbon farming, nature-inclusive farming, conservation agriculture, and organic regenerative agriculture (Newton et al. 2020).

Here are some key principles and practices associated with regenerative agriculture:

(i) Soil Health

Regenerative agriculture places a strong emphasis on improving and maintaining soil health. Healthy soils are essential for productive agriculture and have numerous benefits for the environment. Practices such as minimal or no-till farming, cover cropping, and crop rotation are used to build soil organic matter, improve soil structure, and increase nutrient availability.

(ii) Biodiversity

Promoting biodiversity is a central component of regenerative agriculture. Diverse ecosystems are more resilient and can provide natural pest control, pollination, and enhanced nutrient cycling. Farmers may plant hedgerows, establish wildlife corridors, or create habitat for beneficial insects to support biodiversity.

(iii) Reduced Chemical Inputs

Regenerative agriculture seeks to minimize the use of synthetic pesticides and fertilizers, which can have negative environmental impacts. Instead, it encourages integrated pest management, where natural predators are used to control pests, and nutrient management practices like composting and organic matter incorporation are employed.

(iv) Agroforestry

Integrating trees and other perennial vegetation into agricultural systems is a common practice in regenerative agriculture. Agroforestry can provide multiple benefits, including carbon sequestration, improved soil health, and additional income streams for farmers through products like fruits and nuts.

(v) Water Management

Sustainable water management is crucial in regenerative agriculture. Practices like water harvesting, contour farming, and use of cover crops can help reduce soil erosion, enhance water retention, and improve water quality.

(vi) Holistic Management

Regenerative farmers often employ holistic management approaches, which involve considering the whole ecosystem, including soil, plants, animals, and people, in decision-making processes. This helps ensure that farming practices are sustainable in the long term.

(vii) Carbon Sequestration

One of the significant benefits of regenerative agriculture is its potential to sequester carbon dioxide from the atmosphere and mitigate climate change. Healthy soils can act as carbon sinks, storing carbon in the form of organic matter.

(viii) Local and Sustainable Food Systems

Regenerative agriculture often supports local and sustainable food systems by encouraging the production of food closer to the point of consumption. This reduces the carbon footprint associated with food transportation and fosters community resilience.

(ix) Adaptive Management

Regenerative farmers are encouraged to adapt their practices based on local conditions and feedback from the land. This flexibility allows for a more responsive and sustainable approach to agriculture.

Overall, regenerative agriculture is driven by the goal of creating farming systems that are not only economically viable but also ecologically and socially responsible. It seeks to promote agricultural practices that regenerate and enhance the health of ecosystems, increase resilience to environmental challenges, and provide a foundation for sustainable food production. However, choice of cropping system depends on factors such as climate, soil conditions, available resources, market demand, and farmer preferences. Integrated approaches that combine multiple cropping systems and sustainable practices are increasingly being adopted to optimize productivity, conserve natural resources, and promote agricultural resilience. These cropping systems offer advantages such as increased productivity, risk reduction, efficient resource use, and improved sustainability. However, successful implementation requires careful crop selection, proper planning, effective pest and nutrient management, and knowledge of the specific ecological requirements of the crops involved.

1.2 Global Cropping Systems

Half of the world's habitable land is used for agriculture. One-third of all land is used for cropping or animal husbandry (Fig. 1.4).

Different types of cropping systems exist across the globe based upon climatic conditions and soil properties. Waha et al. (2020) identified top five cropping

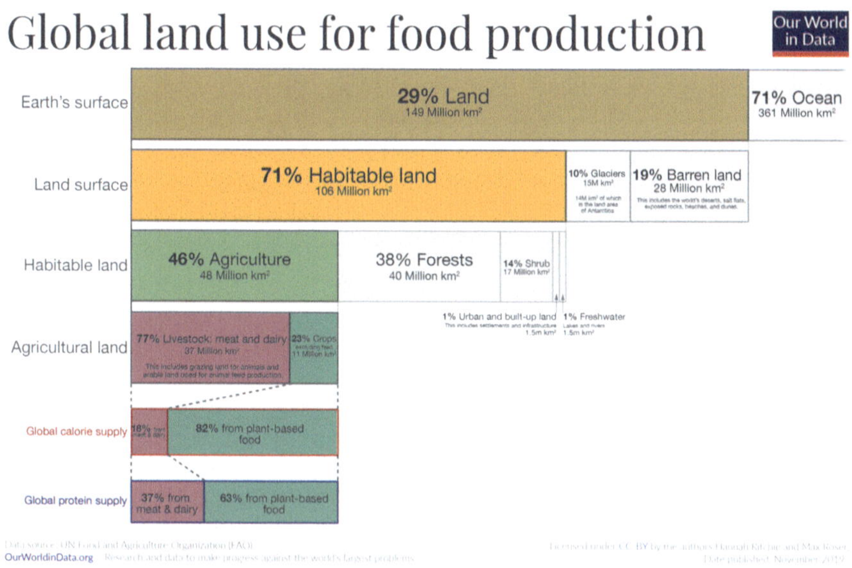

Fig. 1.4 Global land use for food production

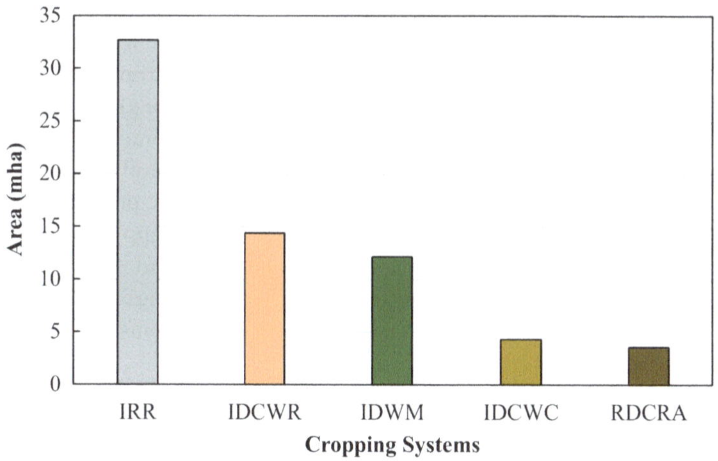

Fig. 1.5 Area of top five cropping systems: (i) irrigated rice-rice (IRR), (ii) irrigated double cropping with wheat and rice (IDCWR), (iii) irrigated double cropping with wheat and maize (IDWM), (iv) irrigated double cropping with wheat and cotton (IDCWC), and (v) rainfed double cropping with rapeseed and another annual crop (RDCRA). (Source: Waha et al. 2020)

systems based on physical area, and it includes (i) irrigated rice-rice (IRR), (ii) irrigated double cropping with wheat and rice (IDCWR), (iii) irrigated double cropping with wheat and maize (IDWM), (iv) irrigated double cropping with wheat and cotton (IDCWC), and (v) rainfed double cropping with rapeseed and another annual crop (RDCRA) (Fig. 1.5). These systems account for greater than 50% of global

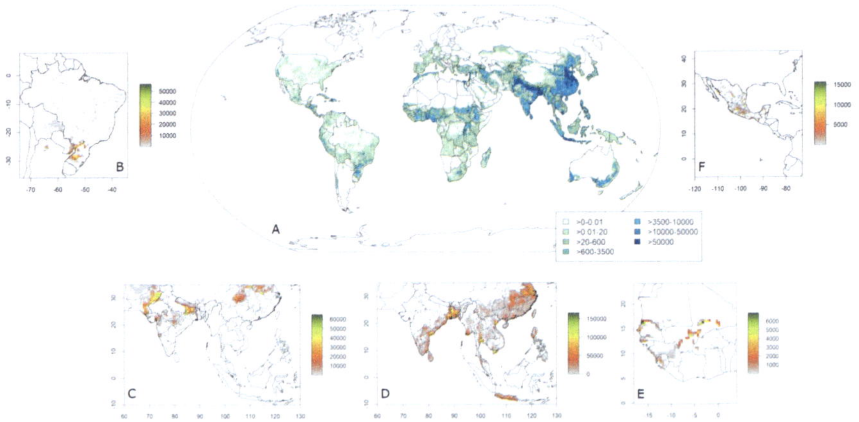

Fig. 1.6 Physical area (hectare) of multiple-cropping systems per 30 arc-min grid cell, 1998–2002. (**a**) Global multiple cropping area. (**b**) Rainfed soybean-wheat double-cropping system in South America. (**c**) Irrigated wheat-rice and rice-rice (**d**) double-cropping system in South, East, and Southeast Asia. (**e**) Irrigated rice-rice double-cropping system in West Africa. (**f**) Irrigated maize-wheat double-cropping system in Central America. White areas indicate locations with total crop area less than or equal to 1% of the grid cell area

cropland when combined with monocultures of wheat, maize, rice, soybean, and pulses, which occupies 468.8 Mha. Furthermore, their estimate reported 134.4 Mha land under multiple cropping, which is 12% of total global crop land as shown in Fig. 1.6. Similarly, 40% of global irrigated crop land and 5% of global rainfed cropland are under multiple cropping. Unsustainable farming practices/systems and urbanization have shown great impacts on natural resources (Hoffmann et al. 2019).

Sustainable agriculture is key to preserving these resources, protecting biodiversity, and producing food feed and bioenergy (Snapp 2017; Snapp and Pound 2017). It is essential to identify new pathways, which can help us to design sustainable agriculture in both temperate and tropical regions under changing climate (Malézieux 2012). A three-step framework for designing cropping systems from nature was proposed by Malézieux (2012) as shown in Fig. 1.7.

The first step is observation of natural ecosystem in the area, and it includes identification of all biodiversity in the target area in consultation with the local farmers. Second step includes experimentation based on the knowledge established in step one. This should answer the following questions:

(i) What are the issues in the existing cropping systems, and what levels of performances/services existing cropping systems fail to deliver?
(ii) What levels of performances/services new cropping systems can achieve?
(iii) What life-forms and species are needed in the new proposed system?

Specific or a combination of practices are required to design new novel cropping systems, and these include rotations, intercropping, mixed cropping, cover crops,

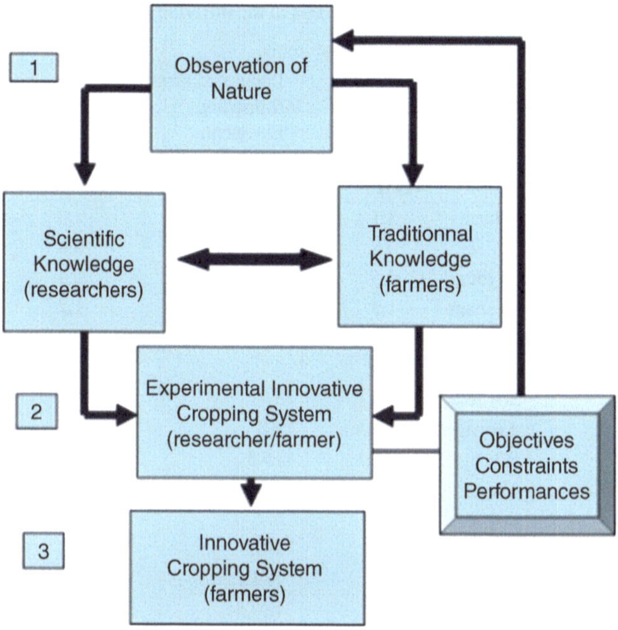

Fig. 1.7 Framework for designing cropping systems from nature. (Source: Malézieux 2012)

service plants, green manuring crops, composting, conservation tillage, perennial cropping, and agroforestry. However, success of all these proposed new cropping systems depends on how well farmers can incorporate or adopt these practices at field scale so that it can satisfy all three indicators, i.e., ecology, economy, and society (Step 3). Furthermore, sustainability in agricultural systems is possible by providing viable solutions to different economic, environmental, and production issues (Fresco 2009; Park and Seaton 1996). Fundamental measures can be reduction in greenhouse gas emissions and improved energy efficiency (Dyer and Desjardins 2003). In general, energy requirement of the agriculture sector is low as compared to other sectors, but to achieve economic sustainability and reduction in greenhouse gas emissions, identification of systems with low energy requirements is needed. Alluvione et al. (2011) estimated the energy flows of wheat-maize-soybean-maize rotation under three different cropping systems, i.e., low input integrated farming (LIIF), integrated farming following European regulations (IFFER), and conventional farming (CF). Results showed that minimum tillage with balance N fertilization can reduce energy inputs by 11% and 65%, respectively. They further highlighted large differences among crops in energy efficiency, i.e., soybean 4.1 MJ kg^{-1} grain, maize 2.2 MJ kg^{-1} grain, and wheat 2.6 MJ kg^{-1} grain, and suggested that crop management in rotation is equally important in determining the energy efficiency of a cropping system. Different energy indicators were used to check the efficiency of cropping systems as shown in Table 1.1. It was observed that

Table 1.1 Energy indicators of the different crops in the differing cropping systems

Energy indicators	Crops	Cropping systems			
		Low input integrated farming (LIIF)	Integrated farming following European regulations (IFFER)	Conventional farming (CF)	Average
Indicator of immediate removal (IIR)	Wheat	3.1	2.7	2.4	2.7
	Maize	0.8	0.9	0.8	0.8
	Soybean	0.5	0.6	0.4	0.5
	Average	**1.5**	**1.4**	**1.2**	**1.4**
Energy intensity (EI) (MJ kg^{-1} grain)	Wheat	2.1	2.6	3.3	2.7
	Maize	1.9	2	2.6	2.2
	Soybean	3.4	3.7	5.3	4.1
	Average	**2.5**	**2.8**	**3.7**	**3.0**
Energy-use efficiency (EUE)	Wheat	18.8	15.5	12.3	15.5
	Maize	10.2	9.8	7.5	9.2
	Soybean	7.3	6.7	4.7	6.2
	Average	**12.1**	**10.7**	**8.2**	**10.3**
Net energy (NE) (GJ ha^{-1} grain)	Wheat	188.3	184.1	174.4	182.3
	Maize	181.4	189.5	189.1	186.7
	Soybean	60.7	64.2	51.4	58.8
	Average	**143.5**	**145.9**	**138.3**	**142.6**
Environmental efficiency of support energy (EESE)		11.7	7	4.3	7.7
Net environmental energy (NEE)		173.8	102.8	74.2	116.9

energy-use efficiency of LIIF and IFFER was increased by 32.7% and 31.4%, respectively, as compared to CF.

Rice-wheat cropping system (RWCS) is feeding a large population of Indo-Gangetic Plains (IGPs) of South Asia. However, sustainability of RWCS has been a big concern since past few decades due to open-field burning of rice residues. This burning leads to environmental pollution due to emissions of greenhouse gases as shown in Fig. 1.8. This also deteriorates soil health and increases C footprints. In Asia, mostly rice is harvested mechanically, which leaves large amounts of straw (~600–800 million tons of rice straw) and stubble in field. Global production of rice straw is around 800–1000 million tons. In the north-western part of India, ~500 million tons of crop residues are produced annually. However, in Pakistan, ~16 million tons of paddy straw is produced annually, of which 60% is burnt. It has been reported that 1.0 Mg dry mass straw releases 1515 kg CO_2, 92 kg CO, 2.7 kg CH_4, and 0.07 kg N_2O. Additionally, 1.0 Mg of rice straw burning leads to loss of 5.5 kg N, 2.3 kg P, 25 kg K, and 1.2 kg S from the soil, thus deteriorating soil health (Singh et al. 2023; Pathak et al. 2011; Andreae 2019). Thus, management of rice straw is an utmost important task to have sustainable cropping system.

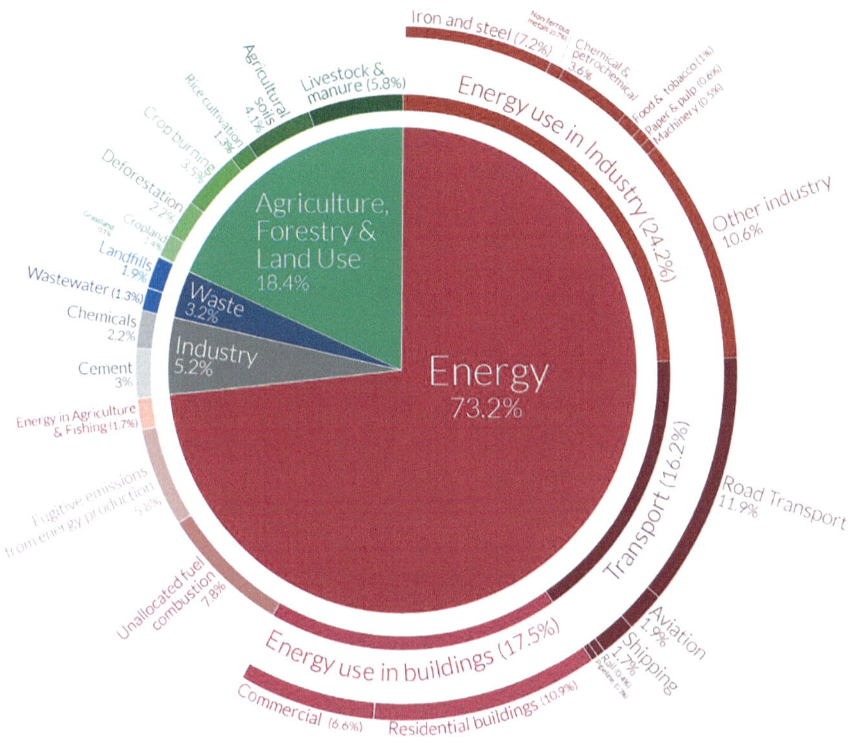

Fig. 1.8 Global greenhouse gas emissions by sector. (Source: Our world in data)

Different ways have been suggested to utilize rice straw and minimize greenhouse gas emissions. These include (i) rice straw mushroom production, (ii) rice straw silage for cattle feed, (iii) mechanized composting of rice straw, (iv) rice straw for improved soil fertility, (v) alternative source of energy and bioethanol production (Swain et al. 2019; Samaddar et al. 2017), (vi) pulping and paper making (Nayeem et al. 2023), (vii) source of silicon (Nayeem et al. 2023; Ma and Takahashi 2002), and (viii) animal feed (Khir and Pan 2019).

1.3 Cropping System Modeling

Models can be used to quantify the efficiency of cropping systems. They can also assess agricultural production and environmental risks. Similarly, crop models can help to design adaptation (e.g., agronomic, nature based, technological, and financial) options under future changing climate. Different process-based cropping system models have been developed to suggest sustainable cropping system. These include Agricultural Production Systems Simulator (APSIM), AquaCrop, CropSyst, Daisy, Decision Support System for Agrotechnology Transfer (DSSAT), DeNitrification-DeComposition (DNDC), EPIC, FarmSim, Farm ASSEssment Tool

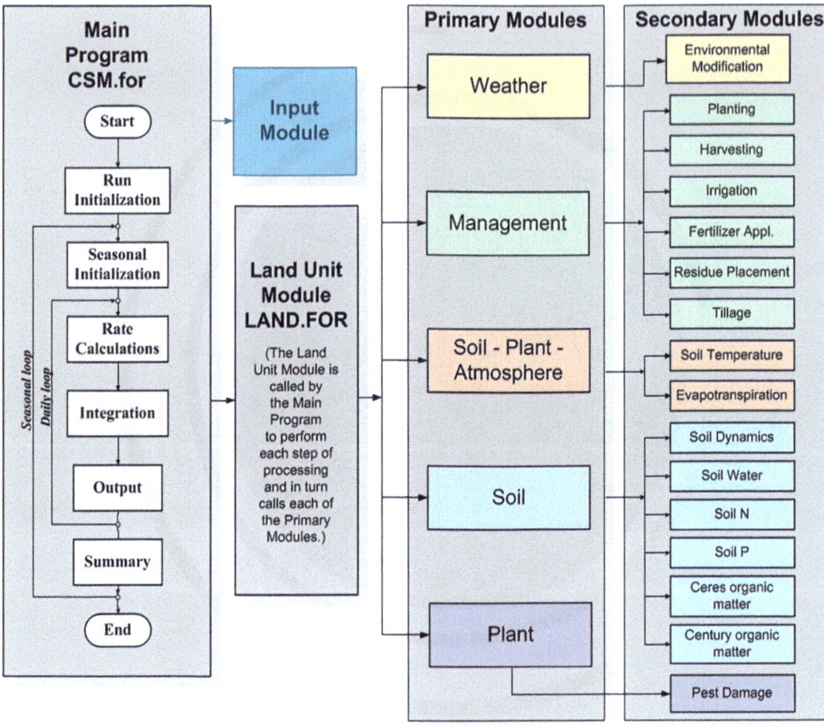

Fig. 1.9 Schematic view of DSSAT cropping system model. (Source: DSSAT.net)

(FASSET), HERMES, STICS, SUCROS, SWAP, and WOFOST. DSSAT is a well-known cropping system model, and it contains more than 40 crop models derived from CERES-Wheat, CERES-Maize, PNUTGRO, and SOYGRO. Schematic diagram of DSSAT illustrates connections between the primary and secondary modules (Fig. 1.9).

Gao et al. (2022) used DSSAT to simulate the impacts of crop rotation on crop evapotranspiration, percolation, water-use efficiency, and yield. Results showed that the model simulated groundwater with good accuracy. Their results suggested that DSSAT is a very useful tool for selecting suitable cropping systems based on the water use for local farmers. Liu et al. (2017) simulated wheat yield and soil organic carbon under a wheat-maize cropping system using DSSAT. They suggested that if the model is calibrated accurately, then it can be a useful tool for assessing and predicting different parameters of cropping system. Irrigation management in the cropping system research is very important as it can help to minimize water losses. DSSAT can help to choose best irrigation management practices under different climates as concluded in earlier work (Malik and Dechmi 2019; Montoya et al. 2020; Amouzou et al. 2019; Mehrabi and Sepaskhah 2020; Shelia et al. 2019; Araya et al. 2017; Attia et al. 2016; Galmarini et al. 2024; Wahab et al. 2024; Ahmed et al. 2019; Liu et al. 2019; Asseng et al. 2019; Ahmed et al. 2017, 2016, 2014a, b and

Ahmed and Hassan 2011; Jiang et al. 2016; He et al. 2013). DSSAT is a very useful tool to improve different agronomic management operations, e.g., irrigation timing under water-limited conditions (Attia et al. 2016; Araya et al. 2017), impact of climate change and planting date (Abbas et al. 2023; Zhang et al. 2023), N fertilizer management (Malik and Dechmi 2020; Amouzou et al. 2019), and productivity of forage-based cropping system (Baath et al. 2021). Incorporation of legume forage in the dryland cropping system could help to provide nutritious forage for livestock as it will help to minimize N application, protect soil erosion, and improve precipitation-use efficiency. Baath et al. (2021) conducted a study using DSSAT to evaluate the impact of forage soybean of different maturity groups on winter wheat and double-cropping systems in comparisons to the fallow-wheat system. The model was calibrated and validated using field data of crop yield and evapotranspiration. Results showed that mid-maturity group soybean gives higher yield and water-use efficiency as compared to late-maturing group. Figure 1.10 shows the simulation performance of DSSAT to simulate biomass and evapotranspiration (ET). However, double-cropping forage soybean and winter wheat resulted in the reduction in winter wheat yield and higher seasonal ET, but it can be compensated due to economic competitiveness and other ecological benefits of double-cropped forage soybean-wheat systems.

He et al. (2021) used three process-based models, i.e., DNDC, DayCent, and DSSAT, to simulate soil carbon sequestration under diverse cropping systems in the semiarid prairies of western Canada. Higher soil organic carbon (SOC) was simulated for the cropping systems where there was higher incorporation of residues or fixation of N as compared to fallow-wheat systems. Better SOC was estimated by DNDC, while DSSAT predicted yield with good accuracy. Furthermore, they suggested that diverse cropping systems, e.g., canola and legume, have higher potential to store SOC as compared to traditional cropping systems. DNDC has been used by different researchers to simulate greenhouse gas emissions and SOC under different agroecosystems (Waldrip et al. 2013; Li et al. 2012; Giltrap et al. 2010; Wang et al. 2022). Based on DNDC's ability to simulate N_2O emissions in response to different agronomic managements, DNDC can be recommended as a valuable tool for designing mitigation strategies.

APSIM is also a well-known widely used cropping system model developed by Agricultural Production Systems Research Unit (APSRU), CSIRO, and state of Queensland Government agencies. Vogeler et al. (2023) applied APSIM model to simulate crop rotation to check N leaching, N uptake, and crop yield and concluded that the model showed good results regarding crop rotation. Their results suggested that APSIM is the best tool to work with crop rotation to reduce the amount of N leaching by reducing fertilization rates and to increase the yield and uptake of N for the crops. He et al. (2023) used pre-validated APSIM to assess the combined influences of cowpea cover crops and three residue retention levels on soil water balance, SOC, N dynamics, crop yield, and gross margin across six crop rotation systems during the historical period (1985–2020), near future (2021–2056), and far future (2057–2092) in southeast Australia. Their results showed that the use of cover crops resulted in higher SOC and yield, reduced N loss, and better uptake of

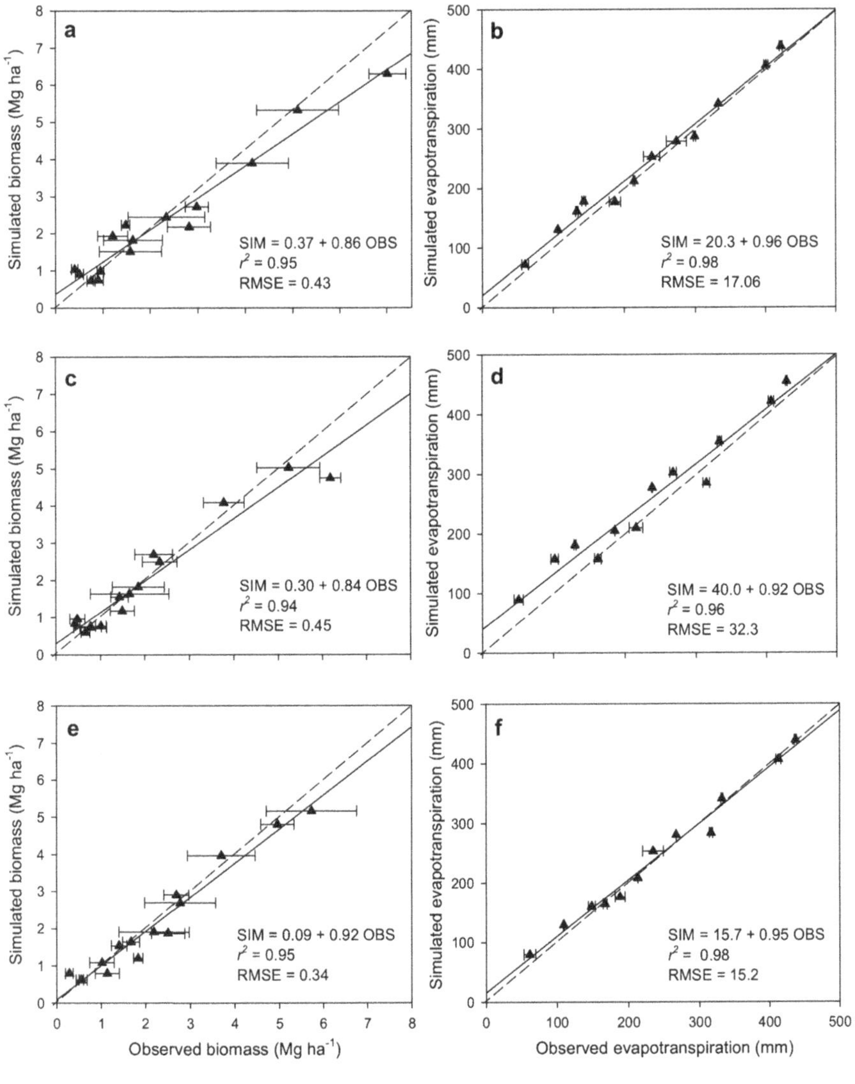

Fig. 1.10 Simulated versus observed biomass and evapotranspiration values for three forage soybean cultivars: (**a, b**) Donegal MGV, (**c, d**) Derry MGVI, and (**e, f**) Tyrone MGVII, using data from field studies conducted at the USDA-ARS Grazinglands Research Laboratory near El Reno, Oklahoma, USA. (Source: Baath et al. 2021)

N in cereals as compared to legume crops. Thus, they concluded that sustainability in crop production with environmental co-benefit is possible by adopting cover crops in the dryland cropping system. Pasley et al. (2023) developed mung bean APSIM next-generation model by using data from 28 diverse fields. They concluded that APSIM is a robust model as it successfully captured the dynamics of crop response to sowing dates, water/irrigation regimes, and climate. Hence, APSIM is a useful tool to evaluate crop sowing dates, total water requirement, and total fertilizer rate of the crop and improve breeding strategies and climate. It is also a useful tool for the farmer to help them to examine options for improving management required for the crop and assess all the production risk across the growing regions. Bana et al. (2023) used APSIM to analyze 37 years (1984–2022) of diverse conservation agriculture (CA) scenarios on productivity, sustainability, and carbon footprints in the rice-wheat cropping system (RWCS). The study highlighted that APSIM was able to capture the impact of CA on SOC, carbon sequestration, and water productivity in RWCS. Yang et al. (2020) used APSIM to evaluate the impact of perennial legumes on the economic profitability, hydrological balance, and agronomic productivity of cropping system of Loess Plateau of China using different climate change scenarios. Five different cropping systems ((i) continuous maize (M), (ii) continuous winter wheat (W), (iii) continuous lucerne (L), (iv) maize-wheat-soybean rotation (MWS), and (v) lucerne (4 years)-winter wheat (2 years) rotation (LW)) were investigated under five series of temperature and precipitation change scenarios. The results showed that LW system has the greatest potential for producing acceptable yield and economic profit under future temperature and precipitation scenarios for this local environment. Similarly, these different process-based models can be used to suggest on-farm different adaptation options as elaborated by Farrell et al. (2023). These include agronomic, nature based, and technological adaptation as shown in Table 1.2.

Pathak et al. (2011) have developed InfoRCT (Information on Use of Resource-Conserving Technologies) that can establish input-output relationships in RWCS. It can simulate GHG emissions and system productivity in response to different crop management practices.

1.4 Conclusion

Land use and climate change are closely interconnected and have significant impacts on each other. Land use refers to how land is utilized, developed, and managed for various purposes, including agriculture, urbanization, forestry, and conservation. Cropping systems: Unsustainable cropping systems are agricultural practices that are detrimental to the long-term health of the environment, the productivity of the land, and often the economic well-being of farmers. These systems may prioritize short-term gains but result in negative consequences over time. Addressing unsustainable cropping systems typically involves adopting more sustainable and environmentally friendly agricultural practices. Sustainable agriculture promotes practices like crop rotation, reduced chemical input use, agroforestry, integrated pest management, and conservation tillage. These approaches aim to protect the environment, maintain or improve soil health, conserve biodiversity, and ensure long-term food security while also considering the economic viability of farming operations. Transitioning to sustainable cropping systems is essential to meet the challenges of feeding a growing global population while protecting natural resources

Table 1.2 On-farm different adaptation options and their inclusion in yield models

Categories	Adaptation options	On-farm examples	Inclusion in yield models (%)
Agronomic (common crop management practices)	Fertilizer management	Improved management of fertilizer in response to decrease or increase in yield potential, e.g., precision fertilizer application	35
	Water management	Water management in response to drought and heat, e.g., deficit irrigation, precision irrigation, and rainwater harvesting	30
	Shift in planting date	Adjustment in sowing date to have better planting and harvesting time and provision of better weather conditions during critical growth stages of crop	20
	Change in cropping systems	Sustainable cropping systems	25
	Cultivar shift/crop ideotypes	Cultivars with different maturity groups to avoid stress and capitalize resources effectively	18
	Crop shift/crop diversification	Shift to different crop species and crop insurance	1
	Shift in planting area	Change in planting location	2
	Changing planting density	Adjusting the planting density to better coincide with future prevailing weather conditions	<1
	Insect pest management	Adjusting the management of crop to minimize the risks of diseases, pest, and pathogen	
	Other	Shade management, windbreaks, etc.	
	Weed management	Adjusting the management of crop to minimize the weed damage	<1
Nature based (agroecological practices usually implemented to sustain or restore agroecosystems)	Conservation agriculture	Soil health management, e.g., minimum tillage, crop residue use, mulching	<1
	Crop diversification	Incorporation of mixed species, crop rotation, and intercropping	<1
	On-farm biodiversity	Provision of ecosystem services	<1
	Agroforestry	Fruit tree and planting of shade tress	<1
Technological (off-farm innovations aiming to enhance farm management)	Technological innovations	Use of biotechnology, plant genetic improvement, and novel agronomy	<1
	Climate services	Weather forecasting and monitoring, reporting, and verification system	<1
	Decision support	Development of apps and decision support tools	<1
	Novel cropping systems	Use of novel systems, e.g., vertical farming and protection agriculture	<1

(continued)

Table 1.2 (continued)

Categories	Adaptation options	On-farm examples	Inclusion in yield models (%)
Financial (farm management decisions to improve financial stability)	Agricultural insurance	Application of insurance to manage risks under changing climate	<1
	Livelihood diversification	Labor adjustments or resource allocation	<1
	Exit farming/ migration	Adjustments to individuals shifting away from agriculture	<1

and mitigating climate change. Crop modeling plays a significant role in the development and implementation of sustainable cropping systems. Sustainable cropping systems aim to maximize agricultural productivity while minimizing negative environmental impacts and ensuring long-term viability. Crop models, which are mathematical representations of crop growth and development, can help farmers, researchers, and policymakers make informed decisions about sustainable agriculture practices. Crop models can simulate the growth of crops under different environmental conditions, including variations in climate, soil, and management practices. By using these models, farmers can optimize resource use, such as water, nutrients, and pesticides, to maximize crop yields while minimizing waste and environmental impact. Similarly, with the increasing uncertainty and variability in climate patterns due to climate change, crop models help farmers adapt to changing conditions. They can provide insights into which crop varieties and planting dates are best suited to specific climate scenarios, helping farmers make decisions that increase resilience. Crop models can assist in selecting the most suitable crops for a given region or climate. They can predict how different crop species and varieties will perform under specific conditions, helping farmers choose crops that are well suited to their local environment. Crop models can simulate the effects of crop rotation and diversification on soil health, pest and disease management, and overall system sustainability. Farmers can use these models to plan crop rotations that enhance soil fertility and reduce the risk of pests and diseases. Efficient water management is crucial for sustainable cropping systems, especially in regions with limited water resources. Crop models can help optimize irrigation scheduling and assess the water requirements of different crops, contributing to water conservation efforts. Crop models assist in nutrient management by predicting crop nutrient uptake and nutrient requirements based on soil conditions. This helps farmers apply fertilizers more efficiently, reducing the risk of nutrient runoff and pollution. Crop models can simulate pest and disease dynamics, helping farmers make informed decisions about pest control measures. Integrated pest management strategies can be developed based on model predictions, reducing the need for chemical pesticides. Crop models can provide yield predictions based on current conditions, allowing farmers to estimate their expected harvest and plan accordingly. This information is valuable for marketing, storage, and distribution. Researchers can use crop models to explore new agricultural practices and technologies in a virtual environment. This accelerates the development and testing of sustainable farming practices and crop varieties. Policymakers and agricultural extension services can use crop models to design and promote sustainable farming practices at regional or national levels. This can help align agricultural policies with sustainability goals. Finally, crop modeling is a valuable tool for designing, implementing, and optimizing sustainable cropping systems. By providing insights into crop growth, environmental interactions, and management strategies, these models contribute to more efficient, resilient, and environmentally friendly agricultural practices, which are essential for addressing global challenges such as food security and climate change mitigation.

References

Abbas G, Ahmed M, Fatima Z, Hussain S, Kheir AMS, Ercişli S, Ahmad S (2023) Modeling the potential impact of climate change on maize-maize cropping system in semi-arid environment and designing of adaptation options. Agric For Meteorol 341:109674. https://doi.org/10.1016/j.agrformet.2023.109674

Ahmed M (2020) Introduction to modern climate change. Andrew E. Dessler: Cambridge University Press, 2011, 252 pp, ISBN-10: 0521173159. Sci Total Environ 734:139397. https://doi.org/10.1016/j.scitotenv.2020.139397

Ahmed M (2023) The science of climate change. In: Brinkmann R (ed) The palgrave handbook of global sustainability. Springer International Publishing, Cham, pp 195–222. https://doi.org/10.1007/978-3-031-01949-4_22

Ahmed M, Ahmad S (eds) (2023) Disaster risk reduction in agriculture. Springer Nature. https://doi.org/10.1007/978-981-99-1763-1. ISBN 978-981-99-1762-4, Published: 03 November 2023

Ahmed M, Hassan FU (2011) APSIM and DSSAT models as decision support tools. In: 19th International Congress on Modelling and Simulation, Perth, Australia, pp 1174–1180. http://mssanz.org.au/modsim2011

Ahmed M, Hassan FU, Aslam M, Aslam MA (2012) Physiological attributes based resilience of wheat to climate change. Int J Agric Biol 14(3):407–412

Ahmed M, Aqeel M, Hassan F-U, Asif M, Hayat R (2014a) Use of APSIM to model nitrogen use efficiency of rain-fed wheat. Int J Agric Biol 16(3):461–470

Ahmed M, Fayyaz Ul H, Van Ogtrop FF (2014b) Can models help to forecast rainwater dynamics for rainfed ecosystem? Weather Climate Extremes 5-6:48–55. https://doi.org/10.1016/j.wace.2014.07.001

Ahmed M, Akram MN, Asim M, Aslam M, Hassan F-u, Higgins S, Stöckle CO, Hoogenboom G (2016) Calibration and validation of APSIM-Wheat and CERES-Wheat for spring wheat under rainfed conditions: models evaluation and application. Comput Electron Agric 123:384–401. https://doi.org/10.1016/j.compag.2016.03.015

Ahmed M, Stöckle CO, Nelson R, Higgins S (2017) Assessment of climate change and atmospheric CO_2 impact on winter wheat in the Pacific Northwest using a multimodel ensemble. Front Ecol Evol 5. https://doi.org/10.3389/fevo.2017.00051

Ahmed M, Stöckle CO, Nelson R, Higgins S, Ahmad S, Raza MA (2019) Novel multimodel ensemble approach to evaluate the sole effect of elevated CO_2 on winter wheat productivity. Sci Rep 9(1):7813. https://doi.org/10.1038/s41598-019-44251-x

Ahmed M, Hayat R, Ahmad M, ul-Hassan M, Kheir AMS, ul-Hassan F, ur-Rehman MH, Shaheen FA, Raza MA, Ahmad S (2022a) Impact of climate change on dryland agricultural systems: a review of current status, potentials, and further work need. Int J Plant Prod. https://doi.org/10.1007/s42106-022-00197-1

Ahmed M, Ahmad S, Kheir AMS (2022b) Climate change: an overview. In: Ahmed M (ed) Global agricultural production - resilience to climate change. Springer Nature Switzerland AG. https://doi.org/10.1007/978-3-031-14973-3_1

Alluvione F, Moretti B, Sacco D, Grignani C (2011) EUE (energy use efficiency) of cropping systems for a sustainable agriculture. Energy 36(7):4468–4481. https://doi.org/10.1016/j.energy.2011.03.075

Amouzou KA, Lamers JPA, Naab JB, Borgemeister C, Vlek PLG, Becker M (2019) Climate change impact on water- and nitrogen-use efficiencies and yields of maize and sorghum in the northern Benin dry savanna, West Africa. Field Crop Res 235:104–117. https://doi.org/10.1016/j.fcr.2019.02.021

Andreae MO (2019) Emission of trace gases and aerosols from biomass burning—an updated assessment. Atmos Chem Phys 19(13):8523–8546. https://doi.org/10.5194/acp-19-8523-2019

Araya A, Kisekka I, Gowda PH, Prasad PVV (2017) Evaluation of water-limited cropping systems in a semi-arid climate using DSSAT-CSM. Agric Syst 150:86–98. https://doi.org/10.1016/j.agsy.2016.10.007

Aslam Z, Ahmed M, Sajad M, Asif M, Akmal M, Awan FK, Ijaz W, Ashraf R, Khattak JZK (2013) A comparison of statistical and dynamic modeling of wheat (*Triticum aestivum* L.) fungal diseases under the climate change. J Food Agric Environ 11(2):1327–1332

Asseng S, Martre P, Maiorano A, Rötter RP, O'Leary GJ, Fitzgerald GJ, Girousse C, Motzo R, Giunta F, Babar MA, Reynolds MP, Kheir AMS, Thorburn PJ, Waha K, Ruane AC, Aggarwal PK, Ahmed M, Balkovič J, Basso B, Biernath C, Bindi M, Cammarano D, Challinor AJ, De Sanctis G, Dumont B, Eyshi Rezaei E, Fereres E, Ferrise R, Garcia-Vila M, Gayler S, Gao Y, Horan H, Hoogenboom G, Izaurralde RC, Jabloun M, Jones CD, Kassie BT, Kersebaum K-C, Klein C, Koehler A-K, Liu B, Minoli S, Montesino San Martin M, Müller C, Naresh Kumar S, Nendel C, Olesen JE, Palosuo T, Porter JR, Priesack E, Ripoche D, Semenov MA, Stöckle C, Stratonovitch P, Streck T, Supit I, Tao F, Van der Velde M, Wallach D, Wang E, Webber H, Wolf J, Xiao L, Zhang Z, Zhao Z, Zhu Y, Ewert F (2019) Climate change impact and adaptation for wheat protein. Glob Chang Biol 25(1):155–173. https://doi.org/10.1111/gcb.14481

Attia A, Rajan N, Xue Q, Nair S, Ibrahim A, Hays D (2016) Application of DSSAT-CERES-Wheat model to simulate winter wheat response to irrigation management in the Texas High Plains. Agric Water Manag 165:50–60. https://doi.org/10.1016/j.agwat.2015.11.002

Baath GS, Northup BK, Rao SC, Kakani VG (2021) Productivity and water use in intensified forage soybean-wheat cropping systems of the US southern Great Plains. Field Crop Res 265:108086. https://doi.org/10.1016/j.fcr.2021.108086

Bana RS, Bamboriya SD, Kumar V, Godara S, Gaydon DS, Laing AM, Dhakar R, Shivay YS, Meena VS, Singh D (2023) Identifying optimum residue levels for stable crop and water productivity and carbon sequestration under a conservation agriculture based rice-wheat system. Soil Tillage Res 232:105745. https://doi.org/10.1016/j.still.2023.105745

Dyer J, Desjardins R (2003) Simulated farm fieldwork, energy consumption and related greenhouse gas emissions in Canada. Biosyst Eng 85(4):503–513

Farrell AD, Deryng D, Neufeldt H (2023) Modelling adaptation and transformative adaptation in cropping systems: recent advances and future directions. Curr Opin Environ Sustain 61:101265. https://doi.org/10.1016/j.cosust.2023.101265

Fresco LO (2009) Challenges for food system adaptation today and tomorrow. Environ Sci Pol 12(4):378–385

Galmarini S, Solazzo E, Ferrise R, Srivastava AK, Ahmed M, Asseng S, Cannon AJ, Dentener F, De Sanctis G, Gaiser T, Gao Y, Gayler S, Gutierrez JM, Hoogenboom G, Iturbide M, Jury M, Lange S, Loukos H, Maraun D, Moriondo M, McGinnis S, Nendel C, Padovan G, Riccio A, Ripoche D, Stockle CO, Supit I, Thao S, Trombi G, Vrac M, Weber TKD, Zhao C (2024) Assessing the impact on crop modelling of multi- and uni-variate climate model bias adjustments. Agric Syst 215:103846. https://doi.org/10.1016/j.agsy.2023.103846

Gao F, Luan X, Yin Y, Sun S, Li Y, Mo F, Wang J (2022) Exploring long-term impacts of different crop rotation systems on sustainable use of groundwater resources using DSSAT model. J Clean Prod 336:130377. https://doi.org/10.1016/j.jclepro.2022.130377

Giltrap DL, Li C, Saggar S (2010) DNDC: a process-based model of greenhouse gas fluxes from agricultural soils. Agric Ecosyst Environ 136(3):292–300. https://doi.org/10.1016/j.agee.2009.06.014

He J, Cai H, Bai J (2013) Irrigation scheduling based on CERES-Wheat model for spring wheat production in the Minqin Oasis in Northwest China. Agric Water Manag 128:19–31. https://doi.org/10.1016/j.agwat.2013.06.010

He W, Grant BB, Jing Q, Lemke R, St. Luce M, Jiang R, Qian B, Campbell CA, VanderZaag A, Zou G, Smith WN (2021) Measuring and modeling soil carbon sequestration under diverse cropping systems in the semiarid prairies of western Canada. J Clean Prod 328:129614. https://doi.org/10.1016/j.jclepro.2021.129614

He Q, Liu DL, Wang B, Cowie A, Simmons A, Waters C, Li L, Feng P, Li Y, Pd V, Huete A, Yu Q (2023) Modelling interactions between cowpea cover crops and residue retention in Australian dryland cropping systems under climate change. Agric Ecosyst Environ 353:108536. https://doi.org/10.1016/j.agee.2023.108536

Hoffmann H, Schomers S, Meyer C, Sander K, Hickey V, Feuerbacher A (2019) Agriculture and ecosystem services. In: Ferranti P, Berry EM, Anderson JR (eds) Encyclopedia of food security and sustainability. Elsevier, Oxford, pp 9–13. https://doi.org/10.1016/B978-0-08-100596-5.22202-6

Jabeen M, Gabriel HF, Ahmed M, Mahboob MA, Iqbal J (2017) Studying impact of climate change on wheat yield by using DSSAT and GIS: a case study of Pothwar Region. In: Ahmed M, Stockle OC (eds) Quantification of climate variability, adaptation and mitigation for agricultural sustainability. Springer International Publishing, Cham, pp 387–411

Jiang Y, Zhang L, Zhang B, He C, Jin X, Bai X (2016) Modeling irrigation management for water conservation by DSSAT-maize model in arid northwestern China. Agric Water Manag 177:37–45. https://doi.org/10.1016/j.agwat.2016.06.014

Khan MA, Tahir A, Khurshid N, Husnain MI, Ahmed M, Boughanmi H (2020) Economic effects of climate change-induced loss of agricultural production by 2050: a case study of Pakistan. Sustain For 12(3):1216. https://doi.org/10.3390/su12031216

Khir R, Pan Z (2019) Rice. In: Pan Z, Zhang R, Zicari S (eds) Integrated processing technologies for food and agricultural by-products. Academic Press, London, pp 21–58. https://doi.org/10.1016/B978-0-12-814138-0.00002-2

Li C, Salas W, Zhang R, Krauter C, Rotz A, Mitloehner F (2012) Manure-DNDC: a biogeochemical process model for quantifying greenhouse gas and ammonia emissions from livestock manure systems. Nutr Cycl Agroecosyst 93(2):163–200. https://doi.org/10.1007/s10705-012-9507-z

Liu H-L, Liu H-B, Lei Q-L, Zhai L-M, Wang H-Y, Zhang J-Z, Zhu Y-P, Liu S-P, Li S-J, Zhang J-S, Liu X-X (2017) Using the DSSAT model to simulate wheat yield and soil organic carbon under a wheat-maize cropping system in the North China Plain. J Integr Agric 16(10):2300–2307. https://doi.org/10.1016/S2095-3119(17)61678-2

Liu B, Martre P, Ewert F, Porter JR, Challinor AJ, Müller C, Ruane AC, Waha K, Thorburn PJ, Aggarwal PK, Ahmed M, Balkovič J, Basso B, Biernath C, Bindi M, Cammarano D, De Sanctis G, Dumont B, Espadafor M, Eyshi Rezaei E, Ferrise R, Garcia-Vila M, Gayler S, Gao Y, Horan H, Hoogenboom G, Izaurralde RC, Jones CD, Kassie BT, Kersebaum KC, Klein C, Koehler A-K, Maiorano A, Minoli S, Montesino San Martin M, Naresh Kumar S, Nendel C, O'Leary GJ, Palosuo T, Priesack E, Ripoche D, Rötter RP, Semenov MA, Stöckle C, Streck T, Supit I, Tao F, Van der Velde M, Wallach D, Wang E, Webber H, Wolf J, Xiao L, Zhang Z, Zhao Z, Zhu Y, Asseng S (2019) Global wheat production with 1.5 and 2.0°C above pre-industrial warming. Glob Chang Biol 25(4):1428–1444. https://doi.org/10.1111/gcb.14542

Liu B, Martre P, Ewert F, Webber H, Waha K, Thorburn PJ, Ruane AC, Aggarwal PK, Ahmed M, Balkovič J, Basso B, Biernath C, Bindi M, Cammarano D, Cao W, Challinor AJ, De Sanctis G, Dumont B, Espadafor M, Eyshi Rezaei E, Fereres E, Ferrise R, Garcia-Vila M, Gayler S, Gao Y, Horan H, Hoogenboom G, Izaurralde RC, Jabloun M, Jones CD, Kassie BT, Kersebaum KC, Klein C, Koehler A-K, Maiorano A, Minoli S, Montesino San Martin M, Müller C, Naresh Kumar S, Nendel C, O'Leary GJ, Eivind Olesen J, Palosuo T, Porter JR, Priesack E, Ripoche D, Rötter RP, Semenov MA, Stöckle C, Stratonovitch P, Streck T, Supit I, Tao F, Van der Velde

M, Wang E, Wolf J, Xiao L, Zhang Z, Zhao Z, Zhu Y, Asseng S (2023) AgMIP-Wheat multi-model simulations on climate change impact and adaptation for global wheat. Open Data J Agric Res 9:10–25. https://doi.org/10.18174/odjar.v9i0.18092
Ma JF, Takahashi E (2002) Chapter 8 - Summary and prospect of silicon research. In: Ma JF, Takahashi E (eds) Soil, fertilizer, and plant silicon research in Japan. Elsevier Science, Amsterdam, pp 181–190. https://doi.org/10.1016/B978-044451166-9/50008-7
Malézieux E (2012) Designing cropping systems from nature. Agron Sustain Dev 32(1):15–29. https://doi.org/10.1007/s13593-011-0027-z
Malik W, Dechmi F (2019) DSSAT modelling for best irrigation management practices assessment under Mediterranean conditions. Agric Water Manag 216:27–43. https://doi.org/10.1016/j.agwat.2019.01.017
Malik W, Dechmi F (2020) Modelling agricultural nitrogen losses to enhance the environmental sustainability under Mediterranean conditions. Agric Water Manag 230:105966. https://doi.org/10.1016/j.agwat.2019.105966
Mehrabi F, Sepaskhah AR (2020) Winter wheat yield and DSSAT model evaluation in a diverse semi-arid climate and agronomic practices. Int J Plant Prod 14:221–243. https://doi.org/10.1007/s42106-019-00080-6
Montoya F, Camargo D, Córcoles JI, Domínguez A, Ortega JF (2020) Analysis of deficit irrigation strategies by using SUBSTOR-Potato model in a semi-arid area. J Agric Sci 157:1–14. https://doi.org/10.1017/S002185961900090X
Nadeem M, Nazer Khan M, Abbas G, Fatima Z, Iqbal P, Ahmed M, Ali Raza M, Rehman A, Ul Haq E, Hayat A, Ali M, Ahmad S (2022) Application of CSM-CANEGRO model for climate change impact assessment and adaptation for sugarcane in semi-arid environment of Southern Punjab, Pakistan. Int J Plant Prod. https://doi.org/10.1007/s42106-022-00192-6
Nayeem J, Rahman MM, Jahan MS, Popy RS (2023) Chapter 14 - Pulping and papermaking of rice straw. In: Ainun ZMA, Sapuan SM, Ilyas RA (eds) Pulping and papermaking of nonwood plant fibers. Academic Press, London, pp 245–265. https://doi.org/10.1016/B978-0-323-91625-7.00001-1
Newton P, Civita N, Frankel-Goldwater L, Bartel K, Johns C (2020) What is regenerative agriculture? A review of scholar and practitioner definitions based on processes and outcomes. Front Sustain Food Syst 4:577723. https://doi.org/10.3389/fsufs.2020.577723
Park J, Seaton R (1996) Integrative research and sustainable agriculture. Agric Syst 50(1):81–100
Pasley H, Wenham K, Bell L, Huth N, Holzworth D, Chaki A, Gaydon D, Collins M (2023) APSIM next generation mungbean model: a tool for advancing mungbean production. Field Crop Res 298:108955. https://doi.org/10.1016/j.fcr.2023.108955
Pathak H, Saharawat YS, Gathala M, Ladha JK (2011) Impact of resource-conserving technologies on productivity and greenhouse gas emissions in the rice-wheat system. Greenh Gases Sci Technol 1(3):261–277. https://doi.org/10.1002/ghg.27
Samaddar A, Mohibbe Azam M, Singaravadivel K, Venkatachalapathy N, Swain BB, Mishra P (2017) Chapter 11 - Postharvest management and value addition of rice and its by-products. In: Mohanty S, Chengappa PG, Mruthyunjaya et al (eds) The future rice strategy for India. Academic Press, London, pp 301–334. https://doi.org/10.1016/B978-0-12-805374-4.00011-7
Shelia V, Hansen J, Sharda V, Porter C, Aggarwal P, Wilkerson CJ, Hoogenboom G (2019) A multi-scale and multi-model gridded framework for forecasting crop production, risk analysis, and climate change impact studies. Environ Model Softw 115:144–154. https://doi.org/10.1016/j.envsoft.2019.02.006
Singh P, Singh G, Gupta A, Sodhi GPS (2023) Data envelopment analysis based energy optimization for improving energy efficiency in wheat established following rice residue management in rice-wheat cropping system. Energy 284:128615. https://doi.org/10.1016/j.energy.2023.128615
Snapp S (2017) Chapter 5 - Designing for the long-term: sustainable agriculture. In: Snapp S, Pound B (eds) Agricultural systems, 2nd edn. Academic Press, San Diego, pp 123–167. https://doi.org/10.1016/B978-0-12-802070-8.00005-0

Snapp S, Pound B (2017) Chapter 4 - Farming systems for sustainable intensification. In: Snapp S, Pound B (eds) Agricultural systems, 2nd edn. Academic Press, San Diego, pp 93–122. https://doi.org/10.1016/B978-0-12-802070-8.00004-9

Swain MR, Singh A, Sharma AK, Tuli DK (2019) Chapter 11 - Bioethanol production from rice-and wheat straw: an overview. In: Ray RC, Ramachandran S (eds) Bioethanol production from food crops. Academic Press, London, pp 213–231. https://doi.org/10.1016/B978-0-12-813766-6.00011-4

Václavík T, Lautenbach S, Kuemmerle T, Seppelt R (2013) Mapping global land system archetypes. Glob Environ Chang 23(6):1637–1647. https://doi.org/10.1016/j.gloenvcha.2013.09.004

Vogeler I, Hansen EM, Thomsen IK (2023) The effect of catch crops in spring barley on nitrate leaching and their fertilizer replacement value. Agric Ecosyst Environ 343:108282. https://doi.org/10.1016/j.agee.2022.108282

Waha K, Dietrich JP, Portmann FT, Siebert S, Thornton PK, Bondeau A, Herrero M (2020) Multiple cropping systems of the world and the potential for increasing cropping intensity. Glob Environ Chang 64:102131. https://doi.org/10.1016/j.gloenvcha.2020.102131

Wahab AA, Hussain M, Fatima Z, Ahmed M, Ahmad S (2024) Evaluating the interactive impact of nitrogen levels and cultivars on yield traits, nitrogen use efficiency, water use efficiency and benefit cost ratio of ratoon rice in Punjab, Pakistan. Int J Plant Prod. https://doi.org/10.1007/s42106-023-00280-1

Waldrip HM, Todd RW, Li C, Cole NA, Salas WH (2013) Estimation of ammonia emissions from beef cattle feedyards using the process-based model manure-DNDC. Trans ASABE 56(3):1103–1114. https://doi.org/10.13031/trans.56.9906

Wang C, Zhao J, Gao Z, Feng Y, Laraib I, Chen F, Chu Q (2022) Exploring wheat-based management strategies to balance agricultural production and environmental sustainability in a wheat–maize cropping system using the DNDC model. J Environ Manag 307:114445. https://doi.org/10.1016/j.jenvman.2022.114445

Yang X, Li Z, Cui S, Cao Q, Deng J, Lai X, Shen Y (2020) Cropping system productivity and evapotranspiration in the semiarid Loess Plateau of China under future temperature and precipitation changes: an APSIM-based analysis of rotational vs. continuous systems. Agric Water Manage 229:105959. https://doi.org/10.1016/j.agwat.2019.105959

Zhang Z, Li Y, Chen X, Wang Y, Niu B, Liu DL, He J, Pulatov B, Hassan I, Meng Q (2023) Impact of climate change and planting date shifts on growth and yields of double cropping rice in southeastern China in future. Agric Syst 205:103581. https://doi.org/10.1016/j.agsy.2022.103581

Sequential Modeling

2

2.1 Introduction

The world's rapidly growing population is the concern of the scientists in the modern era. It is predicted that by 2050, there will be nine billion people on the planet. Food insecurity and malnutrition are still major issues in numerous countries, despite great success in reducing hunger over the past few decades (Ahmad et al. 2023). Unfortunately, the existing levels of food production cannot satisfy the needs of the current population. Around 11% of the people worldwide are chronically malnourished, which means that they lack access to adequate food (FAO 2024). It will be difficult to achieve global food security by 2050; unless we take urgent action to stop severe ripple effects, the agriculture sector will be impacted by climate change and soil erosion, further limiting the efficient use of the world's agricultural resources (Ahmed 2020).

At the same time as the world's population is increasing, the land area is decreasing due to increased urbanization. The development of methods to produce sufficient food utilizing less land while preserving the environment and economic prosperity of rural areas is one of the major challenges facing society in the twenty-first century (Ahmad et al. 2015, 2023). Conventional methods of crop cultivation over the last 50 years have depended largely on manufactured fertilizers and pesticides in order to enhance yields, but they have also deteriorated water quality and posed risks to the health of both people and animals. As a result, there is an enhanced focus on the creation of crop production systems with increased resource usage efficiency and less harmful environmental effects (Bahri et al. 2019).

Modern agriculture, which now accounts for a substantial portion of the world's food supply, is the result of the enormous advancements that have been made in breeding, agronomy, and chemistry over the past few decades. The high productivity of agro-systems relies on a small number of enhanced, high-yielding species, as well as on extensive use of fossil fuels and chemical inputs. Most farmed systems nowadays are made up of just one genetically and biologically enhanced

M. Ahmed et al., *Cropping Systems Modeling Under Changing Climate*,
https://doi.org/10.1007/978-981-97-0331-9_2

single-species crop. Through the advancement of the green revolution over the past 30 years, this crop diversification, along with the use of chemical inputs, such as pesticides and fertilizers, has led to tremendous production gains worldwide (Abbas et al. 2020a, b). On the other hand, cropping systems or multispecies cropping systems are frequently seen as a real-world implementation of ecological concepts based on biodiversity, plant interactions, and other natural regulatory mechanisms. They have potential benefits in terms of profitability, productivity, output stability, tolerance to disruption, and sustainability (Vandermeer 1989).

Cropping systems that are employed year after year lead to complications such as soil compaction, soil degradation, and resultantly reduced productivity (Tariq et al. 2018; Hamza and Anderson 2005). Various cropping systems, such as rotation, intercropping, and crop diversification, as well as associated agronomic techniques employed in agriculture, have an impact on the health and quality of the soil from a variety of geographical and temporal perspectives (Vukicevich et al. 2016).

Cropping systems were first created to maximize yield from agro-systems, but modern agriculture is now more concerned with cropping systems' long-term environmental viability (Fargione et al. 2018). The maintenance of soil health aims to guarantee the cropping systems' long-term stability, high production, and environmental sustainability (Hatfield et al. 2017). Both from an environmental and human points of view, soil is a very complex ecosystem and a very valuable resource for the cropping systems. Due to its numerous vital activities in the cropping system, soil is, without a doubt, one of our most strategic and important resources, for its role in organic matter decomposition, essential nutrient recycling, carbon sequestration, organic contaminant detoxification, supply of quality water, and availability of fuel, fiber, and food (Yang et al. 2020). The best way to evaluate how soil management affects soil sustainability under various cropping systems is to look its productivity over time (Karlen et al. 2001). Targeted crop diversification in a cropping system has a number of advantages for growers as well as the soil environment, as it can increase the viability of chemical nutrients, physical structure of the soil, and functioning of soil microorganisms at various levels, ultimately resulting in a higher crop yield (Maron et al. 2011; Bardgett and van der Putten 2014), and help to reduce the availability of some species of soilborne pathogens (Zhu et al. 2000). Optimized diversified farming produces varied microhabitats in soil that preserve the range, structure, and behavioral complementarity of the beneficial soil microbial population in the cropping system (Pivato et al. 2007).

Crop rotation is a time-tested and effective method for controlling agricultural ecosystems and biodiversity because it improves soil structure, controls outbreaks of diseases and pests, and increases crop yield (Barbieri et al. 2019). In a cropping system, rotation enhances the crop efficiency and value that depends on the crop rotation type, the duration of the rotation period, and the agronomic and soil management history of farmland (Li et al. 2019; Bennett et al. 2012; Tiemann et al. 2015). As a result of enhanced soil nitrogen availability from biological fixation following a legume crop, the rotation of crops with legume grain can increase the performance of wheat as the subsequent crop (Gan et al. 2003) in the cropping system. Short-period rotations have decreased yield when compared to those with a longer duration because they are more susceptible to host-specific diseases (Bennett et al. 2012). By decreasing artificial chemical pollution, preventing soil disease, boosting plant root function, improving soil nutrient and space usage efficiency, and enhancing the behavior of soil microorganisms, intercropping practices can improve soil and crop productivity in the cropping system (Hinsinger et al. 2011; Lemaire et al. 2014; Vukicevich et al. 2016; Bukovsky-Reyes et al. 2019; Sun et al. 2019). There is discussion to move from mono-cropping to multiple cropping, in which various systems are adopted, like sequential cropping, intercropping, mixed intercropping, and relay cropping.

2.2 Types of Multiple-Cropping Systems

Multiple-cropping system is a technique to produce two or more crops on a same piece of land in 1 year to increase the cropping intensity, enhance yield and income, and maintain soil fertility. There are different ways to achieve multiple-cropping systems like sequential, intercropping, relay, and mixed cropping.

2.2.1 Sequential Cropping

It is a form of multiple cropping in which two or more crops are grown on the same piece of land as a rotation in a sequence for 1 or more years, depending on the duration of the single crop in the rotation. After harvesting the first crop, the second crop is planted (Fig. 2.1a).

2.2.2 Intercropping

It is a practice of multiple cropping in which two or more crops are grown simultaneously in a row pattern on the same piece of land (Fig. 2.1b). It increases the farmer's profit and improves the ecosystem. However, it is more complicated to manage, especially the final harvest.

2.2.3 Relay Cropping

It is a technique in which a second crop is sown when a first crop reaches its reproductive phase and is still standing in the field (Fig. 2.1c). It enhances the efficacy of land use and lowers the cost of cultivation.

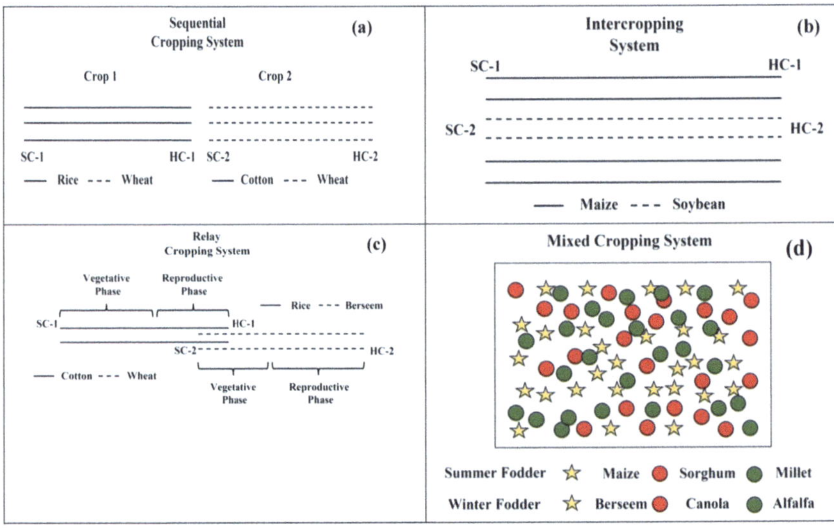

Fig. 2.1 Pictorial view showing sequential cropping system (*SC-1* sowing crop 1, *HC-1* harvesting crop-1, *SC-2* sowing crop-2, *HC-2* harvesting crop-2) (**a**); pictorial view showing intercropping system (*SC-1* sowing crop 1, *HC-1* harvesting crop-1, *SC-2* sowing crop-2, *HC-2* harvesting crop-2) (**b**); pictorial view showing relay cropping system (*SC-1* sowing crop 1, *HC-1* harvesting crop-1, *SC-2* sowing crop-2, *HC-2* harvesting crop-2) (**c**); pictorial view showing mixed cropping system (**d**)

2.2.4 Mixed Cropping

It is a type of farming in which different crops are sown simultaneously on the same piece of land (Fig. 2.1d). In case of mixed cropping, the seeds of different crops are mixed and sown together without having row pattern or row-row or plant-plant distance. It enhances yield, lowers the risk of pest attack, and minimizes crop failure risks, for example, sowing of winter fodders (berseem, alfalfa, oat, canola, toria, etc.) and summer fodders (maize, sorghum, millet, janter, etc.).

2.3 Cropping Systems in the World

A cropping system states the sequence and type of crops grown and practices used for raising them. It includes every cropping sequence that has been used over time and space depending on the crop production technologies that are currently available. Cropping systems have always been designed to increase crop yield. Cropping systems in agriculture, such as crop rotation, intercropping, and crop diversification, and all other agronomic related practices have an impact on soil health and quality from a variety of temporal and spatial perspectives (Vukicevich et al. 2016). Initially, the purpose of designing cropping systems was to increase the production from agro-systems. However, modern agriculture is becoming more concerned with the overall sustainability of the cropping system (Fargione et al. 2018).

Various types of cropping systems are used all over the world, depending upon different factors including available resources, soil type, climate, and cultural practices (Table 2.1).

Table 2.1 Important cropping systems across the world

Cropping system	Country	Factor	References
Rice-wheat	Pakistan	Tillage practices	Shahzad et al. (2017)
	India	Tillage practices + crop residue	Saurabh et al. (2021)
	China	Nitrogen + cropping system	Xing et al. (2002)
	Pakistan	System phenology	Ahmad et al. (2019)
	Pakistan	Climate change	Ahmad et al. (2015)
	Bangladesh	Potassium rate + crop residue	Miah et al. (2008)
Rice-rice	Bangladesh	NPKS + cow dung + rice straw + mung bean residue + poultry manure + dhaincha	Ali et al. (2009)
	Sri Lanka	Cropping system research	Ratnayake et al. (2017)
	Thailand	Cropping system	Polthanee (1997)
Potato-potato	Pakistan	Phenology	Naz et al. (2022)
Wheat-maize	China	Organic and inorganic nutrients	Jiang et al. (2006)
	The Netherlands	Intercropping	Gou et al. (2016)
	Pakistan	N from organic and inorganic sources	Sarwar et al. (2021)
Rice-peanut	Thailand	Cropping system	Polthanee (1997)
Sesame-rice	Thailand	Cropping system	Polthanee (1997)
Early rice-late rice-wheat	China	Nitrogen + cropping system	Xing et al. (2002)
Cotton-wheat	Pakistan	Tillage practices	Shahzad et al. (2017)
	Turkey	Seed rate + cultivars	Kiliç and Gürsoy (2010)
	Uzbekistan	Cropping system	Boboev et al. (2019)
	Pakistan	Climate change	Mubeen et al. (2020)
	Australia	Crop rotation + tillage system	Hulugalle et al. (2005)
	Pakistan	Climate change and adaptation	Ahmad et al.
Sorghum-wheat	Pakistan	Tillage practices	Shahzad et al. (2017)
	India	NPK + Zn + S + FYM	Katkar et al. (2011)
Mung bean-wheat	Pakistan	Tillage practices	Shahzad et al. (2017)

(continued)

Table 2.1 (continued)

Cropping system	Country	Factor	References
Sunflower-groundnut	India	N and P rates and frequencies with and without crop residue	Aulakh and Garg (2007)
Maize-maize	Pakistan	Sowing dates + cultivars; nitrogen rates + hybrids	Abbas et al. (2020a, b)
	China	Double-cropping silage	Meiyun et al. (2010)
Maize-maize-mustard	India	Tillage + mulching	Yadav et al. (2021)
Sunflower-sunflower	India	NPK + FYM	Suresh et al. (2015)
	Pakistan	Climate change + genetics + managements	Tariq et al. (2018)
Soybean-sunflower	India	NPK + FYM	Suresh et al. (2015)
Soybean-soybean	USA	Soybean cultivars resistant to aphids	Morriss et al. (2017)
	Pakistan	Cultivar + sowing dates	Ejaz et al. (2022)
Sugarcane system	Brazil	Erosion	Thomaz et al. (2022)
	Pakistan	Climate change	Nadeem et al. (2022)
	Mauritius	C3-type vegetation and cropping pattern	Umrit et al. (2014)
	Pakistan	Irrigation	Watto and Mugera (2015)
Alfalfa system	Western Australia	Plant density	Dolling et al. (2005)
	New Zealand	Sowing dates	Moot et al. (2015)
	Jammu and Kashmir	Three different genetic cultivars	Iqbal et al. (2018)

2.4 Cereal-Based Cropping Systems

2.4.1 Rice-Wheat Cropping System

The rice-wheat cropping system (RWCS) is an important cropping system across the world because it provides the world's population with basic foods (Banjara et al. 2021; Ahmad et al. 2019). The RWCS is the most technologically advanced system compared to other cropping systems, and it has undergone significant area under cultivation. 13.5 million hectares (mha) are used for RWCS in Asia, with 57% of the total area in South Asia (Wahab et al. 2024; Ahmad et al. 2015). The Indo-Gangetic Plain (IGP) (Banjara et al. 2021) is home to more than 85% of the RWCS used across South Asia. The most significant cropping system in southern and eastern Asia is rice-wheat rotations, which are grown in Pakistan, Bangladesh, China, Myanmar, Nepal, and India over an estimated 21.9 mha (Mann and Garrity 1994). Better agronomic management techniques and heat-tolerant varieties of wheat have allowed the system to be successfully expanded to tropical nations of Southeast Asia, such as the Philippines, Indonesia, and Thailand (Kohli et al. 1991). Since rice is the main crop in the tropics and subtropics, the temperate regions of Australia, China, and Japan produce the highest yields of rice (Yadav et al. 2000). Contrarily, wheat has traditionally been the main crop farmed in temperate regions, despite the fact that it is currently only moderately successful in subtropical and tropical climates (Saunders and Hettel 1994).

2.4.2 Rice-Rice Cropping System

Double-rice cropping is a farming method in which the same plot of land produces two crops of rice in the same calendar year (Chen et al. 2020). This method is mostly used in regions with warm, humid climates, sufficient water for irrigation, and fertile soil to support two crops of rice. Many Southeast Asian countries, including Vietnam (Watanabe et al. 2017), the Philippines (Bouman et al. 2002), Cambodia (Phaloeun et al. 2003), Indonesia (Asnawi et al. 2021), and Thailand (Maneepitak et al. 2019), often practice double-rice cultivation. It is also practiced in some regions of India (Gogoi et al. 2021), China (Chen et al. 2020), Myanmar (Naing et al. 2008), and Bangladesh (Emran et al. 2022). In China, late rice is transplanted at the end of July and harvested at the end of October or the beginning of November under the double-rice cropping method. The early rice is direct-seeded or transplanted at the end of April and harvested at the end of July. Some farmers produce dryland crops (rapeseed, vegetables, green manure, etc.) or leave the land fallow over the winter and early spring following the harvest of late-season rice (Chen et al. 2020).

2.4.3 Wheat-Maize Cropping System

The wheat-maize cropping system is an example of a cropping system used worldwide to optimize food production, including countries with growing populations such as China, India, and Pakistan. In this cropping system, wheat and maize crops are grown in the same field during different seasons in a single year. Wheat-maize system provides high yield and optimizes resource use. In this system, wheat and maize crops are being grown in a sequence.

2.4.4 Rice-Maize Cropping System

The rice-maize cropping system is a major cropping system which is practiced in various places across the world. In Asia alone, the area under the cultivation of rice-maize cropping system is about 3.5 mha (Timisina et al. 2010). Rice-maize system is used in China (Zeng et al. 2021), India (Ghosh et al. 2020), Indonesia (Agus et al. 2019), Vietnam (Le 2014), Thailand (Amnuaylojaroen et al. 2021), the Philippines (Haefele et al. 2013), and Brazil (Heinemann et al. 2008). It is primarily used in lowlands in Asia and in highlands in Latin America. In this cropping system, rice and maize crops are grown in the same field during different seasons in a single year. Because it can provide a high yield and optimize resource use, the system is well-liked. Due to maize's role as a rice pest trap crop, it also offers the benefit of minimizing pest issues. In this system, rice and maize crops are being grown in a sequence.

2.5 Legume-Based Cropping Systems

Due to their extensive array of advantages, legume-based cropping systems are becoming more and more popular on a global scale. Improved soil health, enhanced output potential, and a decreased dependency on outside inputs like fertilizers and pesticides are just a few of these advantages. Furthermore, legume-based cropping systems are environmentally friendly since they support biodiversity and lower greenhouse gas emissions (Azeem et al. 2019). A legume crop's special capability to fix atmospheric nitrogen through symbiotic connections with soil microbes increases the availability of this crucial nutrient for plant growth resulting in a higher crop yield and better soil health. In addition, legume crops use a less water than other conventional crops such as cereals (Gurjar and Swami 2019).

Legume-based cropping systems are practiced in different regions all over the world as a crucial source of income, food, and nutrition. These systems have been extensively adopted in South Asia (India, Bangladesh, and Pakistan) (Shrestha and Neupane 2016), sub-Saharan Africa (Ethiopia and Uganda) (Lal 2017), North America (Canada) (Niu et al. 2018), and Australia (Siddique and Sykes 1997). Legume-based cropping systems are also used for both feed and food purpose in Europe (Anglade et al. 2015) and the Middle East. These cropping systems offer different benefits, like enhancing crop diversity and soil health and reducing the utilization of chemical pesticides and fertilizers.

2.6 Oilseed-Based Cropping Systems

Edible oil is mostly obtained from oilseed crops. Over the last 30 years, the area of oilseed crop cultivation has expanded by 82%, and world production has increased up to 240% (Rahman and de Jimenez 2016) in the oilseed-based cropping systems. In the USA, northern states and areas on Canadian sides having cool conditions for a shorter period of time are suitable for flax, mustard, and canola. However, central and northern Great Plains having warmer growing conditions for longer period of time are well adapted for safflower and sunflower (Johnston et al. 2002). There are different oilseed-based cropping systems such as the sunflower-sunflower (Tariq et al. 2018; Suresh et al. 2015; Sree et al. 2006), soybean-sunflower (Suresh et al. 2015), soybean-soybean (Morriss et al. 2017), sunflower-groundnut (Aulakh and Garg 2007; Mohammad et al. 1993), soybean-mustard (Rashmi et al. 2021), soybean-safflower (Bhutada et al. 2020), and canola-groundnut (Waris 2012).

2.7 Fodder-Based Cropping Systems

There are several fodder-based cropping systems, such as the alfalfa perennial cropping system (Malik et al. 2018), the *Sesbania* perennial system (Heering 1995), and the bajra–napier hybrid grass and perennial *Desmanthus* system (Manoj et al. 2022). The cost of feeding cattle could be reduced by using fodder or green forage, which is abundant in minerals, proteins, micronutrients, and vitamins and contains a significant amount of dry matter for consumption, good digestibility, and little fiber (Chaudhary et al. 2012), and animal production increases (Manoj et al. 2021) with the help of fodder-based cropping systems.

2.8 Cropping System Research

As discussed earlier, scientists are now working on cropping system research instead of individual crops (Ahmed and Ahmad 2023; Ahmed 2023; Abbas et al. 2023; Liu et al. 2023; Ahmed et al. 2012, 2017, 2022a, b; Ahmed 2020; Nadeem et al. 2022; Khan et al. 2020; Asseng et al. 2019; Jabeen et al. 2017; Aslam et al. 2013). Due to climate change, the major focus of the researchers is to feed the burgeoning population of the world. The United Nations is also focusing on sustainable development goals (SDGs) for sustained growth. The limited resources like land and water and costly input resources like nutrients and insect pest management strategies compelled the scientists to shift their research from individual crops to various cropping systems. Therefore, there is dire need of the hour to explore and refine the outcomes based on decision support system for agrotechnology transfer.

2.9 Decision Support System for Cropping System Research

Crop models are collections of mathematical equations that illustrate both the connections among crops and their surroundings as well as mechanisms within a specific plant system (Hoogenboom et al. 2019, 2021). Given the complexity of agricultural systems and the gaps in our current understanding, it is extremely difficult to capture all of the crop system's processes mathematically as long as agricultural models are still idealized representations of reality (Wallach et al. 2014). Obtaining a precise estimate of the net economic output of the cropping system is the ultimate goal of building crop models. Crop models are created at various levels of complexity, based on the knowledge and data that are accessible at the relevant scale (Jones et al. 2017). Since 1970, a number of mathematical simulation models have been created for simulating the growth, development, and yield of crops that are grown at various levels (Galmarini et al. 2024; Wahab et al. 2024; Ahmed et al. 2014a, b, 2016, 2017, 2019; Liu et al. 2019; Asseng et al. 2019; Ahmed and Hassan 2011). Many varied applications, including management of crops, analyses of yield and gap evaluations, and impact of climate change, make substantial use of these models (Jin et al. 2018; Jones et al. 2017; Ritchie and Alagarswamy 2002; Van Ittersum et al. 2013).

2.10 DSSAT Model Description

The CSM-DSSAT is a well-known crop growth model being used by scientific community all over the globe (Jones et al. 2003; Hoogenboom et al. 2021). It has a modular structure and simulates growth and development of crops based on soil-plant-environment interactions and management scenarios (Porter et al. 2000; Hoogenboom et al. 2021). After calibration and evaluation, it can be successfully utilized for model application like seasonal analysis, sequential analysis, spatial analysis, climate change, and yield forecast analysis (DSSAT v4.8; www.dssat.net; Hoogenboom et al. 2021).

2.11 DSSAT Sequence Analysis: Soybean-Chickpea Cropping System—A Case Study

2.11.1 DSSAT Sequence Analysis Program Description

The DSSAT Sequence Analysis program's main goal is to examine the findings of an experiment on cropping sequences that was carried out utilizing yearly crop simulation models. The experiment will follow a specific pattern, such as continuous soybean-chickpea-fallow planted on a piece of land for 25 years, and it may include a number of duplicates produced using various weather patterns. The meteorological information used to power crop models could be historical or could even be generated statistically. The user can use the program to evaluate the sequence's efficiency in terms of both relative and absolute yield levels, yield risk and loss of N due to leaching, or financial benefits over the course of the sequence (Thornton et al. 1995; Hoogenboom et al. 2021).

2.11.2 Input and Output Files for DSSAT Sequence Analysis

Figure 2.2a shows the screenshot of sequence analysis component in DSSAT v4.8, and Fig. 2.2b shows the experiment name, site, and year in it. In the DSSAT model, we selected the fourth experiment (soybean-chickpea rotation) in sequence application. The soybean-chickpea experiment was conducted in India. The experiment's complete information regarding soybean and chickpea was incorporated in DSSAT sequential analysis management files, and then we ran the model (Fig. 2.2c). Figure 2.2d shows the experiment summary (experiment code, starting year, ending year, runs, runs per replication, treatment name, and crops). The biophysical analysis comprised of output for many variables like anthesis and maturity dates, yield and biomass at maturity, and many others.

In the DSSAT Sequence Analysis program, different files are required and developed (Table 2.2). These three files are described briefly below. The sequence experiment results were kept in the output file of a summary model that was saved and

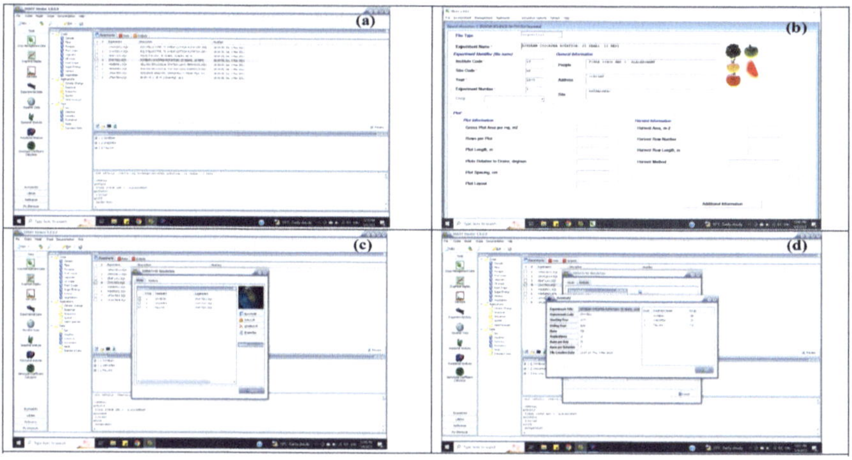

Fig. 2.2 Screenshot showing sequential model component in DSSAT v4.8 (**a**), screenshot showing soybean-chickpea sequential file (ITHY7501.SQX) in DSSAT v4.8 (**b**), screenshot showing soybean-chickpea sequential file showing start of simulation (**c**), screenshot showing soybean-chickpea sequential file with complete simulation (**d**)

Table 2.2 Soybean-chickpea cropping system sequential analysis simulation DSSAT model input and output files

Files	Description
Model summary output	Soybean-chickpea simulation results, stored as one model run per record (input file)
Summary output listing	List of soybean-chickpea model summary output files in current directory (input file)
Price-cost file	Price and cost time series of soybean-chickpea for analyzing experiment (input file)
Graphics	Options for hard-copy graphics output of soybean-chickpea in terms of printer, plotter, port, scaling, plotter language, destination (input file)
Variable abbreviations file	Descriptions and legends for all model output variables available for soybean-chickpea system analysis (input file)
Analysis results file	Output file from soybean-chickpea analysis program containing results and calculations (output file)
Plotter files	Optional file containing plot commands for sending to plotter, printer, or document (output file)

Source: Modified and adopted from Thornton et al. (1995); Singh et al. (1999a, b); DSSAT Sequential Experiment: ITHY7501.SQX; Hoogenboom et al. (2021)

recorded in one model and run. Table 2.3 contains a list of the output variables that were written in this file (Jones et al. 1994).

Both crops' (soybean and chickpea) each component's expenses (cost) and prices for every year are added in the sequence analysis program as time series file (Table 2.4). The cost and price have the potential to be fixed (deterministic), defined by a distribution that is normal, in which the standard deviation was provided by a user and

Table 2.3 Variables in the soybean-chickpea rotation/sequential analysis into the summary output file of the multi-season and multi-year DSSAT model sequential analysis

Variable No.	Biophysical analysis variable description	Variable No.	Biophysical analysis variable description
1	Anthesis date (YRDOY)	37	Physiological maturity date (YRDOY)
2	By-product (kg ha^{-1})	38	Cumulative N_2O emission from soil
3	Tops (N kg ha^{-1})	39	Number of days from planting to harvest (#)
4	Average CO_2 (ppm)	40	N fixed during season (kg ha^{-1})
5	Cumulative CO_2 emitted	41	N application number (#)
6	Tops weight (kg ha^{-1})	42	Inorganic N at maturity (kg [N] ha^{-1})
7	Dry matter-ET productivity	43	Total inorganic N applied (kg [N] ha^{-1})
8	Dry matter-irrigation productivity	44	N leached during season (kg [N] ha^{-1})
9	Dry matter-rain productivity	45	N uptake during season (kg [N] ha^{-1})
10	Dry matter-transpiration productivity	46	Organic soil carbon at maturity (kg ha^{-1})
11	Dry matter-N fertilizer productivity	47	Total organic carbon at maturity (kg ha^{-1})
12	Dry matter-N uptake productivity	48	Organic soil N at maturity (kg ha^{-1})
13	Drainage (mm)	49	Total soil applied organic N at maturity (kg ha^{-1})
14	Sowing weight (kg ha^{-1})	50	Organic soil P at maturity (kg ha^{-1})
15	Average day (h)	51	Total soil applied organic P at maturity (kg ha^{-1})
16	Emergence date (YRDOY)	52	Planting date (YRDOY)
17	Total season transpiration (mm)	53	Number of P application (#)
18	Plant season (sowing-harvest) transpiration (mm)	54	Inorganic P applied (kg ha^{-1})
19	Total season evaporation from soil (mm)	55	Total season precipitation (mm)
20	Plant season evaporation from soil (mm)	56	Seasonal cumulative P uptake (kg ha^{-1})
21	Total season evapotranspiration from soil (mm)	57	Pod/ear/panicle weight at maturity (kg ha^{-1})
22	Plant season evapotranspiration from soil (mm)	58	Residues applied (kg ha^{-1})
23	Grain N at maturity (kg ha^{-1})	59	Season surface runoff (mm)
24	Number (#) at maturity (m^{-2})	60	Simulation start date (YRDOY)
25	Number (#) at maturity/unit	61	Soil K at maturity (kg ha^{-1})

(continued)

Table 2.3 (continued)

Variable No.	Biophysical analysis variable description	Variable No.	Biophysical analysis variable description
26	Harvest date (YRDOY)	62	Soil P at maturity (kg ha^{-1})
27	Harvest index at maturity	63	Average solar radiation (MJ m^{-2} d^{-1})
28	Harvested yield (kg [dm] ha^{-1})	64	Extractable water at maturity (mm)
29	Yield at harvest maturity (kg [dm] ha^{-1})	65	Average minimum air temperature (°C)
30	Unit weight at maturity (g [dm]/unit)	66	Average maximum air temperature (°C)
31	Irrigation application number (#)	67	Yield-ET productivity (kg [yield] ha^{-1})
32	Season applied irrigation (mm)	68	Yield-irrigation productivity (kg [yield] ha^{-1})
33	Number of K application (#)	69	Yield-N fertilizer productivity (kg [yield] ha^{-1})
34	Inorganic K applied (kg ha^{-1})	70	Yield-N uptake productivity (kg [yield] ha^{-1})
35	Seasonal cumulative k uptake (kg ha^{-1})	71	Yield-rainfall productivity (kg [yield] ha^{-1})
36	Leaf area index, maximum	72	Yield-transpiration productivity (kg [yield] ha^{-1})

Source: Modified and adopted from Thornton et al. (1995); Singh et al. (1999a, b); DSSAT Sequential Experiment: ITHY7501.SQX; Hoogenboom et al. (2021)

Table 2.4 Costs and prices used in soybean-chickpea cropping system economic analysis of the multi-season and multi-year DSSAT model sequential analysis

Variable No.	Description of price file variables	Units	Soybean	Chickpea	Fallow
1	Price of grain	\$ t^{-1}	330.00	310.00	0.00
2	Price of by-product	\$ t^{-1}	10.00	10.00	0.00
3	Base production costs	\$ ha^{-1}	395.00	295.00	25.00
4	N fertilizer cost	\$ kg^{-1}	0.50	0.50	0.50
5	Cost per N fertilizer application	\$	12.00	12.00	12.00
6	Irrigation costs	\$ mm^{-1}	0.50	0.50	0.50
7	Cost per irrigation application	\$	15.00	15.00	15.00
8	Seed cost	\$ kg^{-1}	0.50	0.50	0.50
9	Organic amendments	\$ t^{-1}	0.00	0.00	0.00
10	P fertilizer cost	\$ kg^{-1}	0.00	0.00	0.00
11	Cost per P fertilizer application	\$	0.00	0.00	0.00
12	K fertilizer cost	\$ kg^{-1}	0.00	0.00	0.00
13	Cost per K fertilizer application	\$	0.00	0.00	0.00

Source: Modified and adopted from Thornton et al. (1995); Singh et al. (1999a, b); DSSAT Sequential Experiment: ITHY7501.SQX; Hoogenboom et al. (2021)

expressed as a mean percentage. The series of times may include nominal or real dollars; they might be historical or produced in accordance with a pattern that the user has chosen. The in-built price file can be modified or revised as per local conditions based on market value of inputs and outputs (Thornton et al. 1995; Hoogenboom et al. 2021).

2.11.3 DSSAT Sequence Analysis Program Operation

The maximum number of distinct simulation runs that the program may analyze in an output file is unlimited. For example, a sequence of soybean-chickpea was repeated for 25 years. With a three-model run per year, soybean was planted at the start of the growing season on 176–152 days of the year (DOY), and the crop reaches its anthesis date 213–217 DOY and its maturity 274–287 DOY. The crop was harvested from 292 to 298 DOY. For example, in 1975, soybean was planted on 152 DOY (1975152) and harvested on 297 DOY (1975297), while chickpea was planted on 298 DOY (1975298), the day after harvesting the soybean crop. In 1976, the chickpea was harvested on 031 DOY (1976031). In 1976, just to differentiate the two crops, a fallow period was introduced on 032 DOY (1976032, 1 day after harvesting the chickpea crop), and this fallow period was harvested on 175 DOY (1976175), and again soybean crop was sown 1 day later on 176 DOY (1976176). This alternate sowing and harvesting of soybean, chickpea, and a fallow period continued for 25 years. There was a total of 750 runs with 10 replicates having 75 runs per replication and 7 runs per rotation. The user can choose from a variety of settings on the package's main menu, including options for altering how hard duplicates of graphs are made, evaluating the sequence of the crop in regard to economic or biophysical outputs, and quitting the program. The output of 35 generated model variables was listed (Table 2.3) and selected by a user, in which mean, maximum, minimum, and standard deviation values of all replicates were calculated by the model and tabulated by seasons. Figure 2.3 represents the simulated soybean and chickpea yields for India in a soybean-chickpea cropping system (Hoogenboom et al. 2021).

2.11.4 DSSAT Sequence Analysis's Economic Analysis

The economic analysis of the price-cost file selected by a user was carried out. Text editors can edit a file of price-cost in the sequence analysis program and choose the variability of price-cost during economic analysis. Because future cost production and prices for goods cannot be accurately determined, the user may decide to include fluctuation in price and cost. After that, the analysis of user-specified distribution is combined with the distribution of the model simulated (Thornton and Hoogenboom 1994; Hoogenboom et al. 2021).

A distribution through time series for net return and gross margin is the result of economic analysis. Both the returns and the numerous crop components that developed crop sequence can be graphed along with this information. The fallow time may cost money in several circumstances, so the user has the option of including or excluding

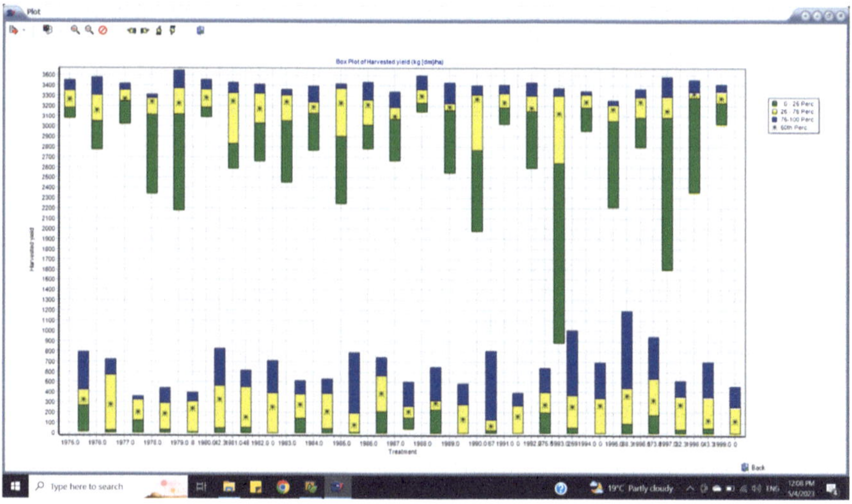

Fig. 2.3 Simulated seed yield of soybean-chickpea cropping system (ITHY7501.SNX) for 25 years; (**a**) 1975–1980, (**b**) 1981–1986, (**c**) 1987–1992, and (**d**) 1993–1999. Box limits represent the 25th and 75th percentiles, box central line represents the median, and outliers represent the minimum and maximum values. The simulated results were obtained from the combination of soybean-chickpea cropping system rotation and historical weather data for 25 years at the ICRISAT center, Patancheru, Andhra Pradesh, India. (Source: Singh et al. 1999a, b; Hoogenboom et al. 2021)

fallow periods from an economic study. The system also determines the probability that critical points of net return will not be reached over time. After that, a time series of probabilities can be plotted (Thornton et al. 1995; Hoogenboom et al. 2021).

2.12 APSIM Model

The APSIM is crop simulation model known to scientists all over the world (Keating et al. 2003). The model was developed by Australian scientists in a modular framework. It was developed for simulation of biophysical processes of various farming systems based on management practices, keeping in view the economic aspects for diverse ecological conditions under prevailing and future climatic scenarios (Keating et al. 2003).

2.13 APSIM Model Rotation Analysis: A Case Study of Wheat-Alfalfa Rotation

2.13.1 APSIM Model Rotational Analysis Program Description

The main objective of the APSIM model rotational program is to determine the findings of sequence cropping experiments that were carried out yearly in crop simulation models. A specific pattern will be followed for the experiment, such as

Table 2.5 Wheat-alfalfa cropping system sequential analysis simulations in APSIM

Number	Description
1	Created a new simulation "Rotation Sample.apsim"
2	Simulation name changed to "Wheat Alfalfa"
3	Selected the "Goondiwindi" met file (1940–1989)
4	"Heavy Clay" soil selected from training soil toolbox
5	Value of "Initial Water" set as 100% filled from the top
6	"Initial Nitrogen" set as 0.1 ppm for NO_3 and NH_4
7	Previous crop simulation and management rule deleted
8	Add "alfalfa" crop and "Alfalfa Ley Management" from Standard and Management toolbox, respectively. Set "must sow" to "yes" in "Alfalfa Ley Management." Same as for wheat
9	Node name changed to "alfalfa" in "Rotation Management component"
10	An arc added from alfalfa node to itself
11	Added "Alfalfa Lay Management.canHarvest" in "Rules" box and "Alfalfa Lay Management.Harvest" in "Actions" box
12	Selected arc toward "alfalfa" and "summer fallow 2" and entered the corresponding "Rules" and "Actions" in both
13	Run model and plotted XY graph

continuous alfalfa ley-summer fallow-wheat-summer fallow-2 planted on a piece of land for 29 years, and in which different weather patterns may be used in replicates. To run the crop model, meteorological data used may be statistically generated or historical. A sequence's efficiency can be evaluated by using the program in terms of yield risks and economic benefits (Table 2.5).

Create a new simulation "Rotation Sample.apsim," change the name of new simulation to "WheatAlfalfa," and save it. Select the met file named "Goondiwindi" from 1940 to 1989 for a simulation run. Select the heavy clay soil from training soil toolbox in which the initial water is set at 100% filled from top and the nitrogen is set at 0.1 ppm in each level of soils. Remove the previous crop and its management rules from simulation, add the alfalfa crop and alfalfa ley management from the standard and management toolbox, and select "must sow" to yes in wheat and alfalfa ley management. After this, in a component of rotation management, change the name of previous node to alfalfa. Right-click on the alfalfa node and added an arc to itself after it selecting this arc "Alfalfa Lay Management.Harvest" added in the rule box and in the action box "Alfalfa Lay Management.Harvest" (Fig. 2.4a). Again, arc toward alfalfa and summer fallow 2, and add the corresponding rules and action in both (Fig. 2.4b). Add the XY graph in paddock, and run the simulation (Fig. 2.4c).

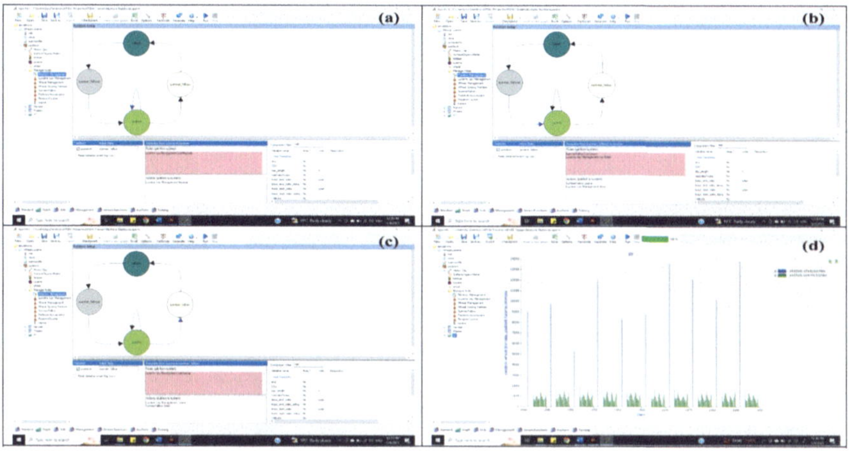

Fig. 2.4 Pictorial view of APSIM model showing multi-cut alfalfa in wheat-alfalfa rotation (**a**); pictorial view of APSIM model showing various commands (summer fallow to alfalfa) in wheat-alfalfa rotation (**b**); pictorial view of APSIM model showing various commands (alfalfa to summer fallow-2) in wheat-alfalfa rotation (**c**); simulated biomass of wheat and alfalfa in wheat-alfalfa rotation using APSIM model (**d**)

2.13.2 APSIM Rotational Program Operation

A sequence of wheat-alfalfa rotation was repeated for 29 years. Alfalfa was planted at the start of the growing season in 1961 and continuously cultivated for 3 years. In 1963, alfalfa was harvested and the land remained fallow in the summer season. In 1964, wheat was planted for one season, and again the land remained fallow in the summer season of 1965. This pattern was continued for 24 years: the alfalfa ley, summer fallow period, wheat, and summer fallow period-2 continued. Figure 2.4d represents the simulated wheat and alfalfa biomass regarding the wheat-alfalfa cropping system (Fig. 2.5).

2.14 Conclusion

The DSSAT and APSIM models can be successfully used for biophysical analysis and economic analysis of diverse cropping systems. These models can be applied for refining the decision based on decision support system for agrotechnology transfer under current and future climate scenarios. A climate-smart production of diverse cropping systems is needed to feed growing populations. Multiple-cropping strategies could be adopted to use existing resources more efficiently, and crop growth models could be successfully used to refine the decision-making process.

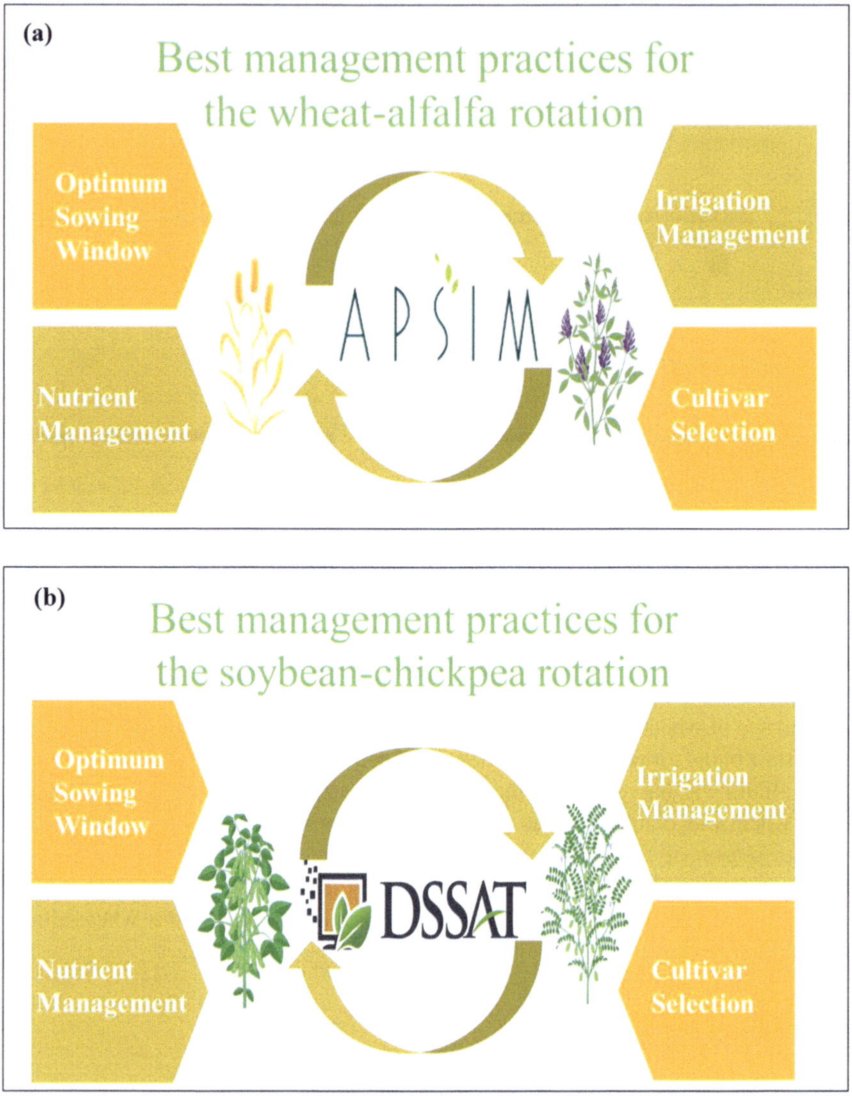

Fig. 2.5 Application of APSIM (**a**) and DSSAT (**b**) model to suggest best management practices

References

Abbas G, Ahmad S, Hussain M, Fatima Z, Hussain S, Iqbal P, Ahmed M, Farooq M (2020a) Sowing date and hybrid choice matters production of maize–maize system. Int J Plant Prod 14:583–595. https://doi.org/10.1007/s42106-020-00104-6

Abbas G, Fatima Z, Hussain M, Hussain S, Atique ur R, Sarwar N, Ahmed M, Ahmad S (2020b) Nitrogen rate and hybrid selection matters productivity of maize–maize cropping system under irrigated arid environment of Southern Punjab, Pakistan. Int J Plant Prod 14:309–320. https://doi.org/10.1007/s42106-020-00086-5

Abbas G, Ahmed M, Fatima Z, Hussain S, Kheir AMS, Ercişli S, Ahmad S (2023) Modeling the potential impact of climate change on maize-maize cropping system in semi-arid environment and designing of adaptation options. Agric For Meteorol 341:109674. https://doi.org/10.1016/j.agrformet.2023.109674

Agus F, Andrade JF, Edreira JI, Deng N, Purwantomo DK, Agustiani N, Aristya VE, Batubara SF, Hosang EY, Krisnadi LY, Makka A (2019) Yield gaps in intensive rice-maize cropping sequences in the humid tropics of Indonesia. Field Crop Res 237:12–22

Ahmad A, Ashfaq M, Rasul G, Wajid SA, Khaliq T, Rasul F, Saeed U, Rahman MH, Hussain J, Ahmad Baig I, Naqvi SA (2015) Impact of climate change on the rice–wheat cropping system of Pakistan. In: Handbook of climate change and agro-ecosystems: the agricultural model intercomparison and improvement project integrated crop and economic assessments, Part 2. World Scientific Publishing, Singapore, pp 219–258. https://doi.org/10.1142/9781783265640_0019

Ahmad S, Abbas G, Ahmed M, Fatima Z, Anjum MA, Rasul G, Khan MA, Hoogenboom G (2019) Climate warming and management impact on the change of phenology of the rice-wheat cropping system in Punjab, Pakistan. Field Crop Res 230:46–61. https://doi.org/10.1016/j.fcr.2018.10.008

Ahmed M (2020) Introduction to modern climate change. Andrew E. Dessler: Cambridge University Press, 2011, 252 pp, ISBN-10: 0521173159. Sci Total Environ 734:139397. https://doi.org/10.1016/j.scitotenv.2020.139397

Ahmed M (2023) The science of climate change. In: Brinkmann R (ed) The Palgrave handbook of global sustainability. Springer International Publishing, Cham, pp 195–222. https://doi.org/10.1007/978-3-031-01949-4_22

Ahmed M, Ahmad S (eds) (2023) Disaster risk reduction in agriculture. Springer Nature. https://doi.org/10.1007/978-981-99-1763-1. ISBN 978-981-99-1762-4, Published: 03 November 2023

Ahmed M, Hassan FU (2011) APSIM and DSSAT models as decision support tools. In: 19th International Congress on Modelling and Simulation, Perth, Australia, pp 1174–1180. http://mssanz.org.au/modsim2011

Ahmed M, Hassan FU, Aslam M, Aslam MA (2012) Physiological attributes based resilience of wheat to climate change. Int J Agric Biol 14(3):407–412

Ahmed M, Aqeel M, Hassan F-U, Asif M, Hayat R (2014a) Use of APSIM to model nitrogen use efficiency of rain-fed wheat. Int J Agric Biol 16(3):461–470

Ahmed M, Fayyaz Ul H, Van Ogtrop FF (2014b) Can models help to forecast rainwater dynamics for rainfed ecosystem? Weather Clim Extremes 5–6:48–55. https://doi.org/10.1016/j.wace.2014.07.001

Ahmed M, Akram MN, Asim M, Aslam M, Hassan F-u, Higgins S, Stöckle CO, Hoogenboom G (2016) Calibration and validation of APSIM-Wheat and CERES-Wheat for spring wheat under rainfed conditions: models evaluation and application. Comput Electron Agric 123:384–401. https://doi.org/10.1016/j.compag.2016.03.015

Ahmed M, Stöckle CO, Nelson R, Higgins S (2017) Assessment of climate change and atmospheric CO_2 impact on winter wheat in the Pacific Northwest using a multimodel ensemble. Front Ecol Evol 5. https://doi.org/10.3389/fevo.2017.00051

Ahmed M, Stöckle CO, Nelson R, Higgins S, Ahmad S, Raza MA (2019) Novel multimodel ensemble approach to evaluate the sole effect of elevated CO2 on winter wheat productivity. Sci Rep 9(1):7813. https://doi.org/10.1038/s41598-019-44251-x

Ahmed M, Hayat R, Ahmad M, ul-Hassan M, Kheir AMS, ul-Hassan F, ur-Rehman MH, Shaheen FA, Raza MA, Ahmad S (2022a) Impact of climate change on dryland agricultural systems: a review of current status, potentials, and further work need. Int J Plant Prod. https://doi.org/10.1007/s42106-022-00197-1

Ahmed M, Ahmad S, Kheir AMS (2022b) Climate change: an overview. In: Ahmed M (ed) Global agricultural production - resilience to climate change. Springer Nature Switzerland AG. https://doi.org/10.1007/978-3-031-14973-3_1

Ahmed M, Rehman A, Sarwar N, Doğan H, Abbas G, Khan MA, Kan M, Raza MA, Hussain S, Ahmad S, Fatima Z (2023) Deciding sowing-window for maize-based cropping system in arid and semiarid environments in Punjab, Pakistan. Turk J Agric For 47(6). https://doi.org/10.55730/1300-011X.3149

Ali ME, Islam MR, Jahiruddin M (2009) Effect of integrated use of organic manures with chemical fertilizers in the rice-rice cropping system and its impact on soil health. Bangladesh J Agric Res 34(1):81–90

Amnuaylojaroen T, Chanvichit P, Janta R, Surapipith V (2021) Projection of rice and maize productions in Northern Thailand under climate change scenario RCP8. 5. Agriculture 11(1):23

Anglade J, Billen G, Garnier J (2015) Relationships for estimating N2 fixation in legumes: incidence for N balance of legume-based cropping systems in Europe. Ecosphere 6(3):1–24

Aslam Z, Ahmed M, Sajad M, Asif M, Akmal M, Awan FK, Ijaz W, Ashraf R, Khattak JZK (2013) A comparison of statistical and dynamic modeling of wheat (*Triticum aestivum* L.) fungal diseases under the climate change. J Food Agric Environ 11(2):1327–1332

Asnawi R, Ratna WA, Reny DT, Made JM (2021) Increasing rice (Oryza sativa L.) productivity and farmer's income through the implementation of modified double rows planting system. Annu Res Rev Biol 36:42–52

Asseng S, Martre P, Maiorano A, Rötter RP, O'Leary GJ, Fitzgerald GJ, Girousse C, Motzo R, Giunta F, Babar MA, Reynolds MP, Kheir AMS, Thorburn PJ, Waha K, Ruane AC, Aggarwal PK, Ahmed M, Balkovič J, Basso B, Biernath C, Bindi M, Cammarano D, Challinor AJ, De Sanctis G, Dumont B, Eyshi Rezaei E, Fereres E, Ferrise R, Garcia-Vila M, Gayler S, Gao Y, Horan H, Hoogenboom G, Izaurralde RC, Jabloun M, Jones CD, Kassie BT, Kersebaum K-C, Klein C, Koehler A-K, Liu B, Minoli S, Montesino San Martin M, Müller C, Naresh Kumar S, Nendel C, Olesen JE, Palosuo T, Porter JR, Priesack E, Ripoche D, Semenov MA, Stöckle C, Stratonovitch P, Streck T, Supit I, Tao F, Van der Velde M, Wallach D, Wang E, Webber H, Wolf J, Xiao L, Zhang Z, Zhao Z, Zhu Y, Ewert F (2019) Climate change impact and adaptation for wheat protein. Glob Chang Biol 25(1):155–173. https://doi.org/10.1111/gcb.14481

Aulakh MS, Garg AK (2007) Yields and nutrient use-efficiency in groundnut-sunflower cropping system in Punjab, India. J Sustain Agric 3(1):89–110

Azeem M, Hayat R, Hussain Q, Ahmed M, Pan G, Ibrahim Tahir M, Imran M, Irfan M, Mehmood ul H (2019) Biochar improves soil quality and N2-fixation and reduces net ecosystem CO2 exchange in a dryland legumecereal cropping system. Soil Tillage Res 186:172–182. https://doi.org/10.1016/j.still.2018.10.007

Bahri H, Annabi M, Cheikh M'Hamed H, Frija A (2019) Assessing the long-term impact of conservation agriculture on wheat-based systems in Tunisia using APSIM simulations under a climate change context. Sci Total Environ 692:1223–1233. https://doi.org/10.1016/j.scitotenv.2019.07.307

Banjara TR, Bohra JS, Kumar S, Singh T, Shori A, Prajapat K (2021) Sustainable alternative crop rotations to the irrigated rice-wheat cropping system of Indo-Gangetic Plains of India. Arch Agron Soil Sci 67:1–18

Barbieri P, Pellerin S, Seufert V, Nesme T (2019) Changes in crop rotations would impact food production in an organically farmed world. Nat Sustain 2:378–385

Bardgett RD, van der Putten WH (2014) Belowground biodiversity and ecosystem functioning. Nature 515:505

Bennett AJ, Bending GD, Chandler D, Hilton S, Mills P (2012) Meeting the demand for crop production: the challenge of yield decline in crops grown in short rotations. Biol Rev 87:52–71

Bhutada PO, Ghuge SB, Kote GM (2020) Influence of mechanization soybean-safflower cropping system in terms of growth, yield and economics. Int J Curr Microbiol Appl Sci 9(8):1303–1306

Boboev H, Djanibekov U, Bekchanov M, Lamers JP, Toderich K (2019) Feasibility of conservation agriculture in the Amu Darya River Lowlands, Central Asia. Int J Agric Sustain 17(1):60–77

Bouman BAM, Castaneda AR, Bhuiyan SI (2002) Nitrate and pesticide contamination of groundwater under rice-based cropping systems: past and current evidence from the Philippines. Agric Ecosyst Environ 92(2–3):185–199

Bukovsky-Reyes S, Isaac ME, Blesh J (2019) Effects of intercropping and soil properties on root functional traits of cover crops. Agric Ecosyst Environ 285:106614

Chaudhary DP, Ashwani K, Sapna SM, Srivastava P, Kumar RS (2012) Maize as fodder? An alternative approach: technical bulletin. Directorate of Maize Research, Pusa Campus, New Delhi

Chen C, Groenigen KJV, Yang H, Hungate BA, Yang B, Tian Y, Chen J, Dong W, Huang S, Deng A, Jiang Y, Zhang W (2020) Global warming and shifts in cropping systems together reduce China's rice production. Glob Food Sec 24:100359

Dolling PJ, Robertson MJ, Asseng S, Ward PR, Latta RA (2005) Simulating lucerne growth and water use on diverse soil types in a Mediterranean-type environment. Aust J Agric Res 56(5):503–515

Ejaz M, Abbas G, Fatima Z, Iqbal P, Raza MA, Kheir AM, Ahmed M, Kakar KM, Ahmad S (2022) Modelling climate uncertainty and adaptations for soybean-based cropping system. Int J Plant Prod 16(2):235–250

Emran SA, Krupnik TJ, Aravindakshan S, Kumar V, Pittelkow CM (2022) Impact of cropping system diversification on productivity and resource use efficiencies of smallholder farmers in south-Central Bangladesh: a multi-criteria analysis. Agron Sustain Dev 42:78

FAO (2024) FAOSTAT database, Food and Agriculture Organization of the United Nations. Available at http://faostat.fao.org/ (Accessed 15 Feb 2024)

Fargione JE, Bassett S, Boucher T, Bridgham SD, Conant RT, Cook-Patton SC, Ellis PW, Falcucci A, Fourqurean JW, Gopalakrishna T, Gu H, Henderson B, Hurteau MD, Kroeger KD, Kroeger T, Lark TJ, Leavitt SM, Lomax G, McDonald RI, Megonigal JP, Miteva DA, Richardson CJ, Sanderman J, Shoch D, Spawn SA, Veldman JW, Williams CA, Woodbury PB, Zganjar C, Baranski M, Elias P, Houghton RA, Landis E, McGlynn E, Schlesinger WH, Siikamaki JV, Sutton-Grier AE, Griscom BW (2018) Natural climate solutions for the United States. Sci Adv 4(11):aat1869. https://doi.org/10.1126/sciadv.aat1869

Galmarini S, Solazzo E, Ferrise R, Srivastava AK, Ahmed M, Asseng S, Cannon AJ, Dentener F, De Sanctis G, Gaiser T, Gao Y, Gayler S, Gutierrez JM, Hoogenboom G, Iturbide M, Jury M, Lange S, Loukos H, Maraun D, Moriondo M, McGinnis S, Nendel C, Padovan G, Riccio A, Ripoche D, Stockle CO, Supit I, Thao S, Trombi G, Vrac M, Weber TKD, Zhao C (2024) Assessing the impact on crop modelling of multi- and uni-variate climate model bias adjustments. Agric Syst 215:103846. https://doi.org/10.1016/j.agsy.2023.103846

Gan YT, Miller PR, McConkey BG, Zentner RP, Stevenson FC, McDonald CL (2003) Influence of diverse cropping sequences on durum wheat yield and protein in the semiarid northern Great Plains. Agron J 95:245–252

Ghosh D, Brahmachari K, Brestic M, Ondrisik P, Hossain A, Skalicky M, Sarkar S, Moulick D, Dinda NK, Das A, Pramanick B (2020) Integrated weed and nutrient management improve yield, nutrient uptake and economics of maize in the rice-maize cropping system of Eastern India. Agronomy 10(12):1906

Gogoi B, Borah N, Baishya A, Nath DJ, Dutta S, Das R, Bhattacharyya D, Sharma KK, Valente D, Petrosillo I (2021) Enhancing soil ecosystem services through sustainable integrated nutrient management in double rice-cropping system of North-East India. Ecol Indic 132:108262

Gou F, van Ittersum MK, Wang G, van der Putten PE, van der Werf W (2016) Yield and yield components of wheat and maize in wheat–maize intercropping in the Netherlands. Eur J Agron 76:17–27

Gurjar G, Swami S (2019) Enhancement of carbon sequestration in soils under climate change scenario in India. Int J Curr Microbiol App Sci 8(2):714–721

Haefele SM, Banayo NP, Amarante ST, Siopongco JD, Mabesa RL (2013) Characteristics and management options for rice–maize systems in the Philippines. Field Crop Res 144:52–61

Hamza MA, Anderson WK (2005) Review soil compaction in cropping systems: a review of the nature, causes and possible solutions. Soil Tillage Res 82:121–145

Hatfield JL, Sauer TJ, Cruse RM (2017) Soil: the forgotten piece of the water, food, energy nexus. In: Sparks DL (ed) Advances in agronomy. Academic Press, London, pp 1–46

Heering JH (1995) Botanical and agronomic evaluation of a collection of Sesbania sesban and related perennial species. Wageningen University and Research, Wageningen

Heinemann AB, Dingkuhn M, Luquet D, Combres JC, Chapman S (2008) Characterization of drought stress environments for upland rice and maize in Central Brazil. Euphytica 162(3):395–410

Hinsinger P, Betencourt E, Bernard L, Brauman A, Plassard C, Shen J, Tang X, Zhang F (2011) P for two, sharing a scarce resource: soil phosphorus acquisition in the rhizosphere of intercropped species. Plant Physiol 156:1078–1086

Hoogenboom G, Porter CH, Boote KJ, Shelia V, Wilkens PW, Singh U, White JW, Asseng S, Lizaso JI, Moreno LP, Pavan W, Ogoshi R, Hunt LA, Tsuji GY, Jones JW (2019) The DSSAT crop modeling ecosystem. In: Boote KJ (ed) Advances in crop modeling for a sustainable agriculture. Burleigh Dodds Science Publishing, Cambridge, pp 173–216. https://doi.org/10.19103/AS.2019.0061.10

Hoogenboom G, Porter CH, Shelia V, Boote KJ, Singh U, White JW, Pavan W, Oliveira FAA, Moreno-Cadena LP, Lizaso JI, Asseng S, Pequeno DNL, Kimball BA, Alderman PD, Thorp KR, Jones MR, Cuadra SV, Vianna MS, Villalobos FJ, Ferreira TB, Batchelor WD, Koo J, Hunt LA, Jones JW (2021) Decision Support System for Agrotechnology Transfer (DSSAT) Version 4.8 (DSSAT.net). DSSAT Foundation, Gainesville

Hulugalle NR, Weaver TB, Scott F (2005) Continuous cotton and cotton-wheat rotation effects on soil properties and profitability in an irrigated Vertisol. J Sustain Agric 27(3):5

Iqbal S, Mushtaq A, Gul Z, Faizan A, Eajaz AD (2018) Alfalfa (Medicago sativa L): a valuable source of forage/fodder crop under cold arid conditions of Ladakh (Jammu and Kashmir). Res J Chem Environ Sci 6(4):74–84

Jabeen M, Gabriel HF, Ahmed M, Mahboob MA, Iqbal J (2017) Studying impact of climate change on wheat yield by using DSSAT and GIS: a case study of Pothwar Region. In: Ahmed M, Stockle OC (eds) Quantification of climate variability, adaptation and mitigation for agricultural sustainability. Springer International Publishing, Cham, pp 387–411

Jiang D, Hengsdijk H, Dai T-B, de Boer W, Jing Q, Cao W-X (2006) Long-term effects of manure and inorganic fertilizers on yield and soil fertility for a winter wheat-maize system in Jiangsu, China1 1Project supported by the National Natural Science Foundation of China (No. 30030090) and the National High-Tech Research and Development Program of China (863 Program) (No. 2003AA209030). Pedosphere 16(1):25–32. https://doi.org/10.1016/S1002-0160(06)60022-2

Jin X, Kumare L, Li L, Feng H, Xu X, Yang G et al (2018) A review of data assimilation of remote sensing and crop models. Eur J Agron 92:141152

Johnston AM, Tanaka DL, Miller PR, Brandt SA, Nielsen DC, Lafond GP, Riveland NR (2002) Oilseed crops for semiarid cropping systems in the northern Great Plains. Agron J 94(2):231–240

Jones JW, Hunt LA, Hoogenboom G, Godwin DC, Singh U, Tsuji et al (1994) General input and output file structures for crop simulation models. DSSAT Version 3.0, vol 1. University of Hawaii Press, Honolulu

Jones JW, Hoogenboom G, Porter CH, Boote KJ, Batchelor WD, Hunt LA, Wilkens PW, Singh U, Gijsman AJ, Ritchie JT (2003) The DSSAT cropping system model. Eur J Agron 18:235–265

Jones JW, Antle JM, Basso B, Boote KJ, Conant RT, Foster I et al (2017) Brief history of agricultural systems modeling. Agric Syst 155:240254

Karlen D, Andrews S, Doran J (2001) Soil quality: current concepts and applications. Adv Agron 74:1–40

Katkar RN, Sonune BA, Kadu PR (2011) Long-term effect of fertilization on soil chemical and biological characteristics and productivity under sorghum (Sorghum bicolor)-wheat (*Triticum aestivum*) system in Vertisol. Indian J Agric Sci 81(8):734

Keating BA, Carberry PS, Hammer GL, Probert ME et al (2003) An overview of APSIM: a model designed for farming systems simulation. Eur J Agron 18:267–288

Khan MA, Tahir A, Khurshid N, Husnain MI, Ahmed M, Boughanmi H (2020) Economic effects of climate change-induced loss of agricultural production by 2050: a case study of Pakistan. Sustain For 12(3):1216. https://doi.org/10.3390/su12031216

Kiliç H, Gürsoy S (2010) Effect of seeding rate on yield and yield components of durum wheat cultivars in cottonwheat cropping system. Sci Res Essays 5(15):2078–2084

Kohli M, Mann CE, Rajaram R (1991) Global status and recent progress in breeding wheat for the warmer areas. In: Saunders DA (ed) Wheat for warmer tropical areas. CIMMYT, Mexico, pp 96–112

Lal R (2017) Improving soil health and human protein nutrition by pulses-based cropping systems. Adv Agron 145:167–204

Le TT (2014) Assessment of the sustainability of the rice-maize cropping system in the Red River Delta of Vietnam and developing reduced tillage practices in rice-maize system in the area. J Vietnamese Environ 5(1):1–7

Lemaire G, Franzluebbers A, Carvalho PCDF, Dedieu B (2014) Integrated crop-livestock systems: strategies to achieve synergy between agricultural production and environmental quality. Agric Ecosyst Environ 190:4–8

Li X, Jousset A, de Boer W, Carrion VJ, Zhang T, Wang X, Kuramae EE (2019) Legacy of land relationship between plant diversity and productivity. Ecol Lett 14:36–41

Liu B, Martre P, Ewert F, Porter JR, Challinor AJ, Müller C, Ruane AC, Waha K, Thorburn PJ, Aggarwal PK, Ahmed M, Balkovič J, Basso B, Biernath C, Bindi M, Cammarano D, De Sanctis G, Dumont B, Espadafor M, Eyshi Rezaei E, Ferrise R, Garcia-Vila M, Gayler S, Gao Y, Horan H, Hoogenboom G, Izaurralde RC, Jones CD, Kassie BT, Kersebaum KC, Klein C, Koehler A-K, Maiorano A, Minoli S, Montesino San Martin M, Naresh Kumar S, Nendel C, O'Leary GJ, Palosuo T, Priesack E, Ripoche D, Rötter RP, Semenov MA, Stöckle C, Streck T, Supit I, Tao F, Van der Velde M, Wallach D, Wang E, Webber H, Wolf J, Xiao L, Zhang Z, Zhao Z, Zhu Y, Asseng S (2019) Global wheat production with 1.5 and 2.0°C above pre-industrial warming. Glob Chang Biol 25(4):1428–1444. https://doi.org/10.1111/gcb.14542

Liu B, Martre P, Ewert F, Webber H, Waha K, Thorburn PJ, Ruane AC, Aggarwal PK, Ahmed M, Balkovič J, Basso B, Biernath C, Bindi M, Cammarano D, Cao W, Challinor AJ, De Sanctis G, Dumont B, Espadafor M, Eyshi Rezaei E, Fereres E, Ferrise R, Garcia-Vila M, Gayler S, Gao Y, Horan H, Hoogenboom G, Izaurralde RC, Jabloun M, Jones CD, Kassie BT, Kersebaum KC, Klein C, Koehler A-K, Maiorano A, Minoli S, Montesino San Martin M, Müller C, Naresh Kumar S, Nendel C, O'Leary GJ, Eivind Olesen J, Palosuo T, Porter JR, Priesack E, Ripoche D, Rötter RP, Semenov MA, Stöckle C, Stratonovitch P, Streck T, Supit I, Tao F, Van der Velde M, Wang E, Wolf J, Xiao L, Zhang Z, Zhao Z, Zhu Y, Asseng S (2023) AgMIP-Wheat multi-model simulations on climate change impact and adaptation for global wheat. Open Data J Agric Res 9:10–25. https://doi.org/10.18174/odjar.v9i0.18092

Malik W, Boote KJ, Hoogenboom G, Cavero J, Dechmi F (2018) Adapting the CROPGRO model to simulate alfalfa growth and yield. Agron J 110:1777–1790

Maneepitak S, Ullah H, Datta A, Shrestha RP, Shrestha S, Kachenchart B (2019) Effects of water and rice straw management practices on water savings and greenhouse gas emissions from a double-rice paddy field in the Central Plain of Thailand. Eur J Agron 107:18–29

Mann RA, Garrity DP (1994) Green manures in rice-wheat cropping systems in Asia. In: Ladha JK, Garrity DP (eds) Green manure production systems for Asian Ricelands. IRRI, Los Banos, pp 27–42

Manoj KN, Shekara BG, Sridhara S, Jha PK, Prasad PV (2021) Biomass quantity and quality from different year-round cereal–legume cropping systems as forage or fodder for livestock. Sustainability 13(16):9414

Manoj KN, Shekara BG, Sridhara S, Mudalagiriyappa, Chikkarugi NM, Gopakkali P, Jha PK, Vara Prasad PV (2022) Carbon footprint assessment and energy budgeting of different annual and perennial forage cropping systems: a study from the semi-arid region of Karnataka, India. Agronomy 12:1783

Maron JL, Marler M, Klironomos JN, Cleveland CC (2011) Soil fungal pathogens and the use history determines reprogramming of plant physiology by soil microbiome. ISME J 13:738–751

Meiyun W, Tianzhi R, Ming Z, Shaokun L, Xiaobo W, Lijuan L, Changli C (2010) Matter production and resources use efficiency of double-cropping silage maize system. Agric Food Sci: 1316–1323

Miah MM, Saha PK, Islam A, Hasan MN, Nosov V (2008) Potassium fertilization in rice-rice and rice-wheat cropping system in Bangladesh. Bangladesh J Agric Environ 4(5):51–67

Mohammad S, Reddy A, Putnaml DH (1993) Evaluation of planting pattern and density effects on sunflower: groundnut sole and intercrop system in a systbmatic ean design. Helia 16(18):1–14
Moot DJ, Hargreaves J, Brown HE, Teixeira EI (2015) Calibration of the APSIM-Lucerne model for 'Grasslands Kaituna' lucerne crops grown in New Zealand. N Z J Agric Res 58(2):190–202
Morriss SC, Studham ME, Tylka GL, MacIntosh GC (2017) Validation of a hairy roots system to study soybean-soybean aphid interactions. PLoS One 12(3):0174914
Mubeen M, Ahmad A, Hammad HM, Awais M, Farid HU, Saleem M, Din MS, Amin A, Ali A, Fahad S, Nasim W (2020) Evaluating the climate change impact on water use efficiency of cotton-wheat in semi-arid conditions using DSSAT model. J Water Climate Change 11(4):1661–1675
Nadeem M, Nazer Khan M, Abbas G, Fatima Z, Iqbal P, Ahmed M, Ali Raza M, Rehman A, Ul Haq E, Hayat A, Ali M, Ahmad S (2022) Application of CSM-CANEGRO model for climate change impact assessment and adaptation for sugarcane in semi-arid environment of Southern Punjab, Pakistan. Int J Plant Prod. https://doi.org/10.1007/s42106-022-00192-6
Naing TAA, Kingsbury AJ, Buerkert A, Finckh MR (2008) A survey of Myanmar rice production and constraints. J Agric Rural Dev Trop Subtrop 109(2):151–168
Naz S, Ahmad S, Abbas G, Fatima Z, Hussain S, Ahmed M, Khan MA, Khan A, Fahad S, Nasim W, Ercisli S (2022) Modeling the impact of climate warming on potato phenology. Eur J Agron 132:126404
Niu Y, Bainard LD, May WE, Hossain Z, Hamel C, Gan Y (2018) Intensified pulse rotations buildup pea rhizosphere pathogens in cereal and pulse based cropping systems. Front Microbiol 9:1909
Phaloeun C, Basnayake J, Ngoy CK, Fukai S, Sarom M (2003) The effect of water availability on rice-based double cropping in rainfed lowlands in Cambodia. Water Agric 116:72–85
Pivato B, Mazurier S, Lemanceau P, Siblot S, Berta G, Mougel C, Van Tuinen D (2007) Medicago species affect the community composition of arbuscular mycorrhizal fungi associated with roots. New Phytol 176:197–210
Polthanee A (1997) Rice-based cropping systems in northeastern Thailand. In: Breeding strategies for rainfed lowland rice in drought-prone environments. Australian Centre for International Agricultural Research, Canberra, pp 13–22
Porter C, Jones JW, Braga R (2000) An approach for modular crop model development. International Consortium for Agricultural Systems Applications, Honolulu
Rahman M, de Jimenez MM (2016) Designer oil crops. In: Breeding oilseed crops for sustainable production. Academic Press, London, pp 361–376
Rashmi I, Ali S, Mina BL, Kumar K, Kumar A, Singh RK (2021) Short-term effect of resource conservation practices on soil loss, productivity and energy budgeting of soybean–mustard system in table lands of Chambal ravine. Agric Res 10:223–233
Ratnayake RR, Perera BM, Rajapaksha RP, Ekanayake EM, Kumara RK, Gunaratne HM (2017) Soil carbon sequestration and nutrient status of tropical rice based cropping systems: Rice-Rice, Rice-Soya, Rice-Onion and Rice-Tobacco in Sri Lanka. Catena 150:17–23
Ritchie JT, Alagarswamy G (2002) Overview of crop models for assessment of crop production. In: Doering OC III, Randolph JC, Southworth J, Pfeifer RA (eds) Effects of climate change and variability on agricultural production systems. Kluwer Academic Publishing, Dordrecht, pp 43–68
Sarwar N, Farooq O, Wasaya A, Hussain M, El-Shehawi AM, Ahmad S, Brestic M, Mahmoud SF, Zivcak M, Farooq S (2021) Integrated nitrogen management improves productivity and economic returns of wheat-maize cropping system. J King Saud Univ Sci 33(5):101475
Saunders DA, Hettel GP (eds) (1994) Wheat in heat-stressed environments: irrigated, dry areas and rice–wheat farming systems. CIMMYT, Mexico
Saurabh K, Rao KK, Mishra JS, Kumar R, Poonia SP, Samal SK, Roy HS, Dubey AK, Choubey AK, Mondal S, Bhatt BP (2021) Influence of tillage based crop establishment and residue management practices on soil quality indices and yield sustainability in rice-wheat cropping system of Eastern Indo-Gangetic Plains. Soil Tillage Res 206:104841

Shahzad M, Hussain M, Farooq M, Farooq S, Jabran K, Nawaz A (2017) Economic assessment of conventional and conservation tillage practices in different wheat-based cropping systems of Punjab, Pakistan. Environ Sci Pollut Res 24:634–643

Shrestha R, Neupane R (2016) Agronomic management and cropping patterns of pulses. In: Gurung TR, Bokhtiar SM (eds) Pulses for sustainable food and nutrition security in SAARC Region. SAARC Agriculture Centre (SAC), Dhaka, pp 33–64

Siddique KHM, Sykes J (1997) Pulse production in Australia past, present and future. Aust J Exp Agric 37(1):103–111

Singh P, Alagarswamy G, Pathak P, Wani SP, Hoogenboom G, Virmani SM (1999a) Soybean-chickpea rotation on Vertic Inceptisols I. Effect of soil depth and landform on light interception, water balance and crop yields. Field Crop Res 63:211–224

Singh P, Alagarswamy G, Hoogenboom G, Pathak P, Wani SP, Virmani SM (1999b) Soybean-chickpea rotation on Vertic Inceptisols II. Long-term simulation of water balance and crop yields. Field Crop Res 63:225–236

Sree PSS, Sridhar V, Sree DS (2006) Production potential and economics of sunflower-based cropping systems under scarce rainfall zone of Andhra Pradesh. Agric Sci Digest 26(1):60–62

Sun F, Pan K, Olatunji OA, Li Z, Chen W, Zhang A, Song D, Sun X, Huang D, Tan X (2019) Specific legumes allay drought effects on soil microbial food web activities of the focal species in agroecosystem. Plant Soil 437:455–471

Suresh G, Babu SS, Qureshi AA (2015) Productivity and nutrient use of sunflower (Helianthus annuus L.) in sequential cropping system in Vertisols. Indian Soc Oilseeds Res 32:139–141

Tariq M, Ahmad S, Fahad S, Abbas G, Hussain S, Fatima Z, Nasim W, Mubeen M, ur Rehman MH, Khan MA, Adnan M (2018) The impact of climate warming and crop management on phenology of sunflower-based cropping systems in Punjab, Pakistan. Agric For Meteorol 256:270–282

Thomaz EL, Marcatto FS, Antoneli V (2022) Soil erosion on the Brazilian sugarcane cropping system: an overview. Geogr Sustain 3:129–138

Thornton PK, Hoogenboom G (1994) A computer program to analyze single season crop model outputs. Agron J 86:860–868

Thornton PK, Hoogenboom G, Wilkens PW, Bowe WT (1995) A computer program to analyze multiple-season crop model outputs. Agron J 87:131–136

Tiemann LK, Grandy AS, Atkinson EE, Marin-Spiotta E, McDaniel MD (2015) Crop rotational diversity enhances belowground communities and functions in an agroecosystem. Ecol Lett 18:761–771

Timisina J, Jat ML, Majumdar K (2010) Rice-maize systems of South Asia: current status, future prospects and research priorities for nutrient management. Plant Soil 335:65–82

Umrit G, Ng Cheong R, Gillabel J, Merckx R (2014) Effect of conventional versus mechanized sugarcane cropping systems on soil organic carbon stocks and labile carbon pools in Mauritius as revealed by 13C natural abundance. Plant Soil 379:177–192

Van Ittersum MK, Cassman KG, Grassini P, Wolf J, Tittonell P, Hochman Z (2013) Yield gap analysis with local to global relevance—a review. Field Crop Res 143:417

Vandermeer JH (1989) The ecology of intercropping. Cambridge University Press, Cambridge

Vukicevich E, Lowery T, Bowen P, Úrbez-Torres JR, Hart M (2016) Cover crops to increase soil microbial diversity and mitigate decline in perennial agriculture. A review. Agron Sustain Dev 36:1–14

Wahab AA, Hussain M, Fatima Z, Ahmed M, Ahmad S (2024) Evaluating the interactive impact of nitrogen levels and cultivars on yield traits, nitrogen use efficiency, water use efficiency and benefit cost ratio of ratoon rice in Punjab, Pakistan. Int J Plant Prod. https://doi.org/10.1007/s42106-023-00280-1

Wallach D, Makowski D, Jones JW, Brun F (eds) (2014) Working with dynamic crop models: methods, tools and examples for agriculture and environment. Academic Press, Waltham

Waris M (2012) Optimizing crop water productivity under center pivot sprinkler irrigation system. Thesis M. Phil. PARC Institute of Advanced Studies in Agriculture (PIASA). 55p

Watanabe T, Luu HM, Inubushi K (2017) Effects of the continuous application of rice straw compost and chemical fertilizer on soil carbon and available silicon under a double rice cropping system in the Mekong Delta, Vietnam. Japan Agric Res Q 51(3):233–239
Watto MA, Mugera AW (2015) Efficiency of irrigation water application in sugarcane cultivation in Pakistan. J Sci Food Agric 95(9):1860–1867
Xing GX, Shi SL, Shen GY, Du LJ, Xiong ZQ (2002) Nitrous oxide emissions from paddy soil in three rice-based cropping systems in China. Nutr Cycl Agroecosyst 64:135–143
Yadav RL, Dwivedi BS, Pandey PS (2000) Rice-wheat cropping system: assessment of sustainability under green manuring and chemical fertilizer inputs. Field Crop Res 65(1):15–30
Yadav GS, Das A, Kandpal BK, Babu S, Lal R, Datta M, Das B, Singh R, Singh VK, Mohapatra KP, Chakraborty M (2021) The food-energy-water-carbon nexus in a maize-maize-mustard cropping sequence of the Indian Himalayas: an impact of tillage-cum-live mulching. Renew Sust Energ Rev 151:111602. https://doi.org/10.1016/j.rser.2021.111602
Yang T, Siddique KHM, Liu K (2020) Cropping systems in agriculture and their impact on soil health—a review. Glob Ecol Conserv 23:e01118
Zeng F, Zuo Z, Mo J, Chen C, Yang X, Wang J, Wang Y, Zhao Z, Chen T, Li Y, Zhang Z (2021) Runoff losses in nitrogen and phosphorus from paddy and maize cropping systems: a field study in Dongjiang Basin, South China. Front Plant Sci 12:675121
Zhu Y, Chen H, Fan J, Wang Y, Li Y, Chen J, Fan J, Yang S, Hu L, Leung H, Mew TW, Teng PS, Wang Z, Mundt CC (2000) Genetic diversity and disease control in rice. Nature 406:718–722

Rice-Wheat System

3

3.1 Introduction

Rice-wheat system is a life-supporting production system as it plays an important role in food security worldwide since it offers basic foods to the worldwide people (Banjara et al. 2021a; Panwar et al. 2019; Lalik et al. 2014). This system feeds the world's 15% population (Ray et al. 2012). This system is mostly being grown in India, Pakistan, Bangladesh, Nepal (Ladha et al. 2003; Timsina and Connor 2001), and some areas of China (Lianzheng and Yixian 1994; Jiaguo 2000) and less practiced in Korea, Iran, Bhutan, Japan, Egypt, Myanmar, and Fuerte River valley in Mexico. However, it is successfully being grown in these countries, i.e., Sri Lanka, Thailand, Indonesia, the Philippines, and Vietnam (Timsina and Connor 2001). The area under rice-wheat system varies significantly, (Paroda 1996) reported 22.4 Mha, while Ladha et al. (2009) reported as 24 Mha. However, the total is 28.8 Mha: 13 Mha in China (Jiaguo 2000; Mahajan and Gupta 2009), 12.3 Mha in India (Kumar et al. 1998), 2.2 Mha in Pakistan (Bhatt et al. 2016), 0.8 Mha in Bangladesh (Ladha et al. 2009), and 0.5 Mha in Nepal (Paroda 1996). This system has been thoroughly cultivated and is the world's most advanced technological system; 13.5 Mha is grown in Asia, with South Asia accounting for 57% of the total (Ladha et al. 2009; Ahmad and Iram 2006). This system accounts for 32% of rice area and 42% of wheat area in Asian countries (Panwar et al. 2019). Moreover, the Indo-Gangetic Plain accounts for more than 85% of all rice-wheat cropping system cultivated in South Asia (Banjara et al. 2021b). The overall area and production of rice-wheat cropping system are presented in Table 3.1, and system overview is presented in Fig. 3.1 Currently, the ratoon rice is also being grown in Pakistan (Wahab et al. 2024).

M. Ahmed et al., *Cropping Systems Modeling Under Changing Climate*,
https://doi.org/10.1007/978-981-97-0331-9_3

Table 3.1 Area and production of rice-wheat systems across various regions in the world

Countries	Rice		Wheat		Rice-wheat system	
	Area (Mha)	Production (MMT)	Area (Mha)	Production (MMT)	Area (Mha)	Production (MMT)
Bangladesh	11.62	35.85				
India	46.38	130.29	31.13	109.59	77.51	239.88
Nepal	01.47	03.73	00.72	02.08	02.19	05.81
Pakistan	03.54	09.10	09.17	27.46	12.71	36.56
Iran	00.56	01.90	06.00	12.00	06.56	13.90
China	29.92	148.99	23.57	136.95	53.49	285.94
Egypt	00.50	02.90	01.40	09.00	01.90	11.90
USA	01.01	06.09	15.03	44.80	16.04	50.89
European Union	00.40	01.71	24.25	138.29	24.65	140.00

Source: USDA (2022)
Mha million hectare, *MMT* million metric ton

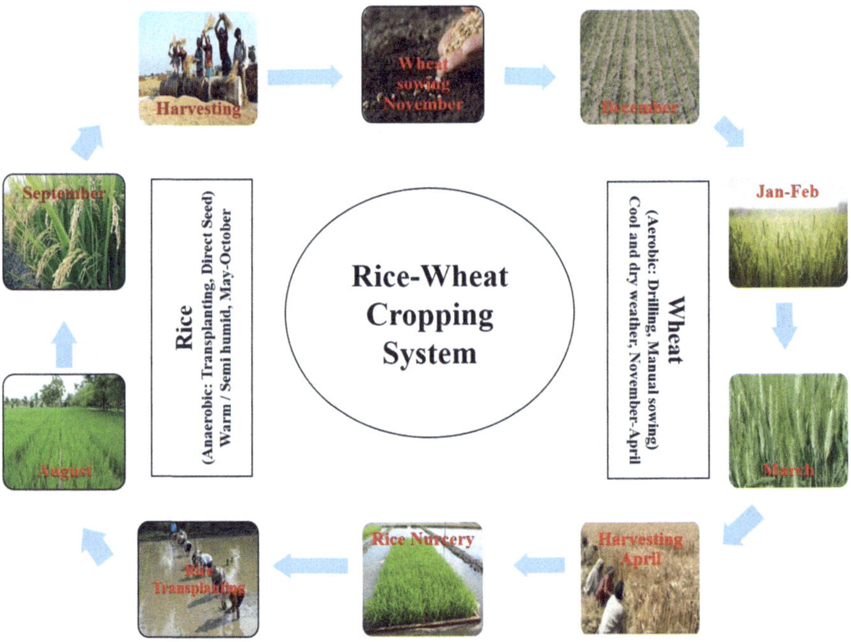

Fig. 3.1 Overview of rice-wheat cropping in South Asia. (Source: Adopted from Qamar et al. 2021)

3.2 Issues of Rice-Wheat Cropping System

Rice-wheat system causes various issues, which are agricultural, technical, environmental, livelihood, and social (Table 3.2): stagnant yield in rice-wheat cropping system due to unattended intervals (Bhatt and Kukal 2015), decreasing subsurface water table (Hira et al. 2004; Humphreys et al. 2010), environmental pollution (Bijay-Singh et al. 2008), and degradation of soil properties (Bhandari et al. 2002). Large quantity of water (rainfall and irrigation) is required for rice crop, as in R-W rotation system, flooded rice is mainly sown from June to October/November (Xu et al. 2019); however, less amount of water is required for wheat as it is grown from October/November to subsequent May (Qiu et al. 2019, 2022). Rice and wheat are nutrient-exhaustive and heavy feeder crops. The productivity of rice-wheat cropping system is under threat due to monotonous practices, nutrient mining, compaction, and productivity fatigue in the same field (Panwar et al. 2019; Qamar et al. 2021). Rice-wheat system disease (Fig. 3.2) and insect attacks have increased due to continuous growing (Bhatt et al. 2016). Industrial growth and the consequent shifting of people to cities, as well as a desire for work except for agriculture, have impacted labor demand at farms (Farooq and Nawaz 2014). In the rice season, puddling changes the porosity of the soil, which affects the moisture-holding capacity (McDonald et al. 2006); bulk density of soil increases (Farooq and Nawaz 2014); shoot-root growth decreases in wheat (McDonald et al. 2006); and high intensity of puddling decreases wheat yield (Arora et al. 2006). Burning of rice residue before

Table 3.2 Agricultural, technical, ecological, livelihood, and social issues of rice-wheat cropping systems

Issues of rice-wheat cropping system				
Agricultural	Technical	Ecological	Livelihood	Social
1. Deteriorating soil health	1. Low mechanization	1. Environmental pollution	1. High energy requirement	1. Population expansion
2. Large management yield gaps	2. Lack of appropriate seeders, especially for small- and medium-scale farmers	2. Declining underground water table	2. Decreased land productivity	2. Farmers' mindset with traditional practices
3. Residue management	3. Lack of adoption rate of new technology	3. Groundwater pollution	3. Decreased water productivity	3. Youth moving away from farming
4. Least attended intervening period		4. Diverse weed flora	4. Decreased efficiency of water use	4. Landholding getting smaller and fragmented
5. Labor shortage		5. New diseases and insect pests	5. Poor incomes	
6. Declining crop response		6. Reduced biodiversity	6. Climatic issues	

Source: Modified and adopted from Kakraliya et al. (2018)

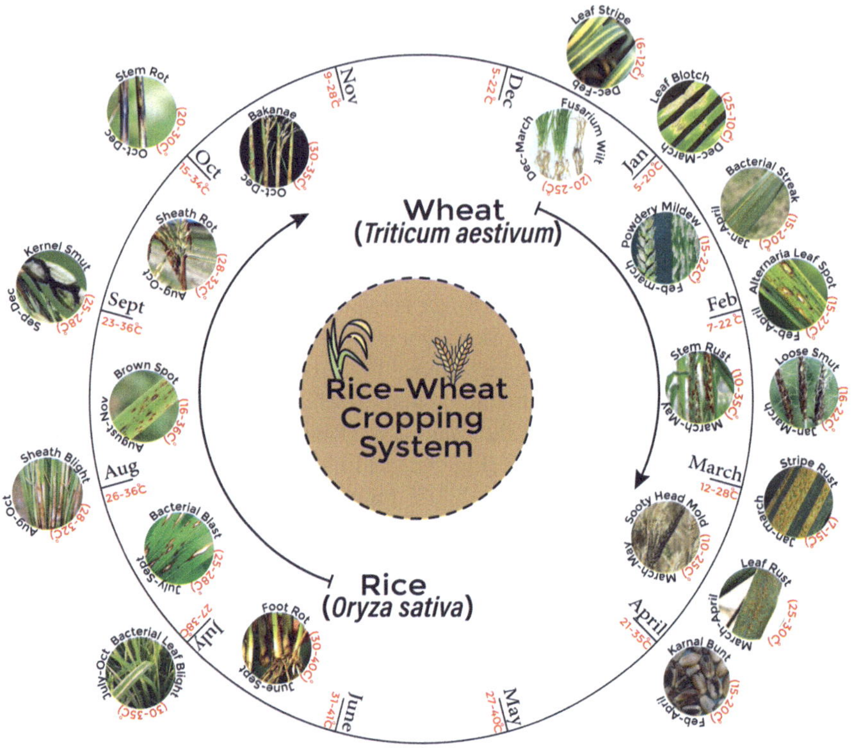

Fig. 3.2 Occurrence of diseases in rice-wheat system in Punjab, Pakistan. (Courtesy: Dr. Ummad ud Din Umar; Mr. Muhammad Fahad Hakim; Department of Plant Pathology, Bahauddin Zakariya University, Multan, Pakistan)

wheat sowing increases the amount of hydrocarbons and greenhouse gases in the atmosphere in the rice-wheat cropping system (Ahmad et al. 2013; Ahmad et al. 2012; Ahmed et al. 2016; FAO 2021; Li et al. 2014; Tao et al. 2013: Jain et al. 2014).

3.3 Rice-Wheat Modeling

The genetic coefficients of rice and wheat are presented in Tables 3.3 and 3.4: the rice-wheat system's observed as well as simulated LAI (Fig. 3.3) of rice (a, b) and wheat (c, d) and biomass (Fig. 3.4) of rice (a, b) and wheat (c, d) at a study site at Punjab, Pakistan.

Table 3.3 Genetic coefficients, description, ranges, and guidelines for model calibration of rice-wheat crops in DSSAT

Genetic coefficient	Description	Genetic coefficient	Description
VAR#	Identification code or number for a specific cultivar		
VAR-Name	Name of cultivar		
EXPNO	Number of experiments used to estimate cultivar parameters		
ECO#	Ecotype code of this cultivar, points to the ecotype in the ECO file		
Rice		*Wheat*	
P1	Time period (expressed as growing degree days [GDDs] in °C-d above a base temperature of 9 °C) from seedling emergence during which the rice plant is not responsive to changes in photoperiod. This period is also referred to as the basic vegetative phase of the plant. **Range:** 150–800 °C-d. **Calibration:** Flexible, compare with observed panicle initiation and flowering dates	VSEN	Sensitivity to vernalization
P2O	Critical photoperiod or the longest day length (in hours) at which the development occurs at a maximum rate. At values higher than P2O, developmental rate is slowed, and hence there is delay due to longer day lengths. **Range:** 11–13 h. **Default:** 12 h. **Calibration:** Do not go below 11 unless data are available	PPSEN	Sensitivity to photoperiod
P2R	Extent to which phasic development leading to panicle initiation is delayed (expressed as GDD in °C-d) for each hour increase in photoperiod above P2O. **Range:** 5–300 °C-d. **Calibration:** Modern rice varieties will have values in the lower range	P1	Thermal time from seedling emergence to the end of the juvenile phase
P5	Time period in GDD °C-d from beginning of grain filling (3–4 days after flowering) to physiological maturity with a base temperature of 9 °C. **Range:** 150–850 °C-d. **Calibration:** Ensure that P1, P2O, and P2R are correctly calibrated for anthesis data. Then calibrate P5 for observed maturity date	P5	Thermal time (base 0 °C) from beginning of grain fill to maturity: range 500–700
G1	Potential spikelet number coefficient as estimated from the number of spikelets per g of main culm dry weight (less leaf blades and sheaths plus spikes) at anthesis. **Range:** 50–75 #/g. A typical value is 55 #/g	PHINT	Phyllochron interval

(continued)

Table 3.3 (continued)

Genetic coefficient	Description	Genetic coefficient	Description
G2	Single grain weight (g) under ideal growing conditions, i.e., non-limiting light, water, nutrients, and absence of pests and diseases. **Range:** 0.015–0.030 g. **Default:** 0.025 g. **Calibration:** Very low flexibility	GRNO	Coefficient of kernel number per stem weight at the beginning of grain filling [kernels (g stem)$^{-1}$]
G3	Tillering coefficient (scalar value) relative to IR64 cultivar under ideal conditions. **Range:** 0.7–1.3. **Calibration:** A higher tillering cultivar would have a coefficient greater than 1.0	MXFIL	Potential kernel growth rate [mg kernel^{-1} day^{-1}]: Values between 1.0 and 3.0
PHINT	Phyllochron interval (°C-d). Time interval in degree-days for each leaf tip to appear under non-stressed conditions. **Range:** 55–90 °C-d. Default 83 °C-d. **Calibration:** Recommend not to change unless field data on leaf numbers are available	STMMX	Potential final dry weight of a single tiller (excluding grain) (g stem^{-1}) values 1.0–3.0
THOT	Temperature (°C) above which spikelet sterility is affected by high temperature. **Range:** 25–34 °C. **Default:** 28 °C. **Calibration:** Recommend not to change unless hot environment data are available. Convert old cultivars THOT = 28/G4	SLAP1	Ratio of leaf area to mass at emergence (cm^2/g)
TCLDP	Temperature (°C) below which panicle initiation is further delayed (other than P1, P2O, and P2R) by low temperature. **Range:** 12–18 °C. Default 15 °C. **Calibration:** Recommend not to change unless cold environment data are available. Convert old cultivars TCLDP = 15.*G5		
TCLDF	Temperature (°C) below which spikelet sterility is affected by low temperature. **Range:** 10–20 °C. Default 15 °C. **Calibration:** Recommend not to change unless cold environment data are available. Convert old cultivars TCLDF = 15.*G5		

Source: DSSAT v4.8; Hoogenboom et al. (2024)
Note: The cultivar coefficient G5 was added to DSSAT Version 4.7.5. The cultivar coefficients G4 and G5 were removed—DSSAT Version 4.8. The cultivar coefficients THOT, TCLDP, and TCLDF were added to DSSAT Version 4.8

Table 3.4 Genetic coefficients and description of wheat crop in APSIM model

Genetic coefficients	Description
photop_sens	Photoperiod sensitivity
vern_sens	Vernailzation sensitivity
tt_end_of_juvenile	Thermal time needed from sowing to the end of juvenile (°C days)
tt_flowering	Thermal time needed in anthesis phase (°C days)
tt_floral_initiation	Thermal time from floral initiation to flowering (°C days)
tt_start_grain_fill	Thermal time from the start of grain filling to maturity (°C days)
max_grain_size	Maximum grain size (G)
potential_grain_growth_rate	Grain growth rate from flowering to grain filling (g $grain^{-1}$ day^{-1})
potential_grain_filling_rate	Potential daily grain filling rate (g $grain^{-1}$ day^{-1})
grains_per_gram_stem	Grain number per stem weight at the start of grain filling (G)

Source: Ahmad et al. (2018)

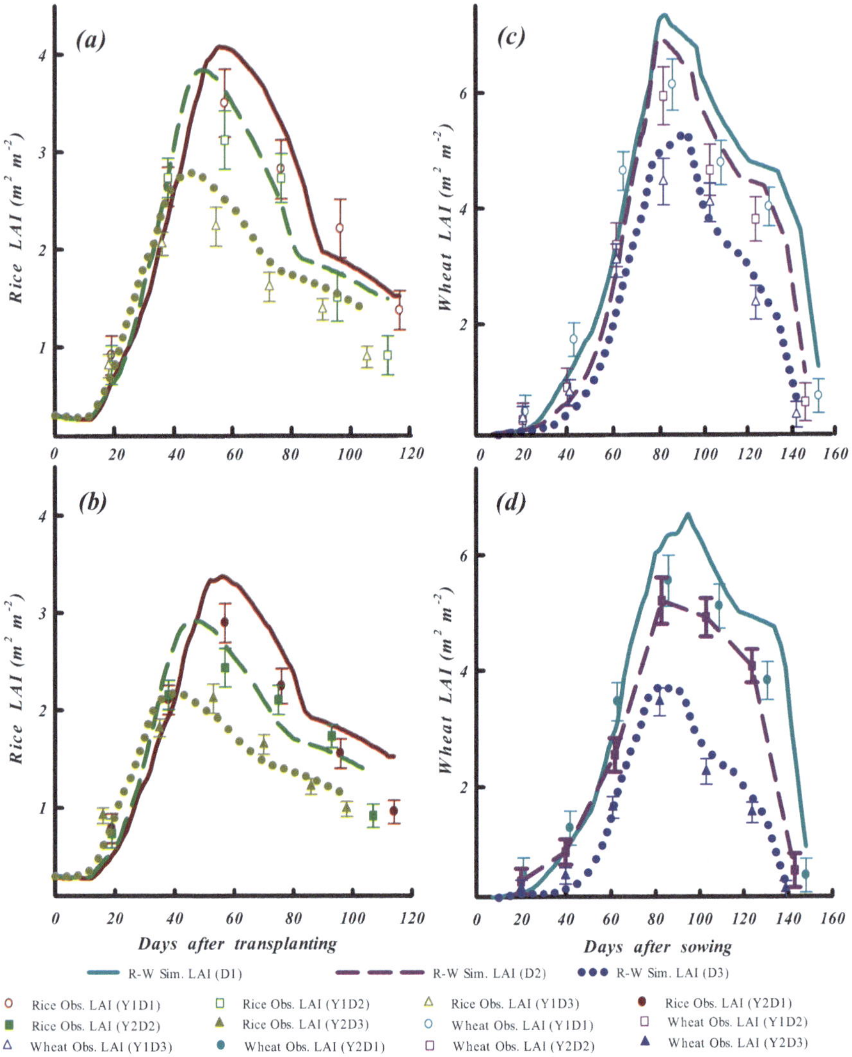

Fig. 3.3 Rice-wheat system observed and simulated leaf area index (LAI) of rice (**a, b**) and wheat (**c, d**) as affected by transplanting/sowing dates under irrigated environment at Multan, Pakistan. (CERES-Rice and CERES-Wheat model evaluation; Source: Adopted from Ahmad et al. 2019)

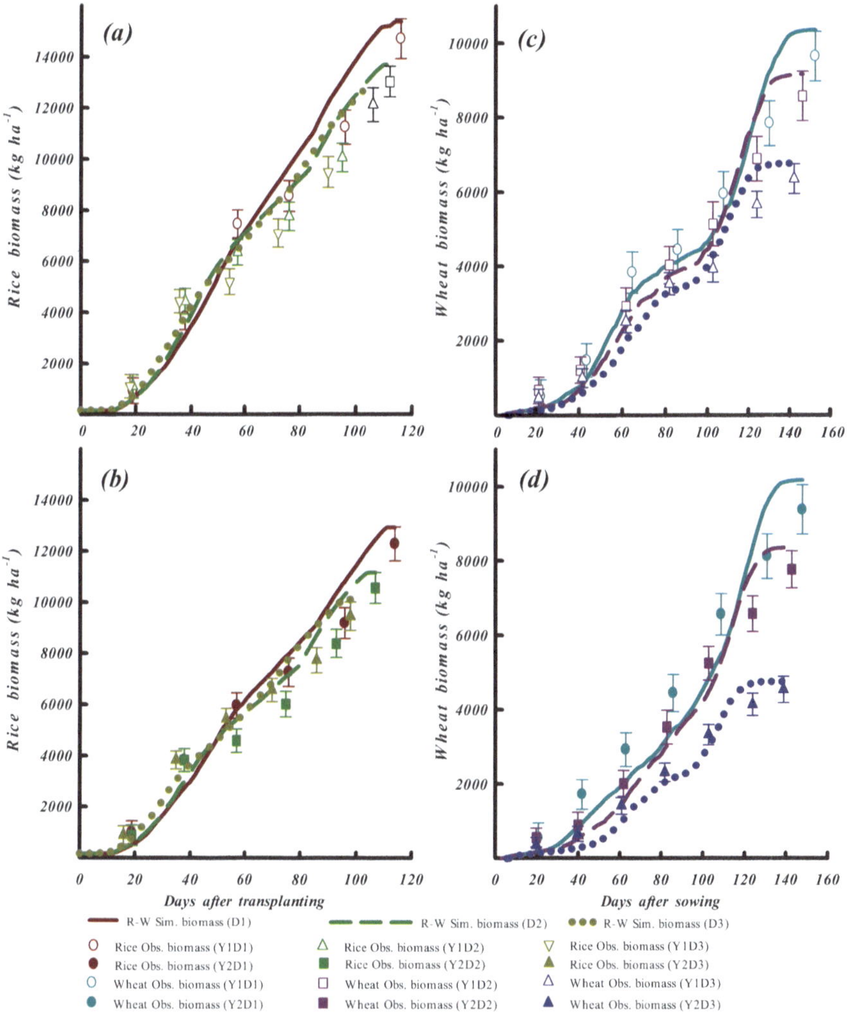

Fig. 3.4 Rice-wheat system observed and simulated biomass of rice (**a, b**) and wheat (**c, d**) as affected by transplanting/sowing dates under irrigated environment at Multan, Pakistan. (CERES-Rice and CERES-Wheat model evaluation; Source: Adopted from Ahmad et al. 2019)

3.4 Climate Change Influence on Rice-Wheat System Phenology

Rice-wheat cropping system phenology is presented in Fig. 3.5. Climate changes intimidate agricultural production locally, regionally, and globally (Abbas et al. 2017; Liu et al. 2018; Ahmad et al. 2015, 2021). Climate change trends recorded average temperature from 0.78 to 1.5 °C in the previous 30 years, while it is projected to be 2–4 °C around 2100 (IPCC 2014; Bokhari et al. 2017). It is anticipated that upsurging temperature trend will have exaggerated effects on agriculture of arid as well as semiarid areas (Amin et al. 2018; Ahmad et al. 2017a, 2017b; Tariq et al. 2018). The rice-wheat system means temperature trends (Fig. 3.6) from 1980 to 2014 and observed and simulated "A" and "M" dates (Fig. 3.7) in Punjab, Pakistan, at ten sites are presented (Ahmad et al. 2019). The phenology of crops in various cropping

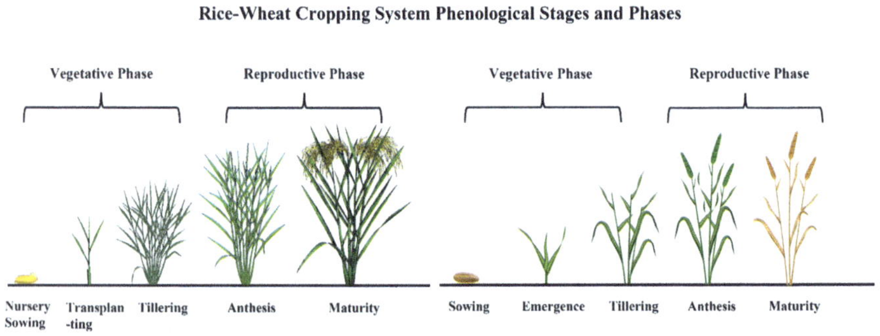

Fig. 3.5 Illustration of rice-wheat cropping system phenology

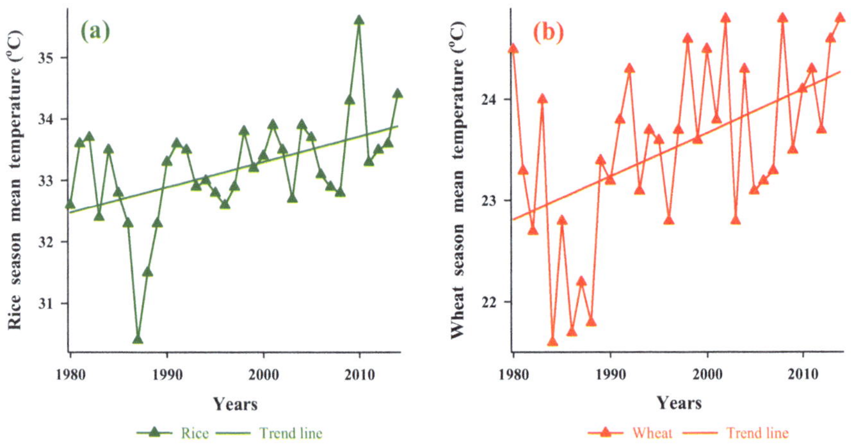

Fig. 3.6 Rice-wheat system mean temperature trends during rice (**a**) and wheat (**b**) seasons from 1980 to 2014 in Punjab, Pakistan. (Source: Modified and adopted from Ahmad et al. 2019)

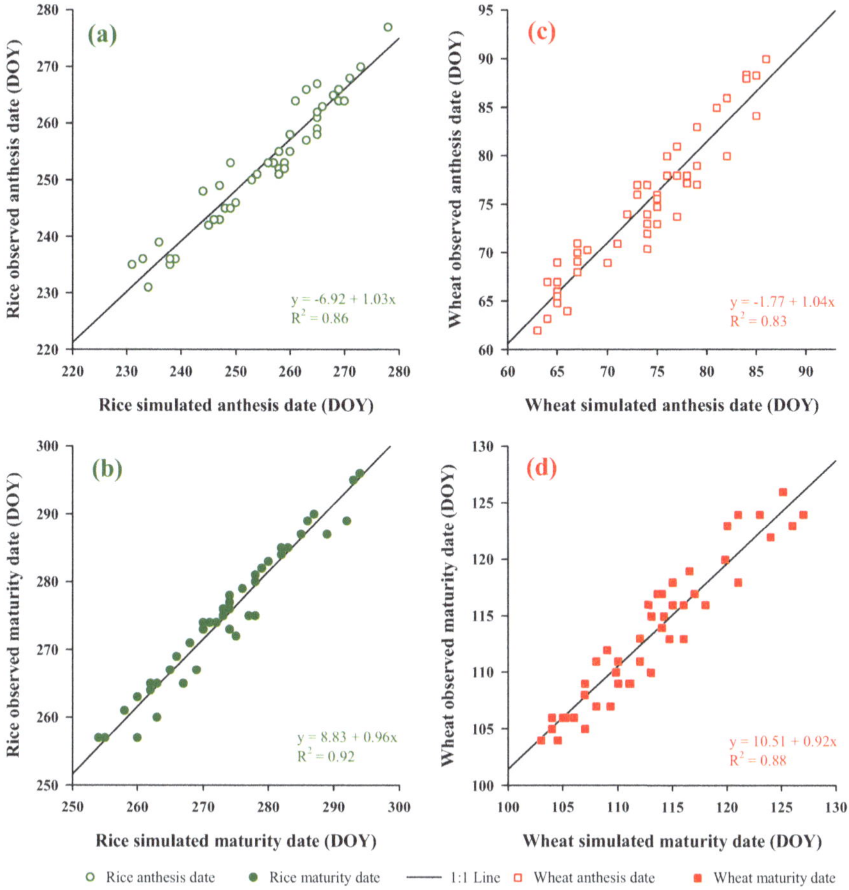

Fig. 3.7 Rice-wheat system observed and simulated anthesis and maturity dates of rice (**a, b**) and wheat (**c, d**) at ten locations in Punjab, Pakistan. (Source: Modified and adopted from Ahmad et al. 2019)

systems directly affects productivity and final biomass of these cropping systems, and therefore adaptation of cropping systems is very imperative to weather vagaries (Mueller et al. 2014; Ahmad et al. 2016; Liu et al. 2018). Figure 3.8 describes the impact of climate change on rice-wheat system phenology for ten locations in Punjab, Pakistan (Ahmad et al. 2019). The management practices (cultivar and sowing date selection; Fig. 3.9), growing degree day's fulfilment, and ecological circumstances have a substantial influence on cropping system's phenology (stages as well as phases) (Liu et al. 2018; Ahmad et al. 2019). Synchronization of weather parameters like temperature and precipitation with critical developmental stages of different cropping systems is very decisive for the full expression of their genetic

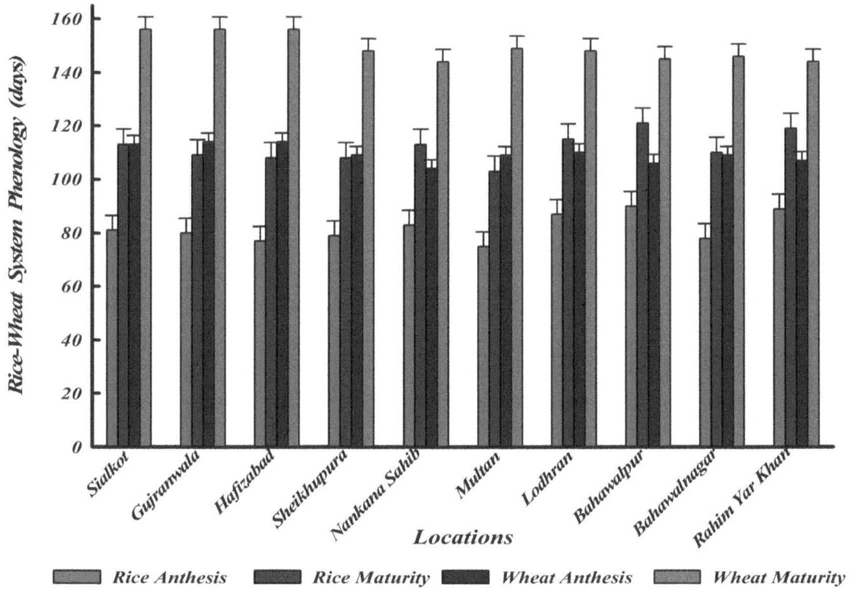

Fig. 3.8 Impact of climate warming on rice-wheat system phenology (mean days to anthesis and maturity from transplanting/sowing) at ten locations during the period of 1980–2014. (Source: Adopted from Ahmad et al. 2019)

potential (Amin et al. 2018; Ahmad et al. 2021). Climatic change-based changes in cropping system stages as well as phases are shrinking the growth cycle of cropping systems, ultimately reducing the time and rate of accumulation for grain yield and aboveground biomass (Figueiredo et al. 2015; Ahmad et al. 2023). Understanding cropping systems' phenological responses to climate change is necessary for better agronomic practices and breeding approaches for mitigation of impacts of climate change (Tariq et al. 2018; Ahmad et al. 2021). The anthesis/flowering date of winter wheat approached earlier in recent seven decades in the USA because of upsurge in temperature due to climate change. Consequently, lesser duration was available for light interception, utilization efficiency, and biomass accumulation (Hu et al. 2005). Wang et al. (2008) reported that in Loess Plateau of China during 1983–2004, due to climate change, growth of wheat was advanced for stem elongation, booting, and flowering, and maturity averaged by 13.2, 9.8, 11.0, and 10.8 days decade^{-1}. Xiao et al. (2013) reported that wheat anthesis and maturity stages approached earlier by 2.7 and 1.4 days decade^{-1} because of climate change during 1981–2009 in Northern China. The sowing and emergence dates were delayed averagely 1.2 and 1.3 days decade^{-1}, and anthesis, maturity, sowing-anthesis, and sowing-maturity advanced averagely by 3.7, 3.1, 4.3, and 5.0 days decade^{-1} in China (He et al. 2015). Rezaei et al. (2015) concluded that cropping systems' growths and developments

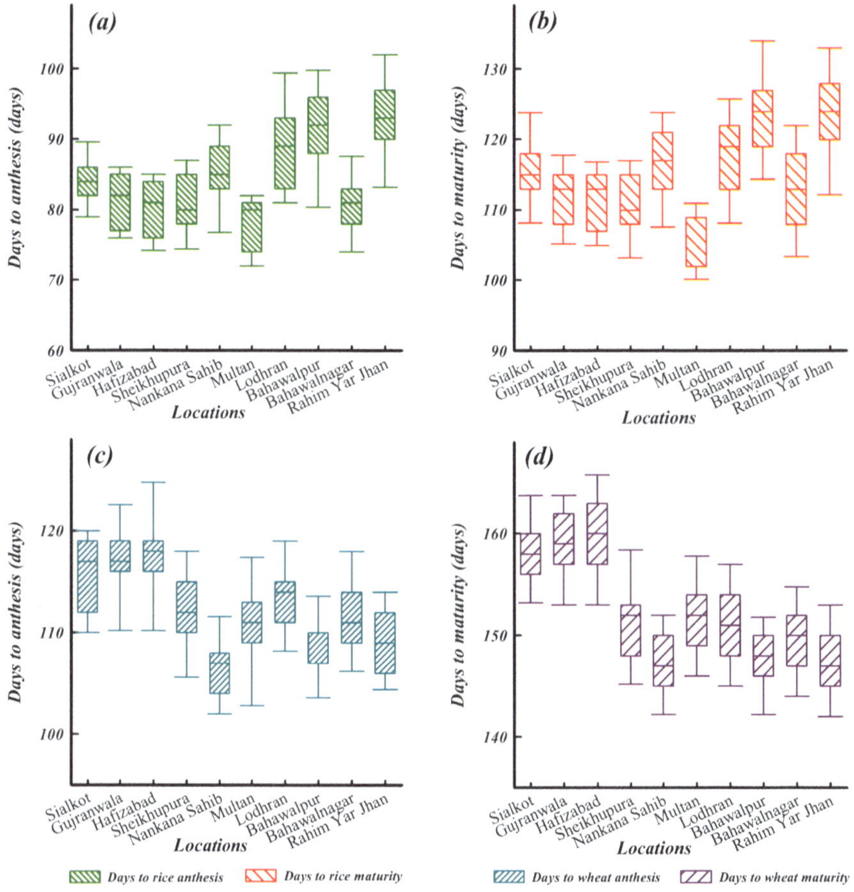

Fig. 3.9 Interactive effect of climate change and crop management practices on rice-wheat system phenology of rice (**a, b**) and wheat (**c, d**) system at ten locations during 1980–2014. Box central line represents median; box lower and upper limits represent 25th and 75th percentiles; whiskers represent minimum and maximum values. (Source: Adopted from Ahmad et al. 2019)

would be augmented because of higher thermal trends, and adverse effects of climate change could be reduced by sowing late and introducing newer cultivars desiring higher thermal times in various cropping systems. (Ahmed and Ahmad 2023; Ahmed 2023; Abbas et al. 2023; Liu et al. 2023; Ahmed et al. 2022a; Ahmed et al. 2022b; Nadeem et al. 2022; Ahmed 2020; Khan et al. 2020; Asseng et al. 2019; Ahmed et al. 2017; Jabeen et al. 2017; Aslam et al. 2013; Ahmed et al. 2012). However, Li et al. (2016) reported that determination of cropping system phenology against thermal trends is a very complicated task due to consistent sowing or transplanting time's adjustments and sowing of new cultivars. The statistical models

Table 3.5 Observed rice-wheat system phenology in Punjab, Pakistan, during the period of 1980–2014 (Source: Adopted from Ahmad et al. 2019)

Locations	Sowing	Transplanting	Panicle[a] initiation	Anthesis[b]	Maturity[c]
Rice					
Sialkot	115 ± 4.8	141 ± 6.0	190 ± 5.8	215 ± 6.3	278 ± 5.3
Gujranwala	122 ± 6.2	148 ± 7.1	185 ± 6.6	219 ± 5.3	287 ± 4.7
Hafizabad	113 ± 5.3	138 ± 4.2	192 ± 3.9	222 ± 4.4	280 ± 6.3
Sheikhupura	120 ± 4.9	146 ± 5.1	196 ± 4.8	228 ± 6.6	274 ± 4.2
Nankana Sahib	124 ± 7.1	150 ± 6.4	194 ± 6.1	223 ± 3.2	286 ± 5.4
Multan	118 ± 5.8	146 ± 5.9	198 ± 5.4	232 ± 3.9	293 ± 5.2
Lodhran	124 ± 3.2	153 ± 4.6	201 ± 4.2	226 ± 4.4	289 ± 6.0
Bahawalpur	121 ± 6.1	147 ± 5.2	205 ± 4.9	234 ± 5.8	298 ± 7.4
Bahawalnagar	121 ± 3.5	156 ± 3.4	199 ± 3.2	231 ± 4.9	302 ± 5.9
Rahim Yar Khan	127 ± 7.2	158 ± 6.2	208 ± 5.8	229 ± 6.8	296 ± 6.3
Locations	**Sowing**	**Emergence**		**Anthesis[a]**	**Maturity[b]**
Wheat					
Sialkot	316 ± 5.6	321 ± 5.2		65 ± 5.3	113 ± 6.4
Gujranwala	319 ± 4.9	324 ± 4.5		72 ± 6.1	118 ± 5.8
Hafizabad	314 ± 3.7	318 ± 3.4		78 ± 5.3	120 ± 4.1
Sheikhupura	326 ± 5.9	330 ± 5.3		70 ± 6.8	123 ± 5.9
Nankana Sahib	322 ± 6.2	326 ± 5.9		74 ± 7.4	122 ± 4.7
Multan	321 ± 5.1	325 ± 4.5		79 ± 4.9	118 ± 6.7
Lodhran	318 ± 6.4	323 ± 6.1		73 ± 6.2	126 ± 4.0
Bahawalpur	326 ± 4.8	330 ± 4.5		78 ± 4.4	122 ± 5.6
Bahawalnagar	324 ± 5.7	328 ± 5.2		80 ± 5.8	124 ± 5.2
Rahim Yar Khan	328 ± 5.3	332 ± 4.8		83 ± 6.4	127 ± 6.1

Source: Ahmad et al. (2019)
± Standard deviation
[a]50% Panicle initiation
[b]50% Anthesis
[c]50% Physiological maturity

could not find mutual interrelationships among agronomic practices, environmental variable variability, and cultivar shifts. The interrelationships could be determined and predicted by processing-based models like DSSAT and APSIM; furthermore, the effect of a single element could be determined by other elements (Jones et al. 2003; Keating et al. 2003). Ahmad et al. (2019) determined the spatiotemporal variability of rice-wheat system phenology, agronomic practices, and cultivar shifts in Punjab, Pakistan, using DSSAT model (Tables 3.5 and 3.6). However, comparative reactions of rice-wheat system phenology against mean temperature using observed and simulated data in Pakistan for 1980–2014 are presented in Fig. 3.10.

Table 3.6 Comparison of responses of rice-wheat system phenology with mean temperature using observed and simulated data in Punjab, Pakistan, during 1980–2014

Phenology	Regression coefficient (days °C^{-1})		Difference between obs. and sim. data (days °C^{-1})	t-Test (*p*-value)
	Obs. data	Sim. data		
Rice				
Sowing-transplanting	−1.08	−1.55	−0.47	0.0029*
Transplanting-maturity	−1.2	−1.75	0.55	0.0012**
Panicle initiation-anthesis	−0.73	−1.05	0.32	0.018*
Panicle initiation-maturity	−1.31	−1.79	0.48	0.0015**
Anthesis-maturity	−0.46	−1.01	0.55	0.0018**
Wheat				
Sowing-anthesis	−1.59	−2.07	0.48	0.0013**
Sowing-maturity	−2.09	−2.38	0.29	0.0002**
Anthesis-maturity	−1.81	−2.55	0.74	0.0009**

Source: Ahmad et al. (2019)
Obs. observed, *Sim.* simulated
*Significant at 0.05 probability level; ** significant at 0.01 probability level

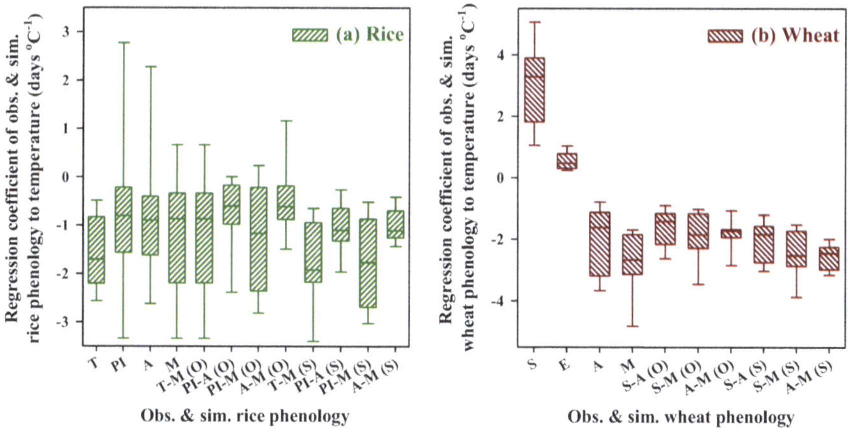

Fig. 3.10 Rice-wheat system observed phenology (stages and phases) and simulated phenology (phases) versus temperature trends using a standard rice (a-part) cultivar and wheat (b-part) cultivar (*S* sowing, *E* emergence, *T* transplanting, *PI* panicle initiation, *A* anthesis, *M* maturity, *S-T* sowing-transplanting, *T-M* transplanting-maturity, *PI-A* panicle initiation-anthesis, *PI-M* panicle initiation-maturity, *S-A* sowing-anthesis, *A-M* anthesis-maturity, *S-M* sowing-maturity, *O* & *Obs.* observed, *S* & *Sim* simulated). Box central line represents median; box lower and upper limits represent 25th and 75th percentiles; whiskers represent minimum and maximum values. (Source: Modified and adopted from Ahmad et al. 2019)

3.5 Climate Change Influence on Rice-Wheat System

3.5.1 Impact of Climate Change on Rice

Reduced rice yield was anticipated in Pakistan, China, India, Vietnam, etc., but the increase was predicted in Myanmar. The CERES-Rice model with five global circulation models and under RCP 8.5 predicted the yield of hybrid, fine, and coarse yield in Pakistan for a period of mid-century. An increase in average rainfall than baseline and future increase in temperature reduced the yield of hybrid rice (24–64%), fine rice (18–54%), and coarse rice (18–54%). Rice production was threatened by dry and hot climatic conditions in Pakistan (Nasir et al. 2020). In India, rice production was also predicted by CERES-Rice Model with eight GCMs and under four RCPs (2.6, 4.5, 6.0, and 8.5). Rice simulated yield for the future showed a reduction of 1.2–8.8%, 0.7–12.6%, and −2.9–17.8% in 2020, 2050, and 2080, respectively, under RCP 8.5, 6.0, and 4.5, but RCP 2.6 showed a positive response to wheat yield (Gupta and Mishra 2019). The negative impact of CO_2 on rice yield and duration of flowering was predicted with one GCM, PRECIS-RCM, and B2 at the end of the century in China (Table 3.7). Rice yield decreased by 0.25–7% at different locations in China (Yao et al. 2007). The effect of climate change on rice production in Thailand was estimated by the econometric panel data model under PRECIS-RCM, with A2 and B2. Rice yield was reduced by 5–34% at the mid of the century (Sinnarong et al. 2019). In other study related to Thailand's climatic conditions' effects on the water footprint of rice yield, the CROPWAT model with five GCMs and RCP (4,5 and 8.5) for the early, mid, and late centuries was used. Simulated results showed that under RCP 4.5, rice yield increased by 1.3–28.9%, but in RCP 8.5, it is the total opposite, with yield reduced by 11.7–28.9% (Arunrat et al. 2020). In Bangladesh, rice yield was reduced by 20–50% at mid-century, and 2070 was predicted by the CERES-RICE model and PRECIS RCM (Basak et al. 2010). Rice yield was affected by water management and climatic conditions. Rice yield was reduced by 23% due to an increase in temperature and decreased rainfall by AquaCrop and RCP 8.5 in Indonesia (Santikayasa et al. 2018). Shrestha et al. (2014) reported that a rice yield increase of 16–40% in 2020, 2050, and 2080 in Myanmar was predicted under AquaCrop, two GCMs, A1, A2, B1, and B2. In Vietnam, the yield was reduced by 0.25–0.49% in 2020, 2030, and 2040, which was predicted under equilibrium displacement model (EDM) (Le 2016).

3.5.2 Impact of Climate Change on Wheat

The CERES-Wheat model was used in Pakistan to know the impact of changing climate on wheat yield (Table 3.8). Climate scenarios RCP 4.5 and RCP 8.5 showed an increasing trend of maximum and minimum temperatures that will decrease the yield at D.I. Khan by 7.9% and 11% under RCP (4.5 and 8.5). At Peshawar, yield will also decrease by 5.5% and 8.4% under RCP (4.5 and 8.5), but the yield will show an increase of 10% and 14% under RCP (4.5 and 8.5) in Chitral for

Table 3.7 Impact of climate change on rice production

Country	Model	Climatic scenarios	Yield	References
Pakistan	CERES-Rice	5 GCMs and RCP (8.5) (1980–2010 and 2040–2069)	Yield decreased up to 10.33–64% due to high temperature	Nasir et al. (2020)
India	CERES-Rice	8 GCMs and 4 RCPs (2.6, 4.5, 6.0, and 8.5) (1975–2005, 2006–2035, 2036–2065, and 2066–2095)	Yield decreased up to 1.2–8.8%, 0.7–12.6%, and 2.9–17.8% in 2020, 2050, and 2080	Gupta and Mishra (2019)
China	CERES-Rice	1 GCM, 1RCM (PRECIS), and B2 (2071–2090)	Yield decreased from 0.25–7% at different locations	Yao et al. (2007)
Thailand	CROPWAT	5 GCMs and 2 RCPs (4.5 and 8.5) (2020–2039, 2040–2059, 2060–2079, and 2080–2099)	Yield increased from 1.3% to 29.8% under RCP 4.5, but it fluctuated under RCP 8.5 from 11.7% to −29.8%	Arunrat et al. (2020)
Thailand	Panel Data	RCM (PRECIS), A2 and B2 (1989–2009, 2020, 2030, and 2040)	Yield decreased from 5% to 34%	Sinnarong et al. (2019)
Indonesia	AquaCrop	GCM and RCP 8.5 (1971–2000, 2021–2050)	Yield decreased by 23% with the increase in temperature and decrease in rainfall	Santikayasa et al. (2018)
Indonesia	CROPWAT	17 GCMs and RCPs (4.5 and 8.5) (1981–2010, 2011–2040, 2041–2070)	Yield decreased 2.8–29.3% under RCP 4.5 and 3.6–30.2% under RCP 8.5	Candradijaya et al. (2014)
Bangladesh	CERES-Rice	RCM (PRECIS) (2008, 2030, 2050, 2070)	Yield decreases up to 20% and 50% for the years 2050 and 2070	Basak et al. (2010)
Myanmar	AquaCrop	2 GCMs, A1, A2, B1, and B2 (1961–1990, 2020, 2050, and 2080)	Under different climatic conditions, yield increased by 16–40%	Shrestha et al. (2014)
Vietnam	Equilibrium displacement model (EDM)	(1980–1999, 2020, 2030, and 2040)	Yield decreased from 0.25% to 0.49%	Le (2016)

Table 3.8 Impact of climate change on wheat production

Country	Model	Climatic scenarios and study periods	Yield	References
Pakistan	CERES-Wheat	HAPPI Scenarios (2030–2040)	Yield decreased due to temperature and precipitation in the province of Punjab (3.2% and 4.7%) and in Sindh (17.8% and 13.8%)	Fahad et al. (2020)
China	CERES-Wheat	2 GCMs and 3 RCPs (4.5, 6.0, and 8.5) (2020–2050)	Yield increased up to 28% in all scenarios	Zheng et al. (2020)
Pakistan	CERES-Wheat	5 GCMs and RCPs (4.5 and 8.5), mid-century (2040–2069)	Yield increases in the district Chitral (10% and 14%) and reduced in D.I. Khan (7.9% and 11%) and in Peshawar (5.5% and 8.4%)	Gul et al. (2020)
China	Cobb-Douglas	5 GCMs and 2 RCPs (4.5 and 8.5) (1981–2005 and 2021–2050)	Yield increased up to 1.47% and 2.16% in RCPs 4.5 and 8.5	Geng et al. (2019)
India	InfoCrop WHEAT	1 GCM and 1 RCM (PRECIS) (2000–2007, 2020–2050, and 2070–2100)	Yield decreased 6–23% and 15–25% by 2050 and 2080	Kumar et al. (2014)
Australia	APSIM-Wheat	9 GCMs and RCMs (1900–1999 and 2080)	Yield decreased up to 13.5–32%	Luo et al. (2005)
Canada	DNDC, DayCent, and DSSAT	20 GCMs and RCP 8.5 (2006–2015, 2025, 2040, 2052, 2063)	Yield increase showed in all models	Qian et al. (2019)
USA	CERES-Wheat	4 GCMs and 2 RCPs (6.0 and 8.5) (1980–2014 and 2040–2060)	Yield increases up to 3–10% and 4–20% in both RCPs	Dhakal et al. (2018)
France	Fixed-effects regression	5 GCMs and 4 RCPs (2.6, 4.5, 6.0, and 8.5) (1976–2005, 2037–2065, 2071–2099)	Yield decreased up to 3–17%	Gammans et al. (2017)
Russia	Fixed-effects regression	4 RCPs (2.6, 4.5, 6.0, and 8.5) (1971–2000, 2046–2065, 2081–2100)	Yield decreased up to 6.7%	Belyaeva and Bokusheva (2018)

mid-century (2040–2069) (Gul et al. 2020), while the use of HAPPI scenarios showed that changes in precipitation and temperature reduced the yield in Sindh (13.8% and 17.8%) and in Punjab (4.7% and 3.2%) province, while in Baluchistan (15.3% and 9.4%) and Khyber Pakhtunkhwa (15.3% and 9.4%), wheat yield will increase (Fahad et al. 2020). The impact of changes in climate was estimated using the CSM-CERES-Wheat model under three RCPs (4.5, 6.0, and 8.5) for a period of 2025–2050. The results showed that anthesis and maturity days were shortened (3–23 and 4–24 days), respectively, but in all scenarios, the wheat yield was enhanced by 28% (Zheng et al. 2020). At the end of the century, the wheat yield will be decreased in India. Climatic scenarios GCM and RCM results showed that yield will reduce by 6–23% and 15–25% by 2050 and 2080, respectively, due to increases in seasonal minimum and maximum temperature (Kumar et al. 2014). The APSIM-Wheat model with nine GCMs assessed the effect of climate change on wheat yield in Australia. Under the changing climate conditions (atmospheric CO_2, regional rainfall and temperature), yield was reduced up to 13.5–32% among all scenarios (Luo et al. 2005). Qian et al. (2019) used three different models (DNDC, DayCent, and DSSAT) to estimate the yield changes with 20 GCMs and 5 different global warming levels (1.5, 2.0, 2.5, and 3.0 °C). With increase in global warming to 2 °C, yield of wheat increased in all crop models and scenarios in Canada (Qian et al. 2019). Influence of climate on future wheat production in Oklahoma, USA, was predicted by the CERES-Wheat model with four general circulation models and two RCP (6.0 and 8.5) for the 2040–2060 future period. Annual rainfall decreases and temperature increases were predicted by the GCM and increase of 3–10% and 4–20% wheat yield by RCP (6.0 and 8.5). Wheat yield will increase in 2040–2060 as compared to the 1980–2014 baseline period (Dhakal et al. 2018). The negative impact of changes in climate on wheat yield in France was predicted by the fixed-effect regression model with five GCPs and four RCP (2.6, 4.5, 6.0, and 8.5). The model predicted that the temperature remaining less (7–12 °C) showed a positive effect on wheat crops rather than the temperature going high (12–32 °C), which showed a negative effect on yield. Yield decreases (3–13%) mid-century and end of the century (17%) (Gammans et al. 2017).

3.6 Rice-Wheat Cropping System Under Changing Climate: AgMIP Pakistan, a Case Study

3.6.1 Sensitivity of Current Rice-Wheat System Productivity to Climate Change

Current productivity at agricultural farms using current technology and perturbed climate is being affected by climate change in Asia, Africa, and all other parts of the world. The semi-subsistence farms, supporting livelihood of resource-poor farmers, are more vulnerable. The economic analysis in terms of socioeconomic indicators is also predicting similar trends (Claessens et al. 2012; Ahmad et al. 2015, 2021).

Ahmad et al. (2015) reported the impact of climate change on rice-wheat cropping system in Punjab, Pakistan. In terms of the average production of rice

(18,349 kg farm^{-1}) and wheat (18,915 kg farm^{-1}) crops, five GCMs predicted decrease in average production for rice (DSSAT, 6–19%; APSIM, 10–12%) and wheat (DSSAT, 8–30%; APSIM, 14–19%) in the rice-wheat system. Subash et al. (2021) also described the impact of climate change on rice-wheat cropping system in India. Due to climate change, mean irrigated rice production decreased marginally by DSSAT, and 3–10% by APSIM under RCP 4.5, in India. However, for RCP 8.5, it was 3% for DSSAT and 13% for APSIM. However, wheat yield reduction due to climate change for hot/dry GCM was 10% for DSSAT and 16% for APSIM under RCP 4.5 in India. However, for all GCMs and RCP 8.5, it was 7–13% for DSSAT and 6–19% for APSIM. The net farm returns under rice-wheat cropping system decreased 4–11% for RCP 4.5 and 5–14% for RCP 8.5. The per capita income also reduced by 3–7% for RCP 4.5.

3.6.2 The Impacts of Climate Change on Future Rice-Wheat System Production

This section describes the results regarding rice-wheat using strategy A (future rice-wheat system production with current climate) and strategy B (future rice-wheat system production with future climate) during mid-century (2040–2069) with five GCMs. This analysis was carried out to determine the individual role of climate impacts on RWS with adaptations. Climate impact on future rice-wheat system was estimated through gains and losses for individual GCM. The predicted rice production is 25,073 kg farm^{-1} for all GCMs (Table 3.8), and reductions in rice mean production range in 8–30% for DSSAT and 14–19% for APSIM. The predicted wheat productivity is 26,473 kg farm^{-1} regarding all GCMs (Table 3.8), and reductions in wheat average productivity range in 6.2–19% for DSSAT and 10.6–12.3% for APSIM. However, for five GCMs, percentage of rice-wheat growers which may lose their farm income ranged in 57–70% for DSSAT simulations and 61–71% for APSIM simulations. Furthermore, percentage gains for DSSAT will be 11–13% and losses would be 16–22%. Similar values for APSIM are 10–15% (gains) and 17–19% (losses). The poverty rate without climate change would be 16% for DSSAT and APSIM; however, with climate change, it would increase slightly ranging in 17–19% for DSSAT and 18–19% for APSIM.

3.7 Conclusion

The major area under RWS is situated in South Asia. Currently, productivity of RWS is stagnant. Among reasons for yield stagnation, climate change is one of the key reasons. Due to climate change, food security in developing countries is at risk. Scientists are quantifying climate change influence on RWS phenology and productivity along with adapting climate-smart practices.

References

Abbas G, Ahmad S, Ahmad A, Nasim W, Fatima Z, Hussain S, Rehman MH, Khan MA, Hasanuzzaman M, Fahad S, Boote KJ, Hoogenboom G (2017) Quantification the impacts of climate change and crop management on phenology of maize-based cropping system in Punjab, Pakistan. Agric For Meteorol 247:42–55

Abbas G, Ahmed M, Fatima Z, Hussain S, Kheir AMS, Ercişli S, Ahmad S (2023) Modeling the potential impact of climate change on maize-maize cropping system in semi-arid environment and designing of adaptation options. Agric For Meteorol 341:109674. https://doi.org/10.1016/j.agrformet.2023.109674

Ahmad I, Iram S (2006) Rice-wheat cropping pattern and resource conservation technologies. Agriculture overview. Pakistan. http://www.pakissan.com/english/agri.overview/rice.wheat.cropping.pattern.html. Accessed 21 Mar 2021

Ahmad S, Ahmad A, Soler CMT, Ali H, Zia-Ul-Haq M, Anothai J, Hussain A, Hoogenboom G, Hasanuzzaman M (2012) Application of the CSM-CERES-Rice model for evaluation of plant density and nitrogen management of fine transplanted rice for an irrigated semiarid environment. Precis Agric 13:200–218

Ahmad S, Ahmad A, Ali H, Hussain A, Garcia y Garcia A, Khan MA, Zia-Ul-Haq M, Hasanuzzaman M, Hoogenboom G (2013) Application of the CSM-CERES-Rice model for evaluation of plant density and irrigation management of transplanted rice for an irrigated semiarid environment. Irrig Sci 31:491–506

Ahmad A, Ashfaq M, Rasul G, Wajid SA, Khaliq T, Rasul F, Saeed U, Rahman MH, Hussain J, Baig IA, Naqvi SAA, Bokhari SAA, Ahmad S, Naseem W, Hoogenboom G, Valdivia RO (2015) Impact of climate change on the rice-wheat cropping system of Pakistan. In: Hillel D, Rosenzweig C (eds) Handbook of climate change and agro-ecosystems, vol 3. Imperial College Press and the American Society of Agronomy, London, pp 219–258

Ahmad S, Nadeem M, Abbas G, Fatima Z, Khan RJZ, Ahmed M, Ahmad A, Rasul G, Khan MA (2016) Quantification of the effects of climate warming and crop management on sugarcane phenology. Clim Res 71:47–61

Ahmad S, Abbas Q, Abbas G, Fatima Z, Atique-ur-Rehman NS, Younis H, Khan RJ, Nasim W, Rehman MH, Ahmad A, Rasul G, Khan MA, Hasanuzzaman M (2017a) Quantification of climate warming and crop management impacts on cotton phenology. Plan Theory 6:1–16

Ahmad S, Abbas G, Fatima Z, Khan RJ, Anjum MA, Ahmed M, Khan MA, Porter CH, Hoogenboom G (2017b) Quantification of the impacts of climate warming and crop management on canola phenology in Punjab, Pakistan. J Agron Crop Sci 203:442–452

Ahmad S, Rehman AU, Ejaz M, Fatima Z, Kan M, Ahmed M (2018) Agricultural land-use change of major field crops in Pakistan. Sci Technol Dev 37(3):113–121

Ahmad A, Abbas G, Ahmed M, Fatima Z, Anjum MA, Rasul G, Khan MA, Hoogenboom G (2019) Climate warming and management impact on the change of phenology of the rice-wheat cropping system in Punjab, Pakistan. Field Crop Res 230:46–61

Ahmad A, Ashfaq M, Rasul G, Wajid SA, Ahmad I, Khaliq T, Nasir J, Rasul F, Riaz F, Ahmad B, Ahmad S, Baig IA, Valdivia RO, Hoogenboom G (2021) Development of climate change adaptation strategies for cotton-wheat cropping system of Punjab Pakistan. In: Rosenzweig C, Mutter CZ, Contreras EM (eds) Handbook of climate change and agroecosystems: climate change and farming system planning in Africa and South Asia: AgMIP stakeholder-driven research. World Scientific Publishing Europe Ltd, London, pp 277–327

Ahmed M (2020) Introduction to modern climate change. Andrew E. Dessler: Cambridge University Press, 2011, 252 pp, ISBN-10: 0521173159. Sci Total Environ 734:139397. https://doi.org/10.1016/j.scitotenv.2020.139397

Ahmed M (2023) The science of climate change. In: Brinkmann R (ed) The Palgrave handbook of global sustainability. Springer International Publishing, Cham, pp 195–222. https://doi.org/10.1007/978-3-031-01949-4_22

Ahmed M, Ahmad S (eds) (2023) Disaster risk reduction in agriculture. Springer Nature. https://doi.org/10.1007/978-981-99-1763-1. ISBN 978-981-99-1762-4, Published: 03 November 2023

Ahmed M, Hassan FU, Aslam M, Aslam MA (2012) Physiological attributes based resilience of wheat to climate change. Int J Agric Biol 14(3):407–412

Ahmed M, Akram MN, Asim M, Aslam M, F-u Hassan Higgins S, Stockle CO, Hoogenboom G (2016) Calibration and validation of APSIM-wheat and CERES-wheat for spring wheat under rainfed conditions: models evaluation and application. Comput Electron Agric 123:384–401

Ahmed M, Stöckle CO, Nelson R, Higgins S (2017) Assessment of climate change and atmospheric CO_2 impact on winter wheat in the Pacific Northwest using a multimodel ensemble. Front Ecol Evol 5. https://doi.org/10.3389/fevo.2017.00051

Ahmed M, Hayat R, Ahmad M, ul-Hassan M, Kheir AMS, ul-Hassan F, ur-Rehman MH, Shaheen FA, Raza MA, Ahmad S (2022a) Impact of climate change on dryland agricultural systems: a review of current status, potentials, and further work need. Int J Plant Prod. https://doi.org/10.1007/s42106-022-00197-1

Ahmed M, Ahmad S, Kheir AMS (2022b) Climate change: an overview. In M. Ahmed (ed.), Global agricultural production - resilience to climate change, Springer Nature Switzerland AG https://doi.org/10.1007/978-3-031-14973-3_1

Ahmed M, Rehman A, Sarwar N, Doğan H, Abbas G, Khan MA, Kan M, Raza MA, Hussain S, Ahmad S, Fatima Z (2023) Deciding sowing-window for maize-based cropping system in arid and semiarid environments in Punjab, Pakistan. Turk J Agric For 47(6):Article 22. https://doi.org/10.55730/1300-011X.3149

Amin A, Nasim W, Mubeen M, Sarwar S, Urich P, Ahmad A, Wajid A, Khaliq T, Rasul F, Hammad HM, Rehmani MIA, Mubarak H, Mirza N, Wahid A, Ahmad S, Fahad S, Ullah A, Khan MN, Ameen A, Amanullah, Shahzad B, Saud S, Alharby H, Ata-Ul-Karim ST, Adnan M, Islam F, Ali QS (2018) Regional climate assessment for temperature and precipitation in Southern Punjab (Pakistan) using SimCLIM climate model for different temporal scales. Theor Appl Climatol 131:121–131

Arora VK, Gajri PR, Uppal HS (2006) Puddling, irrigation, and transplanting-time effects on productivity of rice-wheat system on a sandy loam soil of Punjab, India. Soil Tillage Res 85:212–220

Arunrat N, Pumijumnong N, Sereenonchai S, Chareonwong U, Wang C (2020) Assessment of climate change impact on rice yield and water footprint of large-scale and individual farming in Thailand. Sci Total Environ 726:137864

Aslam Z, Ahmed M, Sajad M, Asif M, Akmal M, Awan FK, Ijaz W, Ashraf R, Khattak JZK (2013) A comparison of statistical and dynamic modeling of wheat (*Triticum aestivum* L.) fungal diseases under the climate change. J Food Agric Environ 11(2):1327–1332

Asseng S, Martre P, Maiorano A, Rötter RP, O'Leary GJ, Fitzgerald GJ, Girousse C, Motzo R, Giunta F, Babar MA, Reynolds MP, Kheir AMS, Thorburn PJ, Waha K, Ruane AC, Aggarwal PK, Ahmed M, Balkovič J, Basso B, Biernath C, Bindi M, Cammarano D, Challinor AJ, De Sanctis G, Dumont B, Eyshi Rezaei E, Fereres E, Ferrise R, Garcia-Vila M, Gayler S, Gao Y, Horan H, Hoogenboom G, Izaurralde RC, Jabloun M, Jones CD, Kassie BT, Kersebaum K-C, Klein C, Koehler A-K, Liu B, Minoli S, Montesino San Martin M, Müller C, Naresh Kumar S, Nendel C, Olesen JE, Palosuo T, Porter JR, Priesack E, Ripoche D, Semenov MA, Stöckle C, Stratonovitch P, Streck T, Supit I, Tao F, Van der Velde M, Wallach D, Wang E, Webber H, Wolf J, Xiao L, Zhang Z, Zhao Z, Zhu Y, Ewert F (2019) Climate change impact and adaptation for wheat protein. Glob Chang Biol 25(1):155–173. https://doi.org/10.1111/gcb.14481

Banjara TR, Bohra JS, Kumar S, Singh T, Shori A, Prajapat K (2021a) Sustainable alternative crop rotations to the irrigated rice-wheat cropping system of Indo-Gangetic Plains of India. Arch Agron Soil Sci 67:1–18

Banjara TR, Bohra JS, Kumar S, Ram A, Pal V (2021b) Diversification of rice-wheat cropping system improves growth, productivity and energetics of rice in the Indo-Gangetic Plains of India. Agric Res 10:1–10

Basak JK, Ali A, Islam MN, Alam MJB (2010) Assessment of the effect of climate change on boro rice production in Bangladesh using CERES-Rice model. J Civ Eng 38(2):95–108

Belyaeva M, Bokusheva R (2018) Will climate change benefit or hurt Russian grain production? A statistical evidence from a panel approach. Clim Chang 149:205–217

Bhandari AL, Ladha JK, Pathak H, Padre AT, Dawe D, Gupta RK (2002) Yield and soil nutrient changes in a long-term rice-wheat rotation in India. Soil Sci Soc of Am J 66:162–170

Bhatt R, Kukal SS (2015) Delineating soil moisture dynamics as affected by tillage in wheat, rice and establishment methods during intervening period. J Appl Nat Sci 7(1):364–368

Bhatt R, Kukal SS, Busari MA, Arora S, Yadav M (2016) Sustainability issues on rice-wheat cropping system. Int Soil Water Conserv Res 4:64–74

Bijay-Singh, Shah YH, Beebout J, Singh Y, Buresh RJ (2008) Crop residue management for low land rice based cropping systems in Asia. Adv Agron 98:117–199

Bokhari SAA, Rasul G, Ruane AC, Hoogenboom G, Ahmad A (2017) The past and future changes in climate of the rice-wheat cropping zone in Punjab, Pakistan. Pak J Meteorol 13(26):9–23

Candradijaya C, Kusmana Y, Syaukat L, Syaufina FA (2014) Climate change impact on rice yield and adaptation response of local farmers in Sumedang District, West Java, Indonesia. Int J Ecosyst 4(5):212–223

Claessens L, Antle JM, Stoorvogel JJ, Valdivia RO, Thornton PK, Herrero M (2012) A method for evaluating climate change adaptation strategies for small-scale farmers using survey, experimental and modeled data. Agric Syst 111:85–95

Dhakal K, Kakani VG, Linde E (2018) Climate change impact on wheat production in the Southern Great Plains of the US using downscaled climate data. Atmos Clim Sci 8(2):143–162

Fahad M, Ahmad I, Athar H (2020) Implications of 1.5 and 2.0 °C additional warming for wheat yield using a gridded modeling approach. Atmosfera 33(4):337–355

FAO (2021) The state of food and agriculture 2021. Making agrifood systems more resilient to shocks and stresses. Food and Agriculture Organization, Rome, pp 157–161

Farooq M, Nawaz A (2014) Weed dynamics and productivity of wheat in conventional and conservation rice-based cropping systems. Soil Tillage Res 141:1–9

Figueiredo N, Carranca C, Trindade H, Pereira J, Goufo P, Coutinho J, Marques P, Maricato R, de Varennes A (2015) Elevated carbon dioxide and temperature effects on rice yield, leaf greenness, and phenological stages duration. Paddy Water Environ 13(4):313–324

Gammans M, Merel P, Ortiz-Bobea A (2017) Negative impacts of climate change on cereal yields: statistical evidence from France. Environ Res Lett 12:054007

Geng X, Wang F, Ren W, Hao Z (2019) Climate change impacts on winter wheat yield in Northern China. Adv Meteorol 2019:2767018

Gul F, Ahmed I, Ashfaq M, Jan D, Fahad S, Lin X, Wang D, Fahad M, Fayyaz M, Shah SA (2020) Use of crop growth model to simulate the impact of climate change on yield of various wheat cultivars under different agro-environmental conditions in Khyber Pakhtunkhwa, Pakistan. Arab J Geosci 13:112

Gupta R, Mishra A (2019) Climate change induced impact and uncertainty of rice yield of agro-ecological zones of India. Agric Syst 173:1–11

He L, Asseng S, Zhao G, Wu D, Yang X, Zhuang W, Jin N, Yu Q (2015) Impacts of recent climate warming, cultivar changes, and crop management on winter wheat phenology across the Loess Plateau of China. Agric For Meteorol 200:135–143

Hira GS, Jalota SK, Arora VK (2004) Efficient management of water resources for sustainable cropping in Punjab, Research bulletin. Department of Soils, Punjab Agricultural University, Ludhiana, p 20

Hoogenboom G, Porter CH, Shelia V, Boote KJ, Singh U, Pavan W, Oliveira FAA, Moreno-Cadena LP, Ferreira TB, White JW, Lizaso JI, Pequeno DNL, Kimball BA, Alderman PD, Thorp KR, Cuadra SV, Vianna MS, Villalobos FJ, Batchelor WD, Asseng S, Jones MR, Hopf A, Dias HB, Hunt LA, Jones JW (2024) Decision Support System for Agrotechnology Transfer (DSSAT) version 4.8.2 (www.DSSAT.net). DSSAT Foundation, Gainesville, FL

Hu Q, Weiss A, Feng S, Baenziger PS (2005) Earlier winter wheat heading dates and warmer spring in the US Great Plains. Agric For Meteorol 135:284–290

Humphreys E, Kukal SS, Christen EW, Hira GS, Singh B, Sudhir-Yadav SRK (2010) Halting the groundwater decline in north-west India - which crop technologies will be winners? Adv Agron 109:156–199
IPCC (2014) Climate change 2014. In: Field CB, Barros VR, Dokken DJ, Mach KJ, Mastrandrea MD, Bilir TE, Chatterjee M, Ebi KL, Estrada YO, Genova RC, Girma B, Kissel ES, Levy AN, MacCracken S, Mastrandrea PR, White LL (eds) Impacts, adaptation, and vulnerability, contribution of working group II to the fifth assessment report of the intergovernmental panel on climate change. Cambridge University Press, New York
Jabeen M, Gabriel HF, Ahmed M, Mahboob MA, Iqbal J (2017) Studying impact of climate change on wheat yield by using DSSAT and GIS: a case study of Pothwar Region. In: Ahmed M, Stockle OC (eds) Quantification of climate variability, adaptation and mitigation for agricultural sustainability. Springer International Publishing, Cham, pp 387–411
Jain N, Bhatia A, Pathak H (2014) Emission of air pollutants from crop residue burning in India. Aerosol Air Qual Res 14:422–430
Jiaguo Z (2000) Rice-wheat cropping system in China. In: Hobbs PR, Gupta RK (eds) Soil and crop management practices for enhanced productivity of the rice-wheat cropping system in the Sichuan Province of China, Rice-wheat consortium paper series, vol 9. Rice-Wheat Consortium for the Indo-Gangetic Plains, New Delhi, pp 1–10
Jones JW, Hoogenboom G, Porter CH, Boote KJ, Batchelor WD, Hunt LA, Wilkens PW, Singh U, Gijsman AJ, Ritchie JT (2003) The DSSAT cropping system model. Eur J Agron 18(3–4):235–265
Kakraliya SK, Kumar S, Kakraliya SS, Choudhary KK, Singh LK (2018) Remedial options for the sustainability of rice wheat cropping system. J Pharmacogn Phytochem 7(2):163–171
Keating BA, Carberry PS, Hammer GL, Probert ME, Robertson MJ, Holzworth D, Huth NI, Hargreaves JNG, Meinke H, Hochman Z, McLean G, Verburg K, Snow V, Dimes JP, Silburn M, Wang E, Brown S, Bristow KL, Asseng S, Chapman S, McCown RL, Freebairn DM, Smith CJ (2003) An overview of APSIM: a model designed for farming systems simulation. Eur J Agron 18:267–288
Khan MA, Tahir A, Khurshid N, Husnain MI, Ahmed M, Boughanmi H (2020) Economic effects of climate change-induced loss of agricultural production by 2050: a case study of Pakistan. Sustain For 12(3):1216. https://doi.org/10.3390/su12031216
Kumar P, Joshi PK, Johansen C, Asokan M (1998) Sustainability of rice-wheat based cropping systems in India. Econ Polit Weekly 33:A182–A188
Kumar SN, Aggarwal PK, Rani DNS, Saxena R, Chauhan N, Jain S (2014) Vulnerability of wheat production to climate change in India. Clim Res 59:173–187
Ladha JK, Dawe D, Pathak H, Padre AT, Yadav RL, Bijay S, Yadvinder-Singh, Singh Y, Singh P, Kundu AL, Sakal R, Ram N, Regmi AP, Gami SK, Bhandari AL, Amin R, Yadav CR, Bhattarai EM, Das S, Aggarwal HP, Gupta RK, Hobbs PR (2003) How extensive are yield declines in long-term rice-wheat experiments in Asia? Field Crop Res 81:159–180
Ladha JK, Kumar V, Alam MM, Sharma S, Gathala MK, Chandna P, Saharawat Y, Balasubramanian V (2009) Integrating crop and resource management technologies for enhanced productivity, profitability and sustainability of the rice-wheat system in South Asia. In: Ladha JK, Singh Y, Erenstein O, Hardy B (eds) Integrated crop and resource management in the rice-wheat system of South Asia. International Rice Research Institute, Los Banos, pp 69–108
Lalik R, Sharma S, Idris M, Singh AK, Singh SS, Bhatt BP, Saharawat Y, Humphreys E, Ladha JK (2014) Integration of conservation agriculture with best management practices for improving system performance of the rice-wheat rotation in the eastern Indo-Gangetic Plains of India. Agric Ecosyst Environ 195:68–82
Le THH (2016) Effects of climate change on rice yield and rice market in Vietnam. J Agric Appl Econ 48(4):366–382
Li Z, Yang P, Tang H, Wu W, Yin H, Liu Z, Zhang L (2014) Response of maize phenology to climate warming in Northeast China between 1990 and 2012. Reg Environ Chang 14(1):39–48
Li K, Yang X, Tian H, Pan S, Liu Z, Lu S (2016) Effects of changing climate and cultivar on the phenology and yield of winter wheat in the North China Plain. Int J Biometeorol 60(1):21–32

Lianzheng W, Yixian G (1994) Rice-wheat systems and their development in China. In: Paroda RS, Woodhead T, Singh RB (eds) Sustainability of rice-wheat production systems in Asia. FAO, Bangkok, pp 160–171

Liu Y, Chen Q, Ge Q, Dai J, Qin Y, Dai L, Zou X, Chen J (2018) Modelling the impacts of climate change and crop management on phenological trends of spring and winter wheat in China. Agric For Meteorol 248:518–526

Liu B, Martre P, Ewert F, Webber H, Waha K, Thorburn PJ, Ruane AC, Aggarwal PK, Ahmed M, Balkovič J, Basso B, Biernath C, Bindi M, Cammarano D, Cao W, Challinor AJ, De Sanctis G, Dumont B, Espadafor M, Eyshi Rezaei E, Fereres E, Ferrise R, Garcia-Vila M, Gayler S, Gao Y, Horan H, Hoogenboom G, Izaurralde RC, Jabloun M, Jones CD, Kassie BT, Kersebaum KC, Klein C, Koehler A-K, Maiorano A, Minoli S, Montesino San Martin M, Müller C, Naresh Kumar S, Nendel C, O'Leary GJ, Eivind Olesen J, Palosuo T, Porter JR, Priesack E, Ripoche D, Rötter RP, Semenov MA, Stöckle C, Stratonovitch P, Streck T, Supit I, Tao F, Van der Velde M, Wang E, Wolf J, Xiao L, Zhang Z, Zhao Z, Zhu Y, Asseng S (2023) AgMIP-Wheat multi-model simulations on climate change impact and adaptation for global wheat. Open Data J Agric Res 9:10–25. https://doi.org/10.18174/odjar.v9i0.18092

Luo Q, Bellotti W, Williams M, Bryan B (2005) Potential impact of climate change on wheat yield in South Australia. Agric Meteorol 132:273–285

Mahajan A, Gupta RD (2009) The rice-wheat cropping system. In: Integrated nutrient management (INM) in a sustainable rice-wheat cropping system. Springer Science + Business Media, Dordrecht, pp 109–117

McDonald AJ, Riha SJ, Duxbury JM, Steenhuis TS, Lauren JG (2006) Soil physical responses to novel rice cultural practices in the rice-wheat system: comparative evidence from a swelling soil in Nepal. Soil Tillage Res 86:163–175

Mueller V, Gray C, Kosec K (2014) Heat stress increases long-term human migration in rural Pakistan. Nat Clim Chang 4:182–185

Nadeem M, Nazer Khan M, Abbas G, Fatima Z, Iqbal P, Ahmed M, Ali Raza M, Rehman A, Ul Haq E, Hayat A, Ali M, Ahmad S (2022) Application of CSM-CANEGRO model for climate change impact assessment and adaptation for sugarcane in semi-arid environment of Southern Punjab, Pakistan. Int J Plant Prod. https://doi.org/10.1007/s42106-022-00192-6

Nasir IR, Rasul F, Ahmad A, Asghar HN, Hoogenboom G (2020) Climate change impacts and adaptations for fine, coarse, and hybrid rice using CERES-Rice. Environ Sci Pollut Res 27:9454–9464

Panwar AS, Shamim M, Babu S, Ravishankar N, Prusty AK, Alam NM, Singh DK, Bindhu JS, Kaur J, Dashora LN et al (2019) Enhancement in productivity, nutrients use efficiency, and economics of rice-wheat cropping systems in India through farmer's participatory approach. Sustain For 11:122. https://doi.org/10.3390/su11010122

Paroda RS (1996) Sustaining the green revolution: new paradigms. B. P. Pal Commemoration Lecture. 2nd International Crop Sci. Congress, November 22, 1996, New Delhi

Qamar R, Atique-ur-Rehman, Javeed HMR, Rehman A, Safdar ME, Ali H, Ahmad S (2021) Tillage systems affecting rice-wheat cropping system. Sains Malaysiana 50(6):1543–1562

Qian B, Zhang X, Smith W, Grant B, Jing Q, Cannon AJ, Neilsen D, McConkey B, Li G, Bonsal B, Wan H, Xue L, Zhao J (2019) Climate change impacts on Canadian yields of spring wheat, canola and maize for global warming levels of 1.5°C 2.0°C, 2.5°C and 3.0°C. Environ Res Lett 14:1–13

Qiu RJ, Liu CW, Cui NB, Wu YJ, Wang ZC, Li G (2019) Evapotranspiration estimation using a modified Priestley-Taylor model in a rice-wheat rotation system. Agric Water Manag 224:105755

Qiu RJ, Li G, Liu CW, Wang ZC, Zhang BZ, Liu ZD (2022) Evapotranspiration estimation using a modified crop coefficient model in a rotated rice-winter wheat system. Agric Water Manag 264:107501

Ray DK, Ramankutty N, Mueller ND, West PC, Foley JA (2012) Recent patterns of crop yield growth and stagnation. Nat Commun 3:1293

Rezaei EE, Siebert S, Ewert F (2015) Intensity of heat stress in winter wheat? Phenology compensates for the adverse effect of global warming. Environ Res Lett 10(2):024012

Santikayasa IP, Amdan W, Perdinan, Sugiarto Y (2018) Quantification the impact of climate change on paddy field yield production in Subang, Indonesia. Earth Environ Sci 187:012059

Shrestha S, Thin NMM, Deb P (2014) Assessment of climate change impacts on irrigation water requirement and rice yield for Ngamoeyeik irrigation project in Myanmar. J Water Clim Chang 5(3):427–442

Sinnarong N, Chen CC, McCar B, Tran BL (2019) Estimating the potential effects of climate change on rice production in Thailand. Paddy Water Environ 17(4):761–769

Subash N, Singh H, Singhm SV, Meena MS, Singh B, Paudel GP, Baigorria G, Bhaskar S, Panwar AS, SP MD, Valdivia RO (2021) Integrated assessment of climate change impacts on rice-wheat farms of IGP-India through multi-climate-crop model approach: a case study Meerut district, Uttar Pradesh, India. In: Rosenzweig C, Mutter CZ, Contreras EM (eds) Handbook of climate change and agroecosystems: climate change and farming system planning in Africa and South Asia: AgMIP stakeholder-driven research. World Scientific Publishing Europe Ltd, London, pp 329–355

Tao F, Zhang Z, Shi W, Liu Y, Xiao D, Zhang S, Zhu Z, Wang M, Liu F (2013) Single rice growth period was prolonged by cultivars shifts, but yield was damaged by climate change during 1981–2009 in China, and late rice was just opposite. Glob Change Biol 19(10):3200–3209

Tariq M, Ahmad S, Fahad S, Abbas G, Hussain S, Fatima Z, Wajid N, Mubeen M, Rehman M, Khan MA, Adnan M, Wilkerson CJ, Hoogenboom G (2018) The impact of climate warming and crop management on phenology of sunflower-based cropping systems in Punjab, Pakistan. Agric For Meteorol 256–257:270–282

Timsina J, Connor DJ (2001) Productivity and management of rice-wheat cropping systems, Issues and challenges. Field Crop Res 69:93–132

USDA (2022) Crop production. National Agricultural Statistics Service, United States Department of Agriculture, Washington, DC

Wahab AA, Hussain M, Fatima Z, Ahmed M, Ahmad S (2024) Evaluating the interactive impact of nitrogen levels and cultivars on yield traits, nitrogen use efficiency, water use efficiency and benefit cost ratio of ratoon rice in Punjab, Pakistan. Int J Plant Prod. https://doi.org/10.1007/s42106-023-00280-1

Wang HL, Gan YT, Wang RY, Niu JY, Zhao H, Yang QG, Li GC (2008) Phenological trends in winter wheat and spring cotton in response to climate changes in northwest China. Agric For Meteorol 148(8–9):1242–1251

Xiao D, Tao F, Liu Y, Shi W, Wang M, Liu F, Zhang S, Zhu Z (2013) Observed changes in winter wheat phenology in the North China Plain for 1981–2009. Int J Biometeorol 57(2):275–285

Xu JZ, Bai WH, Li YW, Wang HY, Yang SH, Wei Z (2019) Modeling rice development and field water balance using AquaCrop model under drying-wetting cycle condition in eastern China. Agric Water Manag 213:289–297

Yao F, Xu Y, Lin E, Yokozawa M, Zhang J (2007) Assessing the impacts of climate change on rice yields in the main rice areas of China. Clim Chang 80:395–409

Zheng Z, Cai H, Wang Z, Wang X (2020) Simulation of climate change impacts on phenology and production of winter wheat in Northwestern China using CERES-Wheat model. Atmos 11:681

Maize-Maize System

4

4.1 Introduction

Maize crops are grown during different growing seasons such as spring-, autumn-, or summer-season maize crop (Abbas et al. 2017, 2019, 2020a, b, c; Abbas and Ahmad 2018). The maize-maize cropping system is a vital rotation system all over the world. In Pakistan, the spring maize season lasts from February to May, while the fall maize season lasts from July to October (Ahmad et al. 2012a, b, 2013, 2015a, b, c, 2016a, b, c, 2017a, b, 2018, 2019, 2021, 2023a, b). Autumn maize grows in warm to moderate climates, whereas spring maize develops in mild to warm climates. Scientific name of maize is *Zea mays*. The family name of maize is Poaceae. In areas of the Central/South America, maize is considered to have originated 55–70 million years ago. Since then, it has evolved into about 10,000 non-domestic relatives (Jones and Thornton 2003; Rouf et al. 2016). Today, maize is produced all across the world and is said to have originated in Mesoamerica. *Zea mays* L., sometimes known as corn or maize, is a member of the Poaceae plant family and is a significant annual grain crop in the globe. Zea, an ancient Greek term that means "supporting life," and Mays, a Taino word that means "life provider," are both related concepts. The simplest way to describe the plant is "maize," which is derived from the Spanish word "maiz" (Nuss and Tanumihardjo 2010). The plant is known by a number of various names, including zea, silk maize, makka, barajovar, etc. In many regions of the world, it is regarded as a staple meal. After rice and wheat, maize crop is the third utmost imperative crop across the globe due to its high nutritional value and abundance of macronutrients (starch, fiber, protein, fat, etc.) as well as micronutrients (vitamin B complex, ß-carotene, etc.) and critical minerals like magnesium, zinc, phosphorus, and copper (Ahmed et al. 2016, 2018a, b, 2019, 2020, 2022; Ahmad and Hasanuzzaman 2012; Ali et al. 2013; Amin et al. 2018a, b, c). The maize crop is a stable cereal that is quite popular all over the world. Additionally, maize has an antioxidant booster that guards against a variety of degenerative disorders (Mehmood et al. 2018). 967 million metric tonnes of

M. Ahmed et al., *Cropping Systems Modeling Under Changing Climate*,
https://doi.org/10.1007/978-981-97-0331-9_4

Table 4.1 Area and production of maize-maize systems across various regions in the world

Countries	Spring maize Area (Mha)	Spring maize Production (MMT)	Autumn maize Area (Mha)	Autumn maize Production (MMT)	Maize-maize system Area (Mha)	Maize-maize system Production (MMT)
Pakistan	0.7	6.1	0.8	4.5	1.6	10.6
India	1.7	7.9	8.2	23.6	9.9	31.5
USA	2.9	54.3	29.8	329.6	32.7	383.9
China	25.7	186.8	17.4	85.6	43.1	272.4
Brazil	10.2	80.5	12.5	34.5	22.7	115.0
Mexico	5.1	21.4	2.1	6.2	7.2	27.6
Argentina	4.6	35.1	2.5	18.9	7.1	54.0
Ukraine	1.4	14.6	3.1	27.5	4.5	42.1

Source: USDA (2022)
Mha million hectare, *MMT* million metric tonne

maize were produced globally. 197 million hectares are used for dry grain production on a worldwide scale (Erenstein et al. 2022). The overall area and production of maize-maize cropping system are presented in Table 4.1. Across the world, maize is used in feed (61%), food (17%), and industrial sectors (22%). 83% of the world maize production is used in the feed, starch, and biofuel sectors; therefore, it has gained a status as an industrial crop (Grote et al. 2021). Shekhar and Singh (2021) mentioned that corn crop is cultivated in India in two growing seasons, which are kharif season and rabi season. The maize growing during kharif season accounts for around 83% of maize area, whereas maize growing during rabi season accounts for 17%. Maize-maize cropping system is an imperative rotation system in the province of Punjab in Pakistan, where spring maize spans from February to May and fall maize spans from July to October. 53% of cultivated maize area is during spring season, while 47% of area is during autumn season in Pakistan (Table 4.1).

4.2 Issues of Maize-Maize System

The main difficulty with the maize-maize farming system is heat stress and water scarcity caused by global climate change (Mahmood et al. 2013; Mubeen et al. 2016; Nasim et al. 2018; Mubarak et al. 2022; Rehman et al. 2015; Rahman et al. 2018, 2021; Sarwar et al. 2013, 2021; Shabbir et al. 2016). A lot of issues are now impeding maize rotation in various areas. Abiotic stresses are one type of challenge. The majority of maize agriculture in the poor countries is dependent on rainfall. As a result, the crop is especially sensitive to drought and heat. According to current estimates, repeated droughts endanger 25% of the output of maize. It is commonly believed that these stressors would worsen due to climate change, with the result that maize output in the maize-growing areas of Africa and Latin America may decline by up to 10% by 2055. Poor/degraded soil quality is another abiotic stressor that is recognized as significantly influencing low yields. Diseases including downy mildew, rust, leaf blight, stripe virus, and recently lethal necrosis have all but destroyed maize crops in some regions of the world. Insects, such as the stem borer, can result in losses of 20–40% (Tariq et al. 2017, 2018a, b; Tarriq et al. 2021; Hussain et al. 2018; Khan et al. 2019, 2020; Zahoor et al. 2022).

4.3 Decision Support System for Agrotechnology Transfer

Crop models are quite effective for assessing crop growth, development, and production in various cropping systems at the field level (Ejaz et al. 2022; Fatima et al. 2018, 2020, 2021; Nadeem et al. 2022; Naz et al. 2022; Jones et al. 2003; Keating et al. 2003; Hoogenboom et al. 2019, 2021). Crop growth models are novel techniques for determining the effects of crop management strategies such as sowing dates and plant genetics, as well as their interactions with the environment (Reidsma et al. 2009). Maize cropping system has been the subject of several modeling studies across the world (Basso et al. 2016). Crop growth simulation as well as multi-linear regression models may be employed to evaluate anticipated possible consequences of climate change on maize growth and production in maize-maize cropping system. AquaCrop, APSIM, DSSAT, EPIC, CropSyst, RZWQM2, ALAMANC, WOFOST, ADEL, GEPIC, Empirical, InfoCrop, and EcoCrop are some examples (Chisanga et al. 2022). Cropping system models make forecasts based on numerical simulations of meteorological, soil, and agricultural data. Climate data may be obtained from a variety of sources, including government meteorological and research agencies, international organizations, and commercial entities. Statistics and dynamical downscaling technologies may also be used to create climate data (Ma et al. 2006). Maize crop growth models, which use biophysical principles of plant development, may offer detailed estimations of crop situations such as phenology status, LAI, and yields for specific crop varieties in maize-maize cropping system. Cropping system models, such as DSSAT model (Jones et al. 2003) and APSIM model (Keating et al. 2003), are built on ecological principles to simulate crop growth of maize in maize-maize cropping system in relation to environmental factors, soil characteristics, and management techniques (Malik et al. 2019). Previously, similar software were employed to examine the consequences of climate warming on the maize crop in maize-maize cropping system, such as high temperatures, rising carbon dioxide levels, and inconsistent rainfall patterns. In order to determine information on short- and long-term management strategies and agricultural production technologies for increased maize yield in maize-based cropping system, many modeling methods are used. Cropping system model includes the CSM-CERES-Maize model, which is a comprehensive system investigation tactic that includes maize crop management evaluation processes such as planting dates, irrigation, fertilizers, and many other variables in cropping system (Babel et al. 2019). It offers the capacity to model the effects and interaction of soil, crop plant genetics, and climate factors on maize crop development, growth, and grain yield in different locales. The DSSAT model's CSM-CERES-Maize is studied for maize crop management such as irrigation, water, and nitrogen in maize-maize cropping system. Furthermore, they are being utilized to evaluate the influence of climate change upon *Zea mays* productivity under maize-maize cropping system under various scenarios of climate change (Rugira et al. 2021). Furthermore, structural development in CSM-CERES-Maize

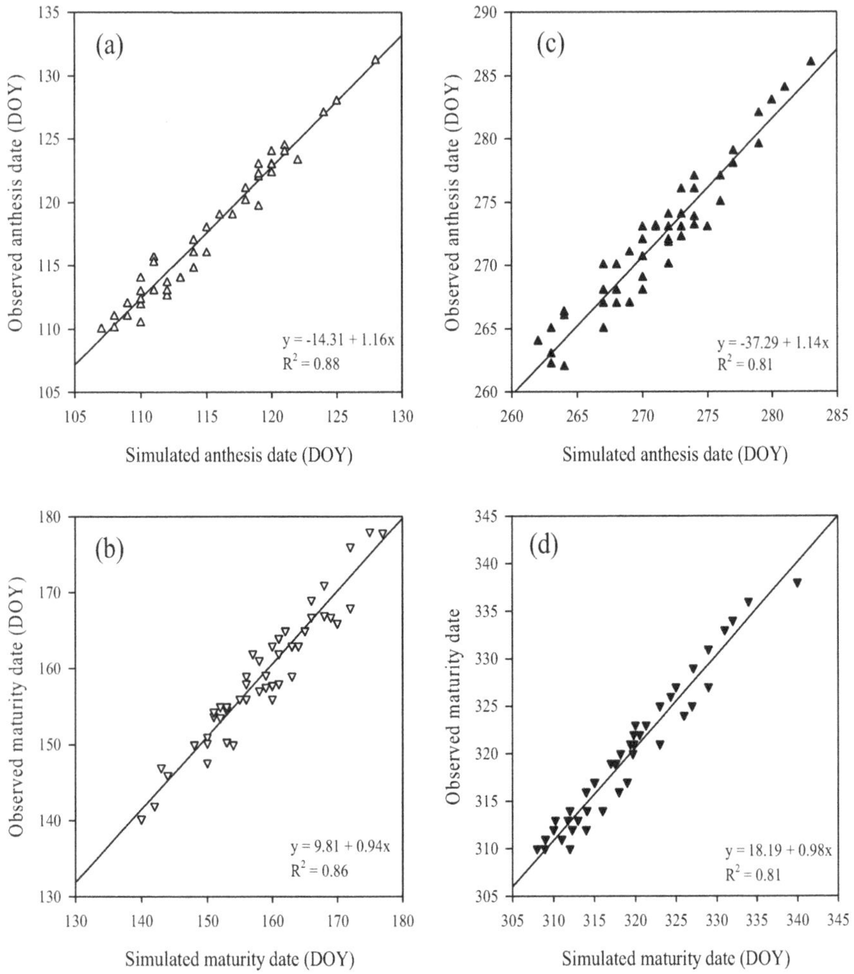

Fig. 4.1 Comparison of simulated and observed anthesis dates and maturity dates for spring (**a, b**; unfilled symbols) and autumn season (**c, d**; filled symbols) maize crop, respectively, during 1980–1985 for ten locations in Punjab, Pakistan (Source: Adopted from Abbas et al. 2017)

was created in relation to leaf area, grain number, cob growth, integration and partitioning, and yield for improved simulation, resulting in the CSM-IXIM-Maize model under the DSSAT. Better maize phenology forecasts in maize-maize cropping system are critical for assessing the effects of changing climate on maize crop production (Menefee et al. 2021). The Agricultural Production Systems Simulator is a model capable of simulating the influence of thermo-temporal fluctuations on maize crop growth, development, and physiology, as well as providing alternative management techniques for sustainable production in maize-maize cropping system. Abbas et al. (2017) used CSM-CERES-Maize model for maize-maize cropping system in Pakistan, in which model performance was good for both spring and autumn maize crop (Fig. 4.1). The genetic coefficients of maize crop in DSSAT and APSIM models are presented in Tables 4.2 and 4.3, respectively.

Table 4.2 Genetic coefficients, description, ranges, and guidelines for model calibration of maize in DSSAT v4.8

Genetic coefficient	Description
VAR#	Identification code or number for a specific cultivar
VAR-NAME	Name of cultivar
EXPNO	Number of experiments used to estimate cultivar parameters
P1	Thermal time from seedling emergence to the end of the juvenile phase (expressed in degree days above a base temperature of 8 °C) during which the plant is not responsive to changes in photoperiod
P2	Extent to which development (expressed as days) is delayed for each hour increase in photoperiod above the longest photoperiod at which development proceeds at a maximum rate (which is considered to be 12.5 h)
P5	Thermal time from silking to physiological maturity (expressed in degree days above a base temperature of 8 °C)
G2	Maximum possible number of kernels per plant
G3	Kernel filling rate during the linear grain filling stage and under optimum conditions (mg/day)
PHINT	Phyllochron interval; the interval in thermal time (degree days) between successive leaf tip appearances

Source: DSSAT v4.8; Hoogenboom et al. (2021)

Table 4.3 Genetic coefficients, description, ranges, and guidelines for model calibration of maize in APSIM

Genetic coefficients	Description
tt_emerg_to_endjuv	Duration from emergence to the end of juvenile °C
est_days_endjuv_to_init	Duration—end of juvenile to flowering initiation °C day
tt_flag_to_flower	Duration—flag leaf to flowering stage °C day
Photoperiod crit_1	Hour—photoperiod critical 1 (h)
Photoperiod crit_2	Hour—photoperiod critical 2 (h)
tt_flower_to_start_grain	Duration—flowering to start of grain filling °C day
tt_flower_to_maturity	Duration, flowering to maturity °C day
tt_maturity_to_ripe	Duration—maturity to seed ripening °C day
head_grain_no_max	Grain maximum number per head
grain_gth_rate	Grain growth rate mg/day
t_base	Base temperature °C day

Source: Beah et al. (2021)

4.4 Climate Change and Temperature Trend

Global climate change is a genuine occurrence, as evidenced by the rise in sea level, reduction in snow and ice cover, and rise in ocean and atmospheric temperatures. The average temperature of the earth's surface has increased during the past three decades due to warmer conditions than in previous centuries. Climate warming is a serious issue in maize crop in maize-maize cropping system (Meza and Silva 2009; Cairns et al. 2013). The global average surface temperature is expected to rise by 1.4–5.8 °C between 1990 and 2100, implying global warming rates of between 0.1 and 0.5 °C every decade (Abbas et al. 2023a, b; Zakir et al. 2023). This is compared to a measured global warming rate of 0.15 °C every decade since the 1970s. As a result, crop growth and production responses to faster warming are of concern (Tokatlidis 2013). Trends in temperature, as well as their impacts on crop development and production, have become significant in maize-growing areas across the world. The observed climate change patterns and their impacts were diverse both spatially and temporally in maize-maize system. The sensitivity of maize crop responses to temperature change is also influenced by other factors such as changes in other climate parameters (e.g., precipitation), and management practices, suggesting a potential role of management for adaptation (Southworth et al. 2000). The global trend for temperature rose under the maize-maize farming scheme. For spring and fall, maize produced in Pakistan showed a warming tendency of 0.95 and 0.89 °C/decade from 1980 to 2014 (Fig. 4.2). In comparison to the fall season, a stronger warming tendency was seen in spring period. For spring and autumn seasons, average temperature tendency rises from sowing to anthesis ranging from 0.7 to 1.1 and 0.5 to 1.1 °C/decade, respectively, and averagely 0.90 and 0.82 °C. During anthesis to maturity phase, the warming trend was also noticeable, ranging from 0.8 to 1.1 °C/decade of maize in the spring period (a mean of 0.95 °C/decade) and 0.6 to 1.1 °C/decade of maize in the fall period (a mean of 0.85 °C/decade). Warming tendency for the sowing to maturity phase was similar, ranging from 0.7 to 1.1 °C/decade for spring period and 0.5 to 1.1 °C/decade for fall period (Fig. 4.3).

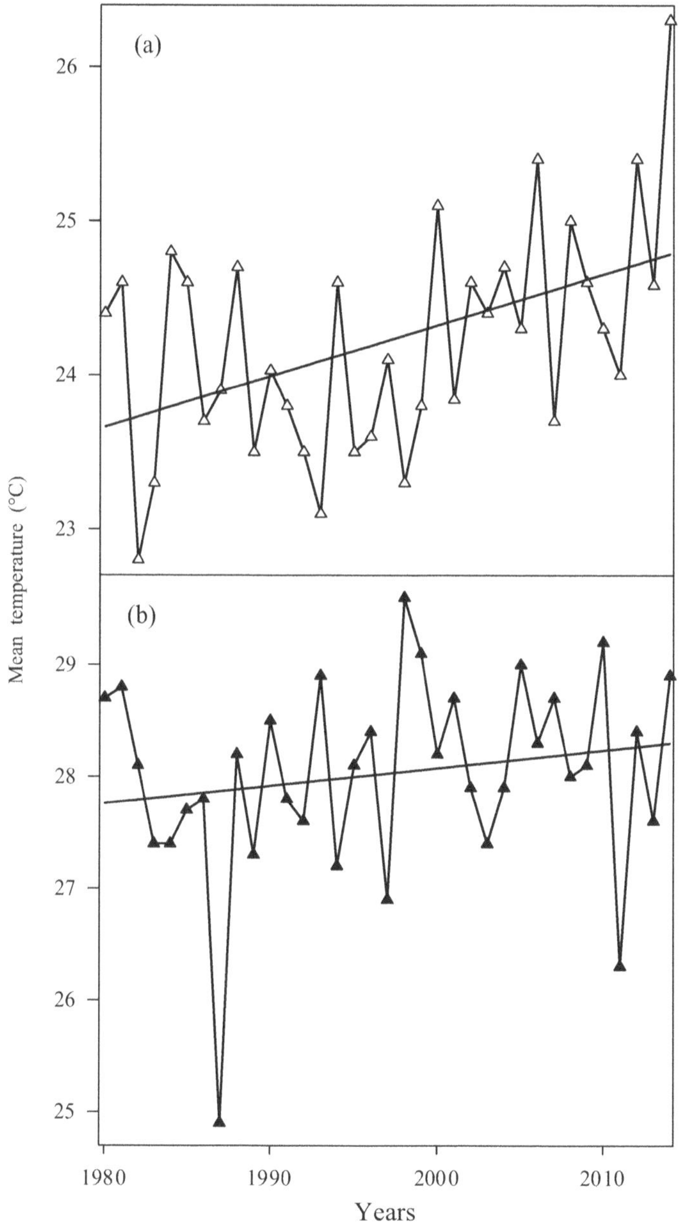

Fig. 4.2 Mean temperature trends in spring (**a**) and autumn (**b**) season from 1980 to 2014 in Punjab, Pakistan (Source: Adopted from Abbas et al. 2017)

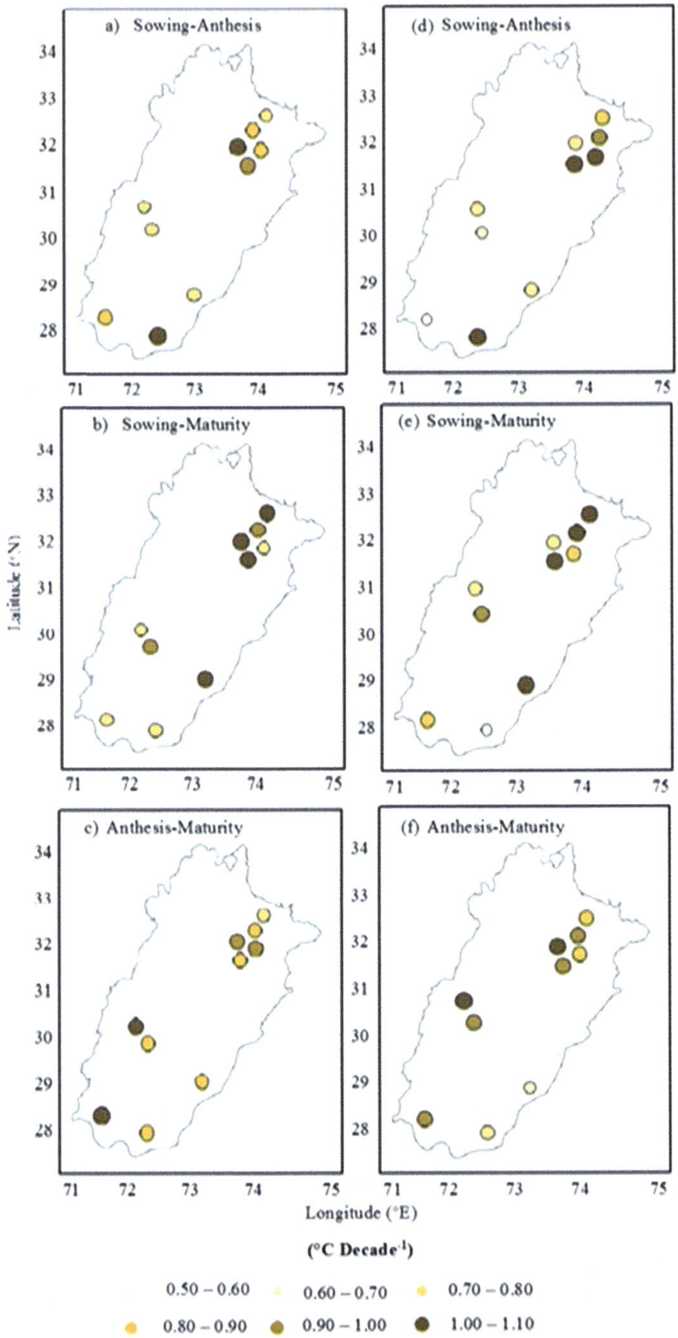

Fig. 4.3 Observed trends in mean temperature during the phenological phases {sowing-anthesis (**a, d**); sowing-maturity (**b, e**); and anthesis-maturity (**c, f**)} for spring (**a–c**; left) and autumn (**d–f**; right) season maize crop from 1980 to 2014 in Punjab, Pakistan (Source: Adopted from Abbas et al. 2017)

4.5 Climate Warming Impact on Maize Morphophysiological Responses

High-temperature stress is critical for maize plant physiological characteristics in maize-maize cropping system. High temperatures lowered the quantity of cobs, grains, and chlorophyll efficiency and increased leaf burning and tassel blasting (Sabagh et al. 2020; Abendroth et al. 2021). Temperatures above 35 °C for an extended length of time are regarded unfavorable for crop growth and development in maize-maize cropping system, while temperatures above 40 °C at flowering as well as grain filling have a significant detrimental influence on grain production in maize-maize cropping system. Maize plants subjected to heat stress had considerably lower stomatal conductance, having a slower photosynthesis rate (Tao and Zhang 2010). Additionally, excessive heat reduces seed weight, leaf area index, net photosynthetic process, and biomass accumulation in maize crop in maize-maize cropping system. The greatest metabolite-producing heat-tolerant corn cultivars, however, are not often high-yielding kinds (Chandra et al. 2023). Lesser plant height, leaves, and a lower LAI are typical characteristics for heat-tolerant types, which eventually result in lower productivity. Hence, while selecting for heat tolerance for maize, a variety of things must be considered (Wang et al. 2011). At the cellular level, climate warming heat stress stimulates the expression of specific genes and boosts the buildup of specific metabolites that may improve maize plants' capacity to withstand heat in maize-based cropping system. In general, considerable genotypic differences in stomatal conductance were found. Stresses have a big influence on critical characteristics of leaf photosynthesis and stomatal conductance. High production of hybrids at regular watering and drought stresses has been observed to bring a delay in canopy senescence because of different interceptions of light with green leaves (Deihimfard et al. 2022). Higher intensity light falling on maize crop plants can permanently harm membrane. Cell membrane is considered as the first vulnerable component to heat stress and turns inactive functionally. Stretching the cell wall and membrane function are mutually exclusive (Rahimi-Moghaddam et al. 2018). Membrane stability might be used to evaluate a plant's capacity to withstand high temperatures. That is suitable along with a practical test, which might observe electrolyte leakages at higher temperatures (Tesfaye et al. 2017). Membrane stability might be used to evaluate a maize plant's capacity to withstand high temperatures (Vučetić 2011; Liu et al. 2013).

4.6 Impact of Climate Change on Crop Phenology

Climate warming is a serious threat on maize phenology in maize-maize cropping system. Changes in crop phenology are crucial for yield formation because they control how long biomass accumulates in different organs of maize plant (Xiao et al. 2016). Climate change would likely decrease crop duration, thus lowering production and endangering future global food security (Ahmed 2020). As a result, influences of climate change upon maize phenology have been a key concern, particularly during the past two decades. Crop phenological variations in response to climate change have become a popular study area with the increase in global warming. Phenology is the study of periodic occurrences in the life cycles of maize plants and how these events are impacted by seasonal and interannual climatic fluctuations as well as habitat-related variables (Tao et al. 2006). Phenology of any crop may be used as an essential biological indicator to indicate the reactions of terrestrial systems to climate change since it will alter with environmental changes. Maize crop phenology (stages and phases) in maize-maize cropping system is often represented in growing degree days' accumulation and is influenced by the current weather conditions (Liu et al. 2021). Maize phenology changes due to climate change influence maize crop management techniques such as cultivar choice and changes in sowing dates. The length of each growth stage and phase in maize is negatively impacted by a rise in temperature during the growing season brought on by global warming. Variations in a crop's phenological stages and phases serve as important indications of changes in the environment and the climate. Maize crop phenological phases in maize-maize cropping system are accelerated by a climate heat trend, whereas cultivars with longer duration can result in slower growth. Li et al. (2014) indicated that in maize crop in maize-maize cropping system, climate change may shorten the growing season and speed up crop growth. If there were no human adaptations to climate change or agronomic advancements, the maize anthesis to maturity may be reduced to some amount. Only three stations in our analysis had considerably shorter maize grain production, and 34% of the stations had significantly higher maize grain production during the previous two decades. More than 40% of the stations in the southern and central parts of Northeast China may have early seedling and heading dates as well as postponed maturity dates, which might be the cause of these alterations. These findings suggest that the majority of farmers may have already shifted the timing of their maize plantings in response to climate change, which has a bigger impact on the early phenological phases. However, certain results revealed that maize phenology varied in response to temperature in a different way;

for example, anthesis to maturity length was significantly inversely linked with T mean at seven locations. Xiao et al. (2019) studied that anthesis along with maturity date in maize-based cropping system was advanced due to climatic warming by 0.2–5.5 and 0.6–11.1 days/decade, respectively. Three growth phases, vegetative (S-A), reproductive (A-M), and overall (S-M), were all reduced by 0.2–5.5, 0.4–5.6, and 0.6–11.1 per decade, respectively, as a result of climate-driven variations in maize phenology crop. Liu et al. (2021) resulted that in contrast to summer maize, the vegetative growth time of spring along with intercropping was shorter. Spring, summer, and intercropping maize's reproductive growth seasons were all prolonged. Vegetative periods of spring, summer, and intercropping were each reduced by 0.19, 1.06, and 3.12 days/decade, respectively, whereas reproductive periods were each lengthened by 0.19, 0.74, and 3.47 days/decade. Žydelis et al. (2021) reveal that for future scenario, maize phenology in maize-maize cropping system will be negatively affected due to climate change. Maize phenological phases will be shortened. RCP 2.6 shows a gradual decline in maize-growing season from 152 days for 2020s to 145 days for 2090s (4.6%), RCP 4.5 shows a decline from 151 to 142 days in the 2090s (5.9%), and RCP 8.5 shows a decline from 149 to 137 days in 2090s (8.1%). It appears that significant changes in climatic circumstances would result in shorter growth periods, since maize duration generally fell more quickly under RCP 8.5 (1.8 days/decade) than under RCP 4.5 (1.4 days/decade) or RCP 2.6 (1 day/decade). Moradi et al. (2014) projected results of response to effects of climate variation for all scenarios and GCM models, which showed that the day to anthesis and anthesis period of various cultivars of maize were reduced; these reductions ranged from 0.5% to 17.5% for day to anthesis and 5% to 33% for anthesis period.

4.7 Case Study in Pakistan

Abbas et al. (2017) studied the impact of climate change on maize stages and phases in spring maize and fall cropping system in Punjab. The CSM-CERES-Maize model was used to assess the spatiotemporal variability of phenology of maize-maize cropping system, agronomic practices, and cultivar alterations. Between 1980 and 2014, variations in spring and fall crop phenology were detected in Punjab. Spring maize seeding was 4.6 days earlier on average basis per decade, whereas autumn maize sowing was 3.0 days delayed on average per decade (Fig. 4.4). Anthesis and maturity dates were observed to be 7.1 and 9.2 days/decade earlier for spring and 2.8 and 4.4 days/decade earlier for fall maize, respectively (Figs. 4.5 and 4.6). Similarly, for

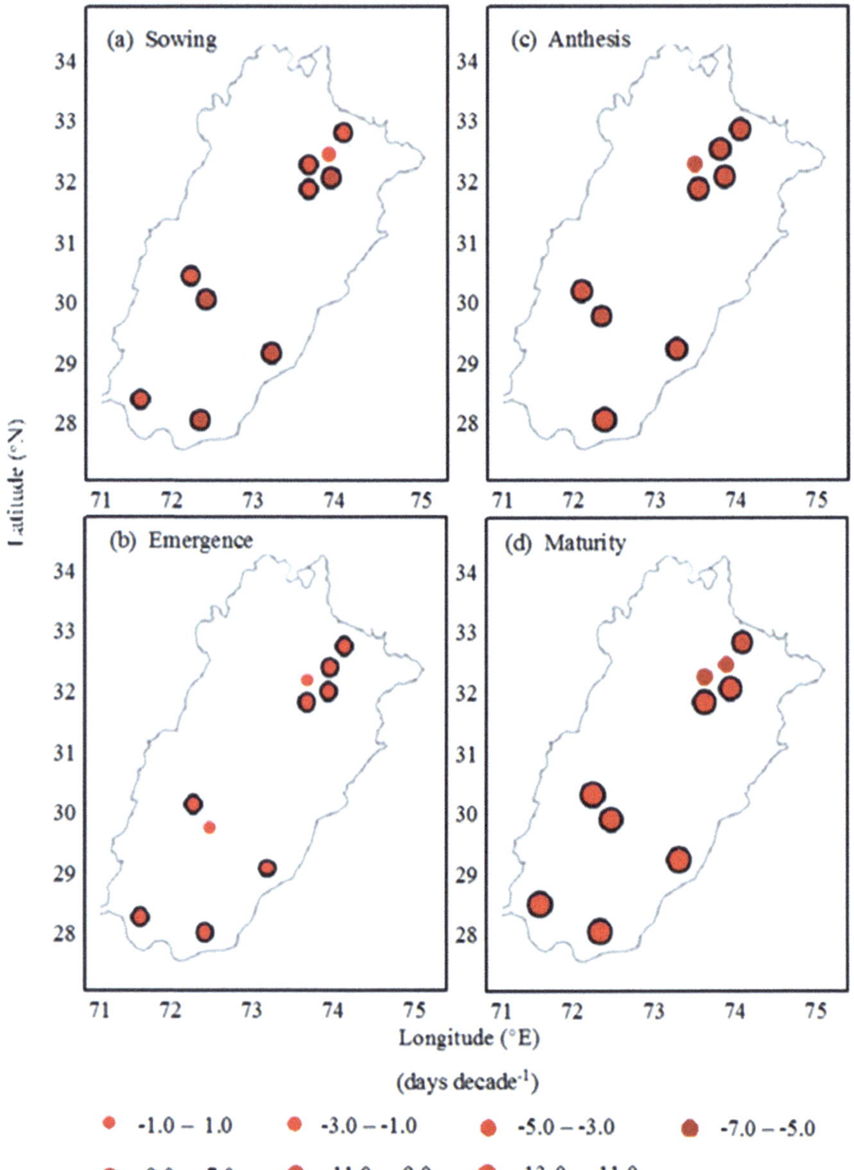

Fig. 4.4 Observed trends in phenology of spring-season maize {sowing (**a**); emergence (**b**); anthesis (**c**); and maturity (**d**); circles with black borderline indicate statistical significant trends at $p = 0.05$ levels} from 1980 to 2014 in Punjab, Pakistan (Source: Adopted from Abbas et al. 2017)

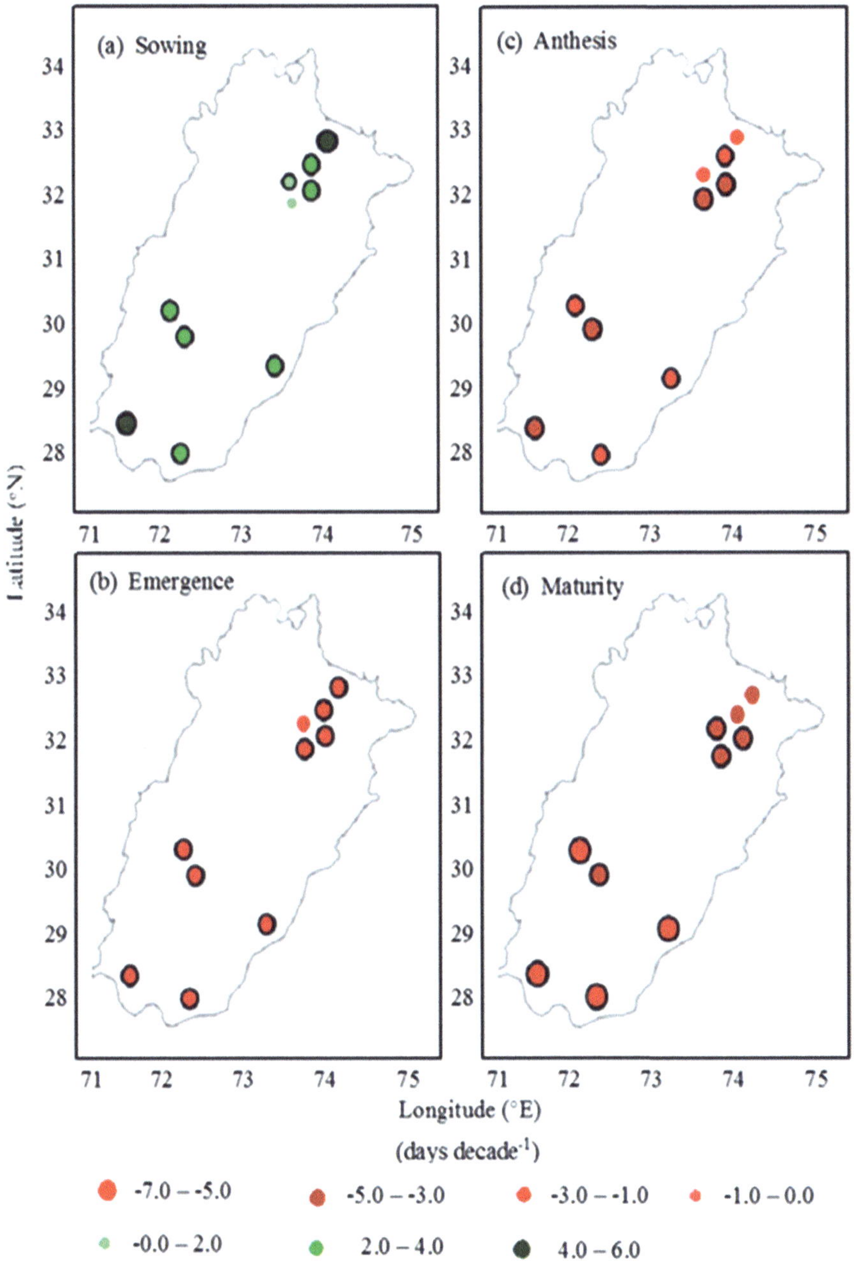

Fig. 4.5 Observed trends in phenology of autumn-season maize {sowing (**a**); emergence (**b**); anthesis (**c**); and maturity (**d**); circles with black borderline indicate statistical significant trends at $p = 0.05$ levels} from 1980 to 2014 in Punjab, Pakistan (Source: Adopted from Abbas et al. 2017)

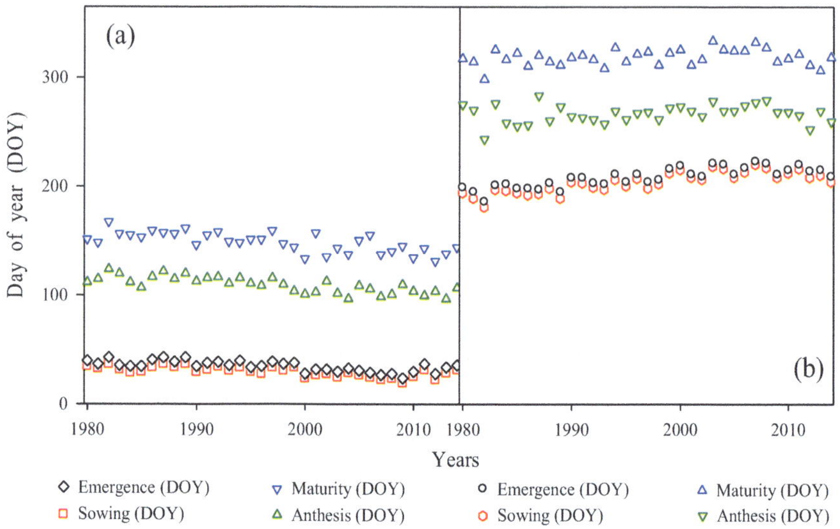

Fig. 4.6 Time series plots of farmer field-observed dates of sowing, emergence, anthesis, and maturity of spring (**a**) and autumn (**b**) maize in Punjab, Pakistan (Source: Adopted from Abbas et al. 2017)

spring and fall in maize-maize cropping system, S-A, sowing to maturity, and A-M phases were reduced by 2.4, 4.6, and 1.9 days/decade and 5.5, 7.8, and 2.2 days/decade, respectively (Fig. 4.7). The variation in phenology periods for spring and fall was significantly related to temperature increases between 1980 and 2014 (Table 4.4). A rise in temperature affects the temporal and spatial variation of spring and fall maize phenological phases (Table 4.5; Fig. 4.8). Table 4.6 indicates that the model-projected stages and phases of phenology were negatively associated with temperature. The temperature response of phenological phases in field-recorded data was lower than in model-simulated phenological data (Table 4.6). Using CERES-Maize with a cultivar for all sites and years revealed that projected phenology has augmented more than in field-observed phenology with climate change (Fig. 4.9). These results indicated that throughout the period of 1980–2014, shifts in cultivars needing high total growing degree days and earlier or late planting may have partially offset the detrimental influences of climate warming on the phenology of maize during spring as well as autumn growing season (Fig. 4.10).

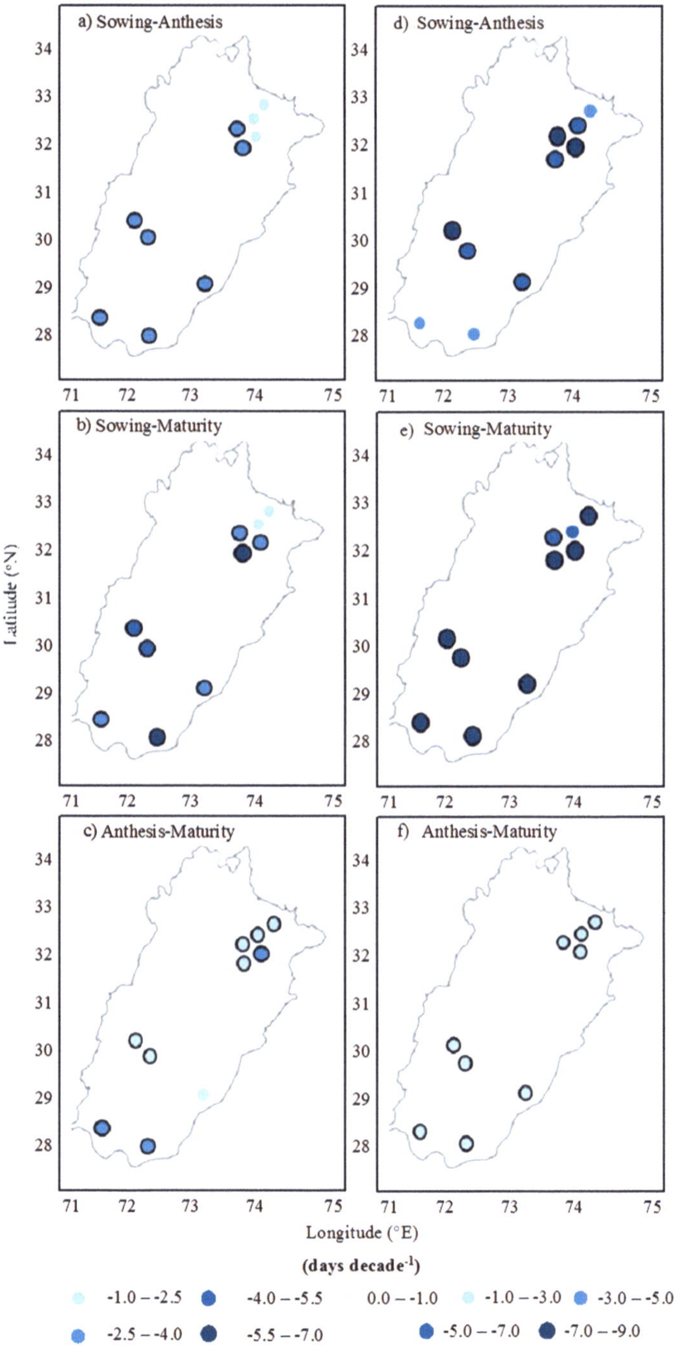

Fig. 4.7 Observed trends in phenology {(**a, d**) sowing-anthesis; (**b, e**) sowing-maturity; and (**c, f**) anthesis to maturity; circles with black borderline indicate statistical significant trends at $p = 0.05$ levels} of spring (**a–c**; left) and autumn (**d–f**; right) season's maize from 1980 to 2014 in Punjab, Pakistan (Source: Adopted from Abbas et al. 2017)

Table 4.4 Average observed phenology (day of year; DOY) of spring and autumn maize in Punjab, Pakistan, during the period of 1980–2014

Location	Sowing	Emergence	Anthesis[a]	Maturity[b]
Spring				
Sialkot	37 ± 6.0	41 ± 5.8	103 ± 6.3	142 ± 5.3
Gujranwala	31 ± 7.1	35 ± 6.6	108 ± 5.3	139 ± 4.7
Hafizabad	39 ± 4.2	44 ± 3.9	102 ± 4.4	147 ± 6.3
Sheikhupura	30 ± 5.1	34 ± 4.8	106 ± 6.6	149 ± 4.2
Nankana Sahib	36 ± 6.4	40 ± 6.1	113 ± 3.2	153 ± 5.4
Multan	32 ± 5.9	36 ± 5.4	107 ± 3.9	149 ± 5.2
Lodhran	40 ± 4.6	44 ± 4.2	113 ± 4.4	144 ± 6.0
Bahawalpur	38 ± 5.2	42 ± 4.9	105 ± 5.8	150 ± 7.4
Bahawalnagar	41 ± 3.4	45 ± 3.2	116 ± 4.9	154 ± 5.9
Rahim Yar Khan	45 ± 6.2	49 ± 5.8	110 ± 6.8	148 ± 6.3
Autumn				
Sialkot	202 ± 5.6	206 ± 5.2	256 ± 5.3	316 ± 6.4
Gujranwala	209 ± 4.9	213 ± 4.5	249 ± 6.1	312 ± 5.8
Hafizabad	213 ± 3.7	217 ± 3.4	258 ± 5.3	310 ± 4.1
Sheikhupura	205 ± 5.9	109 ± 5.3	250 ± 6.8	318 ± 5.9
Nankana Sahib	201 ± 6.2	205 ± 5.9	260 ± 7.4	315 ± 4.7
Multan	204 ± 5.1	208 ± 4.5	264 ± 4.9	317 ± 6.7
Lodhran	207 ± 6.4	211 ± 6.1	257 ± 6.2	324 ± 4.0
Bahawalpur	215 ± 4.8	219 ± 4.5	266 ± 4.4	320 ± 5.6
Bahawalnagar	209 ± 5.7	213 ± 5.2	255 ± 5.8	315 ± 5.2
Rahim Yar Khan	213 ± 5.3	217 ± 4.8	259 ± 6.4	326 ± 6.1

Source: Abbas et al. (2017)
± Standard deviation
[a]50% Anthesis
[b]50% Physiological maturity

Table 4.5 Summary of observed and simulated phenology response to temperature for spring and autumn maize in Punjab, Pakistan, from 1980 to 2014

Phenology	No. of neg.[a]	No. of pos.[b]	No. of sig. neg.[c]	No. of sig. pos.[d]	Reg. mean[e] (days $°C^{-1}$)
Spring stages (observed)					
Sowing	10	0	9	0	−2.08
Emergence	10	0	8	0	−0.38
Anthesis	10	0	9	0	−3.74
Maturity	10	0	8	0	−3.66
Spring phases (observed)					
Sowing-anthesis	10	0	7	0	−2.34
Sowing-maturity	10	0	8	0	−4.19
Anthesis-maturity	10	0	9	0	−1.52
Spring phases (simulated)					
Sowing-anthesis	10	0	9	0	−2.79
Sowing-maturity	10	0	9	0	−4.49
Anthesis-maturity	10	0	10	0	−1.89

(continued)

Table 4.5 (continued)

Phenology	No. of neg.[a]	No. of pos.[b]	No. of sig. neg.[c]	No. of sig. pos.[d]	Reg. mean[e] (days $°C^{-1}$)
Autumn stages (observed)					
Sowing	0	10	0	9	3.52
Emergence	10	0	9	0	−0.46
Anthesis	10	0	8	0	−2.09
Maturity	10	0	8	0	−3.99
Autumn phases (observed)					
Sowing-anthesis	10	0	7	0	−2.19
Sowing-maturity	10	0	9	0	−3.73
Anthesis-maturity	10	0	9	0	−3.10
Autumn phases (simulated)					
Sowing-anthesis	10	0	9	0	−2.41
Sowing-maturity	10	0	10	0	−3.97
Anthesis-maturity	10	0	10	0	−3.42

Source: Abbas et al. (2017)
[a]Number of locations with negative regression coefficients
[b]Number of locations with positive regression coefficients
[c]Number of locations with significant negative regression coefficients
[d]Number of locations with significant positive regression coefficients
[e]Mean of regression coefficients

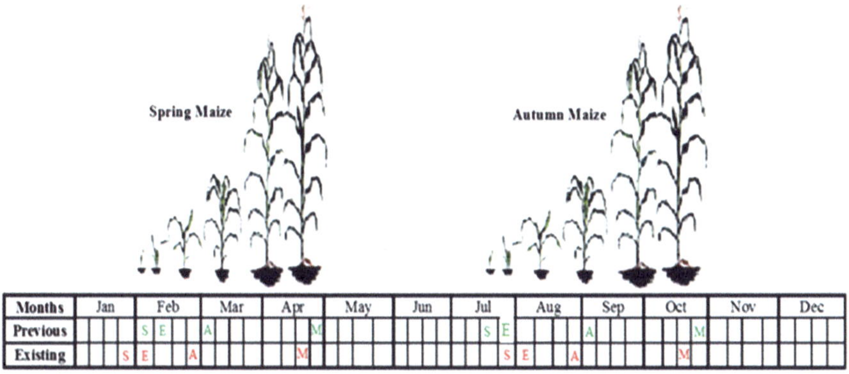

Fig. 4.8 Spatial and temporal variability of spring (left) and autumn (right), respectively, maize phenological phases as effected by an increase in temperature in Punjab, Pakistan. *S* sowing, *E* emergence, *A* anthesis, *M* maturity. Alphabets with green and red colors represent previous and prevailing trends (Source: Adopted from Abbas et al. 2017)

Table 4.6 Comparison of the responses of spring and autumn maize phenology with average temperature using the observed and simulated data in Punjab, Pakistan, during 1980–2014

Phenology	Regression coefficient (days °C^{-1})[a]				Difference between obs. and sim. data (days °C^{-1})		t-Test (p-value)	
	Obs. data		Sim. data					
Phases	Spring	Autumn	Spring	Autumn	Spring	Autumn	Spring	Autumn
Sowing-anthesis	−2.34	−2.19	−2.79	−2.41	0.45	0.22	0.0011**	0.0025**
Sowing-maturity	−4.19	−3.73	−4.49	−3.97	0.30	0.24	0.0095**	0.0180*
Anthesis-maturity	−1.52	−3.10	−1.89	−3.42	0.37	0.24	0.0013**	0.0042**

Source: Abbas et al. (2017)
**Significant at $p = 0.01$
[a]Mean of regression coefficients
[b]Duration of sowing to maturity

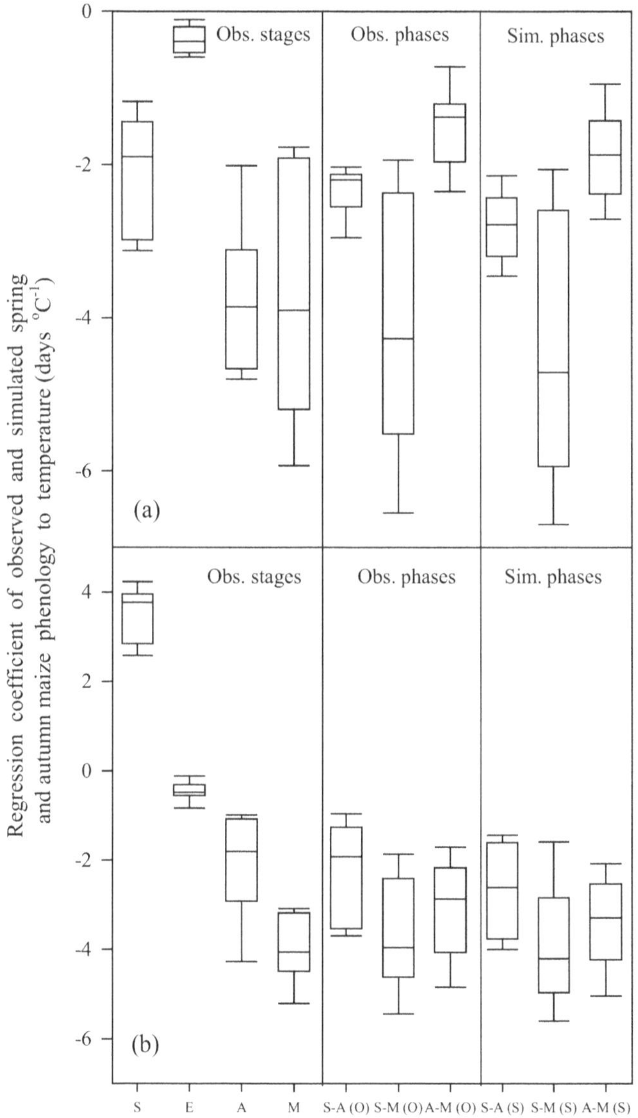

Fig. 4.9 Observed phenology (stages and phases) and simulated phenology (phases) versus temperature trends of spring (**a**) and autumn (**b**) maize cultivar at each location in Punjab, Pakistan, from 1980 to 2014. (*S* sowing, *E* emergence, *A* anthesis, *M* maturity, *S-A* sowing-anthesis, *S-M* sowing-maturity, *A-M* anthesis-maturity, *O* observed, *S* simulated) (Source: Adopted from Abbas et al. 2017)

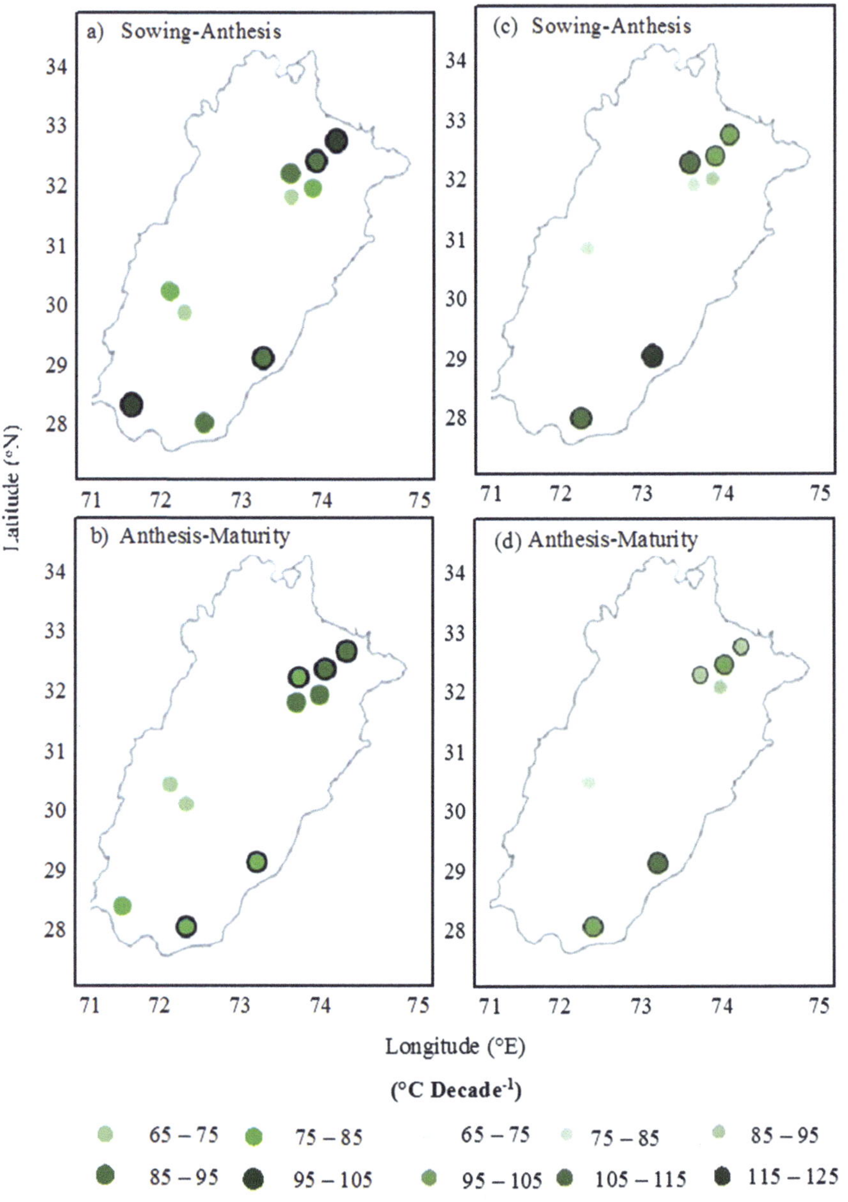

Fig. 4.10 Observed trends in thermal time required for spring (**a, b**; left) and autumn (**c, d**; right) season's maize from sowing-anthesis (**a**) and anthesis to maturity (**b**); circles with black borderline indicate statistically significant trends at $p = 0.05$ levels in Punjab, Pakistan (Source: Adopted from Abbas et al. 2017)

4.8 Climate Change Impact on Current and Future Scenarios of Crop Productivity

Climate change affects maize production in maize-maize cropping system across the world. Climate warming has a significant influence on agricultural productivity, putting food security at risk (Ahmed and Ahmad 2023; Ahmed 2023; Abbas et al. 2023; Liu et al. 2023; Ahmed et al. 2022a; Ahmed et al. 2022b; Nadeem et al. 2022; Ahmed et al. 2020; Khan et al. 2020; Asseng et al. 2019; Ahmed et al. 2017; Jabeen et al. 2017; Aslam et al. 2013; Ahmed et al. 2012). Table 4.7 indicates climate

Table 4.7 Maize yield reduction/increase in maize-maize systems across various regions in the world

Country/ continent	Model	Climatic scenarios	Grain yield reduction/increase (%)	References
Portugal	STICS and AquaCrop	2021–2080; RCP 4.5 and 8.5	−17	Yang et al. (2017)
Ghana	LINTUL5	2016–2045; RCP 4.5 and 8.5	+12	Srivastava et al. (2018)
Africa	APSIM	2039–2099	−20	Rurinda et al. (2015)
USA	Agro-IBIS	1980–2008	−50	Xu et al. (2016)
Ethiopia	CERES	2010–2099; RCP 4.5 and 8.5	−4.7	Araya et al. (2015)
China	APSIM	2039–2099; RCP 8.5	−46	Dai et al. (2016)
Africa	SPEI	1961–2010	−32	Shi and Tao (2014)
Pakistan	CERES	1980–2016	−27	Ahmed et al. (2018)
China	APSIM	1981–2009	−30	Xiao and Tao (2016)
USA	DRAINMOD	2041–2070; RCP 4.5 and 8.5	−27	Gunn et al. (2018)
China	CERES	2006–2050; RCP 4.5 and 8.5	−6	Chen et al. (2020)
Egypt	DNN	2022–2040; RCP 2.6, 4.5, and 8.5	−10	Elbeltagi et al. (2020)
Nepal	VECM	1983 to 2018; RCP 4.5 and 8.5	−26	Chandio et al. (2022)
Somalia	ARDL	1980–2018; RCP 4.5 and 8.5	−12	Warsame et al. (2022)
Mexico	CERES	2012–2100	−30	Murray et al. (2018)
Iran	CERES	2050–2080; RCP 4.5 and 8.5	−42	Moradi et al. (2014)
Malawi	CERES	2011 to 2100; RCP 4.5 and 8.5	−12	Stevens and Madani (2016)

change impact for maize system productivity across the world. Wu et al. (2021) determined climate change effects upon China's maize yields in maize-maize cropping system from 1979 to 2016 and discovered the subsequent consequences that during the years from 1979 to 2016, temperature rises had a detrimental effect on China's maize crop. The yield of maize decreased by 5.19 kg 667 m^{-2} (1.7%) with each 1 °C upsurge of temperature. Since precipitation only slightly increased throughout this time, the effect on maize yield was minimal. The yield of maize rose by a negligible amount of 0.043 kg 667 m^{-2} (0.014%) for every 1 mm increase in precipitation. Preceding research has shown that maize is vulnerable to climate change in all maize-growing seasons. Knox et al. (2012) estimated that by 2050, maize yields in South Asia and Africa will decline by 5% and 16%, respectively. Additionally, it has been shown that the most significant factors affecting maize productivity in France are global temperatures and sun radiation (Ceglar et al. 2016). For instance, it has been found that the majority of the world's main producers of maize, comprising Africa, Brazil, China, and the USA, along with worldwide aggregate output, exhibit negative relationships between productivity and temperatures on average (Tao et al. 2006; Sakurai et al. 2011). The main contributing reason for climate change's negative impact, as shown by empirical research utilizing daily meteorological data, is maize's susceptibility to excessive heat, which is here defined as successive days having temperatures above 30 °C, for temperate and tropical systems (Lobell et al. 2013). Rurinda et al. (2015) stated that in Zimbabwe, under modest radiative forcing, maize cultivars yield dropped by 13% for the years 2010–2039 and 2040–2069 and by 20% for the years 2070–2099, relative to the baseline climate. With significant radiative forcing, larger yield losses of up to 32% can be expected for the period 2070–2099. Lobell et al. (2013) revealed that maize is susceptible to day higher temperatures exceeding 30 °C; along with climate change, expected 2 °C increase in temperatures for much of Africa will impair maize output, further lowering yield in southern Africa despite rising maize demands. Shekhar and Singh (2021) described that climate warming has caused a 10% drop in corn output in Africa. In Zambia, maize yields will fall by 25% in the worst-case scenario, with temperature rises negating the advantages from increasing rainfall. According to Hamududu and Ngoma (2020), climate change would reduce the availability of water for Zambia up to 13% till the end of the century during 2100, posing a far bigger risk to maize crop. Adhikari et al. (2015) indicated that till the end of the twenty-first century, East Africa is expected to lose around 40% of productivity, with climate change affecting maize yield in maize-maize cropping system. Tesfaye et al. (2015) predicted that maize crop yield in Africa will fall by 6–12% and 9–20% in 2050 and 2080, depending on the level of input supplies and GCM estimates. Shakoor et al. (2017) revealed that in Pakistan, the mean temperature will have a detrimental impact on maize crops, resulting in a 6% decline in maize yield by 2030. The average lowest temperature increased maize production by 9% in 2021. An increase in total rainfall would also benefit maize crop productivity. Maize production will grow roughly by 2% by 2030 compared to baseline in 1980. Yang et al. (2017) predicted that in Portugal, the grain yields will show a continuous decrease by 3% in RCP 4.5 or 6% under RCP 8.5 from 2021 to 2040, by 9% under RCP 4.5 or 12.9% under RCP 8.5 from 2041 to 2060, and finally by 10% under RCP 4.5 and 17% under RCP 8.5 from 2061 to 2080. Srivastava et al. (2018)

predicted that in Ghana, there will be a significant increase in mean productivity for all anticipated future climatic scenarios from 2030 to 2080 as compared to baseline because of good effect of CO_2 on crop's RUE paired with relatively mild changes in the quantity of precipitation and incoming radiation for maize growth duration. Xu et al. (2016) stated that across Iowa, USA, in spite of a broader range in expected quantity of warming along with changes in summer rainfall, all predictions estimate a 15–50% decline in maize yield from the late twentieth to the middle and late twenty-first centuries. Araya et al. (2015) simulated that in Ethiopia, the biggest yield decrease (−4.7%) was recorded under RCP 8.5 during the end of the century, while the least yield change (−1.3%) was noticed under RCP 4.5 during the ending of this century. Shi and Tao (2014) revealed that during the maize growth season from 1961 to 2010, each 1 °C elevation in average temperature resulted in yield losses of more than 10% in eight countries, 5–10% in ten countries, and above 5% in four relatively cold nations. Xiao and Tao et al. (2016) predicted that in China, maize productivity is anticipated to decline by roughly 5.36% and 2.55% under RCP 4.5, comparative to baseline considering CO_2 effect. Maize productivity is projected to decrease by nearly 3.0% with a 1 °C increase in average growing period temperature, respectively. Maize yield would be reduced by 2.9% with a 1 °C temperature rise. Moradi et al. (2014) stated that in Iran, during the next 100 years, the predicted grain yields of various cultivars of maize will be steadily reduced across all scenarios by 6.4–42.15% relative to the current climatic circumstances. Jin et al. (2018) reported an average reduction in the yield of corn in the Americas by 12% and 25% under RCP 4.5 and RCP 8.5, respectively.

4.9 Adaption Strategies

Climate change would have a detrimental influence on forthcoming agriculture output as well as food security in all over the world. Adaptive strategies should be implemented to offset the negative impacts on maize-maize cropping system. Heat stress is a major danger to existing and future world maize production (Hellin et al. 2014). Adapting maize to future warmer climates necessitates a better knowledge of crop responses to increased temperatures. Crop management improvements can offset the detrimental impact of climate change, particularly a rise in temperature, on maize crop phenology (Bedeke et al. 2019). When research comes to climate change adaptation in sensitive areas, maize research plays a critical role. Many research have demonstrated that changing cultural practices, for example seed rate, sowing window, phenology changes in varieties, proper management of soil, nutrient management, and irrigation, may favorably or adversely impact grain yield as a result of climatic changes (Wang et al. 2012). Under a climate warming trend, evolving and growing of new varieties with a longer growth duration, for instance, could have an optimistic influence on crop phenological stages and eventually increase yield in maize-maize cropping system. Several investigations have studied climate change impacts on the maize productivity and efficiency, resulting in the promotion of numerous adaptation strategies to alleviate the undesirable influences of climate change (Sutcliffe et al. 2016). Xiao et al. (2016) reported that in the North China Plain, altering variety delayed the maturity and led to a 2.4–3.7 days/decade longer reproductive period. In the Northern China region, maize flowering and maturity were delayed, and overall growth time was extended by an average of 1.5, 6.5, and 6.3 days/decade, respectively. Climate change adaptation techniques in sub-Saharan African maize systems are anticipated to include better germplasm with drought and heat stress tolerance, as well as improved management approaches. Conservation agriculture, for example, has the ability to prevent global warming by lowering greenhouse gas emissions and boosting carbon sequestration. Changing the planting date may be an effective, farmer-friendly, and physiologically feasible practice for avoiding heat stress due to climate change (Meza and Silva 2009). Detrimental influences of climate change on maize production during spring growing season have been mitigated by early sowing dates and prolonged season hybrids. Similar to this, other research results showed that shifting the planting date from the end of April to the end of May resulted in average-based everyday temperature decreases of 1.7 and 4.3 °C and circadian temperature increases of 4.3 and 3.1 °C in the middle and late stages of grain filling, respectively (16–45 days after silking and 45 days afterward from silking to maturity phase). Climate-resilient crop cultivars and adapted farm management practices are obligatory, particularly in zones where irrigation is necessary to gain higher corn yield and irrigation water is reliant on the subversive aquifer, to decrease the undesirable effect of augmented temperatures on growth as well as grain yield of maize crop, especially in lower latitude areas (Meza and Silva 2009). Additionally, it was shown that subsurface drip irrigation pumped at nighttime dropped temperature in the root zone of the crop, promoting corn plant development and output. Optimized water application has the ability to upsurge maize's water usage efficiency, raising its capability to endure heat under climate change circumstances (Southworth et al. 2000). Development of maize in maize-maize cropping system could be delayed, and the reproductive stage could be shortened via soil water scarcity stress along with higher temperature throughout vegetative growing phase, but these conditions are better while moisture contents in maize field are kept at 65% of the field capacity using drip irrigation. In order to develop heat tolerance for maize under high-temperature stress, growth hormones along with exogenous compounds such as ABA and $CaCl_2$ are beneficial to maize plant (Tokatlidis 2013). An appropriate potassium fertilizer boosts the integrity of cell membranes, turgor pressure, as well as water potential during water scarcity due to climate warming (Cairns et al. 2013). The maize-growing region and beyond might be protected against the detrimental impacts of a warming environment on maize crop output and food security by cultivating maize varieties with longer growth periods and greater thermal needs (Liu et al. 2013).

4.10 Conclusion

Global climate change has a considerable impact on the maize-maize agricultural system. In various places, the area used to cultivate maize has experienced an average temperature increase of about 1 °C in the past. The key phenological phases of flowering and crop maturity of the spring season/winter season and autumn season/summer season maize occurred earlier or delayed due to the rise in temperature, and the length of phenology phase from planting to anthesis, planting to maturity, and anthesis to maturity was shortened, potentially lowering grain yield. New hybrids that required more growing degree days were adapted, somewhat reducing the harmful effects of the temperature increase. With adaptation of newly evolved hybrid varieties which had higher growing degree days demand for vital phenology including stages and phases and ultimately more production in maize-maize cropping systems, an increased temperature influence on the phenology and yield of maize during the spring and autumn growing seasons, respectively, can be offset. Future precipitation, temperature, and soil fertility changes would have an impact on maize output and growth. The primary factor affecting maize production in a maize-maize cropping system is a rise in temperature.

References

Abbas Q, Ahmad S (2018) Effect of different sowing times and cultivars on cotton fiber quality under stable cotton-wheat cropping system in southern Punjab, Pakistan. Pak J Life Soc Sci 16(2):77–84

Abbas G, Ahmad S, Ahmad A, Nasim W, Fatima Z, Hussain S, Rehman MH, Khan MA, Hasanuzzaman M, Fahad S, Boote KJ, Hoogenboom G (2017) Quantification the impacts of climate change and crop management on phenology of maize-based cropping system in Punjab, Pakistan. Agric For Meteorol 247:42–55

Abbas G, Fatima Z, Riaz F, Iqbal P, Hussain S, Zakir I, Iqbal M, Haque E, Ahmad S (2019) Study of sowing dates and hybrids effect in maize-based cropping system under arid conditions of Southern Punjab, Pakistan. Pak J Life Soc Sci 17(1):24–31

Abbas G, Ahmad S, Hussain M, Fatima Z, Hussain S, Iqbal P, Ahmed M, Farooq M (2020a) Sowing date and hybrid choice matters production of maize-maize system. Int J Plant Prod 14:583–595

Abbas G, Fatima Z, Hussain M, Hussain S, Atique-ur-Rehman SN, Ahmed M, Ahmad S (2020b) Nitrogen rate and hybrid selection matters productivity of maize-maize cropping system under irrigated arid environment of southern Punjab, Pakistan. Int J Plant Prod 14:309–320

Abbas Q, Atique-ur-Rehman, Sarwar N, Hussain S, Abbas G, Khan MN, Fatima Z, Naz S, Younis H, Ahmed M, Ullah H, Iqbal P, Iqbal M, Kan M, Ahmad S (2020c) Effect of different sowing times and cultivars on wheat grain quality under cotton-wheat cropping system in Southern Punjab, Pakistan. Pak J Life Soc Sci 18(1):41–50

Abbas G, Atique-ur-Rehman, Sarwar N, Fatima Z, Hussain S, Ahmed M, Raza MA, Kan M, DoĞan H, Khan MA, Ahmad S (2023a) Deciding sowing-window for maize-based cropping system in arid and semi-arid environments in Punjab, Pakistan. Turk J Agric For 47(6):1078–1098

Abbas G, Ahmed M, Fatima Z, Hussain S, Kheir AMS, Ercişli S, Ahmad S (2023b) Modeling the potential impact of climate change on maize-maize cropping system in semi-arid environment and designing of adaptation options. Agric For Meteorol 341:109674. https://doi.org/10.1016/j.agrformet.2023.109674

Abendroth LJ, Miguez FE, Castellano MJ, Carter PR, Messina CD, Dixon PM, Hatfield JL (2021) Lengthening of maize maturity time is not a widespread climate change adaptation strategy in the US Midwest. Glob Chang Biol 27(11):2426–2440

Adhikari U, Nejadhashemi AP, Woznicki SA (2015) Climate change and eastern Africa: a review of impact on major crops. Food Energy Security 4(2):110–132

Ahmad S, Hasanuzzaman M (2012) Integrated effect of plant density, N rates and irrigation regimes on the biomass production, N content, PAR use efficiencies and water productivity of rice under irrigated semiarid environment. Notulae Botanicae Horti Agrobotanici Cluj-Napoca 40(1):201–211

Ahmad S, Ahmad A, Soler CMT, Ali H, Zia-Ul-Haq M, Anothai J, Hussain A, Hoogenboom G, Hasanuzzaman G (2012a) Application of the CSM-CERES-Rice model for evaluation of plant density and nitrogen management of fine transplanted rice for an irrigated semiarid environment. Precis Agric 13(2):200–218

Ahmad S, Ali H, Ismail M, Shahzad MI, Nadeem M, Anjum MA, Zia-Ul-Haq M, Firdous N, Khan MA (2012b) Radiation and nitrogen use efficiencies of C_3 winter cereals to nitrogen split application. Pak J Bot 44(1):139–149

Ahmad S, Ahmad A, Ali H, Hussain A, Garcia y Garcia A, Khan MA, Zia-Ul-Haq M, Hasanuzzaman M, Hoogenboom G (2013) Application of the CSM-CERES-Rice model for evaluation of plant density and irrigation management of transplanted rice for an irrigated semiarid environment. Irrig Sci 31(3):491–506

Ahmad A, Ashfaq M, Rasul G, Wajid SA, Khaliq T, Rasul F, Saeed U, Rahman MH, Hussain J, Baig IA, Naqvi SAA, Bokhari SAA, Ahmad S, Naseem W, Hoogenboom G, Valdivia RO (2015a) Impact of climate change on the rice-wheat cropping system of Pakistan. In: Hillel D, Rosenzweig C (eds) Handbook of climate change and agro-ecosystems, vol 3. Imperial College Press and the American Society of Agronomy, London, pp 219–258

Ahmad S, Ali H, Atique-ur-Rehman, Khan RJ, Ahmad W, Fatima Z, Abbas G, Irfan M, Ali H, Khan MA, Hasanuzzaman M (2015b) Measuring leaf area of winter cereals by different techniques: a comparison. Pak J Life Soc Sci 13(2):117–125

Ahmad S, Raza I, Muhammad D, Ali H, Hussain S, Dogan H, Zia-Ul-Haq M (2015c) Radiation, water and nitrogen use efficiencies of *Gossypium hirsutum* L. Turk J Agric For 39(5):825–837

Ahmad S, Ali H, Farooq U, Khan SU, Atique-ur-Rehman, Sarwar N, Shahzad AN, Doğan H, Hussain S, Sultan MT, Waheed A, Zia-ul-Haq M, Hussain K, Khan MA (2016a) Improving nitrogen and radiation-use-efficiencies of C_4 summer cereals by split nitrogen applications under irrigated arid environment. Turk J Agric For 40(2):280–289

Ahmad S, Hussain S, Fatima Z, Abbas G, Atique-ur-Rehman KMR, Younis H, Naz S, Sohail M, Ajmal M, Abbas N, Akhtar M, Rauf A, Khan M, Ali Z, Hassan M, Rizwan M, Bajwa RS, Ajmal A, Naz S, Ali H, Khan AA, Ali M, Sarwar G, Khan MA, Hasanuzzaman M (2016b) Application of DSSAT model for sowing date management of C_4 summer cereals for fodder and grain crops under irrigated arid environment. Pak J Life Soc Sci 14(2):104–114

Ahmad S, Nadeem M, Abbas G, Fatima Z, Khan RJ, Ahmed M, Ahmad A, Rasul G, Khan MA (2016c) Quantification of the effects of climate warming and crop management on sugarcane phenology. Clim Res 71(1):47–61

Ahmad S, Abbas G, Fatima Z, Khan RJ, Anjum MA, Ahmed M, Khan MA, Porter CH, Hoogenboom G (2017a) Quantification of the impacts of climate warming and crop management on canola phenology in Punjab, Pakistan. J Agron Crop Sci 203(5):442–452

Ahmad S, Abbas Q, Abbas G, Fatima Z, Atique-ur-Rehman, Naz S, Younis H, Khan RJ, Nasim W, Rehman MH, Ahmad A, Rasul G, Khan MA, Hasanuzzaman M (2017b) Quantification of climate warming and crop management impacts on cotton phenology. Plan Theory 6:1–16

Ahmad S, Atique-ur-Rehman, Ejaz M, Fatima Z, Kan M, Ahmad M (2018) Agricultural land-use change of major field crops in Pakistan (1961–2014). Sci Technol Dev 37(3):113–121

Ahmad A, Abbas G, Ahmed M, Fatima Z, Anjum MA, Rasul G, Khan MA, Hoogenboom G (2019) Climate warming and management impact on the change of phenology of the rice-wheat cropping system in Punjab, Pakistan. Field Crop Res 230:46–61

Ahmad A, Ashfaq M, Rasul G, Wajid SA, Ahmad I, Khaliq T, Nasir J, Rasul F, Riaz F, Ahmad B, Ahmad S, Baig IA, Valdivia RO, Hoogenboom G (2021) Development of climate change adaptation strategies for cotton–wheat cropping system of Punjab Pakistan. In: Rosenzweig C, Mutter CZ, Contreras EM (eds) Handbook of climate change and agroecosystems: climate change and farming system planning in Africa and South Asia: AgMIP stakeholder-driven research. World Scientific Publishing Europe Ltd, London, pp 277–327

Ahmad S, Ahmad I, Ahmad B, Ahmad A, Wajid A, Khaliq T, Abbas G, Wilkerson CJ, Hoogenboom G (2023a) Regional integrated assessment of climate change impact on cotton production in a semi-arid environment. Clim Res 89:113–132

Ahmad S, Raza MA, Hussain S, Abbas G, Fatima Z, Ahmed M, Goheer MA, Garcia y Garcia A, Wilkerson CJ, Hoogenboom G (2023b) Identification of weak links in the production technology of farmer for bridging canola yield-gap. J Agric Sci 161(2):241–253. https://doi.org/10.1017/S0021859623000187

Ahmed M (2020) Introduction to modern climate change. Andrew E. Dessler: Cambridge University Press, 2011, 252 pp, ISBN-10: 0521173159. Sci Total Environ 734:139397. https://doi.org/10.1016/j.scitotenv.2020.139397

Ahmed M (2023) The science of climate change. In: Brinkmann R (ed) The Palgrave handbook of global sustainability. Springer International Publishing, Cham, pp 195–222. https://doi.org/10.1007/978-3-031-01949-4_22

Ahmed M, Ahmad S (eds) (2023) Disaster risk reduction in agriculture. Springer Nature. https://doi.org/10.1007/978-981-99-1763-1. ISBN 978-981-99-1762-4, Published: 03 November 2023

Ahmed M, Hassan FU, Aslam M, Aslam MA (2012) Physiological attributes based resilience of wheat to climate change. Int J Agric Biol 14(3):407–412

Ahmed M, Akram MN, Asim M, Aslam M, F-u Hassan Higgins S, Stockle CO, Hoogenboom G (2016) Calibration and validation of APSIM-wheat and CERES-wheat for spring wheat under rainfed conditions: models evaluation and application. Comput Electron Agric 123:384–401

Ahmed M, Stöckle CO, Nelson R, Higgins S (2017) Assessment of climate change and atmospheric CO_2 impact on winter wheat in the Pacific Northwest using a multimodel ensemble. Front Ecol Evol 5. https://doi.org/10.3389/fevo.2017.00051

Ahmed I, Ahmed S, Hussain J, Ullah A, Judge J (2018a) Assessing the impact of climate variability on maize using simulation modeling under semi-arid environment of Punjab, Pakistan. Environ Sci Pollut Res 25(28):28413–28430

Ahmed M, Ijaz W, Ahmad S (2018b) Adapting and evaluating APSIM-SoilP-Wheat model for response to phosphorus under rainfed conditions of Pakistan. J Plant Nutr 41(16):2069–2084

Ahmed M, Stockle CO, Nelson R, Higgins S, Ahmad S, Raza M (2019) Novel multimodal ensemble approach to evaluate the sole effect of elevated CO_2 on winter wheat productivity. Sci Rep 9:7813

Ahmed M, Ahmad S, Tariq M, Fatima Z, Aslam Z, Raza MA, Iqbal N, Akmal M, Fayyaz-ul-Hassan, Abbasi NA, Hayat R (2020) Wastes to be the source of nutrients and energy to mitigate climate change and ensure future sustainability: options and strategies. J Plant Nutr 43(6):896–920

Ahmed M, Hayat R, Ahmad M, Mahmood-ul-Hassan, Kheir AMS, Fayyaz-ul-Hassan, Rehman MH, Shaheen FA, Raza MA, Ahmad S (2022) Impact of climate change on dryland agricultural systems: a review of current status, potentials, and further work need. Int J Plant Prod 16:341–363

Ahmed M, Hayat R, Ahmad M, ul-Hassan M, Kheir AMS, ul-Hassan F, ur-Rehman MH, Shaheen FA, Raza MA, Ahmad S (2022a) Impact of climate change on dryland agricultural systems: a review of current status, potentials, and further work need. Int J Plant Prod. https://doi.org/10.1007/s42106-022-00197-1

Ahmed M, Ahmad S, Kheir AMS (2022b) Climate change: an overview. In: Ahmed M (ed) Global agricultural production - resilience to climate change. Springer Nature Switzerland AG. https://doi.org/10.1007/978-3-031-14973-3_1

Ali H, Iqbal N, Ahmad S, Shahzad AN, Sarwar N (2013) Performance of late sown wheat crop under different planting geometries and irrigation regimes in arid climate. Soil Tillage Res 130:109–119

Amin A, Nasim W, Fahad S, Ali S, Ahmad S, Rasool A, Saleem N, Hammad HM, Sultana SR, Mubeen M, Bakhat HF, Ahmad N, Shah GM, Adnan M, Noor M, Basir A, Saud S, Rahman MH, Paz JO (2018a) Evaluation and analysis of temperature for historical (1996–2015) and projected (2030–2060) climates in Pakistan using SimCLIM climate model: ensemble application. Atmos Res 213:422–436

Amin A, Nasim W, Mubeen M, Ahmad A, Nadeem M, Urich P, Fahad S, Ahmad S, Wajid A, Tabassum F, Hammad HM, Sultana SR, Anwar S, Baloch SK, Wahid A, Wilkerson CJ, Hoogenboom G (2018b) Simulated CSM-CROPGRO-cotton yield under projected future climate by SimCLIM for southern Punjab, Pakistan. Agric Syst 167:213–222

Amin A, Nasim W, Mubeen M, Sarwar S, Urich P, Ahmad A, Wajid A, Khaliq T, Rasul F, Hammad HM, Rehmani MIA, Mubarak H, Mirza N, Wahid A, Ahmad S, Fahad S, Ullah A, Khan MN, Ameen A, Amanullah, Shahzad B, Saud S, Alharby H, Ata-Ul-Karim ST, Adnan M, Islam F, Ali QS (2018c) Regional climate assessment for temperature and precipitation in Southern Punjab (Pakistan) using SimCLIM climate model for different temporal scales. Theor Appl Climatol 131:121–131

Araya A, Girma A, Getachew F (2015) Exploring impacts of climate change on maize yield in two contrasting agro-ecologies of Ethiopia. Asian J Appl Sci Eng 4(1):26–36

Aslam Z, Ahmed M, Sajad M, Asif M, Akmal M, Awan FK, Ijaz W, Ashraf R, Khattak JZK (2013) A comparison of statistical and dynamic modeling of wheat (*Triticum aestivum* L.) fungal diseases under the climate change. J Food Agric Environ 11(2):1327–1332

Asseng S, Martre P, Maiorano A, Rötter RP, O'Leary GJ, Fitzgerald GJ, Girousse C, Motzo R, Giunta F, Babar MA, Reynolds MP, Kheir AMS, Thorburn PJ, Waha K, Ruane AC, Aggarwal PK, Ahmed M, Balkovič J, Basso B, Biernath C, Bindi M, Cammarano D, Challinor AJ, De Sanctis G, Dumont B, Eyshi Rezaei E, Fereres E, Ferrise R, Garcia-Vila M, Gayler S, Gao Y, Horan H, Hoogenboom G, Izaurralde RC, Jabloun M, Jones CD, Kassie BT, Kersebaum K-C, Klein C, Koehler A-K, Liu B, Minoli S, Montesino San Martin M, Müller C, Naresh Kumar S, Nendel C, Olesen JE, Palosuo T, Porter JR, Priesack E, Ripoche D, Semenov MA, Stöckle C, Stratonovitch P, Streck T, Supit I, Tao F, Van der Velde M, Wallach D, Wang E, Webber H, Wolf J, Xiao L, Zhang Z, Zhao Z, Zhu Y, Ewert F (2019) Climate change impact and adaptation for wheat protein. Glob Chang Biol 25(1):155–173. https://doi.org/10.1111/gcb.14481

Babel MS, Deb P, Soni P (2019) Performance evaluation of AquaCrop and DSSAT-CERES for maize under different irrigation and manure application rates in the Himalayan region of India. Agric Res 8(2):207–217

Basso B, Liu L, Ritchie JT (2016) A comprehensive review of the CERES-wheat, -maize and -rice models' performances. Adv Agron 136:27–132

Beah A, Kamara AY, Jibrin JM, Akinseye FM, Tofa AI, Ademulegun TD (2021) Simulation of the optimum planting windows for early and intermediate-maturing maize varieties in the Nigerian Savannas using the APSIM model. Front Sustain Food Syst 5:624886

Bedeke S, Vanhove W, Gezahegn M, Natarajan K, Van Damme P (2019) Adoption of climate change adaptation strategies by maize-dependent smallholders in Ethiopia. NJAS-Wageningen J Life Sci 88:96–104

Cairns JE, Hellin J, Sonder K, Araus JL, MacRobert JF, Thierfelder C, Prasanna BM (2013) Adapting maize production to climate change in sub-Saharan Africa. Food Security 5(3):345–360

Ceglar A, Toreti A, Lecerf R, Van der Velde M, Dentener F (2016) Impact of meteorological drivers on regional inter-annual crop yield variability in France. Agric For Meteorol 216:58–67

Chandio AA, Akram W, Bashir U, Ahmad F, Adeel S, Jiang Y (2022) Sustainable maize production and climatic change in Nepal: robust role of climatic and non-climatic factors in the long-run and short-run. Environ Dev Sustain 25:1614–1644

Chandra AK, Joshi A, Tripathi A, Kumar A, Pandey S, Singh A, Lal D, Bharati A, Adhikari S, Dinkar V (2023) Climate-Resilience maize: heat stress, signaling, and molecular interventions. J Plant Growth Regul 42:6349–6366. https://doi.org/10.1007/s00344-022-10844-6

Chen X, Wang L, Niu Z, Zhang M, Li J (2020) The effects of projected climate change and extreme climate on maize and rice in the Yangtze River Basin, China. Agric For Meteorol 282:107867

Chisanga CB, Moombe M, Elijah P (2022) Modelling climate change impacts on maize. CABI Rev. https://doi.org/10.1079/cabireviews202217008

Dai T, Wang J, He D, Wang N (2016) Modeling the impacts of climate change on spring maize yield in Southwest China using the APSIM model. Resour Sci 38:155–165

Deihimfard R, Rahimi-Moghaddam S, Azizi K, Haghighat M (2022) Increased heat stress risk for maize in arid-based climates as affected by climate change: threats and solutions. Int J Biometeorol 66:1365–1378

Ejaz M, Abbas G, Fatima Z, Iqbal P, Raza MA, Kheir AMS, Ahmed M, Kakar KM, Ahmad S (2022) Modelling climate uncertainty and adaptations for soybean-based cropping system. Int J Plant Prod 16:235–250

Elbeltagi A, Aslam MR, Malik A, Mehdinejadiani B, Srivastava A, Bhatia AS, Deng J (2020) The impact of climate changes on the water footprint of wheat and maize production in the Nile Delta, Egypt. Sci Total Environ 743:140770

Erenstein O, Jaleta M, Sonder K, Mottaleb K, Prasanna BM (2022) Global maize production, consumption and trade: trends and R&D implications. Food Security 14:1295–1319

Fatima Z, Abbas Q, Khan A, Hussain S, Ali MA, Abbas G, Younis H, Naz S, Ismail M, Shahzad MI, Nadeem M, Farooq U, Khan SU, Javed K, Khan AA, Ahmed M, Khan MA, Ahmad S (2018) Resource-use-efficiencies of C_3 and C_4 cereals under split nitrogen regimes. Agronomy 8(5):69

Fatima Z, Ahmad M, Hussain M, Abbas G, Ul-Allah S, Ahmad S, Ahmed N, Ali MA, Sarwar G, Haque E, Iqbal P, Hussain S (2020) The fingerprints of climate warming on cereal crops phenology and adaptation options. Sci Rep 10:18013

Fatima Z, Atique-ur-Rehman AG, Iqbal P, Zakir I, Khan MA, Kamal GM, Ahmed M, Ahmad S (2021) Quantification of climate warming and crop management impacts on phenology of pulses-based cropping systems. Int J Plant Prod 15:107–123

Grote U, Fasse A, Nguyen TT, Erenstein O (2021) Food security and the dynamics of wheat and maize value chains in Africa and Asia. Front Sustain Food Syst 4:617009

Gunn KM, Baule WJ, Frankenberger JR, Gamble DL, Allred BJ, Andresen JA, Brown LC (2018) Modeled climate change impacts on subirrigated maize relative yield in northwest Ohio. Agric Water Manag 206:56–66

Hamududu BH, Ngoma H (2020) Impacts of climate change on water resources availability in Zambia: implications for irrigation development. Environ Dev Sustain 22(4):2817–2838

Hellin J, Bellon MR, Hearne SJ (2014) Maize landraces and adaptation to climate change in Mexico. J Crop Improv 28(4):484–501

Hoogenboom G, Porter CH, Boote KJ, Shelia V, Wilkens PW, Singh U, White JW, Asseng S, Lizaso JI, Moreno LP, Pavan W, Ogoshi R, Hunt LA, Tsuji GY, Jones JW (2019) The DSSAT crop modeling ecosystem. In: Boote KJ (ed) Advances in crop modeling for a sustainable agriculture. Burleigh Dodds Science Publishing, Cambridge, pp 173–216. https://doi.org/10.19103/AS.2019.0061.10

Hoogenboom G, Porter CH, Shelia V, Boote KJ, Singh U, White JW, Pavan W, Oliveira FAA, Moreno-Cadena LP, Lizaso JI, Asseng S, Pequeno DNL, Kimball BA, Alderman PD, Thorp KR, Jones MR, Cuadra SV, Vianna MS, Villalobos FJ, Ferreira TB, Batchelor WD, Koo J, Hunt LA, Jones JW (2021) Decision support system for agrotechnology transfer (DSSAT) version 4.8 (DSSAT.net). DSSAT Foundation, Gainesville

Hussain M, Ahmad S, Hussain S, Lal R, Ul-Allah S, Nawaz A (2018) Rice in saline soils: physiology, biochemistry, genetics and management. Adv Agron 148:231–287

Jabeen M, Gabriel HF, Ahmed M, Mahboob MA, Iqbal J (2017) Studying impact of climate change on wheat yield by using DSSAT and GIS: a case study of Pothwar Region. In: Ahmed M, Stockle OC (eds) Quantification of climate variability, adaptation and mitigation for agricultural sustainability. Springer International Publishing, Cham, pp 387–411

Jin X, Kumar L, Li Z, Feng H, Xu X, Yang G, Wang J (2018) A review of data assimilation of remote sensing and crop models. Eur J Agron 92:141–152

Jones PG, Thornton PK (2003) The potential impacts of climate change on maize production in Africa and Latin America in 2055. Glob Environ Chang 13(1):51–59

Jones J, Hoogenboom G, Porter CH, Boote KJ, others (2003) The DSSAT cropping system model. Eur J Agron 18:235–265

Khan MA, Tahir A, Khurshid N, Husnain MI, Ahmed M, Boughanmi H (2020) Economic effects of climate change-induced loss of agricultural production by 2050: a case study of Pakistan. Sustain For 12(3):1216. https://doi.org/10.3390/su12031216

Keating BA, Carberry PS, Hammer GL, Probert ME et al (2003) An overview of APSIM: a model designed for farming systems simulation. Eur J Agron 18:267–288

Khan A, Fahad S, Khan A, Saud S, Adnan M, Wahid F, Noor M, Nasim W, Hammad HM, Bakhat HF, Ahmad S, Rehman MH, Wang D, Sönmez O (2019) Managing tillage operations and manure to restore soil carbon stocks in wheat-maize cropping system. Agron J 111:1–10

Khan MN, Hussain M, Abbas G, Fatima Z, Iqbal P, Khan A, Zakir I, Ahmed M, Mateen A, Khan MA, Ahmad S (2020) Improving resource use efficiencies of sugarcane at farmer field under arid environment. Int J Agric Biol 24(5):1279–1285

Knox J, Hess T, Daccache A, Wheeler T (2012) Climate change impacts on crop productivity in Africa and South Asia. Environ Res Lett 7(3):034032

Li Z, Yang P, Tang H, Wu W, Yin H, Liu Z, Zhang L (2014) Response of maize phenology to climate warming in Northeast China between 1990 and 2012. Reg Environ Chang 14(1):39–48

Liu Z, Hubbard KG, Lin X, Yang X (2013) Negative effects of climate warming on maize yield are reversed by the changing of sowing date and cultivar selection in Northeast China. Glob Chang Biol 19(11):3481–3492

Liu Y, Zhang J, Pan T, Ge Q (2021) Assessing the adaptability of maize phenology to climate change: the role of anthropogenic-management practices. J Environ Manag 293:112874

Liu B, Martre P, Ewert F, Webber H, Waha K, Thorburn PJ, Ruane AC, Aggarwal PK, Ahmed M, Balkovič J, Basso B, Biernath C, Bindi M, Cammarano D, Cao W, Challinor AJ, De Sanctis G, Dumont B, Espadafor M, Eyshi Rezaei E, Fereres E, Ferrise R, Garcia-Vila M, Gayler S, Gao Y, Horan H, Hoogenboom G, Izaurralde RC, Jabloun M, Jones CD, Kassie BT, Kersebaum KC, Klein C, Koehler A-K, Maiorano A, Minoli S, Montesino San Martin M, Müller C, Naresh Kumar S, Nendel C, O'Leary GJ, Eivind Olesen J, Palosuo T, Porter JR, Priesack E, Ripoche D, Rötter RP, Semenov MA, Stöckle C, Stratonovitch P, Streck T, Supit I, Tao F, Van der Velde M, Wang E, Wolf J, Xiao L, Zhang Z, Zhao Z, Zhu Y, Asseng S (2023) AgMIP-Wheat multi-model simulations on climate change impact and adaptation for global wheat. Open Data J Agric Res 9:10–25. https://doi.org/10.18174/odjar.v9i0.18092

Lobell DB, Hammer GL, McLean G, Messina C, Roberts MJ, Schlenker W (2013) The critical role of extreme heat for maize production in the United States. Nat Clim Chang 3(5):497–501

Ma L, Hoogenboom G, Ahuja LR, Ascough Ii JC, Saseendran SA (2006) Evaluation of the RZWQM-CERES-Maize hybrid model for maize production. Agric Syst 87(3):274–295

Mahmood A, Ullah H, Shahzad AN, Ali H, Ahmad S, Zia-Ul-Haq M, Honermeier B, Hasanuzzaman M (2013) Dry matter yield and chemical composition of sorghum cultivars with varying planting density and sowing date. Sains Malaysiana 42(10):1529–1538

Malik W, Isla R, Dechmi F (2019) DSSAT-CERES-maize modelling to improve irrigation and nitrogen management practices under Mediterranean conditions. Agric Water Manag 213:298–308

Mehmood K, Husain M, Aslam M, Ahmedani MS, Aulakh AM, Shaheen FA (2018) Changes in the nutritional composition of maize flour due to Tribolium castaneum infestation and application of carbon dioxide to manage this pest. Environ Sci Pollut Res 25(19):18540–18547

Menefee D, Rajan N, Cui S, Bagavathiannan M, Schnell R, West J (2021) Simulation of dryland maize growth and evapotranspiration using DSSAT-CERES-Maize model. Agron J 113(2):1317–1332

Meza FJ, Silva D (2009) Dynamic adaptation of maize and wheat production to climate change. Clim Chang 94(1):143–156

Moradi R, Koocheki A, Nassiri Mahallati M (2014) Adaptation of maize to climate change impacts in Iran. Mitig Adapt Strateg Glob Chang 19(8):1223–1238

Mubarak MK, Hussain K, Abbas G, Altaf MT, Baloch FS, Ahmad S (2022) Productivity of sorghum (*Sorghum bicolor* L.) at diverse irrigation regimes and sowing dates in semi-arid and arid environment. Turk J Agric For 46(1):1–18

Mubeen M, Ahmad A, Wajid A, Khaliq T, Hammad HM, Sultana SR, Ahmad S, Fahad S, Nasim W (2016) Application of CSM-CERES-Maize model in optimizing irrigated conditions. Outlook Agric 45(3):173–184

Murray GN, Jaramillo VJ, Larsen J (2018) Food security and climate change: the case of rainfed maize production in Mexico. Agric For Meteorol 253:124–131

Nadeem M, Nazer Khan M, Abbas G, Fatima Z, Iqbal P, Ahmed M, Ali Raza M, Rehman A, Ul Haq E, Hayat A, Ali M, Ahmad S (2022) Application of CSM-CANEGRO model for climate change impact assessment and adaptation for sugarcane in semi-arid environment of Southern Punjab, Pakistan. Int J Plant Prod. https://doi.org/10.1007/s42106-022-00192-6

Nasim W, Amin A, Fahad S, Awais M, Khan N, Mubeen M, Wahid A, Turan V, Rehman MH, Ihsan MZ, Ahmad S, Hussain S, Mian IA, Khan B, Jamal Y (2018) Future risk assessment by estimating historical heat wave trends with projected heat accumulation using SimCLIM climate model in Pakistan. Atmos Res 205:118–133

Naz S, Ahmad S, Abbas G, Fatima Z, Hussain S, Khan MA, Ahmed M, Khan A, Fahad S, Nasim W, Ercisli S, Wilkerson CJ, Hoogenboom G (2022) Modeling the impact of climate warming on potato phenology. Eur J Agron 132:126404

Nuss ET, Tanumihardjo SA (2010) Maize: a paramount staple crop in the context of global nutrition. Compr Rev Food Sci Food Saf 9(4):417–436

Rahimi-Moghaddam S, Kambouzia J, Deihimfard R (2018) Risk assessment of heat stress to grain maize in Khuzestan province under climate change. Environ Stress Crop Sci 11(3):749–764

Rahman MH, Ahmad A, Wang X, Wajid A, Nasim W, Hussain M, Ahmad B, Ahmad I, Ali Z, Ishaque W, Awais M, Muddasir M, Shelia V, Ahmad S, Fahad S, Alam M, Ullah H, Hoogenboom G (2018) Multi-model projections of future climate and climate change impacts uncertainty assessment for cotton production in Pakistan. Agric For Meteorol 253–254:94–113

Rahman MH, Ahmad I, Wang D, Fahad S, Afzal M, Ghaffar A, Saddique Q, Khan MA, Saud S, Hassan S, Fahad M, Din M, Ahmad M, Yue Z, Shah RA, Sönmez O, Nasim W (2021) Influence of semi-arid environment on radiation use efficiency and other growth attributes of lentil crop. Environ Sci Pollut Res 28:13697–13711

Rehman A, Ashfaq M, Naqvi SAA, Adil SA, Bashir K, Ahmad A, Ahmad S, Ali A, Imran A (2015) Is climate change worsening the poverty of maize growers? Evidence from Punjab province of Pakistan. Ciencia e Tecnica Vitivinicola 30(1):105–116

Reidsma P, Ewert F, Boogaard H, van Diepen K (2009) Regional crop modelling in Europe: the impact of climatic conditions and farm characteristics on maize yields. Agric Syst 100(1–3):51–60

Rouf T, Prasad K, Kumar P (2016) Maize—a potential source of human nutrition and health: a review. Cogent Food Agric 2(1):1166995

Rugira P, Ma J, Zheng L, Wu C, Liu E (2021) Application of DSSAT CERES-maize to identify the optimum irrigation management and sowing dates on improving maize yield in Northern China. Agronomy 11(04):674

Rurinda J, van Wijk MT, Mapfumo P, Descheemaeker K, Supit I, Giller KE (2015) Climate change and maize yield in southern Africa: what can farm management do? Glob Chang Biol 21(12):4588–4601

Sabagh AE, Hossain A, Iqbal MA, Barutçular C, Islam MS, Çiğ F, Erman M, Sytar O, Brestic M, Wasaya A, Jabeen T (2020) Maize adaptability to heat stress under changing climate. In: Hossain A (ed) Plant stress physiology. IntechOpen, London

Sakurai G, Iizumi T, Yokozawa M (2011) Varying temporal and spatial effects of climate on maize and soybean affect yield prediction. Clim Res 49(2):143–154

Sarwar N, Ali H, Maqsood M, Ullah E, Shahzad AN, Shahzad M, Mubeen K, Shahid MA, Ahmad S (2013) Phenological response of rice plants to different micronutrients application under water saving paddy fields on calcareous soil. Turkish J Field Crops 18(1):52–57

Sarwar N, Atique-ur-Rehman FO, Wasaya A, Hussain M, El-Shehawi AM, Ahmad S, Brestic M, Mahmoud SF, Zivcak M, Farooq S (2021) Integrated nitrogen management improves productivity and economic returns of wheat-maize cropping system. J King Saud Univ Sci 33:101475

Shabbir RN, Waraich EA, Ali H, Nawaz F, Ashraf MY, Ahmad R, Awan MI, Ahmad S, Irfan M, Hussain S, Ahmad Z (2016) Supplemental exogenous NPK application alters biochemical processes to improve yield and drought tolerance in wheat (*Triticum aestivum* L.). Environ Sci Pollut Res 23(3):2651–2662

Shakoor U, Rashid M, Saboor A, Khurshid N, Husnain Z, Rehman A (2017) Maize production response to climate change in Pakistan: a time series assessment. Sarhad J Agric 33(2):320–330

Shekhar M, Singh N (2021) The impact of climate change on changing pattern of maize diseases in Indian Subcontinent: a review. In: El-Esawi MA (ed) Maize genetic resources-breeding strategies and recent advances. IntechOpen, London
Shi W, Tao F (2014) Vulnerability of African maize yield to climate change and variability during 1961–2010. Food Security 6(4):471–481
Southworth J, Randolph JC, Habeck M, Doering OC, Pfeifer RA, Rao DG, Johnston JJ (2000) Consequences of future climate change and changing climate variability on maize yields in the midwestern United States. Agric Ecosyst Environ 82(1–3):139–158
Srivastava AK, Mboh CM, Zhao G, Gaiser T, Ewert F (2018) Climate change impact under alternate realizations of climate scenarios on maize yield and biomass in Ghana. Agric Syst 159:157–174
Stevens T, Madani K (2016) Future climate impacts on maize farming and food security in Malawi. Sci Rep 6(1):1–14
Sutcliffe C, Dougill AJ, Quinn CH (2016) Evidence and perceptions of rainfall change in Malawi: do maize cultivar choices enhance climate change adaptation in sub-Saharan Africa? Reg Environ Chang 16(4):1215–1224
Tao F, Zhang Z (2010) Adaptation of maize production to climate change in North China Plain: quantify the relative contributions of adaptation options. Eur J Agron 33(2):103–116
Tao F, Yokozawa M, Xu Y, Hayashi Y, Zhang Z (2006) Climate changes and trends in phenology and yields of field crops in China, 1981–2000. Agric For Meteorol 138(1–4):82–92
Tao F, Zhang Z, Zhang S, Rötter RP, Shi W, Xiao D, Liu Y, Wang M, Liu F, Zhang H (2016) Historical data provide new insights into response and adaptation of maize production systems to climate change/variability in China. Field Crop Res 185:1–11
Tariq M, Yasmeen A, Ahmad S, Hussain N, Afzal MN, Hasanuzzaman M (2017) Shedding of fruiting structures in cotton: factors, compensation and prevention. Trop Subtrop Agroecosyst 20(2):251–262
Tariq M, Afzal MN, Muhammad D, Ahmad S, Shahzad AN, Kiran A, Wakeel A (2018a) Relationship of tissue potassium content with yield and fiber quality components of Bt cotton as influenced by potassium application methods. Field Crop Res 229:37–43
Tariq M, Ahmad S, Fahad S, Abbas G, Hussain S, Fatima Z, Wajid N, Mubeen M, Rehman M, Khan MA, Adnan M, Wilkerson CJ, Hoogenboom G (2018b) The impact of climate warming and crop management on phenology of sunflower-based cropping systems in Punjab, Pakistan. Agric For Meteorol 256–257:270–282
Tarriq M, Fatima Z, Iqbal P, Nahar K, Ahmad S, Hasanuzzaman M (2021) Sowing dates and cultivars mediated changes in phenology and yield traits of cotton-sunflower cropping system in arid environment. Int J Plant Prod 15:291–302
Tesfaye K, Gbegbelegbe S, Cairns JE, Shiferaw B, Prasanna BM, Sonder K, Boote KJ, Makumbi D, Robertson R (2015) Maize systems under climate change in sub-Saharan Africa: potential impacts on production and food security. Int J Clim Change Strateg Manage 7:247–271
Tesfaye K, Zaidi PH, Gbegbelegbe S, Boeber C, Rahut DB, Getaneh F, Seetharam K, Erenstein O, Stirling C (2017) Climate change impacts and potential benefits of heat-tolerant maize in South Asia. Theor Appl Climatol 130(3):959–970
Tokatlidis IS (2013) Adapting maize crop to climate change. Agron Sustain Dev 33(1):63–79
USDA National Agricultural Statistics Service (2022) Census of Agriculture. Complete data available at www.nass.usda.gov/AgCensus
Vučetić V (2011) Modelling of maize production in Croatia: present and future climate. J Agric Sci 149(2):145–157
Wang M, Li Y, Ye W, Bornman JF, Yan X (2011) Effects of climate change on maize production, and potential adaptation measures: a case study in Jilin Province, China. Clim Res 46(3):223–242
Wang J, Wang E, Yang X, Zhang F, Yin H (2012) Increased yield potential of wheat-maize cropping system in the North China Plain by climate change adaptation. Clim Chang 113(3):825–840
Warsame AA, Sheik-Ali IA, Barre GM, Ahmed A (2022) Examining the effects of climate change and political instability on maize production in Somalia. Environ Sci Pollut Res 30:3293–3306

Wu JZ, Zhang J, Ge ZM, Xing LW, Han SQ, Chen SHEN, Kong FT (2021) Impact of climate change on maize yield in China from 1979 to 2016. J Integr Agric 20(1):289–299

Xiao D, Tao F (2016) Contributions of cultivar shift, management practice and climate change to maize yield in North China Plain in 1981–2009. Int J Biometeorol 60(7):1111–1122

Xiao D, Qi Y, Shen Y, Tao F, Moiwo JP, Liu J, Wang R, Zhang H, Liu F (2016) Impact of warming climate and cultivar change on maize phenology in the last three decades in North China Plain. Theor Appl Climatol 124(3):653–661

Xiao D, Zhao Y, Bai H, Hu Y, Cao J (2019) Impacts of climate warming and crop management on maize phenology in northern China. J Arid Land 11(6):892–903

Xu H, Twine TE, Girvetz E (2016) Climate change and maize yield in Iowa. PLoS One 11(5):e0156083

Yang C, Fraga H, Van Ieperen W, Santos JA (2017) Assessment of irrigated maize yield response to climate change scenarios in Portugal. Agric Water Manag 184:178–190

Zahoor A, Ahmed M, Fayyaz-ul-Hassan, Shabbir G, Ahmad S (2022) Ontogeny growth and radiation use efficiency of canola (*Brassica napus* L.) under various nitrogen management strategies and contrasting environments. Int J Plant Prod 16(1):195–208

Zakir I, Wahab AA, Abbas T, Haider ST, Irum S, Sabir S, Hussain S, Ahmad S (2023) Cereal-vegetable intercropping: hindrances and strategies to increase intercropping. Turk J Agric For 47(6):1115–1129

Žydelis R, Weihermüller L, Herbst M (2021) Future climate change will accelerate maize phenological development and increase yield in the Nemoral climate. Sci Total Environ 784:147175

Sunflower-Sunflower System 5

5.1 Introduction

Sunflower-sunflower cropping system is important in various regions of the world (Tariq et al. 2017, 2018a, b; Tarriq et al. 2021; Hussain et al. 2018; Khan et al. 2019, 2020a; Zahoor et al. 2022). Sunflower is cultivated in spring and autumn growing seasons in sunflower-sunflower cropping system. Endemic Americans have been using sunflower (*Helianthus annuus*, Asteraceae). The sunflower scientific name is *Helianthus annuus* and belongs to the family Asteraceae. This has been originated from areas of North America, since 3000 BCE (Guo et al. 2017; Gurkan et al. 2020). One of the most significant oilseed crops farmed worldwide is the sunflower (*Helianthus annuus* L.). The name *Helianthus annuus*, which is derived from the Greek words helios which means sun, anthus which means flower, and annuus which means annual, refers to a plant that faces its blooms and leaves toward the sun (Radanović et al. 2018). It was firstly commercialized in Russia. The sunflower plant has just lately made a comeback to North America as a crop. But the plant was initially domesticated by American Indians, who turned it into a single-headed plant with a range of seed colors (Adeleke and Babalola 2020). Sunflower is primarily grown for the extraction of oil from its seeds, making it a significant oil crop. After palm, soybean, and rapeseed, sunflower is now recognized as the fourth most significant oil crop in the world. Cooking oil, food, and confections are all made from sunflowers. Different types of sunflowers are used for confectionery and oil applications. The primary application of meal, a by-product of the oil extraction process, is as a component of animal feed rations. Stock and poultry are fed on sunflower oil cake. Additionally, the oil is utilized as a lubricant, in soaps and paints (Kane et al. 2013; Tajik et al. 2023). The seeds are frequently found in bird seed blends and can be consumed dry, roasted, or mashed into nut butter. According to agricultural statistics data, the world's sunflower production in 2022–2023 was 50.70 million metric tonnes, or almost 0.60 million tonnes less than that was anticipated last month. The amount of sunflowers produced in 2017 was 57.32 million tonnes. The

M. Ahmed et al., *Cropping Systems Modeling Under Changing Climate*,
https://doi.org/10.1007/978-981-97-0331-9_5

Table 5.1 Area and production of maize-maize systems across various regions in the world

Countries	Spring sunflower		Autumn sunflower		Sunflower-sunflower system	
	Area (Mha)	Production (MMT)	Area (Mha)	Production (MMT)	Area (Mha)	Production (MMT)
Russian Federation	5.43	10.21	2.98	6.3	8.41	16.51
Ukraine	3.24	6.59	2.71	3.45	5.95	10.04
European Union	1.97	5.78	1.86	3.64	3.83	9.42
Argentina	1.04	2.93	0.83	1.33	1.87	4.26
China	0.58	1.78	0.27	0.83	0.85	2.61
Turkey	0.39	1.26	0.36	0.66	0.75	1.92
USA	0.32	0.87	0.19	0.47	0.51	1.34
Kazakhstan	0.53	0.64	0.28	0.49	0.81	1.13

Source: USDA (2022)
Mha million hectare, *MMT* million metric tonne

anticipated 50.70 million tonnes of sunflowers produced this year might indicate a global production decline of 6.62 million tonnes or 11.55% reduction compared to last year. The annual production of sunflowers is 56.07 million metric tonnes. The Russian Federation produces 16.51 million tonnes of sunflowers annually, making it the greatest producer in the world (Table 5.1). With an annual production of 10.04 million tonnes, Ukraine ranks in second. Together, the Russian Federation and Ukraine account for more than 50% of the world's sunflower production. In Pakistan and India, sunflower is also grown during spring season from January to February and during autumn growing season from July to August (Tariq et al. 2018; Mahmood et al. 2013; Mubeen et al. 2016; Nasim et al. 2018; Mubarak et al. 2022; Rehman et al. 2015; Rahman et al. 2018, 2021; Sarwar et al. 2013, 2021; Shabbir et al. 2016).

5.2 Issues of Sunflower-Sunflower Cropping System

Significant economic losses are brought on by biotic and abiotic stressors in sunflower-sunflower cropping system, which also serve to limit sunflower yield (Abbas et al. 2017, 2019, 2020a, b, c; Abbas and Ahmad 2018; Ahmad et al. 2012a, b, 2013, 2015a, b, c, 2016a, b, c, 2017a, b, 2018, 2019, 2021, 2023a, b). Sunflower (*Helianthus annuus* L.), a rainfed spring crop, is more subject to the detrimental effects of climate change, particularly to high temperatures and drought stress (Lee et al. 2011; Miladinović et al. 2019). The sunflower crop's sowing season falls with wheat crop cultivation. Wheat is preferred by farmers because it has a strong marketing structure and greater support prices. Sunflower maturity coincides with cotton's sowing season, delaying cotton planting. Farmers are therefore hesitant to grow sunflower since a delay in cotton seeding results in a large decrease in production. Sunflower is a labor-intensive crop that requires a lot of fertilizer (Akimowicz et al. 2022; Ahmed et al. 2016, 2018a, b, 2019, 2020, 2022; Ahmad and

Hasanuzzaman 2012; Ali et al. 2013; Amin et al. 2018a, b, c). Furthermore, cotton grown after sunflower harvest is more susceptible to pests, insects, and diseases. These limits are impeding the government's efforts and deterring farmers from increasing sunflower acreage. Sunflower cultivars have a limited yield potential when compared to maize and other crops. Many unscrupulous vendors provide poor-quality seeds to naïve farmers, resulting in poor crop and yields (Blackman et al. 2011; Saha and Mondal 2016). Similarly, with variations, farmers frequently utilize their own seeds, causing seed quality to deteriorate. Although an acceptable number of breeder seed is generated, the quantity of certified seeds produced is insufficient due to a weak seed multiplication chain (Debaeke et al. 2017).

5.3 Decision Support System for Agrotechnology Transfer

Growth prediction models for sunflower in sunflower-sunflower cropping system are progressively utilized as a research tool to help tactical and premeditated verdict to be achieved in a variety of agroclimatic and socioeconomic scenarios (Nasim et al. 2016; Ejaz et al. 2022; Fatima et al. 2018, 2020, 2021; Nadeem et al. 2022; Naz et al. 2022; Jones et al. 2003; Keating et al. 2003; Hoogenboom et al. 2019, 2021). Crop simulation models may be used to swiftly and cheaply analyze variability and complex interconnections. These methods may be used to measure the impact of climate change and assess adaptation choices to lessen its negative consequences. Simple models based on growing degree days have been used to map crop suitability (Arshad et al. 2020). In the past, simple evapotranspiration models, empirical data, and, more recently, process-based models were all employed to estimate yield. Crop simulation models are presently the only method used to examine how climatic variability and change affect grain production and quality. This is because recent developments and improvements (algorithms, databases, climate forecasts) have made them more accurate and efficient (García-López et al. 2014). Sunflower response to climate change may be anticipated in sunflower simulation models by simulating physiological processes as a function of carbon dioxide along with higher temperatures. The following sunflower crop growth simulation models are currently used to simulate the response of sunflower yield in sunflower-sunflower cropping system to various environments: i) generic (one mode for a variety of species) and (ii) specific to the sunflower crop: STICS, CropSyst, EPIC/EPIC-Phase, AquaCrop, AqYield, or WOFOST QSUN (APSIM sunflower) is an oil crop (Arshad et al. 2020). The jeopardy of significantly undervaluing yield reductions in probable circumstances is made possible by the fact that these solitary models incorporate mathematical equations for simulation of the sunflower crop responsiveness to raised temperature and carbon dioxide level. However, the influence of higher carbon dioxide on photosynthesis and transpiration process of the sunflower has been included in the extensive use of the CropSyst, EPIC, and STICS models (Debaeke et al. 2017). Therefore, additional work has to be done in these useful growth simulation models for designing adaptations to climate warming as well as assessing their impacts. Some solitary models obviously take into account seed oil content; therefore, sunflower simulation models need to be updated to incorporate additional physiology-based information on individual as well as combined impacts of carbon dioxide, drought, and temperature on sunflower yield as well as novel outputs. García-López et al. (2014) demonstrated that more accurately than more complex models, modest empirical models that solitarily use climate, edaphic, and field trial-observed data can estimate sunflower yields by identifying key factors like influence of high-temperature stress on flowering period during spring and autumn growing seasons. The applicability of empirical models in regions with climate, crop, and soil circumstances that are significantly diverse from those of the calibration region, however, is constrained. Zhang et al. (2021) disclosed that RZWQM2 model and CSM-CROPGRO-Sunflower component accurately predicted LAI, biomass, phenology, seed yield, and water usage in the east region of Colorado under a variety of water deficiency circumstances. As was also informed in the published data for the CSM-OILCROP-Sunflower model, the findings of the climate change simulations revealed that irrigated water regimes had a far bigger influence on seed yield of sunflower than climate change scenarios. While water potential rose with higher carbon dioxide, the CO_2 fertilization impact was not substantial for sunflower. Parent and Tardieu (2012) determined how recent model for sunflower crop might accurately estimate genetical capriciousness of seed yield in conditions of water scarcity or extreme temperatures. When it comes to temperature, process retorts are frequently approached by a linear function of temperatures, which restricts the model's scope of cogency to the temperature range in which the reaction is roughly linear. Utilizing curvilinear functions for sunflower models is a straightforward initial step that is anticipated to enhance crop development prediction under high-temperature situations. Tariq et al. (2018) applied the calibrated DSSAT-CROPGRO model in sunflower-sunflower cropping system in Pakistan (Figs. 5.1 and 5.2). They indicated that sunflower growth simulation model performed good for entirely studied sites in the Punjab, Pakistan, as revealed through the accord among the simulated flowering and maturity days and field-recorded phenology information for flowering and maturity for the sunflower during spring growing season (slope = 1.05, $R^2 = 0.82$, for calibration; slope = 1.04, $R^2 = 0.80$, for evaluation) and autumn growing season (slope = 1.02, $R^2 = 0.81$, for calibration; slope = 1.04, $R^2 = 0.78$, for evaluation) (Fig. 5.3). The genetic coefficients of sunflower crop in the DSSAT v4.8 are presented in Table 5.2. The sunflower model was applied for phenology simulation for both spring and autumn sunflower.

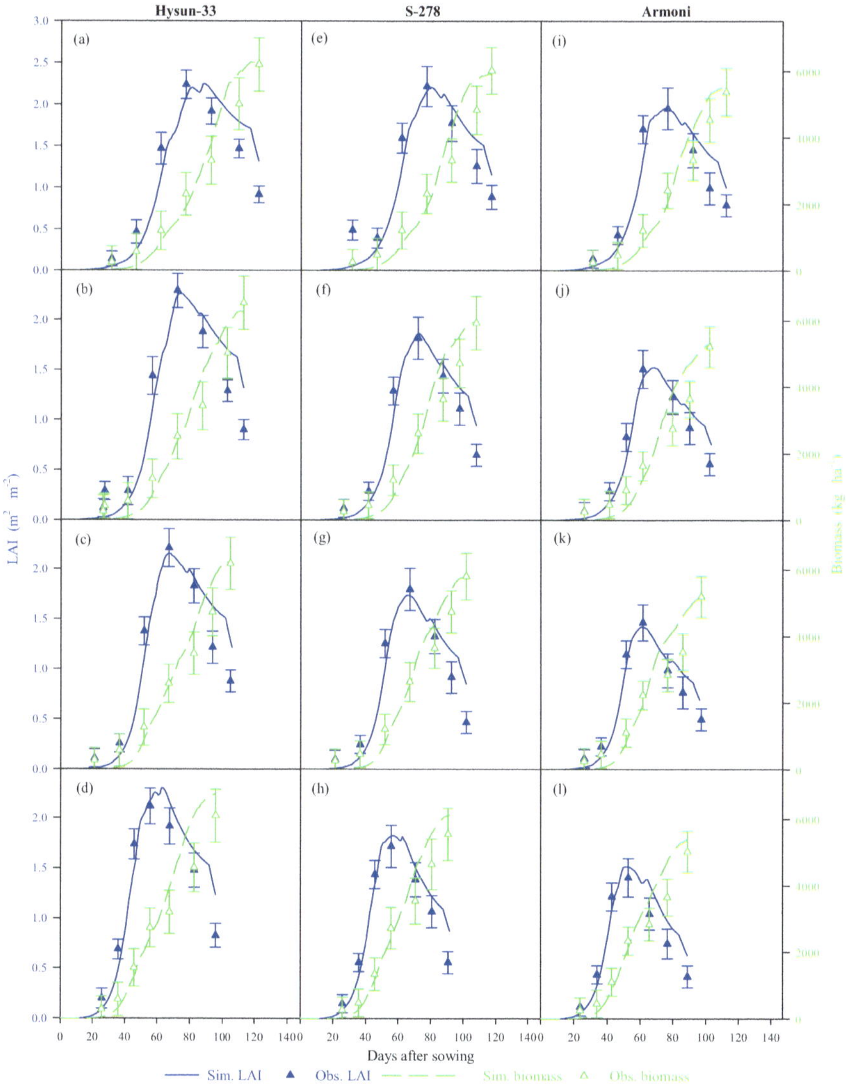

Fig. 5.1 Observed and simulated LAI and biomass as of sunflower hybrids and sowing dates during the first year at study site

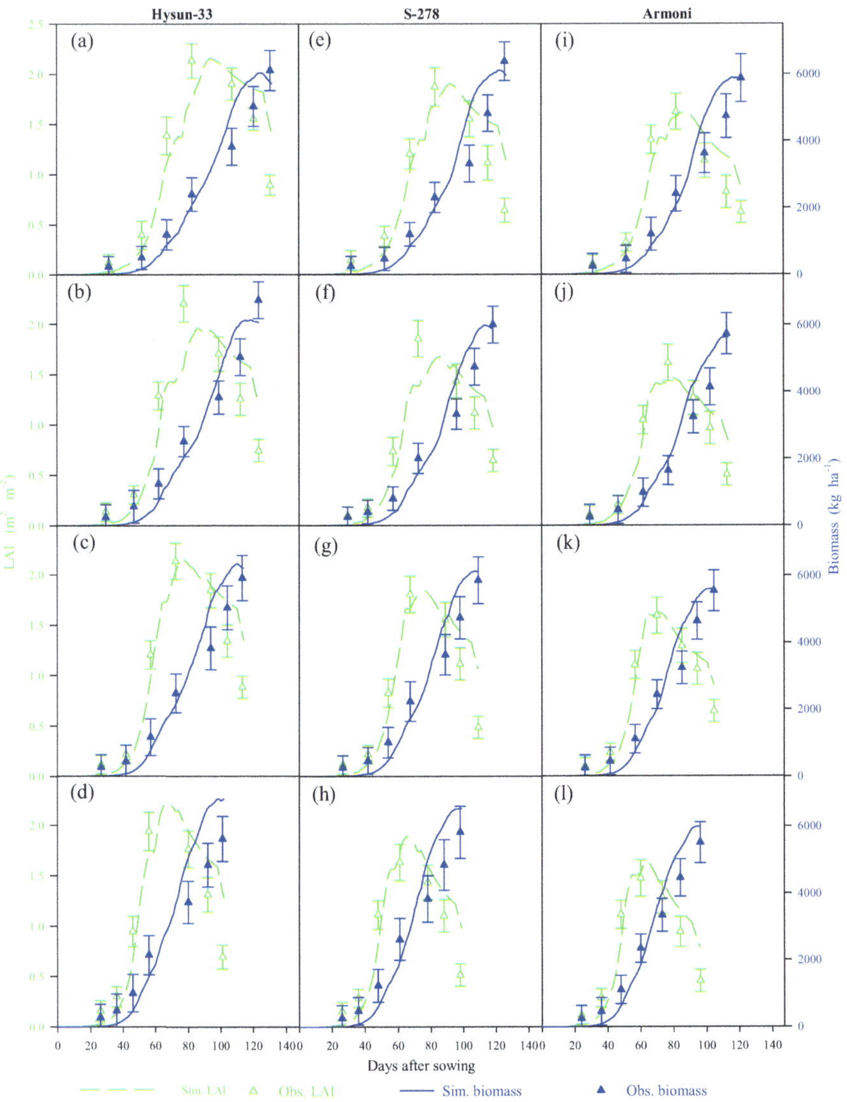

Fig. 5.2 Observed and simulated LAI and biomass as of sunflower hybrids and sowing dates during the second year at study site

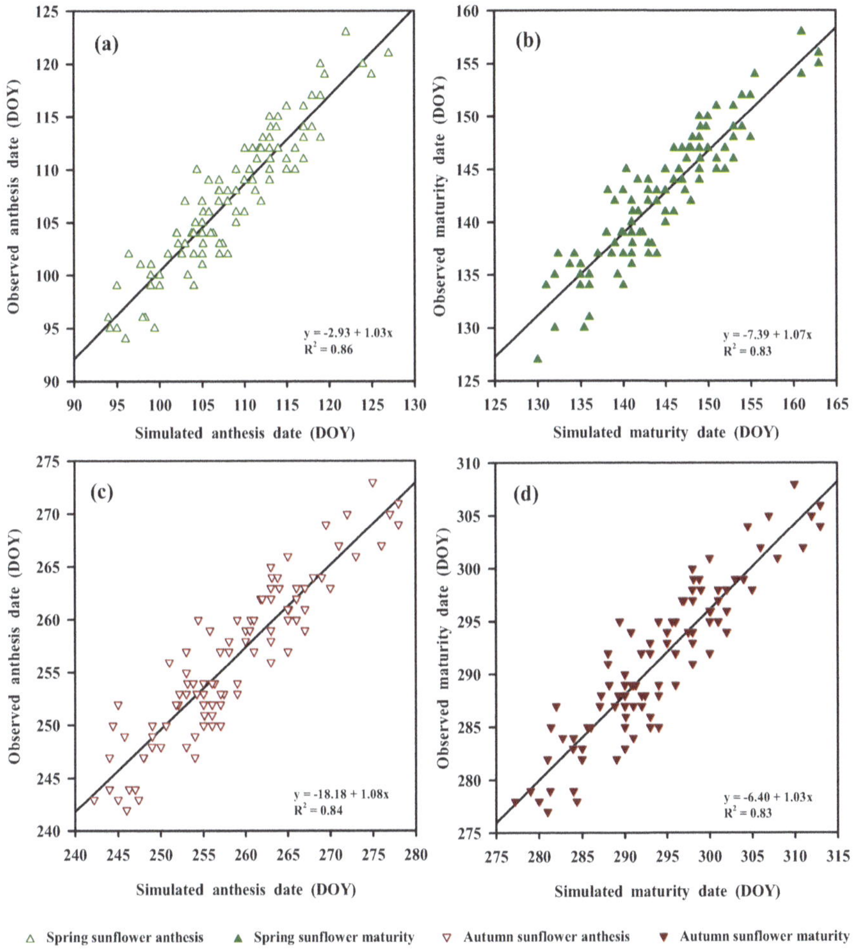

Fig. 5.3 Comparison of simulated and observed anthesis and maturity dates for spring (**a, b**; upward unfilled and filled triangles) and autumn season (**c, d**; downward unfilled and filled triangles) sunflower crop, respectively, during 1980–1985 for 20 locations in Punjab, Pakistan (Source: Adopted from Tariq et al. 2018)

Table 5.2 Genetic coefficients, description, ranges, and guidelines for model calibration of sunflower in DSSAT

Genetic coefficient	Description
VAR#	Identification code or number for a specific cultivar
VAR-NAME	Name of cultivar
EXPNO	Number of experiments used to estimate cultivar parameters
ECO#	Ecotype code of this cultivar, points to the ecotype in the ECO file
CSDL	Critical long day length above which reproductive development progresses with no day length effect (for long-day plants) (h)
PPSEN	Slope of the relative response of development to photoperiod with time (negative for long-day plants) (1/h)—slower at day length less than CSDL
EM-FL	Time between plant emergence and flower appearance (R1) (photothermal days)
FL-SH	Time between starburst and start of the thalamus (R3) (photothermal days)
FL-SD	Time between starburst and anthesis/start of the seed (R5) (photothermal days)
SD-PM	Time between anthesis/start of the seed (R5) and physiological maturity (R7) (photothermal days)
FL-LF	Time between starburst (R1) and end of leaf expansion (photothermal days). Duration should be relatively close to FL-SD, affects SLA
LFMAX	Maximum leaf photosynthesis rate at 30 °C, 350 vpm CO_2, and high light (mg CO_2/m^2-s)
SLAVR	Specific leaf area of cultivar under standard growth conditions (cm^2/g)
SIZLF	Maximum size of full leaf (cm^2)
XFRT	Maximum fraction of daily growth that is partitioned to thalamus, achene, +seed
WTPSD	Maximum weight per seed (g)
SFDUR	Seed filling duration for pod cohort at standard growth conditions (photothermal days)
SDPDV	Average seed per pod under standard growing conditions (#/pod)
PODUR	Time required for cultivar to reach the final pod load under optimal conditions (photothermal days). Must be short for sunflower
THRSH	Threshing percentage. The maximum ratio of (seed/(seed+shell)) at maturity. Causes seeds to stop growing as their dry weight increases until shells are filled in a cohort. Sunflower, achene as % of thalamus
SDPRO	Fraction protein in seeds (g(protein)/g(seed))
SDLIP	Fraction oil in seeds (g(oil)/g(seed))

Source: DSSAT v4.8; Hoogenboom et al. (2021)

5.4 Climate Change and Temperature Trend

Change in climatic conditions is a serious hazard for the sunflower crop in the sunflower-sunflower cropping system and other crop-based cropping systems. Rendering to IPCC report data, the present CO_2 concentration of 400 ppm might rise to 660 ppm during 2060 and 790 ppm during 2090 (Deveci et al. 2019; Awaad 2022). Due to CO_2's capability to absorb infrared light, this is predicted to increase global temperatures and perhaps alter rainfall patterns. The average yearly temperature in Europe climbed by 0.9 °C between 1901 and 2005. Since the end of the 1980s, an increase in air temperature has been clearly documented throughout Europe, and climatologists are referring to a climatic trend rather than spontaneous interannual climatic fluctuation (Ahmed and Ahmad 2023; Ahmed 2023; Abbas et al. 2023; Liu et al. 2023; Ahmed et al. 2012, 2017, 2022b, c; Nadeem et al. 2022; Ahmed 2020; Khan et al. 2020b; Asseng et al. 2019; Jabeen et al. 2017; Aslam et al. 2013). Under hot dry climate conditions, maximum and lowest temperatures rose through 3.5 and 4 °C, respectively, with 30% reduced rainfall (Donatelli et al. 2015). Brouziyne et al. (2018) stated that both scenarios RCP 4.5 and RCP 8.5 are likely to have large rainfall decreases, with RCP 8.5 climate model having a 26% decline in yearly precipitation with a variation of 12%. In all scenarios, the major growth seasons for the sunflower crop, winter and spring, will be the most affected by the drop in rainfall. A significant increase in average temperature is projected to be +1.1 and 1.7 °C for RCP 4.5 and RCP 8.5, respectively (standard deviations of 0.19 and 0.35 °C). Rabi season, an imperative growing season of sunflower, can utmost be influenced, with a considerable shift of increased temperature of 3.4 °C (with standard deviations of 0.52 °C) during the month of January. Conroy et al. (1988) demonstrated that increasing carbon dioxide level in air is helping sunflowers adapt to water shortages in sunflower-sunflower cropping system. When irrigated water volume was inoperative under circumstances that encourage adjustment of the osmotic process, such as when water shortages were gradually executed or when carbon dioxide level was more than 340 ppm, the researchers observed that sunflower crop plants showed more tolerance against water shortage condition. Both leaf conductance to carbon dioxide level and mesophyll's ability to repair carbon dioxide decrease as the water deficit rises. Protracted leaves that had been repeatedly in CO_2 concentration of 660 ppm or had already become used to water scarcity underwent osmotic adjustment. Impact was minimal in non-acclimated sunflower crop plants growing at 340 ppm of carbon dioxide level and was highest when the CO_2 concentrations and water scarcity conditions were combined. Killi et al. (2017) planted two cultivars of sunflower crop (one is water scarcity resistant and the other is sensitive to water shortage stress) for 4 weeks preceding to the application of a water shortage under modest temperature (25 °C) or increased temperature of 35 °C conditions. They discovered that after being exposed to high temperatures, the drought-sensitive variety's net photosynthesis rate and stomatal conductance were more influenced due to soil dryness than after being acclimatized to modest temperature condition of 25 °C. As a result, higher temperatures may enhance the influence of water scarcity stress, having genetic based variations to investigate. Tariq et al. (2018) investigated climate change impact on sunflower-sunflower cropping system in Pakistan. Figure 5.4 shows that temperature tendencies were augmented for total crop period of the spring and autumn sunflower crop, with recorded higher thermal tendency for spring as compared to autumn season. Figure 5.5 indicates that the temperature change from planting to flowering was ranging from 0.4 to 0.8 °C/decade during spring growing period and 0.2 to 0.6 °C/decade during the autumn growing season (statistical significant in which p value is less than 0.05). Rise in temperature from A to M ranged from 0.5 to 1.0 °C/decade (significance in which p value is less than

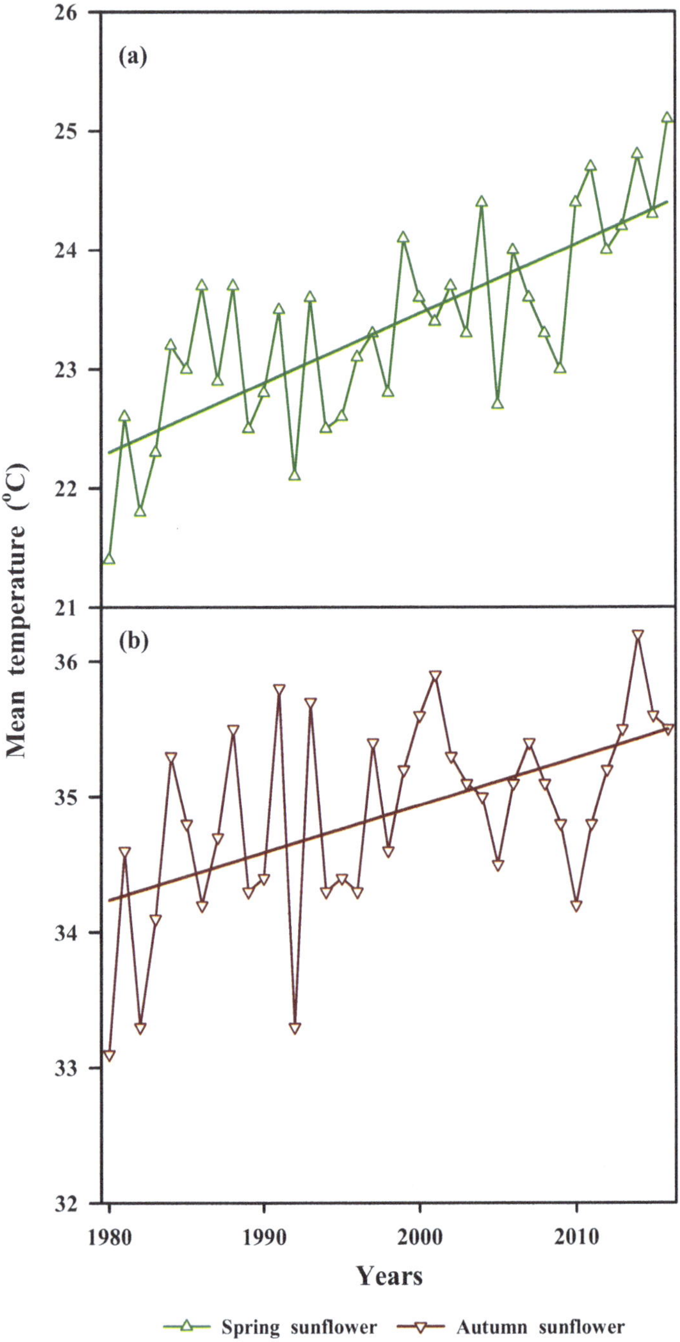

Fig. 5.4 Mean temperature trends in spring (**a**) and autumn (**b**) sunflower season from 1980 to 2016 in Punjab, Pakistan (Source: Adopted from Tariq et al. 2018)

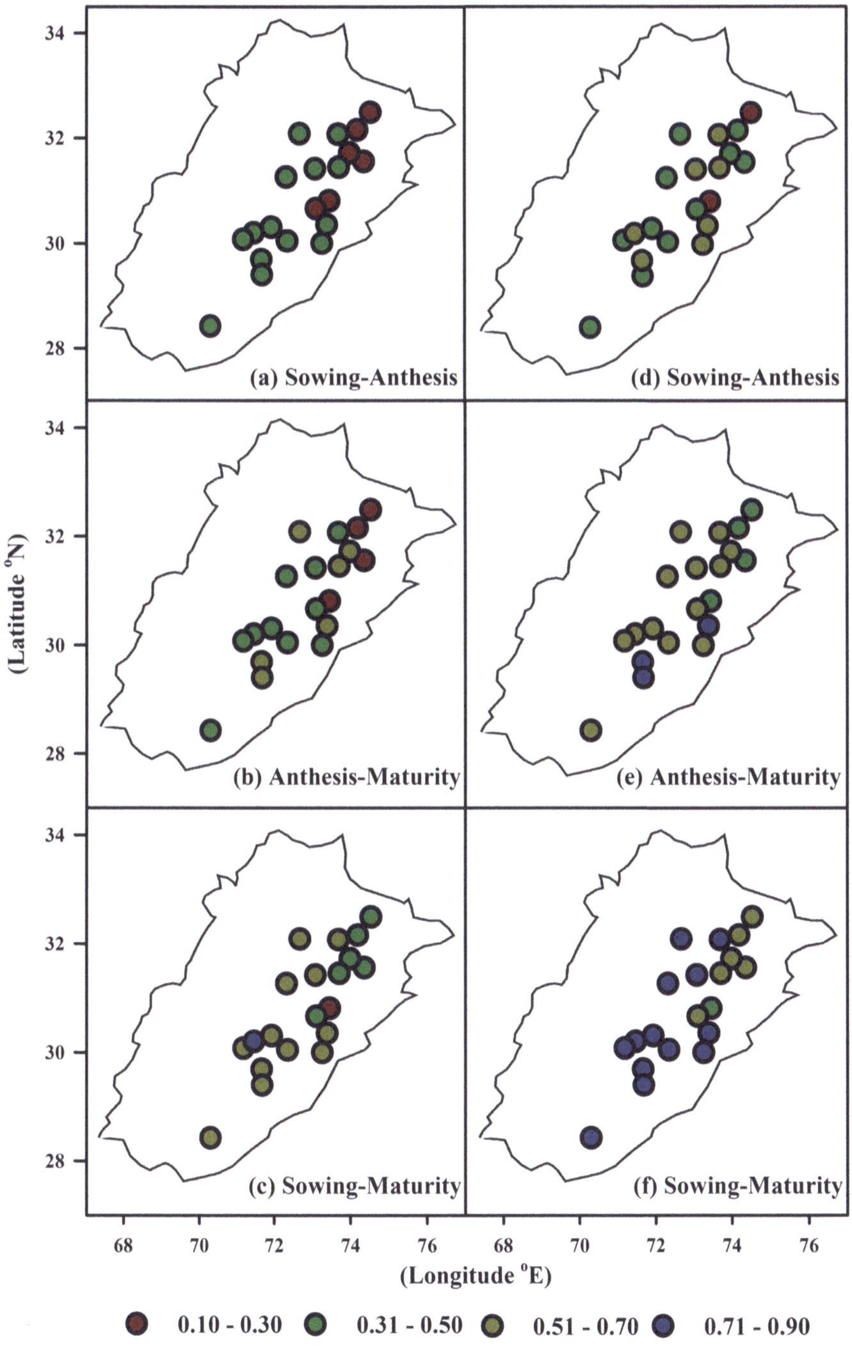

Fig. 5.5 Observed trends in mean temperature during the phenological phases {sowing-anthesis (**a, d**); anthesis-maturity (**b, e**); sowing-maturity (**c, f**)} for spring (**a–c**; left) and autumn (**d–f**; right) season sunflower crop from 1980 to 2016 in Punjab, Pakistan (Source: Adopted from Tariq et al. 2018)

0.05) during the spring season and 0.4 to 0.9 °C/decade (statistically significant, in which p value is less than 0.05) during the fall season. For the time of sowing to maturity, the temperature change ranged from 0.6 to 1.1 °C/decade during the spring growing season (statistical significance, in which p value is less than 0.05) and 0.5 to 0.9 °C/decade during autumn growing season.

5.5 Climate Change Impact on Sunflower Morphophysiological Responses

Increased carbon dioxide levels reduced the quantity of chlorophyll pigments (chlorophyll a and b, as well as carotenoids) and enhanced oxidative stress in sunflower plants by raising H_2O_2 concentrations and lowering the activity of antioxidant enzymes like catalase and ascorbate peroxidase (Agüera and de la Haba 2021). The weakening of antioxidant defenses in the plant most likely raises the concentrations of reactive oxygen species in the chloroplast, lowering the quantity of chlorophyll contents. Biofertilization with fungus (*Rhizophagus irregularis*) on sunflower plants cultivated in higher carbon dioxide conditions results in reduced hydrogen peroxide level and higher antioxidant enzymatic activities (catalase and APX). These findings imply that sunflower symbiosis with *Rhizophagus irregularis* reduces oxidative stress in the sunflower plant (Khan et al. 2017). It appears that a strategy that can be exploited to lessen the impact of carbon dioxide on sunflower plants is an increase in antioxidant defenses. To increase crop output in a CO_2-rich environment, a deeper comprehension of these mechanisms during leaf growth is necessary. Reactive oxygen species (ROSs), such as superoxide radicals, hydrogen peroxide, and hydroxyl radicals, are formed when temperatures are raised, which has been reported to lower antioxidant enzyme activity levels and cause oxidative stress in sunflower plants (Van der Merwe et al. 2015). The buildup of reactive oxygen species not only harms cells but also activates the stress signaling pathways, causing the creation of transcription factors for heat-shock proteins. It was proposed that heat stress may disconnect enzymes and metabolic pathways that result in a buildup of reactive oxygen species, which are responsible for oxidative stress, similar to other kinds of abiotic stress (Killi et al. 2017). The large rise in hydrogen peroxide and the obvious reduction in antioxidant enzyme activity in sunflower plants cultivated at increased temperatures during leaf development indicate that there was severe oxidative stress in these plants compared to sunflower plants grown at normal temperatures. Heat-tolerant sugarcane leaf genotypes that express more CAT and APX at high temperatures may offer defense against reactive oxygen species, hydrogen peroxide, superoxide, and hydroxyl radicals produced by plants when exposed to higher temperatures (Hernández et al. 2018). Elevated temperatures have a detrimental impact on markers typically used to track the development of leaves, inhibiting the growth of primary leaves that are 42 days old on sunflower plants, and increase their oxidative condition. Recently, it was shown that the protein profiles of sunflower leaves indicated distinct variations in the expression of proteins between sunflower plants growing under the two temperature situations (lowered and raised temperatures) (Amutha et al. 2007). These molecular variations seen in primary leaves of sunflower at high temperatures may suggest that sunflower plants may withstand these stressful circumstances better (Janni et al. 2020). Numerous biochemical and physiological properties of plants are impacted by high temperatures. Compared to cereals, less research has been done on the effects of climate warming stress on sunflower, despite the fact that the sunflower crop can be harmed by high temperatures at some critical periods of growth. De la Haba et al. (2014) revealed that when compared to controls, the sunflower leaves subjected to higher temperatures showed decreased growth as seen by lower specific leaf mass and smaller leaf areas. High temperature appears to encourage the breakdown of soluble proteins in leaves, as seen by a greater drop in soluble protein content over leaf life span in plants cultivated at high temperatures compared to control plants. Additionally, high temperatures decrease net photosynthetic rate, potentially through lowering the concentration of pigments of photosynthesis and stomatal conductance. In sunflower leaves exposed to higher temperatures, nitrate reductase along with glutamine synthetase activity dropped, whereas glutamate dehydrogenase deaminating activity rose. According to their findings, high temperatures may have caused early senescence in the sunflower leaves, most likely as a result of the buildups of soluble sugars and corresponding decline in starch levels. Accelerated senescence of primary leaves of sunflower at high temperatures may also have been caused by oxidative damage brought on by increased hydrogen peroxide buildups and a loss in antioxidant activity.

5.6 Impact of Climate Change on Crop Phenology

Crop phenology is affected due to climate change. Phenological stages of sunflower are accelerated due to climate warming across the world. Duration of sunflower phenological phases is reduced as a result of climate change in sunflower-sunflower cropping system in different regions of developed and developing countries. It has been proven that sunflower yield is negatively impacted by temperatures above 31 °C at the anthesis stage as they reduce pollen productivity and floret fertility. Even in well-watered situations, temperatures exceeding 26 °C were supraoptimal for sunflower pollen development. Awais et al. (2018) indicated that the duration of crops was reduced by 3.96% for every 1 °C that is projected to rise in temperature by 2025, and the same effects were observed for higher carbon dioxide concentrations (430 and 550 ppm). With a 2 °C rise in temperature, this crop duration reduction was greater (7.92%) (expected in 2050). The interplay between temperature and carbon dioxide had little influence on crop phenology. The leaf area index (LAI) was negatively impacted by the rising temperature, and these impacts were more pronounced given the current carbon dioxide concentration (393 ppm). At greater carbon dioxide concentrations (430 and

550 ppm), the detrimental impacts of increased temperature were somewhat mitigated.

5.7 Case Study in Pakistan

Tariq et al. (2018) conducted a study to determine the comprehending influences of increasing temperature trends, cultivars, and husbandry practices on sunflower phenological stages and phases in the sunflower-sunflower cropping system in Punjab, Pakistan. In order to develop adaption measures against the trend of rising temperatures in the future decades, this type of research is very essential. The objective of the study was to measure the influence of historical weather data and crop management practices on the phenological stages and phases of sunflower. For this study, spring and autumn phenology and meteorological data were acquired for 20 sites in Pakistan (Table 5.3). From 1980 to 2016, they researched the effect of temperatures' increasing tendency on phenology length of spring and autumn. Results revealed that average temperature is augmented by 0.9 and 0.8 °C/decade, respectively, for spring and autumn sunflower. Phenology stages like S, E, A, and M dates were advanced during the spring growing season by averagely 6.6, 6.3, 3.8, and 2.2 days/decade and delayed during the autumn growing by an average of 5.7, 5.4, 3.1, and 1.8 days/decade (Figs. 5.6 and 5.7). The measured phenology phases S-A, A-M, and S-M were shortened by 2.8, 1.6, and 4.4 days/decade during the spring growing season and by 2.5, 1.2, as well as 3.8 days/decade during the autumn growing season, respectively (Fig. 5.8). There was a negative relationship between model-predicted phenology and temperature during both seasons of sunflower (Table 5.4). Total degree days required for growing spring and autumn cultivars from S to A increased with an average of 80 and 71 °C days/decade, respectively, and ranged from 65 to 99 °C days/decade (significantly at 14 sites) and 57 to 87 °C days/decade (significantly at 13 sites) (Fig. 5.9). Similarly, the requirement of total growing degree days for the period of A to M of spring and autumn sunflower was augmented by 74–106 days/decade (significantly at 16 sites) and 64–92 days/decade (significantly at 15 sites), with an average of 91 and 79 °C days/decade for spring and autumn season cultivars, respectively. From 1980 to 2016, changes in sunflower phenology for spring and fall were significantly related by an increase in the warming tendency. The disparities in phenology between predicted and observed models of spring and autumn types were noteworthy, along with from 1980 to 2016, many unique hybrids with greater needs for total growing degree days were introduced (Table 5.5). The CSM-DSSAT-CROPGRO-Sunflower model, which used standardized cultivars for all sites as well as years, revealed that projected phenology accelerated owing to climate change, with predicted phenological dates being earlier than observed dates in sunflower-sunflower cropping system in Punjab, Pakistan (Fig. 5.10).

Table 5.3 Average observed phenology day of year (DOY) of spring and autumn sunflower in Punjab, Pakistan, during the period of 1980–2016 (Source: Adopted from Tariq et al. 2018)

Location	S (DOY)	E (DOY)	A[a] (DOY)	M[b] (DOY)	S-A (days)	A-M (days)	S-M (days)
Spring							
Sialkot	42 ± 4.5	47 ± 4.4	117 ± 6.4	158 ± 4.2	75 ± 3.2	41 ± 5.3	116 ± 5.0
Lahore	39 ± 6.2	44 ± 6.1	112 ± 5.2	150 ± 4.7	73 ± 4.1	38 ± 5.1	111 ± 6.1
Okara	30 ± 3.7	35 ± 3.2	108 ± 4.3	138 ± 6.2	78 ± 5.3	30 ± 4.1	108 ± 7.5
Gujranwala	26 ± 5.2	31 ± 4.9	106 ± 5.1	137 ± 5.2	80 ± 4.9	31 ± 6.3	111 ± 4.0
Hafizabad	33 ± 8.2	38 ± 7.2	101 ± 3.5	140 ± 6.8	68 ± 2.7	39 ± 5.2	107 ± 5.3
Sheikhupura	31 ± 3.2	36 ± 2.9	103 ± 5.3	133 ± 7.2	72 ± 3.9	30 ± 5.9	102 ± 4.6
Nankana Sahib	26 ± 5.2	31 ± 5.1	100 ± 6.2	137 ± 6.0	74 ± 3.1	37 ± 4.3	111 ± 6.1
Sargodha	34 ± 4.1	39 ± 3.8	104 ± 4.3	142 ± 4.1	70 ± 5.5	38 ± 6.8	108 ± 3.9
Faisalabad	32 ± 3.5	37 ± 3.3	102 ± 5.5	145 ± 3.3	70 ± 6.2	43 ± 4.7	113 ± 5.3
Jhang	29 ± 4.2	34 ± 4.1	104 ± 3.6	139 ± 3.9	75 ± 5.1	35 ± 3.2	110 ± 6.2
Sahiwal	27 ± 4.4	31 ± 4.0	100 ± 5.3	138 ± 4.2	73 ± 6.8	38 ± 4.1	111 ± 6.5
Khanewal	24 ± 3.9	28 ± 3.5	95 ± 6.2	136 ± 5.1	71 ± 4.6	41 ± 3.2	112 ± 3.9
Multan	15 ± 2.8	19 ± 2.6	92 ± 5.9	128 ± 3.9	77 ± 3.9	36 ± 5.2	113 ± 4.4
Vehari	16 ± 3.7	20 ± 3.4	85 ± 6.1	126 ± 4.4	69 ± 4.4	41 ± 3.1	110 ± 5.2
Muzaffargarh	18 ± 5.1	22 ± 4.7	93 ± 4.4	122 ± 3.6	75 ± 7.2	29 ± 4.5	104 ± 4.3
Lodhran	16 ± 2.5	20 ± 2.0	87 ± 3.4	128 ± 3.0	71 ± 3.4	41 ± 5.0	112 ± 5.2
Pakpattan	11 ± 3.1	15 ± 2.8	80 ± 5.3	121 ± 4.9	69 ± 4.6	41 ± 6.3	110 ± 6.7
Bahawalpur	19 ± 5.6	23 ± 4.9	99 ± 4.2	128 ± 6.3	80 ± 5.1	29 ± 6.8	109 ± 4.3
Bahawalnagar	17 ± 5.4	21 ± 4.2	96 ± 4.0	125 ± 7.1	79 ± 6.7	29 ± 4.2	108 ± 4.9
Rahim Yar Khan	14 ± 4.3	18 ± 3.9	85 ± 6.2	119 ± 5.3	71 ± 5.1	34 ± 5.6	105 ± 5.6
Autumn							
Sialkot	202 ± 4.6	207 ± 4.1	275 ± 5.2	314 ± 6.1	73 ± 5.1	39 ± 3.9	112 ± 5.2
Lahore	199 ± 3.2	204 ± 2.9	270 ± 4.8	312 ± 5.5	71 ± 6.3	42 ± 4.4	113 ± 4.9
Okara	190 ± 5.1	195 ± 4.9	266 ± 6.3	305 ± 4.2	76 ± 3.7	39 ± 4.1	115 ± 7.0
Gujranwala	186 ± 2.8	191 ± 2.3	264 ± 7.2	296 ± 5.8	78 ± 5.3	32 ± 4.0	110 ± 6.8

Location	S (DOY)	E (DOY)	A[a] (DOY)	M[b] (DOY)	S-A (days)	A-M (days)	S-M (days)
Hafizabad	193 ± 4.1	198 ± 3.4	259 ± 4.9	298 ± 6.2	66 ± 3.6	39 ± 5.1	105 ± 5.3
Sheikhupura	191 ± 7.1	196 ± 4.1	261 ± 5.1	294 ± 4.8	70 ± 2.7	33 ± 5.5	103 ± 4.7
Nankana Sahib	186 ± 3.9	191 ± 3.6	258 ± 6.2	293 ± 3.9	72 ± 7.3	35 ± 4.8	107 ± 3.1
Sargodha	194 ± 5.2	199 ± 4.9	262 ± 5.5	296 ± 4.1	68 ± 5.7	34 ± 3.7	102 ± 3.8
Faisalabad	192 ± 4.6	197 ± 4.0	260 ± 6.0	293 ± 3.8	68 ± 4.2	33 ± 4.3	101 ± 4.7
Jhang	189 ± 3.3	194 ± 3.1	262 ± 4.4	296 ± 8.2	73 ± 3.6	34 ± 6.3	107 ± 5.2
Sahiwal	187 ± 5.2	192 ± 5.0	258 ± 3.9	290 ± 7.0	71 ± 7.2	32 ± 5.5	103 ± 4.8
Khanewal	184 ± 3.8	189 ± 3.5	253 ± 7.1	292 ± 4.5	69 ± 5.2	39 ± 3.9	108 ± 3.6
Multan	185 ± 3.1	190 ± 2.8	260 ± 6.2	290 ± 3.2	75 ± 4.3	30 ± 7.6	105 ± 7.4
Vehari	186 ± 4.0	191 ± 3.9	253 ± 5.8	288 ± 5.1	67 ± 6.2	35 ± 5.0	102 ± 4.6
Muzaffargarh	188 ± 5.4	193 ± 5.1	261 ± 3.9	297 ± 4.8	73 ± 4.9	36 ± 4.4	109 ± 3.9
Lodhran	186 ± 2.9	191 ± 2.6	255 ± 5.9	293 ± 3.9	69 ± 3.6	38 ± 3.6	107 ± 5.6
Pakpattan	181 ± 3.7	186 ± 3.5	248 ± 4.1	289 ± 5.2	67 ± 2.7	41 ± 6.1	108 ± 5.0
Bahawalpur	189 ± 4.1	194 ± 3.8	267 ± 5.0	295 ± 6.3	78 ± 3.1	28 ± 5.8	106 ± 4.2
Bahawalnagar	187 ± 3.1	192 ± 2.9	264 ± 4.1	290 ± 5.7	77 ± 3.6	26 ± 4.6	103 ± 3.7
Rahim Yar Khan	186 ± 5.7	191 ± 5.4	255 ± 3.7	292 ± 4.9	69 ± 3.9	37 ± 4.0	106 ± 4.9

[a]50% Anthesis
[b]50% Physiological maturity

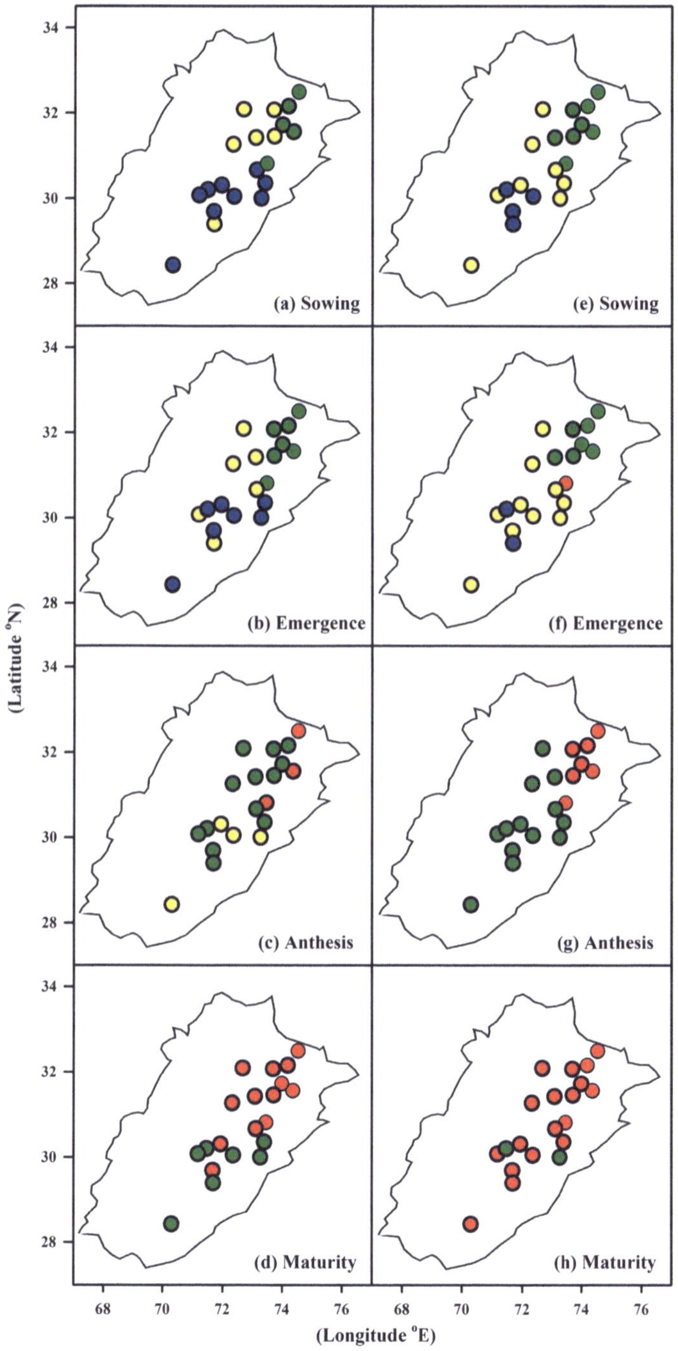

Fig. 5.6 Observed trends in phenology of spring-season sunflower {sowing (**a**); emergence (**b**); anthesis (**c**); and maturity (**d**); circles with black borderline indicate statistically significant trends at $p = 0.05$ levels} from 1980 to 2016 in Punjab, Pakistan. Observed trends in phenology of autumn-season sunflower {sowing (**e**); emergence (**f**); anthesis (**g**); and maturity (**h**); circles with black borderline indicate statistically significant trends at $p = 0.05$ levels} from 1980 to 2016 in Punjab, Pakistan (Source: Adopted from Tariq et al. 2018)

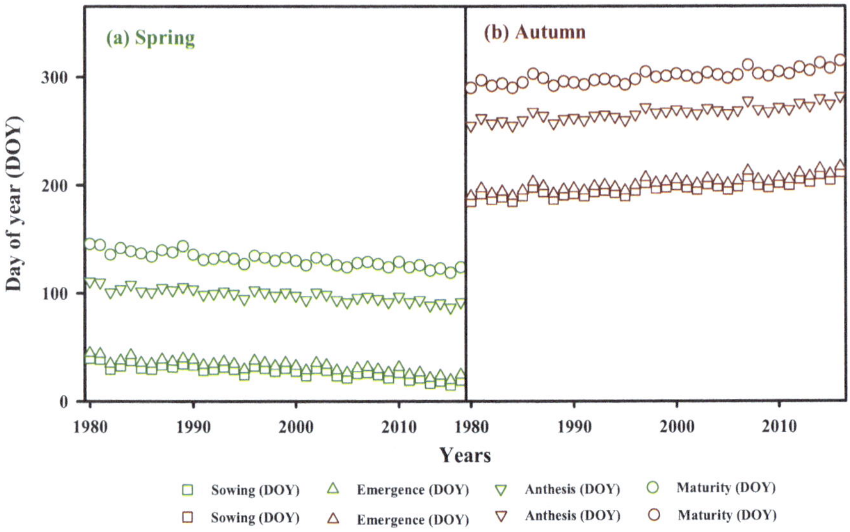

Fig. 5.7 Time series plots of farmer field-observed dates of sowing, emergence, anthesis, and maturity of spring (**a**) and autumn (**b**) season sunflower in Punjab, Pakistan (Source: Adopted from Tariq et al. 2018)

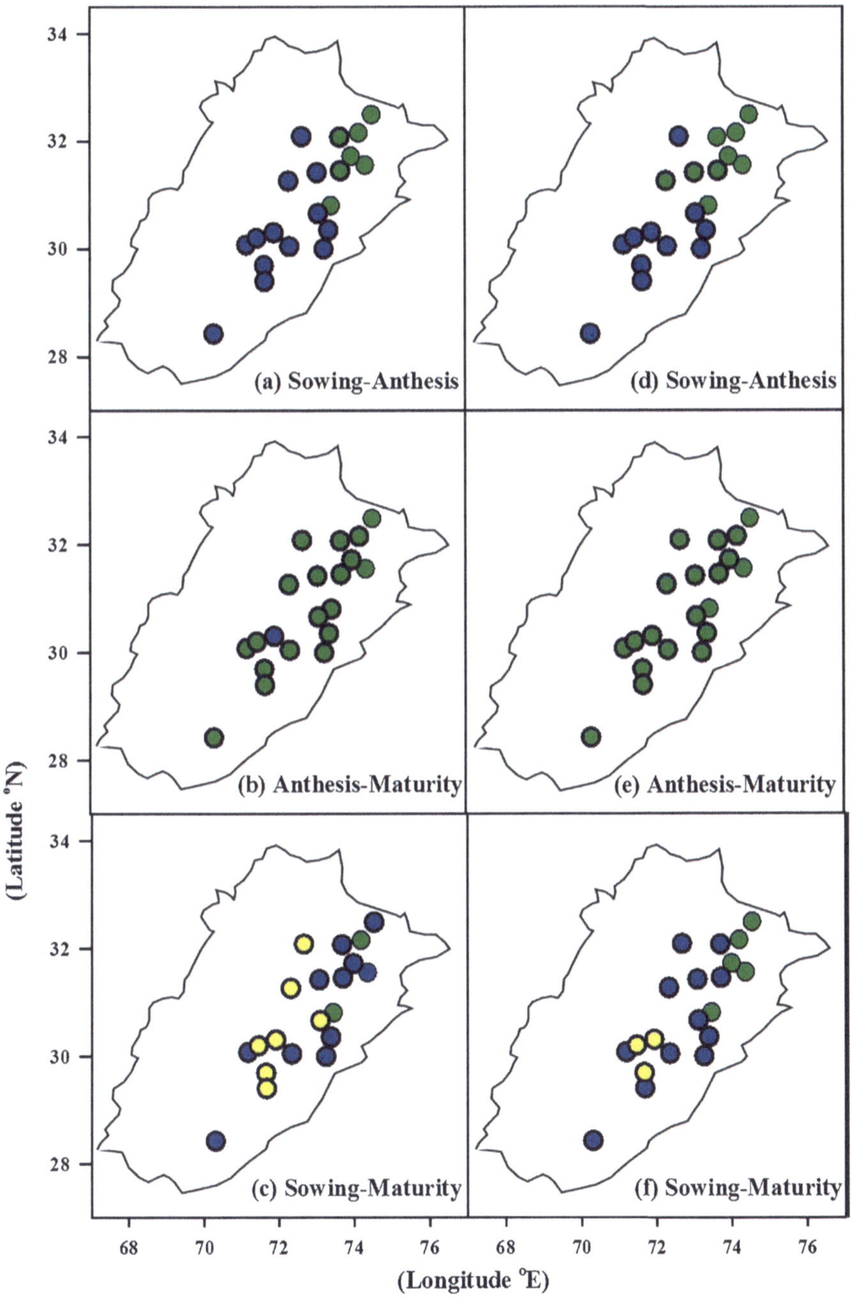

Fig. 5.8 Observed trends in phenology {(**a, d**) sowing-anthesis; (**b, e**) sowing-maturity; and (**c, f**) anthesis to maturity; circles with black borderline indicate statistically significant trends at $p = 0.05$ levels} of spring (**a–c**; left) and autumn (**d–f**; right) season's sunflower from 1980 to 2016 in Punjab, Pakistan (Source: Adopted from Tariq et al. 2018)

Table 5.4 Summary of observed and simulated phenology response to temperature for spring and autumn sunflower in Punjab, Pakistan, for 1980–2016 (Source: Adopted from Tariq et al. 2018)

Phenology	No. of neg.[a]	No. of pos.[b]	No. of sig. neg.[c]	No. of sig. pos.[d]	Reg. mean[e] (days °C^{-1})
Spring stages (observed)					
Sowing	20	0	18	0	−3.0
Emergence	20	0	17	0	−2.8
Anthesis	20	0	19	0	−1.7
Maturity	20	0	16	0	−1.0
Spring phases (observed)					
Sowing-anthesis	20	0	15	0	−1.3
Anthesis-maturity	20	0	17	0	−0.8
Sowing-maturity	20	0	18	0	−2.1
Spring phases (simulated					
Sowing-anthesis	20	0	19	0	−1.8
Anthesis-maturity	20	0	18	0	−1.6
Sowing-maturity	20	0	19	0	−3.0
Autumn stages (observed)					
Sowing	20	0	16	0	−2.3
Emergence	20	0	15	0	−2.2
Anthesis	20	0	17	0	−1.2
Maturity	20	0	16	0	−0.7
Autumn phases (observed)					
Sowing-anthesis	20	0	14	0	−1.1
Anthesis-maturity	20	0	15	0	−0.5
Sowing-maturity	20	0	17	0	−1.6
Autumn phases (simulated)					
Sowing-anthesis	20	0	18	0	−1.5
Anthesis-maturity	20	0	16	0	−1.1
Sowing-maturity	20	0	17	0	−2.3

[a]Number of locations with negative regression coefficients
[b]Number of locations with positive regression coefficients
[c]Number of locations with significant negative regression coefficients
[d]Number of locations with significant positive regression coefficients
[e]Mean of regression coefficients

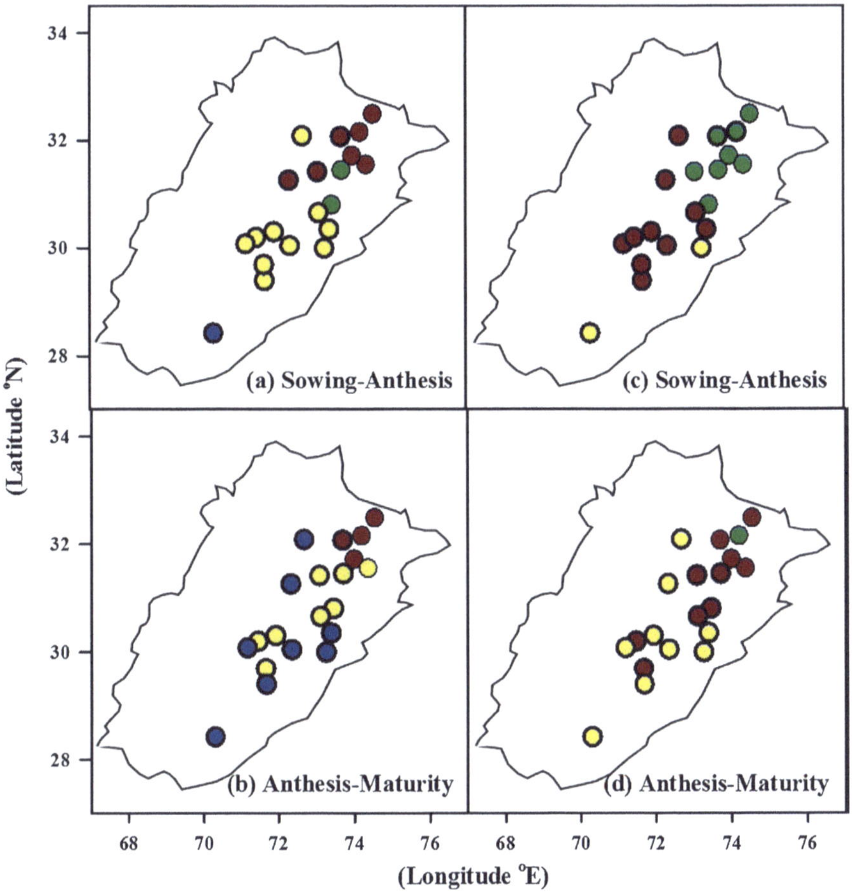

Fig. 5.9 Observed trends in thermal time required for spring (**a, b**; left) and autumn (**c, d**; right) season's sunflower from sowing-anthesis (**a**) and anthesis to maturity (**b**); circles with black borderline indicate statistically significant trends at $p = 0.05$ levels in Punjab, Pakistan (Source: Adopted from Tariq et al. 2018)

Table 5.5 Comparison of the responses of spring and autumn sunflower phenology with average temperature using the observed and simulated data in Punjab, Pakistan, during 1980–2016 (Source: Adopted from Tariq et al. 2018)

Phenology	Regression coefficient (days °C^{-1})[a] Obs. data		Sim. data		Difference between obs. and sim. data (days°C^{-1})		*t*-Test (*p*-value)	
Phases	Spring	Autumn	Spring	Autumn	Spring	Autumn	Spring	Autumn
Sowing-anthesis	−1.3	−1.1	−1.8	−1.5	−0.5	−0.4	0.0061**	0.0038**
Sowing-maturity	−0.8	−0.5	−1.6	−1.1	−0.8	−0.6	0.0019**	0.0021**
Anthesis-maturity	−2.1	−1.6	−3.0	−2.3	−0.9	−0.7	0.0036**	0.0024**

Obs. observed, *Sim.* simulated

** Significant at the 0.01 probability level

[a]Mean of regression coefficients

[b]Duration of sowing to maturity

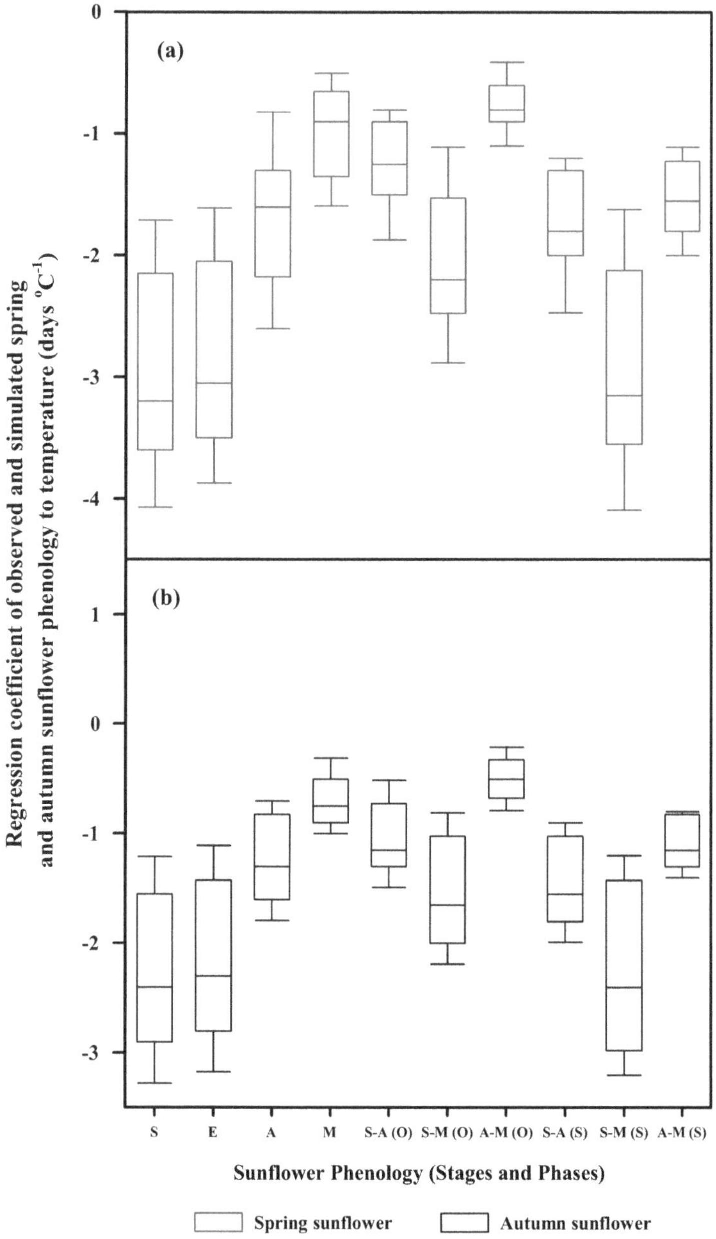

Fig. 5.10 Observed phenology (stages and phases) and simulated phenology (phases) versus temperature trends of spring (**a**) and autumn (**b**) maize cultivar at 20 locations in Punjab, Pakistan, from 1980 to 2016. (*S* sowing, *E* emergence, *A* anthesis, *M* maturity, *S-A* sowing-anthesis, *S-M* sowing-maturity, *A-M* anthesis-maturity, *O* observed, *S* simulated). The central line represents median, box limits are 25th and 75th percentiles, and whiskers represent the minimum and maximum values (Source: Adopted from Tariq et al. 2018)

5.8 Impact of Climate Change on Current and Future Scenarios of Crop Production

Sunflower yield is negatively affected in sunflower-sunflower cropping systems due to climate change across various regions in the world (Table 5.6). At the European level, Moriondo et al. (2011) investigated the direct effects of extreme climatic conditions (i.e., heat stress during flowering stage) by employing the outputs of HadCM3P regional climate model as drivers of a modified version of CropSyst model. The researchers noted that the rise in maximum temperatures under the A2 and B2 scenarios for the period 2071–2100 would result in an upsurge in occurrence of heat stress at "A" compared to baseline. Average productivity loss in Mediterranean region increased from 14% to 34%, while the risk of lower productivity (less than 1.8 t ha^{-1}) increased from 8% to 24%, with greatest changes reported for northeastern as well as southeastern areas, as well as for flat regions. Valverde et al. (2015) projected productivity loss of 6–10% for 2011–2041 timeframe and 11–19% for 2041–2070 duration

Table 5.6 Sunflower yield reduction/increase sunflower-sunflower cropping systems across various regions in the world

Country/ continent	Model	Climatic scenarios	Grain yield reduction/increase (%)	References
Pakistan	OILCROP-SUN	2039–2069 RCP4.5 and 8.5	−15	Nasim et al. (2016)
France	CropSyst	2071–2100 SRES A2 and B2	−34	Moriondo et al. (2011)
Portugal	ISAREG	2011–2041	−10	Valverde et al. (2015)
Russia	AquaCrop	2046–2065	−50	Debaeke et al. (2017)
Turkey	CROPGRO	2041–2098	−39	Gurkan et al. (2020)
Pakistan	CROPGRO	2040–2069	−28	Awais et al. (2020)
Bangladesh	AquaCrop	2015–2050 SRES A1 and B2	+6	Saha and Mondal (2016)
Italy	CropSyst	2010–2050	−40	Tubiello et al. (2000)
America	DayCent	1950–2099 SRES A2 and B1	−24	Lee et al. (2011)
Morocco	SWAT	2031–2050	−26	Brouziyne et al. (2018)
Turkey	AquaCrop	2020–2030 IPCC A2	−22	Deveci et al. (2019)
European Union	CropSyst	2020–2030	−30	Donatelli et al. (2015)
China	Statistical method	1981–2010	+13	Tang et al. (2016)

using water balance technique paired in Portugal. A recent research employed the FAO's AquaCrop model to forecast productivity and output for sunflower crops in Kazakhstan, Ukraine, and Russia during 2046–2065 time span. Debaeke et al. (2017) tested two climatic scenarios: moderate warming (2.4 °C in 2065) as per CGM31 climate model and severe warming (4.3 °C in 2065) according to the MIROC3.2 climatic model. Merely, temperature as well as evapotranspiration was evaluated, with precipitation remaining constant from the baseline and no consideration given to atmospheric CO_2 change. As per worst-case situation, total output would drop till half, having a major influence on the global market since Ukraine would be severely damaged. According to Nasim et al. (2016), an increase in carbon dioxide concentrations from 360 to 550 ppm (varying from 4% in a subhumid environment to 2% in a dry climate) had a substantial impact on sunflower output in Pakistan. Gurkan et al. (2020) projected that in Turkey, both the RCP 4.5 and RCP 8.5 scenarios predict a drop in sunflower production. Sunflower production is expected to fall by 19–31% under the RCP 4.5 scenario and by 3–24% in the RCP 8.5 scenario from 2019 to 2040. Sunflower production is expected to fall by 15–31% for RCP 4.5 and by 24–33% for RCP 8.5 between 2041 and 2070. Sunflower production is estimated to fall by 15–34% for RCP 4.5 and 21–40% for RCP 8.5 between 2071 and 2098. Awais et al. (2020) reported that sunflower productivity reduced by 6–24% and 8–28% using CROPGRO and APSIM model, respectively, for 2040–2069. Saha and Mondal (2016) showed a slightly increasing trend in Bangladesh. Tubiello et al. (2000) predicted that in Italy, in the future, from 2040 to 2080, sunflower seed yield might be reduced by up to 40%. Lee et al. (2011) simulated that in the USA, climatic change will reduce sunflower production by approximately 40% in the long run unless emission of greenhouse gases and the consequent climate change are reduced and/or new management strategies and improved cultivars are adopted. Deveci et al. (2019) demonstrated a 22% drop in sunflower production using the AquaCrop model and the IPCC A2 scenario applied to the Thrace area of Turkey. Donatelli et al. (2015) determined the impact of climate change on sunflower in sunflower-sunflower cropping system in Europe. The results reveal an increase in sunflower production in Spain, Italy, Romania, as well as Bulgaria in 2020, with minor patches of reduction in France and Germany. The increased precipitation can be directly attributed to the improvements. Southern European nations predict moderate gains by 2030, whereas Eastern European countries see yield declines of 10–30%. Adaptation for sunflower was not totally effective under the 2030 time horizon, particularly in northern Europe, where a 10–20% improvement is predicted under the ECHAM5 realization, with occasional circumstances reported under HadleyCM3.

5.9 Adaption Strategies

The only way to prevent climate change from having a significant detrimental influence on crop adaptability for sunflowers is to use climate-smart sunflowers. It may be possible to create, evaluate, and combine adaptations through crop management (planting dates), breeding (earliness, stress tolerance), and relocation of growing regions to partially offset these adverse effects (Moriondo et al. 2011; Nasim et al. 2016). Adaptation solutions are preferred to increase crop productivity for smallholder farmers' economic sustainability and food security due to the anticipated negative effects of climate change on sunflower crop production and yield. Sunflower yield can be increased by using a variety of agronomic management techniques, such as adjustments to fertilizer and irrigation management, as well as soil water conservation techniques like mulching, crop rotation, and minimal or no tillage; site-specific technologies include adjusting the sowing date and planting density, as well as better cultivars (Valverde et al. 2015; Debaeke et al. 2017). It is sometimes recommended that the primary long-term adaptation be breeding for new genotypes that are better adaptable to thermal shocks. Enhanced heat stress resistance, enhanced reactivity of increasing atmospheric carbon dioxide levels, and improved water efficiency are all goals of the breeding techniques (Gurkan et al. 2020). Changing planting date; switching cultivars (earliness); reviewing soil management, fertilization, and integrated pest management; and increasing irrigation water productivity are strategies to minimize the adverse impact of climate warming on sunflower crop. Still, there are several ways to deal with drought-prone regions through crop management (Brouziyne et al. 2018). Early spring sunflower crop sowing aids in preventing plant heat stress and drought during the hotter, drier summer months anticipated under climate change. Winter cultivars that mature more slowly are required to make up for the anticipated crop yield decline brought on by warmer climates' rapid phenological development. No tillage and an early planting date are two sustainable farming practices that can help R'dom be more climate-smart in the face of future climate change impacts (Brouziyne et al. 2018). Donatelli et al. (2015) revealed that with regard to the baseline sowing date, the sowing dates of a sunflower were moved by either moving the sowing ahead or pushing it back by either 10 or 20 calendar days. The duration of the biological cycle was another element that was taken into account as an adaptation approach, leading to simulations of varieties with both a shorter and longer maturity period in comparison to the reference current average varieties. To generate a variance in blooming and physiological maturity, growing degree days for the sunflower parameters were either decreased or increased. The assumption that the temporal horizon of interest is too short to create significantly better varieties led to the exclusion of efficiency differences in genetically based grain production. According to Tariq et al. (2018), in order to ensure sustainable productivity and food security, the adversative effects of high temperature due to climate change on the spring and autumn season-cultivated sunflower phenology have been somewhat mitigated from 1980 to 2016 by advancement of planting dates for spring, later planting dates for autumn season crop, and growing of genotypes that require higher thermal time requirement.

5.10 Conclusion

Significant economic losses are brought on by biotic and abiotic stressors in sunflower-sunflower cropping system, which also serve to limit sunflower yield. Sunflower (*Helianthus annuus* L.), a rainfed spring crop, is more subject to the detrimental effects of climate change, particularly to high temperatures and drought stress. In the sunflower-sunflower cropping systems, models of sunflower crop growth are increasingly employed as instruments to help strategic and tactical decision-making under a variety of agroclimatic and socioeconomic situations. The quantity of photosynthetic pigments in the sunflower plants' leaves dropped due to elevated CO_2 levels and global warming, and oxidative stress increased as a result. If climate change adaptation is not taken into consideration, significant declines in sunflower yield are predicted globally. The increased atmospheric carbon dioxide might offset the negative effects of higher temperature stress and shorter sunflower crop length in the near future (2050s), but the CO_2 fertilization effect will not stop yield decline at the end of this century. The only way to prevent climate change from having a significant detrimental influence on crop adaptability for sunflowers is to adopt climate-smart sunflowers. There are still a lot of questions about biodiversity and biotic elements (such as fungus, birds, insects, and soil macrofauna) that could have an impact on sunflower production in sunflower-sunflower cropping system in the future.

References

Abbas Q, Ahmad S (2018) Effect of different sowing times and cultivars on cotton fiber quality under stable cotton-wheat cropping system in Southern Punjab, Pakistan. Pak J Life Soc Sci 16(2):77–84

Abbas G, Ahmad S, Ahmad A, Nasim W, Fatima Z, Hussain S, Rehman MH, Khan MA, Hasanuzzaman M, Fahad S, Boote KJ, Hoogenboom G (2017) Quantification the impacts of climate change and crop management on phenology of maize-based cropping system in Punjab, Pakistan. Agric For Meteorol 247:42–55

Abbas G, Fatima Z, Riaz F, Iqbal P, Hussain S, Zakir I, Iqbal M, Haque E, Ahmad S (2019) Study of sowing dates and hybrids effect in maize-based cropping system under arid conditions of Southern Punjab, Pakistan. Pak J Life Soc Sci 17(1):24–31

Abbas G, Ahmad S, Hussain M, Fatima Z, Hussain S, Iqbal P, Ahmed M, Farooq M (2020a) Sowing date and hybrid choice matters production of maize-maize system. Int J Plant Prod 14:583–595

Abbas G, Fatima Z, Hussain M, Hussain S, Atique-ur-Rehman, Sarwar N, Ahmed M, Ahmad S (2020b) Nitrogen rate and hybrid selection matters productivity of maize-maize cropping system under irrigated arid environment of Southern Punjab, Pakistan. Int J Plant Prod 14:309–320

Abbas Q, Atique-ur-Rehman, Sarwar N, Hussain S, Abbas G, Khan MN, Fatima Z, Naz S, Younis H, Ahmed M, Ullah H, Iqbal P, Iqbal M, Kan M, Ahmad S (2020c) Effect of different sowing times and cultivars on wheat grain quality under cotton-wheat cropping system in Southern Punjab, Pakistan. Pak J Life Soc Sci 18(1):41–50

Abbas G, Ahmed M, Fatima Z, Hussain S, Kheir AMS, Ercişli S, Ahmad S (2023) Modeling the potential impact of climate change on maize-maize cropping system in semi-arid environment and designing of adaptation options. Agric For Meteorol 341:109674. https://doi.org/10.1016/j.agrformet.2023.109674

Adeleke BS, Babalola OO (2020) Oilseed crop sunflower (Helianthus annuus) as a source of food: nutritional and health benefits. Food Sci Nutr 8(9):4666–4684

Agüera E, de la Haba P (2021) Climate change impacts on sunflower (Helianthus annus L.) plants. Plants 10(12):2646

Ahmad S, Hasanuzzaman M (2012) Integrated effect of plant density, N rates and irrigation regimes on the biomass production, N content, PAR use efficiencies and water productivity of

rice under irrigated semiarid environment. Notulae Botanicae Horti Agrobotanici Cluj-Napoca 40(1):201–211
Ahmad S, Ahmad A, Soler CMT, Ali H, Zia-Ul-Haq M, Anothai J, Hussain A, Hoogenboom G, Hasanuzzaman G (2012a) Application of the CSM-CERES-Rice model for evaluation of plant density and nitrogen management of fine transplanted rice for an irrigated semiarid environment. Precis Agric 13(2):200–218
Ahmad S, Ali H, Ismail M, Shahzad MI, Nadeem M, Anjum MA, Zia-Ul-Haq M, Firdous N, Khan MA (2012b) Radiation and nitrogen use efficiencies of C_3 winter cereals to nitrogen split application. Pak J Bot 44(1):139–149
Ahmad S, Ahmad A, Ali H, Hussain A, Garcia y Garcia A, Khan MA, Zia-Ul-Haq M, Hasanuzzaman M, Hoogenboom G (2013) Application of the CSM-CERES-Rice model for evaluation of plant density and irrigation management of transplanted rice for an irrigated semiarid environment. Irrig Sci 31(3):491–506
Ahmad A, Ashfaq M, Rasul G, Wajid SA, Khaliq T, Rasul F, Saeed U, Rahman MH, Hussain J, Baig IA, Naqvi SAA, Bokhari SAA, Ahmad S, Naseem W, Hoogenboom G, Valdivia RO (2015a) Impact of climate change on the rice-wheat cropping system of Pakistan. In: Hillel D, Rosenzweig C (eds) Handbook of climate change and agro-ecosystems, vol 3. Imperial College Press and the American Society of Agronomy, London, pp 219–258
Ahmad S, Ali H, Atique-ur-Rehman, Khan RJ, Ahmad W, Fatima Z, Abbas G, Irfan M, Ali H, Khan MA, Hasanuzzaman M (2015b) Measuring leaf area of winter cereals by different techniques: a comparison. Pak J Life Soc Sci 13(2):117–125
Ahmad S, Raza I, Muhammad D, Ali H, Hussain S, Dogan H, Zia-Ul-Haq M (2015c) Radiation, water and nitrogen use efficiencies of *Gossypium hirsutum* L. Turk J Agric For 39(5):825–837
Ahmad S, Ali H, Farooq U, Khan SU, Atique-ur-Rehman, Sarwar N, Shahzad AN, Doğan H, Hussain S, Sultan MT, Waheed A, Zia-ul-Haq M, Hussain K, Khan MA (2016a) Improving nitrogen and radiation-use-efficiencies of C_4 summer cereals by split nitrogen applications under irrigated arid environment. Turk J Agric For 40(2):280–289
Ahmad S, Hussain S, Fatima Z, Abbas G, Atique-ur-Rehman KMR, Younis H, Naz S, Sohail M, Ajmal M, Abbas N, Akhtar M, Rauf A, Khan M, Ali Z, Hassan M, Rizwan M, Bajwa RS, Ajmal A, Naz S, Ali H, Khan AA, Ali M, Sarwar G, Khan MA, Hasanuzzaman M (2016b) Application of DSSAT model for sowing date management of C_4 summer cereals for fodder and grain crops under irrigated arid environment. Pak J Life Soc Sci 14(2):104–114
Ahmad S, Nadeem M, Abbas G, Fatima Z, Khan RJ, Ahmed M, Ahmad A, Rasul G, Khan MA (2016c) Quantification of the effects of climate warming and crop management on sugarcane phenology. Clim Res 71(1):47–61
Ahmad S, Abbas G, Fatima Z, Khan RJ, Anjum MA, Ahmed M, Khan MA, Porter CH, Hoogenboom G (2017a) Quantification of the impacts of climate warming and crop management on canola phenology in Punjab, Pakistan. J Agron Crop Sci 203(5):442–452
Ahmad S, Abbas Q, Abbas G, Fatima Z, Atique-ur-Rehman, Naz S, Younis H, Khan RJ, Nasim W, Rehman MH, Ahmad A, Rasul G, Khan MA, Hasanuzzaman M (2017b) Quantification of climate warming and crop management impacts on cotton phenology. Plan Theory 6:1–16
Ahmad S, Atique-ur-Rehman EM, Fatima Z, Kan M, Ahmad M (2018) Agricultural land-use change of major field crops in Pakistan (1961–2014). Sci Technol Dev 37(3):113–121
Ahmad A, Abbas G, Ahmed M, Fatima Z, Anjum MA, Rasul G, Khan MA, Hoogenboom G (2019) Climate warming and management impact on the change of phenology of the rice-wheat cropping system in Punjab, Pakistan. Field Crop Res 230:46–61
Ahmad A, Ashfaq M, Rasul G, Wajid SA, Ahmad I, Khaliq T, Nasir J, Rasul F, Riaz F, Ahmad B, Ahmad S, Baig IA, Valdivia RO, Hoogenboom G (2021) Development of climate change adaptation strategies for cotton–wheat cropping system of Punjab, Pakistan. In: Rosenzweig C, Mutter CZ, Contreras EM (eds) Handbook of climate change and agroecosystems: climate change and farming system planning in Africa and South Asia: AgMIP stakeholder-driven research. World Scientific Publishing Europe Ltd, London, pp 277–327

Ahmad S, Ahmad I, Ahmad B, Ahmad A, Wajid A, Khaliq T, Abbas G, Wilkerson CJ, Hoogenboom G (2023a) Regional integrated assessment of climate change impact on cotton production in a semi-arid environment. Clim Res 89:113–132

Ahmad S, Raza MA, Hussain S, Abbas G, Fatima Z, Ahmed M, Goheer MA, Garcia y Garcia A, Wilkerson CJ, Hoogenboom G (2023b) Identification of weak links in the production technology of farmer for bridging canola yield-gap. J Agric Sci 161(2):241–253. https://doi.org/10.1017/S0021859623000187

Ahmed M (2020) Introduction to modern climate change. Andrew E. Dessler: Cambridge University Press, 2011, 252 pp, ISBN-10: 0521173159. Sci Total Environ 734:139397. https://doi.org/10.1016/j.scitotenv.2020.139397

Ahmed M (2023) The science of climate change. In: Brinkmann R (ed) The Palgrave handbook of global sustainability. Springer International Publishing, Cham, pp 195–222. https://doi.org/10.1007/978-3-031-01949-4_22

Ahmed M, Ahmad S (eds) (2023) Disaster risk reduction in agriculture. Springer Nature. https://doi.org/10.1007/978-981-99-1763-1. ISBN 978-981-99-1762-4, Published: 03 November 2023

Ahmed M, Hassan FU, Aslam M, Aslam MA (2012) Physiological attributes based resilience of wheat to climate change. Int J Agric Biol 14(3):407–412

Ahmed M, Akram MN, Asim M, Aslam M, F-u Hassan Higgins S, Stockle CO, Hoogenboom G (2016) Calibration and validation of APSIM-wheat and CERES-wheat for spring wheat under rainfed conditions: models evaluation and application. Comput Electron Agric 123:384–401

Ahmed M, Stöckle CO, Nelson R, Higgins S (2017) Assessment of climate change and atmospheric CO_2 impact on winter wheat in the Pacific northwest using a multimodel ensemble. Front Ecol Evol 5. https://doi.org/10.3389/fevo.2017.00051

Ahmed I, Rahman MH, Ahmad A, Ahmad S, Hussain J, Ullah A, Judge J (2018a) Assessing the impact of climate variability on maize using simulation modeling under semi-arid environment of Punjab-Pakistan. Environ Sci Pollut Res 25(23):28413–28430

Ahmed M, Ijaz W, Ahmad S (2018b) Adapting and evaluating APSIM-SoilP-Wheat model for response to phosphorus under rainfed conditions of Pakistan. J Plant Nutr 41(16):2069–2084

Ahmed M, Stockle CO, Nelson R, Higgins S, Ahmad S, Raza M (2019) Novel multimodal ensemble approach to evaluate the sole effect of elevated CO_2 on winter wheat productivity. Sci Rep 9:7813

Ahmed M, Ahmad S, Tariq M, Fatima Z, Aslam Z, Raza MA, Iqbal N, Akmal M, Fayyaz-ul-Hassan, Abbasi NA, Hayat R (2020) Wastes to be the source of nutrients and energy to mitigate climate change and ensure future sustainability: options and strategies. J Plant Nutr 43(6):896–920

Ahmed M, Hayat R, Ahmad M, Mahmood-ul-Hassan, Kheir AMS, Fayyaz-ul-Hassan, Rehman MH, Shaheen FA, Raza MA, Ahmad S (2022a) Impact of climate change on dryland agricultural systems: a review of current status, potentials, and further work need. Int J Plant Prod 16:341–363

Ahmed M, Ahmad S, Kheir AMS (2022b) Climate change: an overview. In: Ahmed M (ed) Global agricultural production - resilience to climate change. Springer Nature Switzerland AG. https://doi.org/10.1007/978-3-031-14973-3_1

Ahmed M, Hayat R, Ahmad M, ul-Hassan M, Kheir AMS, ul-Hassan F, ur-Rehman MH, Shaheen FA, Raza MA, Ahmad S (2022c) Impact of climate change on dryland agricultural systems: a review of current status, potentials, and further work need. Int J Plant Prod. https://doi.org/10.1007/s42106-022-00197-1

Akimowicz M, Del Corso JP, Gallai N, Képhaliacos C (2022) The leader, the keeper, and the follower? A legitimacy perspective on the governance of varietal innovation systems for climate changes adaptation. The case of sunflower hybrids in France. Agric Syst 203:103498

Ali H, Iqbal N, Ahmad S, Shahzad AN, Sarwar N (2013) Performance of late sown wheat crop under different planting geometries and irrigation regimes in arid climate. Soil Tillage Res 130:109–119

Amin A, Nasim W, Fahad S, Ali S, Ahmad S, Rasool A, Saleem N, Hammad HM, Sultana SR, Mubeen M, Bakhat HF, Ahmad N, Shah GM, Adnan M, Noor M, Basir A, Saud S, Rahman MH, Paz JO (2018a) Evaluation and analysis of temperature for historical (1996–2015) and

projected (2030–2060) climates in Pakistan using SimCLIM climate model: ensemble application. Atmos Res 213:422–436
Amin A, Nasim W, Mubeen M, Ahmad A, Nadeem M, Urich P, Fahad S, Ahmad S, Wajid A, Tabassum F, Hammad HM, Sultana SR, Anwar S, Baloch SK, Wahid A, Wilkerson CJ, Hoogenboom G (2018b) Simulated CSM-CROPGRO-cotton yield under projected future climate by SimCLIM for Southern Punjab, Pakistan. Agric Syst 167:213–222
Amin A, Nasim W, Mubeen M, Sarwar S, Urich P, Ahmad A, Wajid A, Khaliq T, Rasul F, Hammad HM, Rehmani MIA, Mubarak H, Mirza N, Wahid A, Ahmad S, Fahad S, Ullah A, Khan MN, Ameen A, Amanullah, Shahzad B, Saud S, Alharby H, Ata-Ul-Karim ST, Adnan M, Islam F, Ali QS (2018c) Regional climate assessment for temperature and precipitation in Southern Punjab (Pakistan) using SimCLIM climate model for different temporal scales. Theor Appl Climatol 131:121–131
Amutha R, Muthulaksmi S, Baby Rani W, Indira K, Mareeswari P (2007) Studies on biochemical basis of heat tolerance in sunflower (Helianthus annus L.). Res J Agric Biol Sci 3(4):234–238
Arshad A, Ghani MU, Qamar H, Zubair M (2020) Sunflower modelling: a review. In: Ahmed M (ed) Systems modeling. Springer, Cham, pp 307–326
Aslam Z, Ahmed M, Sajad M, Asif M, Akmal M, Awan FK, Ijaz W, Ashraf R, Khattak JZK (2013) A comparison of statistical and dynamic modeling of wheat (*Triticum aestivum* L.) fungal diseases under the climate change. J Food Agric Environ 11(2):1327–1332
Asseng S, Martre P, Maiorano A, Rötter RP, O'Leary GJ, Fitzgerald GJ, Girousse C, Motzo R, Giunta F, Babar MA, Reynolds MP, Kheir AMS, Thorburn PJ, Waha K, Ruane AC, Aggarwal PK, Ahmed M, Balkovič J, Basso B, Biernath C, Bindi M, Cammarano D, Challinor AJ, De Sanctis G, Dumont B, Eyshi Rezaei E, Fereres E, Ferrise R, Garcia-Vila M, Gayler S, Gao Y, Horan H, Hoogenboom G, Izaurralde RC, Jabloun M, Jones CD, Kassie BT, Kersebaum K-C, Klein C, Koehler A-K, Liu B, Minoli S, Montesino San Martin M, Müller C, Naresh Kumar S, Nendel C, Olesen JE, Palosuo T, Porter JR, Priesack E, Ripoche D, Semenov MA, Stöckle C, Stratonovitch P, Streck T, Supit I, Tao F, Van der Velde M, Wallach D, Wang E, Webber H, Wolf J, Xiao L, Zhang Z, Zhao Z, Zhu Y, Ewert F (2019) Climate change impact and adaptation for wheat protein. Glob Chang Biol 25(1):155–173. https://doi.org/10.1111/gcb.14481
Awaad HA (2022) Approaches in sunflower to mitigate impact of climate change. In: Sustainable agriculture in Egypt. Springer, Cham, pp 243–274
Awais M, Wajid A, Saleem MF, Nasim W, Ahmad A, Raza MAS, Bashir MU, Mubeen M, Hammad HM, Rahman MH, Saeed U (2018) Potential impacts of climate change and adaptation strategies for sunflower in Pakistan. Environ Sci Pollut Res 25(14):13719–13730
Awais M, Anwar-ul-Haq M, Mubeen M, Akhtar J, Wajid SA (2020) Adaptation strategies for sunflower production under climate change for arid environment by using CROPGRO and APSIM-Sunflower models. Pak J Agric Sci 57(6):1529–1538
Blackman BK, Michaels SD, Rieseberg LH (2011) Connecting the sun to flowering in sunflower adaptation. Mol Ecol 20(17):3503–3512
Brouziyne Y, Abouabdillah A, Hirich A, Bouabid R, Zaaboul R, Benaabidate L (2018) Modeling sustainable adaptation strategies toward a climate-smart agriculture in a Mediterranean watershed under projected climate change scenarios. Agric Syst 162:154–163
Conroy JP, Virgona JM, Smillie RM, Barlow EW (1988) Influence of drought acclimation and CO_2 enrichment on osmotic adjustment and chlorophyll a fluorescence of sunflower during drought. Plant Physiol 86(4):1108–1115
De la Haba P, De la Mata L, Molina E, Agüera E (2014) High temperature promotes early senescence in primary leaves of sunflower (Helianthus annuus L.) plants. Can J Plant Sci 94(4):659–669
Debaeke P, Casadebaig P, Flenet F, Langlade N (2017) Sunflower crop and climate change: vulnerability, adaptation, and mitigation potential from case-studies in Europe. OCL Oilseeds Fats Crops Lipids 24(1):15
Deveci H, Konukcu F, Altürk B (2019) Effect of climate change on wheat grown soil moisture profile in Thrace district. J Tekirdag Agric Fac 16(2):202–218

Donatelli M, Srivastava AK, Duveiller G, Niemeyer S, Fumagalli D (2015) Climate change impact and potential adaptation strategies under alternate realizations of climate scenarios for three major crops in Europe. Environ Res Lett 10(7):075005

Ejaz M, Abbas G, Fatima Z, Iqbal P, Raza MA, Kheir AMS, Ahmed M, Kakar KM, Ahmad S (2022) Modelling climate uncertainty and adaptations for soybean-based cropping system. Int J Plant Prod 16:235–250

Fatima Z, Abbas Q, Khan A, Hussain S, Ali MA, Abbas G, Younis H, Naz S, Ismail M, Shahzad MI, Nadeem M, Farooq U, Khan SU, Javed K, Khan AA, Ahmed M, Khan MA, Ahmad S (2018) Resource-use-efficiencies of C_3 and C_4 cereals under split nitrogen regimes. Agronomy 8(5):69

Fatima Z, Ahmad M, Hussain M, Abbas G, Ul-Allah S, Ahmad S, Ahmed N, Ali MA, Sarwar G, Haque E, Iqbal P, Hussain S (2020) The fingerprints of climate warming on cereal crops phenology and adaptation options. Sci Rep 10:18013

Fatima Z, Atique-ur-Rehman AG, Iqbal P, Zakir I, Khan MA, Kamal GM, Ahmed M, Ahmad S (2021) Quantification of climate warming and crop management impacts on phenology of pulses-based cropping systems. Int J Plant Prod 15:107–123

García-López J, Lorite IJ, García-Ruiz R, Domínguez J (2014) Evaluation of three simulation approaches for assessing yield of rainfed sunflower in a Mediterranean environment for climate change impact modelling. Clim Chang 124(1):147–162

Guo S, Ge Y, Na Jom K (2017) A review of phytochemistry, metabolite changes, and medicinal uses of the common sunflower seed and sprouts (Helianthus annuus L.). Chem Cent J 11(1):1–10

Gurkan H, Shelia V, Bayraktar N, Yildirim YE, Yesilekin N, Gunduz A, Boote KJ, Porter CH, Hoogenboom G (2020) Estimating the potential impact of climate change on sunflower yield in the Konya province of Turkey. J Agric Sci 158(10):806–818

Hernández F, Poverene M, Presotto A (2018) Heat stress effects on reproductive traits in cultivated and wild sunflower (Helianthus annuus L.): evidence for local adaptation within the wild germplasm. Euphytica 214(8):1–15

Hoogenboom G, Porter CH, Boote KJ, Shelia V, Wilkens PW, Singh U, White JW, Asseng S, Lizaso JI, Moreno LP, Pavan W, Ogoshi R, Hunt LA, Tsuji GY, Jones JW (2019) The DSSAT crop modeling ecosystem. In: Boote KJ (ed) Advances in crop modeling for a sustainable agriculture. Burleigh Dodds Science Publishing, Cambridge, pp 173–216. https://doi.org/10.19103/AS.2019.0061.10

Hoogenboom G, Porter CH, Shelia V, Boote KJ, Singh U, White JW, Pavan W, Oliveira FAA, Moreno-Cadena LP, Lizaso JI, Asseng S, Pequeno DNL, Kimball BA, Alderman PD, Thorp KR, Jones MR, Cuadra SV, Vianna MS, Villalobos FJ, Ferreira TB, Batchelor WD, Koo J, Hunt LA, Jones JW (2021) Decision Support System for agrotechnology Transfer (DSSAT) Version 4.8 (DSSAT.net). DSSAT Foundation, Gainesville

Hussain M, Ahmad S, Hussain S, Lal R, Ul-Allah S, Nawaz A (2018) Rice in saline soils: physiology, biochemistry, genetics and management. Adv Agron 148:231–287

Jabeen M, Gabriel HF, Ahmed M, Mahboob MA, Iqbal J (2017) Studying impact of climate change on wheat yield by using DSSAT and GIS: a case study of Pothwar Region. In: Ahmed M, Stockle OC (eds) Quantification of climate variability, adaptation and mitigation for agricultural sustainability. Springer International Publishing, Cham, pp 387–411

Janni M, Gullì M, Maestri E, Marmiroli M, Valliyodan B, Nguyen HT, Marmiroli N (2020) Molecular and genetic bases of heat stress responses in crop plants and breeding for increased resilience and productivity. J Exp Bot 71(13):3780–3802

Jones JW, Hoogenboom G, Porter CH, Boote KJ, Batchelor WD, Hunt LA, Wilkens PW, Singh U, Gijsman AJ, Ritchie JT (2003) The DSSAT cropping system model. Eur J Agron 18(3–4):235–265

Kane NC, Burke JM, Marek L, Seiler G, Vear F, Baute G, Knapp SJ, Vincourt P, Rieseberg LH (2013) Sunflower genetic, genomic and ecological resources. Mol Ecol Resour 13(1):10–20

Keating BA, Carberry PS, Hammer GL, Probert ME, Robertson MJ, Holzworth D, Huth NI, Hargreaves JNG, Meinke H, Hochman Z, McLean G, Verburg K, Snow V, Dimes JP, Silburn M, Wang E, Brown S, Bristow KL, Asseng S, Chapman S, McCown RL, Freebairn DM, Smith

CJ (2003) An overview of APSIM: a model designed for farming systems simulation. Eur J Agron 18:267–288

Khan M, Rauf S, Munir H, Kausar M, Hussain MM, Ashraf E (2017) Evaluation of sunflower (Helianthus annuus L.) single cross hybrids under heat stress condition. Arch Agron Soil Sci 63(4):525–535

Khan A, Fahad S, Khan A, Saud S, Adnan M, Wahid F, Noor M, Nasim W, Hammad HM, Bakhat HF, Ahmad S, Rehman MH, Wang D, Sönmez O (2019) Managing tillage operations and manure to restore soil carbon stocks in wheat-maize cropping system. Agron J 111:1–10

Khan MN, Hussain M, Abbas G, Fatima Z, Iqbal P, Khan A, Zakir I, Ahmed M, Mateen A, Khan MA, Ahmad S (2020a) Improving resource use efficiencies of sugarcane at farmer field under arid environment. Int J Agric Biol 24(5):1279–1285

Khan MA, Tahir A, Khurshid N, Husnain MI, Ahmed M, Boughanmi H (2020b) Economic effects of climate change-induced loss of agricultural production by 2050: a case study of Pakistan. Sustain For 12(3):1216. https://doi.org/10.3390/su12031216

Killi D, Bussotti F, Raschi A, Haworth M (2017) Adaptation to high temperature mitigates the impact of water deficit during combined heat and drought stress in C3 sunflower and C4 maize varieties with contrasting drought tolerance. Physiol Plant 159(2):130–147

Lee J, De Gryze S, Six J (2011) Effect of climate change on field crop production in California's Central Valley. Clim Chang 109(1):335–353

Liu B, Martre P, Ewert F, Webber H, Waha K, Thorburn PJ, Ruane AC, Aggarwal PK, Ahmed M, Balkovič J, Basso B, Biernath C, Bindi M, Cammarano D, Cao W, Challinor AJ, De Sanctis G, Dumont B, Espadafor M, Eyshi Rezaei E, Fereres E, Ferrise R, Garcia-Vila M, Gayler S, Gao Y, Horan H, Hoogenboom G, Izaurralde RC, Jabloun M, Jones CD, Kassie BT, Kersebaum KC, Klein C, Koehler A-K, Maiorano A, Minoli S, Montesino San Martin M, Müller C, Naresh Kumar S, Nendel C, O'Leary GJ, Eivind Olesen J, Palosuo T, Porter JR, Priesack E, Ripoche D, Rötter RP, Semenov MA, Stöckle C, Stratonovitch P, Streck T, Supit I, Tao F, Van der Velde M, Wang E, Wolf J, Xiao L, Zhang Z, Zhao Z, Zhu Y, Asseng S (2023) AgMIP-Wheat multi-model simulations on climate change impact and adaptation for global wheat. Open Data J Agric Res 9:10–25. https://doi.org/10.18174/odjar.v9i0.18092

Mahmood A, Ullah H, Shahzad AN, Ali H, Ahmad S, Zia-Ul-Haq M, Honermeier B, Hasanuzzaman M (2013) Dry matter yield and chemical composition of sorghum cultivars with varying planting density and sowing date. Sains Malaysiana 42(10):1529–1538

Miladinović D, Hladni N, Radanović A, Jocić S, Cvejić S (2019) Sunflower and climate change: possibilities of adaptation through breeding and genomic selection. In: Kole C (ed) Genomic designing of climate-smart oilseed crops. Springer, Cham, pp 173–238

Moriondo M, Giannakopoulos C, Bindi M (2011) Climate change impact assessment: the role of climate extremes in crop yield simulation. Clim Chang 104(3):679–701

Mubarak MK, Hussain K, Abbas G, Altaf MT, Baloch FS, Ahmad S (2022) Productivity of sorghum (*Sorghum bicolor* L.) at diverse irrigation regimes and sowing dates in semi-arid and arid environment. Turk J Agric For 46(1):1–18

Mubeen M, Ahmad A, Wajid A, Khaliq T, Hammad HM, Sultana SR, Ahmad S, Fahad S, Nasim W (2016) Application of CSM-CERES-Maize model in optimizing irrigated conditions. Outlook Agric 45(3):173–184

Nadeem M, Khan MN, Abbas G, Fatima Z, Iqbal P, Ahmed M, Raza MA, Rehman A, Haq EU, Hayat A, Ali M, Ahmad S (2022) Application of CSM-CANEGRO model for climate change impact assessment and adaptation for sugarcane in semi-arid environment of Southern Punjab, Pakistan. Int J Plant Prod 16:443–466. https://doi.org/10.1007/s42106-022-00192-6

Nasim W, Belhouchette H, Ahmad A, Rahman MH, Jabran K, Ullah K, Fahad S, Shakeel M, Hoogenboom G (2016) Modelling climate change impacts and adaptation strategies for sunflower in Pakistan. Outlook Agric 45(1):39–45

Nasim W, Amin A, Fahad S, Awais M, Khan N, Mubeen M, Wahid A, Turan V, Rehman MH, Ihsan MZ, Ahmad S, Hussain S, Mian IA, Khan B, Jamal Y (2018) Future risk assessment by estimating historical heat wave trends with projected heat accumulation using SimCLIM climate model in Pakistan. Atmos Res 205:118–133

Naz S, Ahmad S, Abbas G, Fatima Z, Hussain S, Khan MA, Ahmed M, Khan A, Fahad S, Nasim W, Ercisli S, Wilkerson CJ, Hoogenboom G (2022) Modeling the impact of climate warming on potato phenology. Eur J Agron 132:126404

Parent B, Tardieu F (2012) Temperature responses of developmental processes have not been affected by breeding in different ecological areas for 17 crop species. New Phytol 194(3):760–774

Radanović A, Miladinović D, Cvejić S, Jocković M, Jocić S (2018) Sunflower genetics from ancestors to modern hybrids—a review. Genes 9(11):528

Rahman MH, Ahmad A, Wang X, Wajid A, Nasim W, Hussain M, Ahmad B, Ahmad I, Ali Z, Ishaque W, Awais M, Muddasir M, Shelia V, Ahmad S, Fahad S, Alam M, Ullah H, Hoogenboom G (2018) Multi-model projections of future climate and climate change impacts uncertainty assessment for cotton production in Pakistan. Agric For Meteorol 253–254:94–113

Rahman MH, Ahmad I, Wang D, Fahad S, Afzal M, Ghaffar A, Saddique Q, Khan MA, Saud S, Hassan S, Fahad M, Din M, Ahmad M, Yue Z, Shah RA, Sönmez O, Nasim W (2021) Influence of semi-arid environment on radiation use efficiency and other growth attributes of lentil crop. Environ Sci Pollut Res 28:13697–13711

Rehman A, Ashfaq M, Naqvi SAA, Adil SA, Bashir K, Ahmad A, Ahmad S, Ali A, Imran A (2015) Is climate change worsening the poverty of maize growers? Evidence from Punjab province of Pakistan. Ciencia e Tecnica Vitivinicola 30(1):105–116

Saha D, Mondal MS (2016) Assessing the impact of climate change on sunflower yield using the AQUACROP model. In: Proceedings of the 3rd International Conference on Civil Engineering for Sustainable Development (ICCESD 2016), 12–14 February 2016, KUET, Khulna

Sarwar N, Ali H, Maqsood M, Ullah E, Shahzad AN, Shahzad M, Mubeen K, Shahid MA, Ahmad S (2013) Phenological response of rice plants to different micronutrients application under water saving paddy fields on calcareous soil. Turk J Field Crop 18(1):52–57

Sarwar N, Atique-ur-Rehman, Farooq O, Wasaya A, Hussain M, El-Shehawi AM, Ahmad S, Brestic M, Mahmoud SF, Zivcak M, Farooq S (2021) Integrated nitrogen management improves productivity and economic returns of wheat-maize cropping system. J King Saud Univ Sci 33:101475

Shabbir RN, Waraich EA, Ali H, Nawaz F, Ashraf MY, Ahmad R, Awan MI, Ahmad S, Irfan M, Hussain S, Ahmad Z (2016) Supplemental exogenous NPK application alters biochemical processes to improve yield and drought tolerance in wheat (*Triticum aestivum* L.). Environ Sci Pollut Res 23(3):2651–2662

Tajik A, Farhadian A, Khelkhal MA, Rezaeisadat M, Petrov SM, Eskin AA, Vakhin AV, Golafshani MB, Lapuk SE, Buzurov AE, Kiiamov A (2023) Sunflower oil as renewable biomass source to develop highly effective oil-soluble catalysts for in-situ combustion of heavy oil. Chem Eng J 453:139813

Tang J, Wang J, He D, Huang M, Pan Z, Pan X (2016) Comparison of the impacts of climate change on potential productivity of different staple crops in the agro-pastoral ecotone of North China. J Meteorol Res 30(6):983–997

Tariq M, Yasmeen A, Ahmad S, Hussain N, Afzal MN, Hasanuzzaman M (2017) Shedding of fruiting structures in cotton: factors, compensation and prevention. Trop Subtrop Agroecosyst 20(2):251–262

Tariq M, Ahmad S, Fahad S, Abbas G, Hussain S, Fatima Z, Nasim W, Mubeen M, Rehman MH, Khan MA, Adnan M (2018) The impact of climate warming and crop management on phenology of sunflower-based cropping systems in Punjab, Pakistan. Agric For Meteorol 256:270–282

Tariq M, Afzal MN, Muhammad D, Ahmad S, Shahzad AN, Kiran A, Wakeel A (2018a) Relationship of tissue potassium content with yield and fiber quality components of Bt cotton as influenced by potassium application methods. Field Crop Res 229:37–43

Tariq M, Ahmad S, Fahad S, Abbas G, Hussain S, Fatima Z, Wajid N, Mubeen M, Rehman M, Khan MA, Adnan M, Wilkerson CJ, Hoogenboom G (2018b) The impact of climate warming and crop management on phenology of sunflower-based cropping systems in Punjab, Pakistan. Agric For Meteorol 256–257:270–282

Tarriq M, Fatima Z, Iqbal P, Nahar K, Ahmad S, Hasanuzzaman M (2021) Sowing dates and cultivars mediated changes in phenology and yield traits of cotton-sunflower cropping system in arid environment. Int J Plant Prod 15:291–302

Tubiello FN, Donatelli M, Rosenzweig C, Stockle CO (2000) Effects of climate change and elevated CO_2 on cropping systems: model predictions at two Italian locations. Eur J Agron 13(2–3):179–189

USDA National Agricultural Statistics Service (2022) Census of Agriculture. Complete data available at www.nass.usda.gov/AgCensus

Valverde P, de Carvalho M, Serralheiro R, Maia R, Ramos V, Oliveira B (2015) Climate change impacts on rainfed agriculture in the Guadiana river basin (Portugal). Agric Water Manag 150:35–45

Van der Merwe R, Labuschagne MT, Herselman L, Hugo A (2015) Effect of heat stress on seed yield components and oil composition in high-and mid-oleic sunflower hybrids. S Afr J Plant Soil 32(3):121–128

Zahoor A, Ahmed M, Fayyaz-ul-Hassan SG, Ahmad S (2022) Ontogeny growth and radiation use efficiency of canola (*Brassica napus* L.) under various nitrogen management strategies and contrasting environments. Int J Plant Prod 16(1):195–208

Zhang J, Zhang H, Sima MW, Trout TJ, Malone RW, Wang L (2021) Simulated deficit irrigation and climate change effects on sunflower production in Eastern Colorado with CSM-CROPGRO-Sunflower in RZWQM2. Agric Water Manag 246:106672

Cotton-Wheat System

6

6.1 Introduction

Cotton-wheat cropping system is the second most important cropping system after the rice-wheat cropping system (Shah et al. 2016). This cropping system is practiced in Australia (Shabani and Kotey 2016), Central Asia like Uzbekistan (Kienzler et al. 2011), China (Yuan et al. 2021), Egypt (Sultan et al. 2012), Brazil (Balota et al. 2004), some parts of Europe like Turkey (Basal et al. 2020), and South Asia like Pakistan and India (Tariq et al. 2022; Singh et al. 2014). In a significant cropping pattern, cotton-wheat used throughout the world, cotton is seeded after the wheat is harvested at the beginning of summer. The BT cotton varieties have been added recently to resist bollworms more effectively (Jabran et al. 2020). Cotton-wheat cropping system ensures food security as well as farmer's profit by providing food from wheat and fiber from cotton (Tariq et al. 2022). It is also a main source of foreign exchange especially for developing countries (Javed et al. 2009). Wheat is farmed on a large area globally with an estimated annual yield of more than 781.1 million tonnes in 2022 (FAO 2024). The area and production of cotton-wheat cropping system are presented in Table 6.1, and literature published examples are presented in Table 6.2.

In the whole world, food security is under threat due to climate change. However, due to burgeoning population, developing nations are more vulnerable to it (Funk and Brown 2009). This issue is multifaceted, and adaptation to climate change is needed for reducing the risk on agricultural production (Ahmed et al. 2022b; Conrad et al. 2013; Tanveer et al. 2017). Rosenzweig et al. (2013) reported that AgMIP aimed for improving food security and enhancing adaptative capacity in developing and developed nations under climate change. Cotton is a major cash crop in Pakistan, and cotton-wheat cropping system is being adopted in Southern Punjab and several parts of Sindh, Pakistan. Cotton requires higher temperature during vegetative phase and lower temperature during reproductive phase (Ahmad et al. 2023b); extreme

M. Ahmed et al., *Cropping Systems Modeling Under Changing Climate*,
https://doi.org/10.1007/978-981-97-0331-9_6

Table 6.1 Area and production under cotton-wheat cropping system across various countries

Country	Cotton		Wheat		Cotton-wheat	
	Area (Mha)	Production (MMT)	Area (Mha)	Production (MMT)	Area (Mha)	Production (MMT)
Uzbekistan	1.06	1.06	1.36	5.99	2.42	7.05
Pakistan	2.00	2.60	9.17	27.46	11.17	30.06
Australia	0.60	1.73	13.04	36.35	13.64	38.08
Turkey	0.45	1.24	7.05	16.00	7.5	17.24
India	12.15	10.40	31.13	109.59	43.28	119.99
Egypt	0.09	0.09	1.40	9.00	1.49	9.09
Brazil	1.60	3.89	2.74	7.70	4.34	11.59
Mexico	0.15	0.42	0.55	3.28	0.70	0.70

Table 6.2 Literature reported examples of cotton-wheat cropping system

Country name	Continent	Field studies	Results		Economic benefit	References
			Cotton	Wheat		
Pakistan	Asia	Relay intercropping of insect-resistant transgenic cotton	3899 kg ha^{-1}	6854 kg ha^{-1}		Shah et al. (2016)
Pakistan	Asia	Relay cropping Cotton: C1 = fallow, C2 = 75 cm, C3 = 150 cm, C4 = after wheat Wheat: W1 = 22.5 cm, W2 = 4th line cotton, W3 = 7th line cotton	C2 and C3 improved yield by 86.07% and 64.89% and suppressed CLCV by 68.14% and 65.86%, respectively, than C4	W3 and W2 produced 19.81% and 8.16% lesser yield than W1	From 2013 to 2016, W2C2 gave 6704$ more profit than C1	Tariq et al. (2022)
India	Asia	Sowing date, tillage, and residue effect	1820 kg ha^{-1}	3586 kg ha^{-1}	Min. tillage saved 31.0 US$ in cotton and 59.6 US$ in wheat	Jalota et al. (2008)
China	Asia	Plant density and mepiquat chloride application on cotton	4429.5 kg ha^{-1}			Yuan et al. (2021)
India	Asia	Complex fertilizer N(Urea), K(MOP), P(DAP), S(Gypsum), Zn(ZnSO4), N + P + S + Zn (APSZ)	22.2 q ha^{-1}	45.6 q ha^{-1}	Net return = Ind Rs. 15,417.5	Brar et al. (2019)
China	Asia	Early relay intercropping	Lint = 1138 kg/hm^2	8821 kg/hm^2		Wang et al. (2021)
India	Asia	Conservation agriculture	4.0 Mg ha^{-1}	4.98 Mg ha^{-1}		Das et al. (2014)
Pakistan	Asia	Planting technique	2983 kg ha^{-1}	5231 kg ha^{-1}	Net return 1478.6 US $	Ahmad et al. (2023b)
Turkey	Asia + Europe	Effect of seeding rate on wheat		5032 kg ha^{-1}		Kilic and Gursoy (2010)
Pakistan	Asia	Wheat as relay crop	3275 kg ha^{-1}	5166 kg ha^{-1}	Net income = Pak Rs. 370,833	Nasrullah et al. (2017)

weather conditions like heat wave adversely affect productivity, leading to economic crises.

Geographical area of Pakistan is 79.6 Mha; on area basis, it is the 2nd largest country in South Asia and 36th in the world. Two seasons, i.e., winter (rabi) and summer (kharif), are key components of Pakistani agricultural economy. The rabi crops are being sown from November to April and kharif crops from May to October. *Gossypium hirsutum* L. is the main kharif-season crop, and *Triticum aestivum* L is a rabi crop. The six major crops include wheat, rice, cotton, maize, sugarcane, and chickpea contributing 29% to value-added agriculture and 6% to gross domestic product. There are three main wheat-based schemes: RW, CW, and mixed wheat. The rice-wheat system is common in semiarid areas in the central part (Ahmad et al. 2019) and cotton-wheat system in arid areas in the southern part of the country. Rice-wheat and cotton-wheat systems fulfill the food and fiber requirements of the people (Rahman et al. 2018).

Farooqi et al. (2005) concluded that developing countries are at higher risk of current and future climatic extreme events due to their geographical locations, usual poverty, burgeoning population, and reliance on agricultural and natural resources. Eckstein et al. (2018) declared that Pakistan ranks at seventh position among climate change-affected countries and with natural disaster-associated risks. Farooqi et al. (2005) reported that developing countries are vulnerable to sea-level rise, more frequency in heavy flash floods, melting of glaciers, high temperature, and higher drought frequency. These vulnerabilities directly and indirectly affect decision-making and have drastic impacts on the agricultural industry and also threaten water, energy, and food security.

Upsurge in maximum and minimum temperatures during winter will condense in the winter season and prolong in the summer season in South Asia. Therefore, late-onsetting and earlier ending will reduce the length of the crop growth season or duration, and due to phenotypic plasticity, winter crops will try to complete their biological cycle speedily resulting in reduced economic yield (Ahmed et al. 2019, 2022a). Due to earlier ending of winter, temperature will increase in February, and at this stage, wheat is at grain formation and grain-filling duration. This upsurge in temperature will shrink the grain-filling duration with reduced accumulation of photosynthates, and finally the grain size will be reduced (Ahmad et al. 2023b).

The productivity of certain crops would be decreased around 43% and millet productivity around 10% due to changing climate in South Asia (Ullah et al. 2019). Impacts of changing climate in South Asia are previously evident; therefore, there is a prerequisite for quantification of potential climate change impacts along with developing adaptation options to minimize the negative effects.

This case study used by AgMIP quantifies the changing climate impacts on the CW system of Punjab at farm scale using regional integrated assessment (RIA) approach developed by AgMIP associating climate, crop, and economic modeling procedures. The main aim of AgMIP RIA is to share technical information with stakeholders, which might be helpful for decision-making (Rosenzweig et al. 2016; Hillel and Rosenzweig 2015). These aims are to determine the potential climate change impacts for various socioeconomic scenarios and agricultural systems and identify adaptation strategies for improving the farmers/growers' livelihood.

Sharing outcomes with stakeholders like farmers, academia, researchers, and policymakers is significant for ensuring that study outcomes contribute toward evidence-based decision-making in South Asia (Hillel and Rosenzweig 2015).

6.2 Description of Cropping System Investigated

Cotton-wheat cropping system is a time-tested and established cropping system in South Asia having significant contribution to the agricultural sector of China, India, and Pakistan. This is grain-plus-cash enterprise, contributing toward the livelihood of growers by cotton cultivation as an industry by-product and wheat as a staple food and primary constituent of food security. It is extremely remunerative to have secure returns from a cash-grain cropping system, contributing mainly toward the economic welfare of a significant number of families being associated with farming, value-chain enterprises, and textile manufacturing. In Pakistan, area for CWS is 8.83 Mha, that is, around 37% of total cropping area.

Cotton is being sown from April to June and harvested during October to November; however, wheat is planted in winter season from November to April on residual soil moisture with additional supplementary irrigation water mainly because of favorable weather conditions. This area receives a total rainfall of around 400–600 mm during July to October; however, in certain areas, 5–10% precipitation is also received during November to April. The major constraint of cotton-wheat cropping is the late harvesting of cotton due to last picking from the start of December to the first week of January having more economic value that results in late sowing of wheat, which leads to lesser wheat productivity. Khokhar et al. (2010) reported that sowing wheat after mid-November reduces wheat productivity by 1% day^{-1}. The reported optimum time of sowing for wheat is from the first week of November to the first week of December, and sowing at optimum times reduces the risk of hot-weather exposure at critical stages of anthesis, grain formation, and grain filling. Similarly, late sowing of cotton also results in poor early growth, which leads to insect pest attack along with the incidence of cotton leaf curl virus (CLCV). Adaptation of management practices like adjusting sowing times, optimizing balanced nutrients along with efficient nutrient application strategies for cotton-wheat cropping system could be fruitful for harnessing higher yields under changing climatic conditions.

6.3 Issues of Cotton-Wheat Cropping System

In CWS, cultivation of cotton is followed by wheat. The main issue of this system is the no fallow period (Tariq et al. 2022). Because of the overlapping harvesting season of the prior crop, sowing of the subsequent crop is delayed (Buttar et al. 2013; Shah et al. 2016; Sajjad et al. 2018). During various phases of growth and development, late-sown cotton crops are subjected to a variety of challenges. In cotton-wheat cropping system, the growing period of Bt cotton is longer than that of conventional cotton varieties. The farmers were forced by this circumstance to choose between growing cotton or wheat crop in 1 year. This could affect the food security of the agriculture region and cause financial loses for farmers (Jabran et al. 2020). Some of the problems connected with current agricultural production techniques are soil deterioration, compaction, salt and sodicity, soil erosion, decreased soil organic matter, and waterlogging (Soule et al. 1990). Reduced soil fertility, soil degradation, and soil structure disintegration are severe issues in cotton that require rapid attention (Kidron et al. 2010). Heavy tillage causes soil structure deterioration and increases compaction and erosion (Gupta and Allmaras 1987) in cotton-wheat cropping system. Cotton cultivation under the no-till method is a best solution to prevent late sowing of cotton, but it is tough due to several operations such as residue incorporation and the need for tillage for flood irrigation and interrow cultivation (CRDC 2017). Remaining stubbles may disrupt the machinery operation, increase the frequency of pests and illnesses, and increase the reliance on chemical herbicides for weed management (Scott et al. 2010).

6.4 Cotton-Wheat Modeling

The genetic coefficients of cotton and wheat are presented in Tables 6.3 and 6.4. The cotton observed as well as simulated data for LAI and biomass is presented in Figs. 6.1 and 6.2 for first and second years at the study site at Punjab, Pakistan. The exceeding probability of cotton-wheat system productivity is presented in Fig. 6.3.

Table 6.3 Genetic coefficients, description, ranges, and guidelines for model calibration of cotton-wheat crops in DSSAT

Genetic coefficient	Description	Genetic coefficient	Description
EXPNO	Number of experiments used to estimate cultivar parameters		
ECO#	Ecotype code of this cultivar, points to the ecotype in the ECO file		
Cotton		*Wheat*	
CSDL	Critical short-day length below which reproductive development progresses with no day length effect (for short-day plants) (hours)	P1V	Days, optimum vernalizing temperature, required for vernalization
PPSEN	Slope of the relative response of development to photoperiod with time (positive for short-day plants) (1/h)	P1D	Photoperiod response (% reduction in rate/10 h drop in pp)
EM-FL	Time between plant emergence and flower appearance (R1) (photothermal days)	P5	Grain filling (excluding lag) phase duration (oC.d)

(continued)

Table 6.3 (continued)

Genetic coefficient	Description	Genetic coefficient	Description
FL-SH	Time between first flower and first pod (R3) (photothermal days)	G1	Kernel number per unit canopy weight at anthesis (#/g)
FL-SD	Time between first flower and first seed (R5) (photothermal days)	G2	Standard kernel size under optimum conditions (mg)
SD-PM	Time between first seed (R5) and physiological maturity (R7) (photothermal days)	G3	Standard, non-stressed mature tiller wt (incl grain) (g dwt)
FL-LF	Time between first flower (R1) and end of leaf expansion (photothermal days)	PHINT	Interval between successive leaf tip appearances (oC.d)
FLMAX	Maximum leaf photosynthesis rate at 30 °C, 350 vpm CO_2, and high light (mg CO_2/m^2-s)		
SLAVR	Specific leaf area of cultivar under standard growth conditions (cm^2/g)		
SIZLF	Maximum size of full leaf (three leaflets) (cm^2)		
XFRT	Maximum fraction of daily growth that is partitioned to seed + shell		
WTPSD	Maximum weight per seed (g)		
SFDUR	Seed filling duration for pod cohort at standard growth conditions (photothermal days)		
SDPDV	Average seed per pod under standard growing conditions (#/pod)		
PODUR	Time required for cultivar to reach final pod load under optimal conditions (photothermal days)		
THRSH	Threshing percentage. The maximum ratio of (seed/(seed+shell)) at maturity. Causes seeds to stop growing as their dry weight		
SDPRO	Fraction protein in seeds (g(protein)/g(seed))		
SDLIP	Fraction oil in seeds (g(oil)/g(seed))		

Source: DSSAT v4.8; Hoogenboom et al. (2021)

Table 6.4 Genetic coefficients and their description of wheat crop in APSIM model

Genetic coefficients	Description
photop_sens	Photoperiod sensitivity
vern_sens	Vernailzation sensitivity
tt_end_of_juvenile	Thermal time needed from sowing to end of juvenile (°C days)
tt_flowering	Thermal time needed in anthesis phase (°C days)
tt_floral_initiation	Thermal time from floral initiation to flowering (°C days)
tt_start_grain_fill	Thermal time from start of grain filling to maturity (°C days)
max_grain_size	Maximum grain size (G)
potential_grain_growth_rate	Grain growth rate from flowering to grain filling (g $grain^{-1}$ day^{-1})
potential_grain_filling_rate	Potential daily grain filling rate (g $grain^{-1}$ day^{-1})
grains_per_gram_stem	Grain number per stem weight at the start of grain filling (G)

Source: Ahmed et al. (2018)

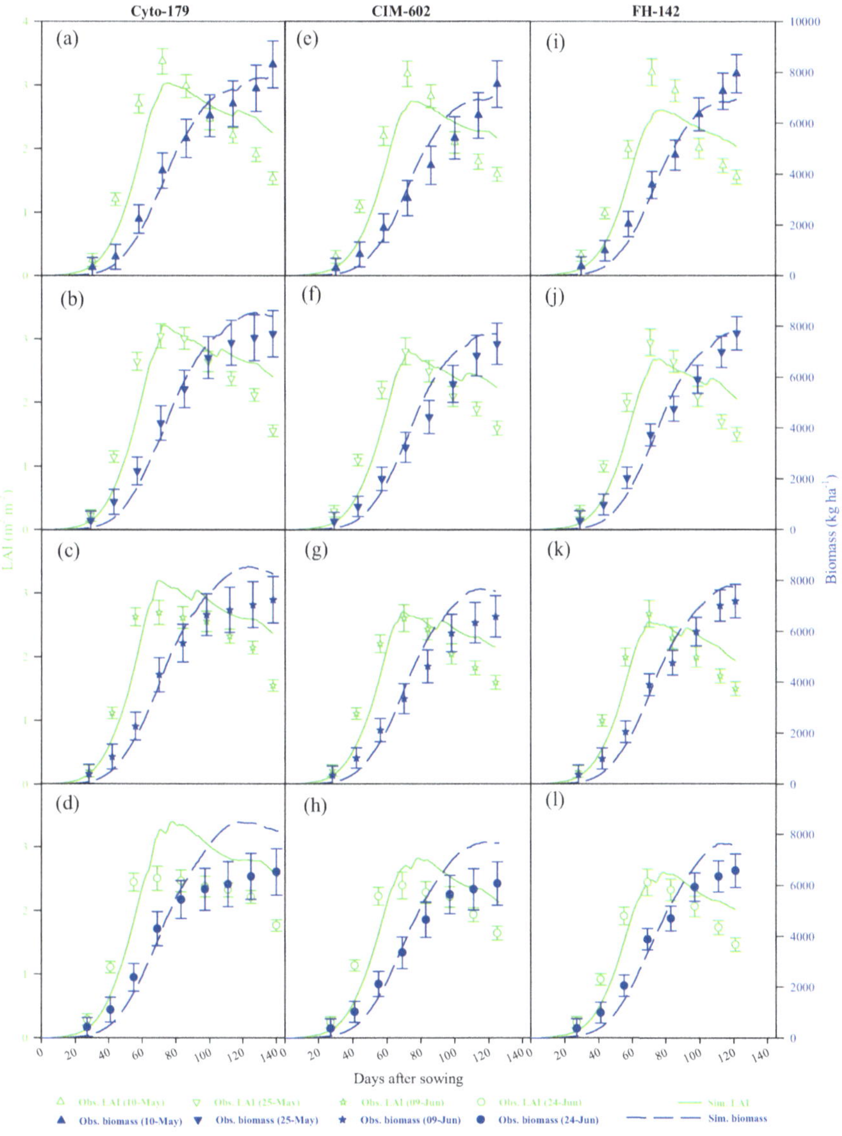

Fig. 6.1 The sowing date's effect on leaf area index and biomass of cotton crop at study site during the first year

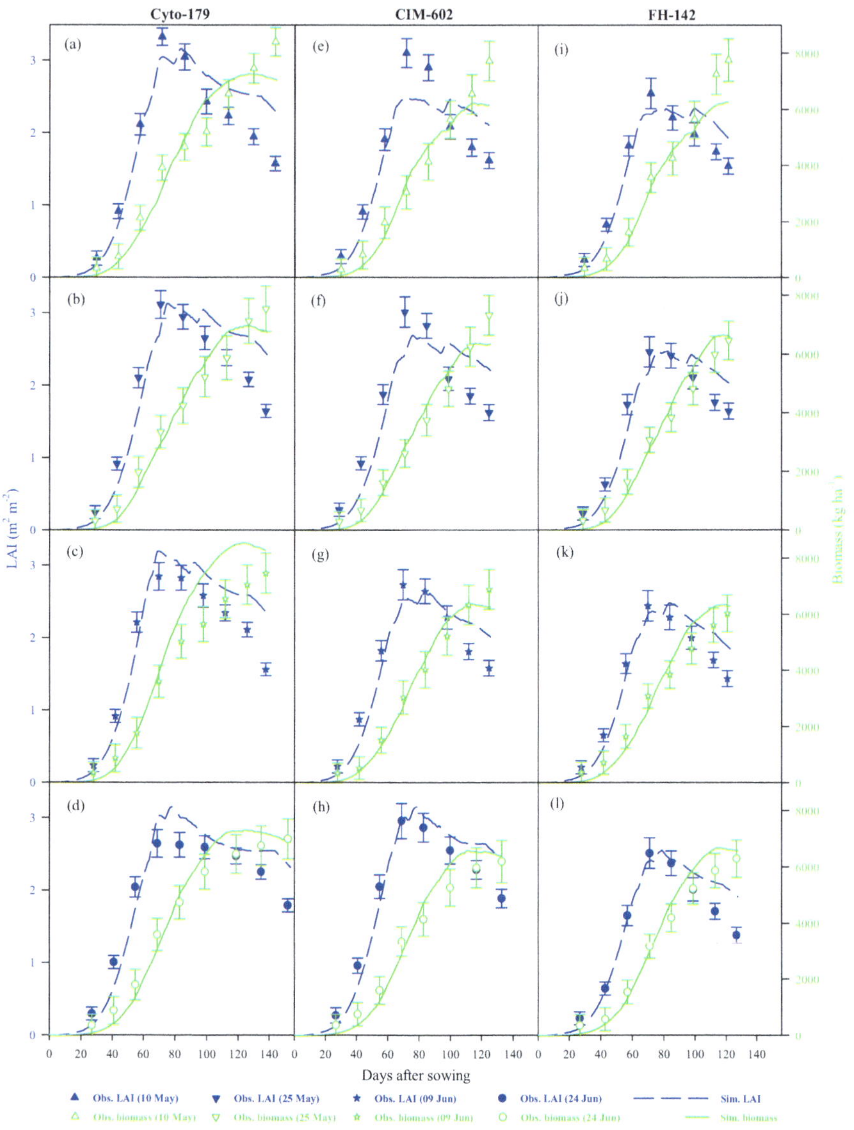

Fig. 6.2 The sowing date's effect on leaf area index and biomass of cotton crop at study site during the second year

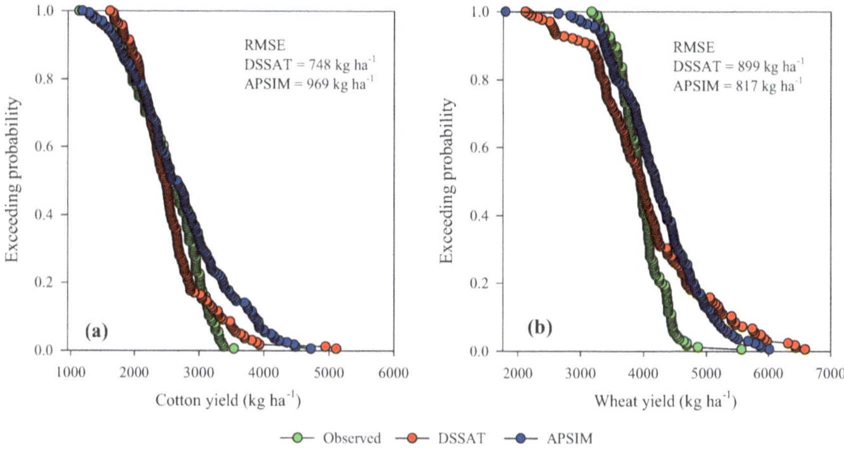

Fig. 6.3 Exceedance probability of cotton-wheat cropping system (**a** and **b**) observed yields at farmer fields for DSSAT and APSIM models

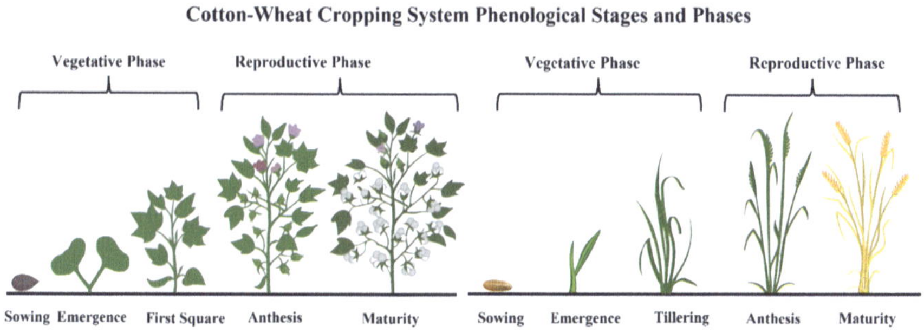

Fig. 6.4 Illustration of cotton-wheat system phenology

6.5 Climate Change Influence on Cotton-Wheat System Phenology

Cotton-wheat cropping system phenology is presented in Fig. 6.4, and recorded and predicted dates of onset of phenology are presented in Fig. 6.5. Climate variability and change threaten agro-based productivity at local, regional, as well as global levels (Liu et al. 2023). During the previous 30 years, temperature has increased by 0.78–1.5 C and is projected to increase by 2–4 C near 2100 (IPCC 2023; Bokhari et al. 2018). It is estimated that the rising trend of temperature might exaggerate the situation in arid and semiarid environments (Tariq et al. 2022). Cotton season's mean temperature trends for cotton system from 1980 to 2015 are presented in Fig. 6.6. Observed temperature trends during cotton

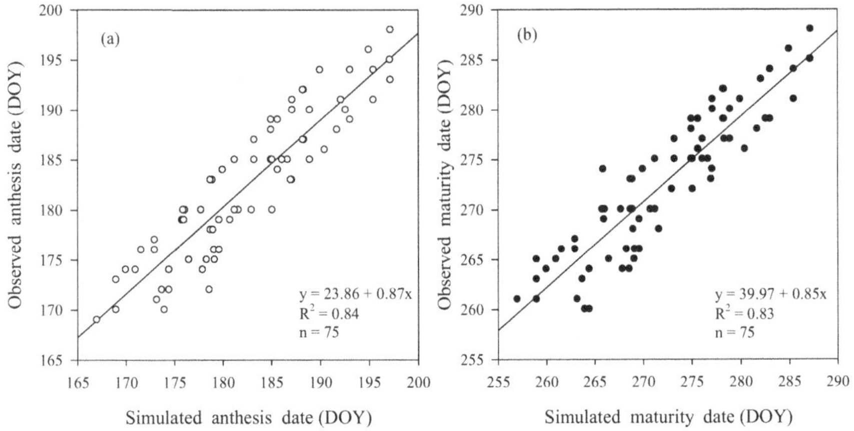

Fig. 6.5 Observed and simulated dates of onset of phenological stages of anthesis (**a**; unfilled symbol; left) and maturity (**b**; filled symbols; right) for cotton in Punjab, Pakistan, during 1983–1985; *DOY* day of year (Source: Adopted from Ahmad et al. 2017)

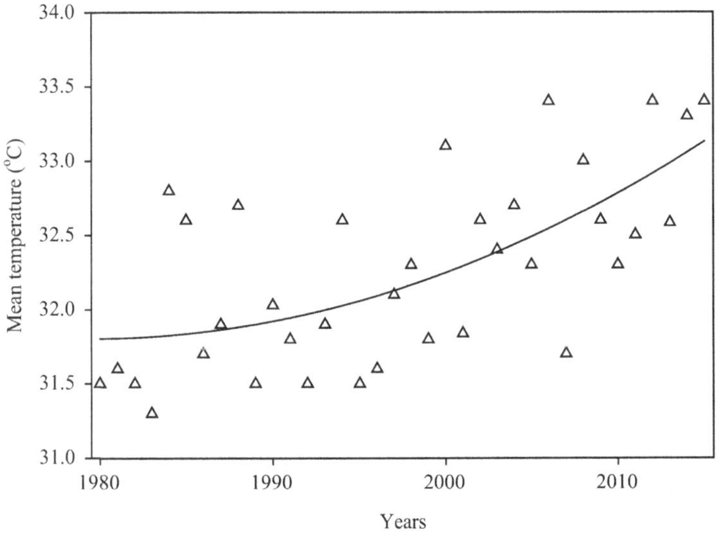

Fig. 6.6 Mean total temperature during the cotton growing season from 1980 to 2015 at 15 sites in Punjab, Pakistan (Source: Adopted from Ahmad et al. 2017)

phenological phases since 1980–2015 for 15 sites are presented in Figs. 6.7, 6.8, and 6.9 (Ahmad et al. 2017). Mean recorded cotton phenology in Punjab, observed as well as predicted phenology as influenced by temperature, along with phenology responses to temperature for 1980–2015 are presented in Tables 6.5, 6.6, and 6.7, respectively. The trends for wheat crop are published earlier (Ahmad et al. 2019) and also presented in rice-wheat system in this book. Due to climate

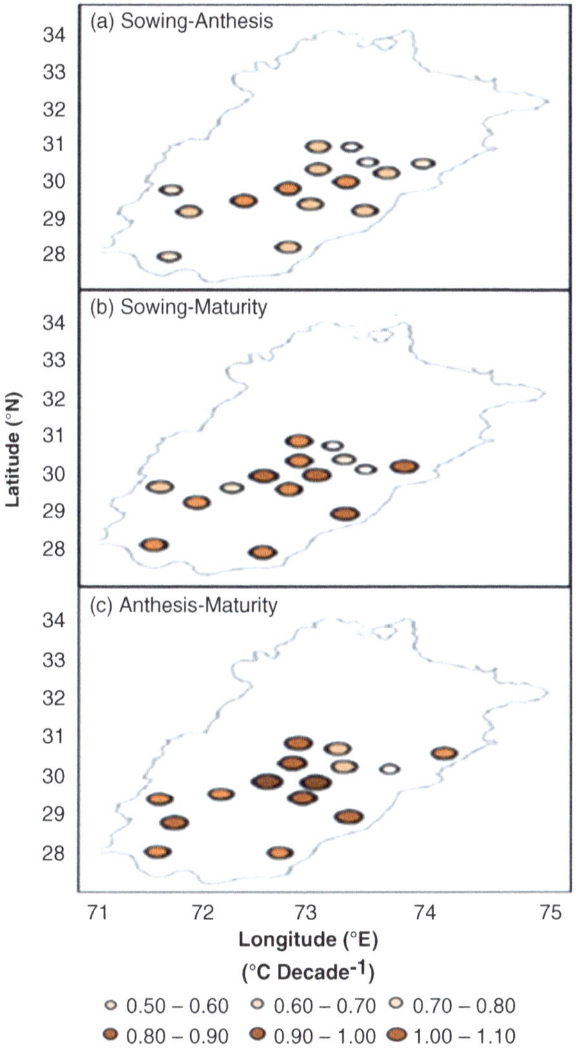

Fig. 6.7 Observed trends in mean temperature during the phenological phases, (**a**) sowing-anthesis, (**b**) sowing-maturity, and (**c**) anthesis-maturity, for cotton from 1980 to 2015 at 15 locations in Punjab, Pakistan (Source: Adopted from Ahmad et al. 2017)

variability and climate change, phenology of cropping systems has been affected adversely and yield as well as biomass production decreased and cropping systems' adaptations are very much needed against weather vagaries. Figures 6.10, 6.11, and 6.12 depict climate change influence on cotton system stages and phases for various sites in Punjab (Ahmad et al. 2019). The husbandry practices like cultivars and sowing dates, GDD requirements, and ecological conditions have a significant influence on the system's phenology (Ahmad et al. 2023b).

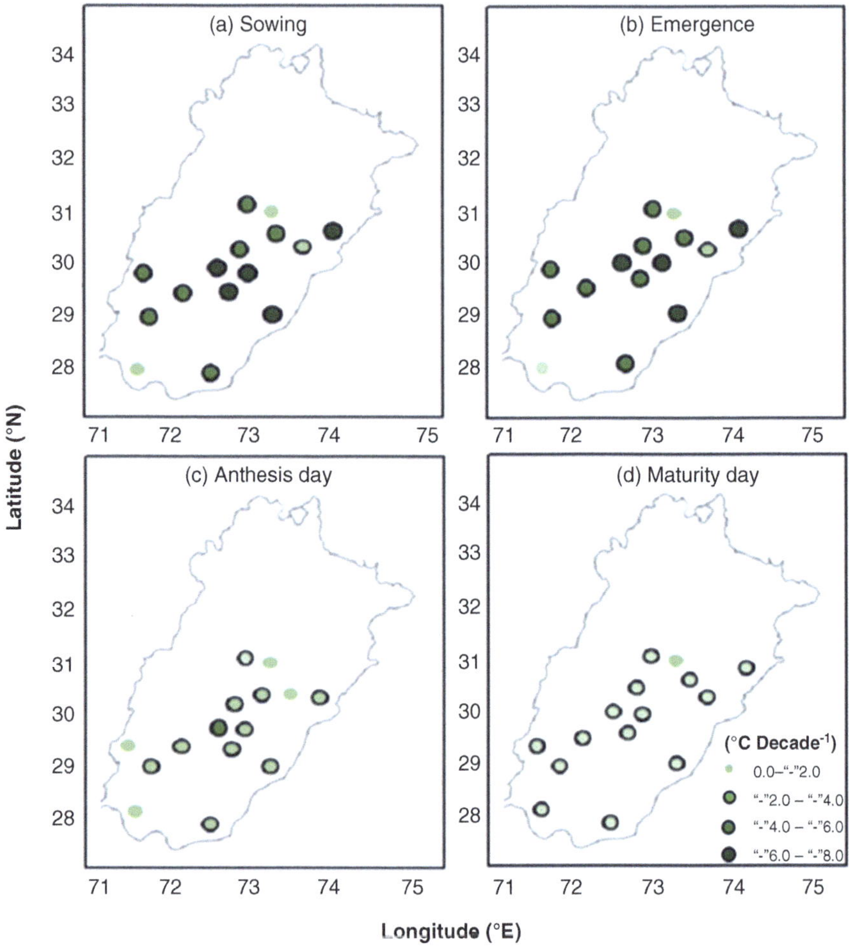

Fig. 6.8 Observed trends in phenological stages of cotton sown from 1980 to 2015 in Punjab, Pakistan; (**a**) sowing, (**b**) emergence, (**c**) anthesis, and (**d**) maturity. Circles with black border indicate statistically significant trends at $p = 0.05$ probability level (Source: Adopted from Ahmad et al. 2017)

Harmonization weather variables like temperature and precipitation with important growth stages of cropping systems are very crucial for fully expressing their genetics for diverse systems (Amin et al. 2018; Ahmad et al. 2023). Climate change-based changes in phenology result in decreasing of the life cycle of crops across various cropping systems like maize-maize sunflower-sunflower (Tariq et al. 2022), rice-wheat (Ahmad et al. 2019), cotton-wheat, soybean-soybean (Ejaz et al. 2022), and pulse-based (Fatima et al. 2020) that finally reduces the time and rate of accumulation of economic yield and biomass (Rezaei et al. 2015; Ahmad et al. 2023b). Therefore, considering the system's response to climate variability and change for better agronomy as well as breeding strategies for

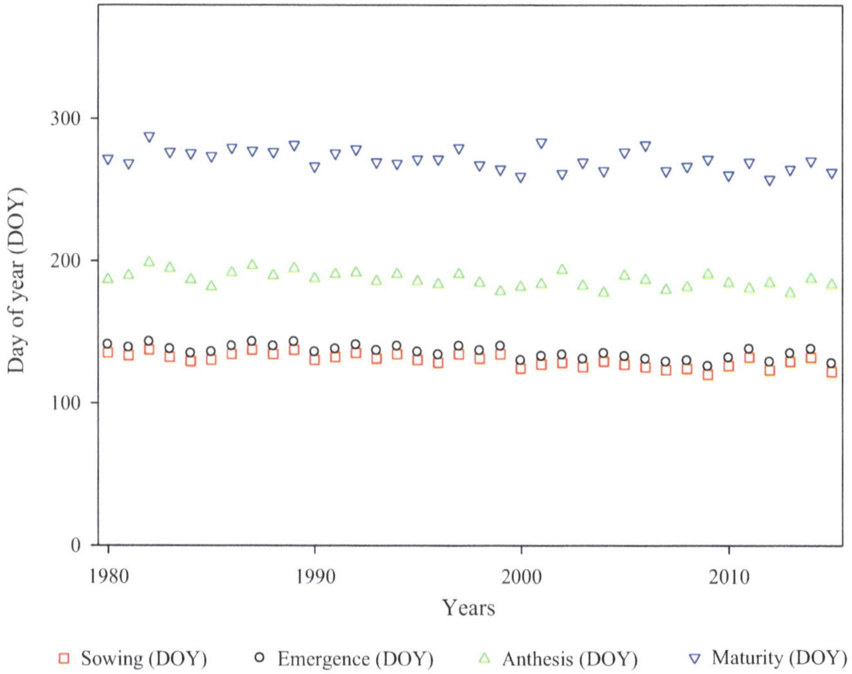

Fig. 6.9 Time series plots of observed dates of onset of phenological stages (sowing, emergence, anthesis, and maturity) for cotton from 1980 to 2015 in Punjab, Pakistan (Source: Adopted from Ahmad et al. 2017)

Table 6.5 Average observed phenology of cotton in Punjab, Pakistan, during the period of 1980–2015 (Source: Adopted from Ahmad et al. 2017)

Districts	Sowing DOY	Emergence DOY	Anthesis[a] DOY	Maturity[b] DOY
Toba Tek Singh	119 ± 7.0	124 ± 6.4	190 ± 5.4	274 ± 9.3
Jhang	121 ± 6.2	126 ± 5.7	188 ± 3.5	281 ± 7.3
Okara	128 ± 5.7	133 ± 5.3	191 ± 7.2	273 ± 5.4
Sahiwal	123 ± 7.2	128 ± 6.9	182 ± 6.1	278 ± 8.3
Pakpattan	130 ± 5.2	135 ± 5.0	189 ± 5.9	270 ± 7.8
Multan	120 ± 4.3	125 ± 4.1	191 ± 7.2	281 ± 6.7
Khanewal	125 ± 5.0	130 ± 4.8	193 ± 8.4	276 ± 5.5
Vehari	128 ± 6.7	133 ± 6.2	197 ± 9.2	285 ± 7.2
Lodhran	125 ± 4.4	130 ± 4.1	189 ± 6.2	278 ± 8.1
Bahawalnagar	131 ± 3.8	136 ± 3.5	192 ± 8.1	272 ± 9.5
Bahawalpur	130 ± 5.1	135 ± 4.5	187 ± 4.9	282 ± 8.5
DG Khan	126 ± 4.2	131 ± 3.9	183 ± 5.8	276 ± 6.7
Rajanpur	122 ± 6.9	127 ± 6.3	190 ± 7.3	280 ± 7.6
Muzaffargarh	127 ± 5.2	132 ± 4.7	194 ± 6.1	276 ± 7.1
Rahim Yar Khan	123 ± 4.8	128 ± 3.9	197 ± 6.8	270 ± 8.6

DOY day of year
[a]50% Anthesis
[b]90% Physiological maturity

Table 6.6 Summary of observed and simulated phenology response to temperature for cotton in Punjab, Pakistan, for 1980–2015 (Source: Adopted from Ahmad et al. 2017)

Phenology	No. of Neg.[a]	No. of Pos.[b]	No. of Sig. Neg.[c]	No. of Sig. Pos.[d]	Reg. Mean[e] (days °C^{-1})
Cotton stages and phases (observed)					
Sowing	15	0	13	0	−2.03
Emergence	15	0	13	0	−1.93
Anthesis	15	0	11	0	−1.09
Maturity	15	0	14	0	−0.42
Sowing-anthesis	15	0	12	0	−0.94
Anthesis-maturity	15	0	11	0	−0.67
Sowing-maturity	15	0	13	0	−1.61
Cotton phases (simulated)					
Sowing-anthesis	15	0	14	0	−1.42
Anthesis-maturity	15	0	13	0	−1.06
Sowing-maturity	15	0	14	0	−1.97

[a]Number of locations with negative regression coefficients
[b]Number of locations with positive regression coefficients
[c]Number of locations with significant negative regression coefficients
[d]Number of locations with significant positive regression coefficients
[e]Mean of regression coefficients

Table 6.7 Comparison of the responses of cotton phenology with average temperature using the observed and simulated data in Punjab, Pakistan, during 1980–2015 (Source: Adopted from Ahmad et al. 2017)

Phenology (phases)	Regression coefficient[a] (days °C^{-1})		Difference between obs. and sim. regression correlations (days °C^{-1})	*t*-Test (*p*-value)
	Observed data	Simulated data		
Sowing-anthesis	−0.94	−1.42	0.48	0.0012**
Anthesis-maturity	−0.67	−1.06	0.39	0.0053**
Sowing-maturity	−1.61	−1.97	0.36	0.0021**

Obs. observed, *Sim.* simulated
** Significant at the 0.01 probability level
[a]Mean of regression coefficients

adaptation and mitigations is the need of the time (Ahmad et al. 2023). Hu et al. (2005) described that in the USA, wheat reached earlier at anthesis owing to changing climate-based rise of temperature. In China, growth stages of wheat

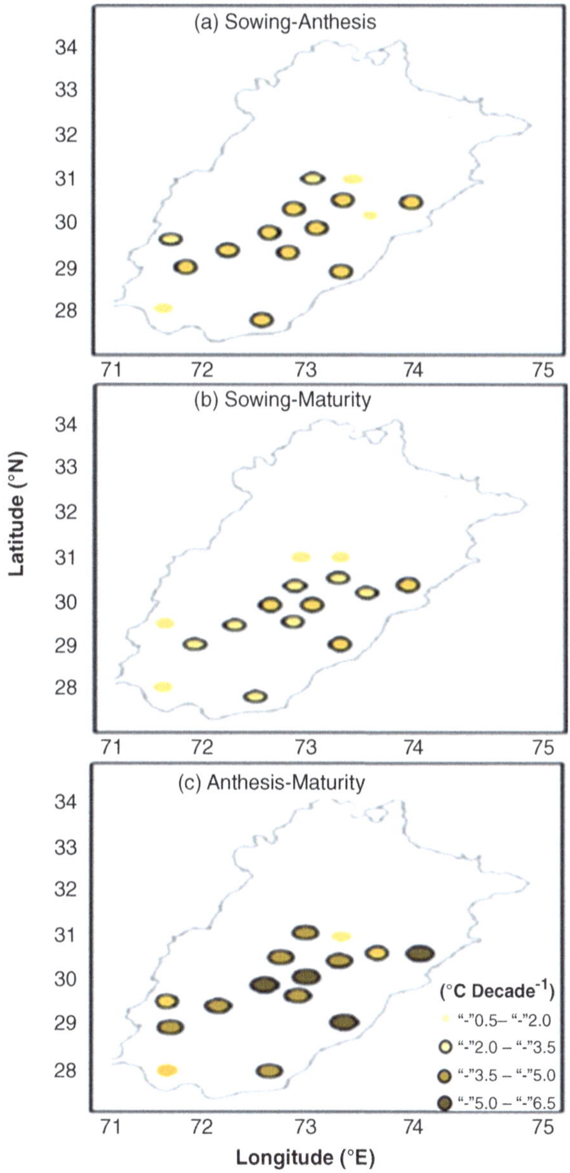

Fig. 6.10 Observed trends in the length of phenological phases for cotton from 1980 to 2015 in Punjab, Pakistan: (**a**) sowing-anthesis, (**b**) anthesis-maturity, and (**c**) sowing-maturity. Circles with black border indicate statistically significant trend at $p = 0.05$ probability level (Source: Adopted from Ahmad et al. 2017)

approached earlier owing to changing climate (He et al. 2015). The spatiotemporal variabilities of cotton phenology (Ahmad et al. 2017) and wheat phenology (Ahmad et al. 2019) are presented in tables and figures, and overall impact on cotton-wheat system due to climate changes and adaptations are described employing DSSAT and APSIM models.

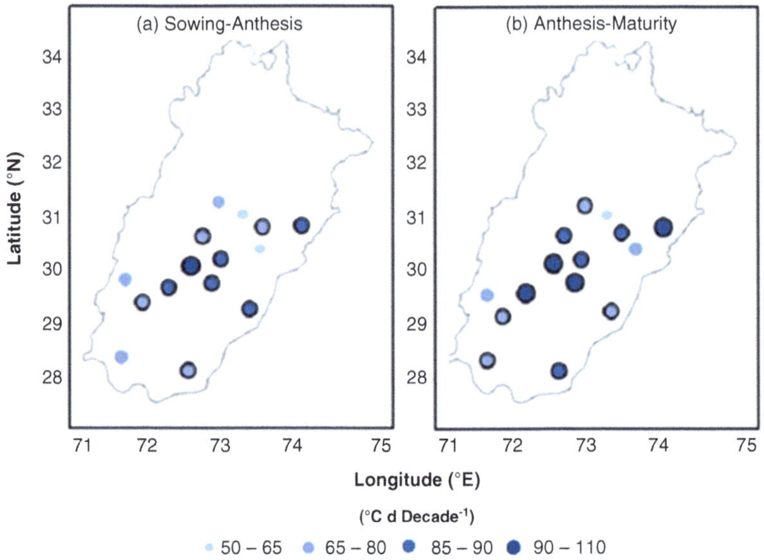

Fig. 6.11 Observed trends in thermal time required for cotton in Punjab, Pakistan, to advance from (**a**) sowing-anthesis and (**b**) anthesis-maturity. Circles with black border indicate statistically significant trend at $p = 0.05$ probability level (Source: Adopted from Ahmad et al. 2017)

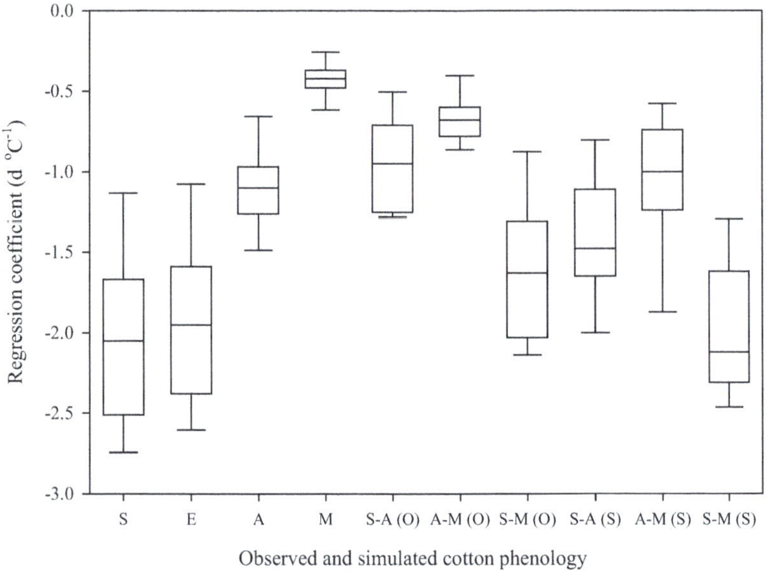

Fig. 6.12 Observed phenology (stages and phases) and simulated phenology (phases) versus temperature trends of cotton, based on data from 15 locations in Punjab, Pakistan, from 1980 to 2015. *S* sowing, *E* emergence, *A* anthesis, *M* maturity, *S-A* sowing-anthesis, *A-M* anthesis-maturity, *S-M* sowing-maturity, *O* observed, *S* simulated. Central horizontal line: median; box limits: 25th and 75th percentiles; whiskers: minimum and maximum values (Source: Adopted from Ahmad et al. 2017)

6.6 Impact of Climate Change on Cotton Production

DSSAT-CSM model was employed to check the impact of climate change upon the cotton productivity in Pakistan under RCP 4.5 and 8.5 climatic scenarios from baseline period 1980–2010 to future prediction periods 2010–2039 and 2014–2069 (Table 6.8). This study revealed that for the period of 2010–2039, the seed cotton yield declined by 8% and 12% for RCP 4.5 and 8.5 climatic scenarios. Same as 20% and 30% decrease in yield was observed during 2040–2069 under RCP 4.5 and RCP 8.5 climatic scenarios, respectively (Rahman et al. 2018). In India, InfoCrop-cotton model was used with three climatic scenarios A2, B2, and A1B. Present climate scenario was based on historical data 1969–1990, while future data was based on forecasts from 2071 to 2100 for A2 and B2, as well as 2031 to 2060 for the A1B scenario. In A2 scenario, average cotton yield decreased up to 477 kg ha^{-1}, and in B2, declined by 268 kg ha^{-1} (Hebbar et al. 2013). For climate and yield research, the fixed-influence model was employed to analyze the effects of climatic factors like Tmax and Tmin, along with precipitation for cotton from 1980 to 2020 in Southeastern region of the USA. The model predicted that a 1 °C upsurge in T (min) improved productivity of the cotton up to 20.8%, whereas 1 °C rise in T (max) decreased the yield of cotton by 20.8%, whereas the rainfall has no significant effect on yield (Sharma et al. 2022). In China, a study was conducted under full (638 mm) and deficit (478 mm) irrigation regimes. RZWQM2 model having a combination of six GCMs for two RCPs was employed to study the effects of changing climate for cotton in the periods 2022–2047, 2048–2073, and 2073–2099 as compared to 1975–2000. Results of this study showed that the yield was declined up to 13–18% and 14–18% under fully irrigated conditions and 10–14% and 11–19.6% for deficit condition (Ahmad et al. 2023b). In Southern Queensland and Northern NSW, eastern Australia influence of changing climate for cotton yield was studied by employing the APSIM model. Climate projections for 30 years' time span of 2030 and 2050 were attained through CCCS. The results showed that in 2030, yield will increase by 6% and in 2050, it will decrease by 3.6% (Williams et al. 2015). In Israel, a study was conducted by applying climate projections from HadCM3, forecast for 2070–2099 as compared to baseline 1960–1990. Average increase in temperature was 5.3 °C in A2 and 3.6 °C in B2 scenario, which decreased the cotton yield by 52% and 58%, respectively (Haim et al. 2008) (Table 6.8).

Table 6.8 Impact of climate change on cotton production

Country	Model	Climatic scenarios and study periods	Yield	References
Pakistan	CROPGRO-Cotton	RCP 4.5 and RCP 8.5 (1980–2010, 2010–2039, 2040–2069)	In 2010–2039, decline by 8% (RCP 4.5) and 12% (RCP 8.5); in 2040–2069, decline by 20% (RCP 4.5) and 30% (RCP 8.5)	Rahman et al. (2018)
India	InfoCrop-cotton	A2, B2, and A1B were derived from PRECIS RCM, which has HADCM3 as its GCM (1969–1990, 2031–2060, 2071–2100)	Declined by 477 kg ha^{-1} in A2, 268 kg ha^{-1} in B2, in A1B decline was not significant	Hebbar et al. (2013)
USA	Fixed effect	Daily T (max), T (min), and rainfall (1980–2020)	Rise in Tmin increased yield by 20.8%, rise in Tmin decreased yield by 10.3%, rainfall nonsignificant	Sharma et al. (2022)
China	RZWQM2 + 6 GCM	Full (638 mm) and deficit (478 mm) irrigation regimes, RCPs 4.5 and 8.5 (1975–2000, 2022–2047, 2048–2073, 2073–2099)	Decrease by 13–18% (RCP 4.5) and 14–18% (RCP 8.5) with full irrigation, 10–14% (RCP 4.5) and 11–19.6% (RCP 8.5) under deficit irrigation	Ahmad et al. (2023b)
Australia	APSIM	Consistent Climate Change Scenarios +6 GCMs 1975–2004, 2030–2050	In 2030, yield will increase by 6% and in 2050 decrease by 3.6%	Williams et al. (2015)
Israel	HadCM3 + SRES	Increase in temperature to 5.3 °C in A2 and 3.6 °C in B2 1960–1990 and 2070–2099	−52% in A2 and −38% in B2	Haim et al. (2008)
Southern Mali	Linear regression	Rainfall, Tmax, Tmin (1965–2005)	Decline in rainfall resulted in −17 kg ha^{-1} $year^{-1}$, for Tmax 4 kg ha^{-1} $year^{-1}$ and for Tmin −4 kg ha^{-1} $year^{-1}$	Traore et al. (2013)
Cameron	CROPGRO + GCM + RCM + CERES + SUBSTOR	ENSEMBLES regional climate changes scenarios (1950–2010, 2010–2050)	+ Yield 1.3 kg ha^{-1} $year^{-1}$	Gérardeaux et al. (2013)
Texas	CROPGRO-Cotton	NARCCAP CMIP3 (1971–2000, 2041–2070)	With increase in CO_2 concentration, yield increases by 14–29%, by water reduction yield reduces by 37–39%	Adhikari et al. (2016)
Greece	AquaCrop + (HadRM3, C4I, REMO-MPI, ETHZ, CNRM, DMI-HIRHAM, KNMI, SMHI)	A1B emission scenario (1961–1990, 2021–2050, 2071–2100)	+43% by REMO-MPI, +70% by ETHZ	Voloudakis et al. (2015)

6.7 Impact of Climate Change on Wheat Production

The AFRCWHEAT2 simulation model with three GCMs was employed to judge the influence of changing climatic conditions upon the growth and productivity of winter wheat and durum winter wheat in the UK and France during 1959–1989 baseline, 2010, 2030, and 2050 in the future (Table 6.9). The predicted wheat yield increased by 12% in the UK, which was higher than in France in 2050, i.e. 9–10% (Semenov et al. 1993). Impact of changing climate in Southwest Australia was estimated by the APSIM-Wheat Model for a baseline period 1954–2003, for mid-century, and end-century. Five different rainfall and three temperature scenarios with APSIM-N wheat predicted the wheat yield reduction due to a rise in temperature and a decrease in rainfall, but wheat yield would be increased by 38–48% under the application of various nitrogen fertilizers (Ludwig and Asseng 2006). The impact of climate change in European Russia was predicted by Climate-Soil-Yield for a baseline period of 1990–1999, and in the future (2030–2039, 2050–2059, and 2090–2099) with Voeikov Main Geophysical Observatory (MGO)-RCM and RCP 8.5. The projected wheat yield dropped to 15.8% and 32.9% in 2050–2059 and 2090–2099 in some parts of European Russia (Pavlova et al. 2019). In China, the APSIM-Wheat model was applied with 28 GCMs under RCPs 4.5 and 8.5 from baseline (1981–2010) and for future periods (2031–2060 and 2071–2100). The wheat yield increased (0.7–22.7%) under a constant concentration of CO_2. In the future, it is anticipated that rising CO2 concentration will benefit wheat by increasing water production at a lower cost of evapotranspiration (Xiao et al. 2018). DSSAT and APSIM models predicted the influence of changing climate upon wheat yield in Pakistan under RCPs 4.5 and 8.5 with three GCMs for the future periods 2017–2040, 2040–2069, and 2070–2100. Model results predicted that increase in minimum and maximum temperature (5.81 and 5.15 °C) and precipitation (17.14–61.72%) decreases the wheat yield by 2.81–31.0% (Munir et al. 2022). CERES-Wheat model predicted the influence of changing climate on the wheat productivity for Turkey with three GCMs and under RCPs 4.5 and 8.5 from the future periods of 2036–2065 and 2066–2095. The results revealed that a rise in minimum temperature from 1.3 °C under RCP 4.5 to 3.1 under RCP 8.5 and maximum temperature from 2 °C for RCP 4.5 to 4 °C for RCP 8.5 has a negative impact on the wheat yield, reduced by 13–14.4% (Vanli et al. 2019). In China (Yellow River basin), CERES-Wheat model with RCP 8.5 predicted that the northern part river wheat yield will decrease by 9.9% but there will be 35.3% increase in the western part of the river (Long et al. 2022) (Table 6.9).

Table 6.9 Impact of climate change on wheat production

Country	Model	Climatic scenarios and study periods	Yield	References
UK and France	AFRCWHEAT2	3 GCMs (1959–1989, 2010, 2030, 2050)	Wheat yield will be increased in France (9–10%) and in the UK (12%) in 2050	Semenov et al. (1993)
Australia	APSIM-Wheat	3 Temperature and 5 rainfall (1954–2003, 2050, 2100)	Wheat yield will be reduced due to a decrease in rainfall and a rise in temperature	Ludwig and Asseng (2006)
European Russia	Climate-soil-yield	RCM, RCP 8.5 (1990–1999, 2030–2039, 2050–2059, and 2090–2099)	Yield will be decreased by 15.8–32.9% in different areas of European Russia	Pavlova et al. (2019)
China	APSIM-Wheat	28 GCMs, RCP (4.5, 8.5) (1981–2010, 2031–2060, and 2071–2100)	Wheat yield will decrease by 0.7–22.7% in Northern China	Xiao et al. (2018)
Pakistan	APSIM and DSSAT	3 GCM Resolutions, RCPs (4.5 and 8.5), 3 RCMs (2017–2040, 2040–2059, 2070–2100)	Increase in rainfall and temperature, and wheat yield will reduce by 2.81–31.0%	Munir et al. (2022)
India	DSSAT	WRF model 3.7.1, RCP 8.5 (1986–2005, 2046–2065)	Reduction of −1% to −8% in wheat yield due to changes in rainfall and temperature	Daloz et al. (2021)
Northern part of Serbia and Southeast Central Europe	SIRIUS-Wheat Model and AGRICLIM Software	4 GCMs (2040–2080)	Heat waves and drought have a negative effect on wheat yield	Lalici et al. (2013)
South-eastern Turkey	CERES-Wheat	3 GCMs, RCPs (4.5 and 8.5) (2036–2065, 2066–2095)	Yield will be reduced by 13–14.4% with the effect of rising temperature	Vanli et al. (2019)
Europe (Finland, Spain, Germany)	26 Crop growth models	1981–2010	In anomalous weather conditions, wheat yield decreases up to 40–70%	Fronzek et al. (2018)
Yellow River basin, China	CERES-Wheat	RCP 8.5 (1975–2005, 2020–2050)	In the western part of YRB, 35.4% increase in yield, but in the northern part, yield decreased by 9.9%	Long et al. (2022)

6.8 AgMIP Case Study

The AgMIP quantified the changing climate influence on cotton-wheat cropping system in Pakistan (Fig. 6.13). Results revealed that temperature will increase in the cotton zone during the cotton season in future. There will be variations in temperatures and in the rainfall distribution (Fig. 6.14). The yield of cotton-wheat system

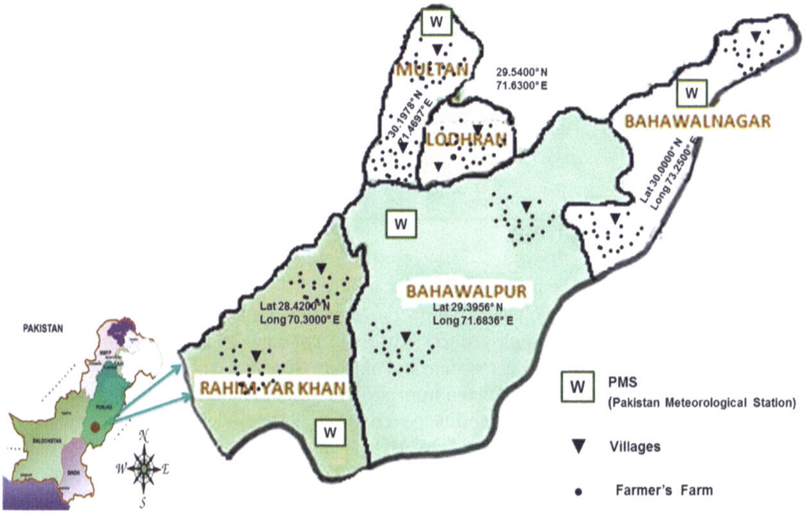

Fig. 6.13 Geographical location of the study site in Punjab, Pakistan

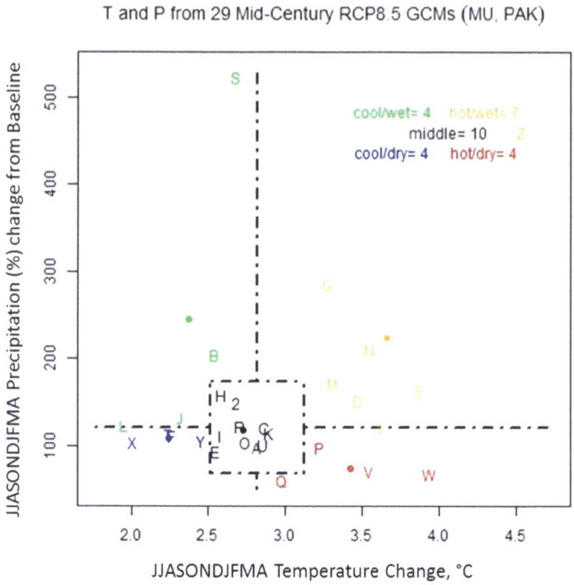

Fig. 6.14 Uncertainty arising from projected change in precipitation (%)

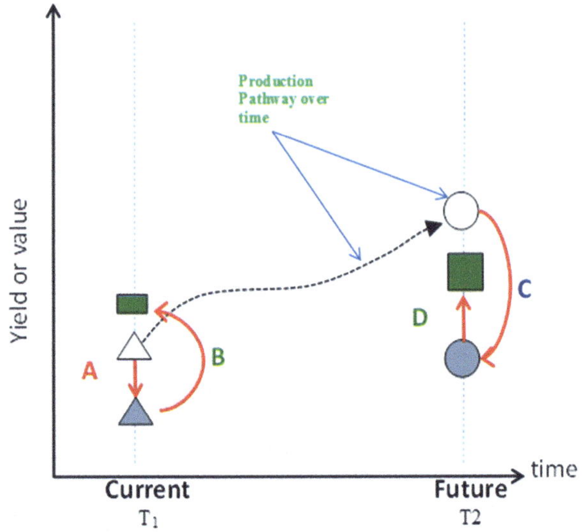

Fig. 6.15 Framework for climate change impact assessment and adaptation benefits (Claessens et al. 2012). The TOA-MD was employed to study socioeconomic impacts of climate change on farming communities for cotton-wheat system in Pakistan. Survey data were collected from 165 farms in five districts. **System 1** is calculated from survey data, and **System 2** uses simulated yields from DSSAT and ASIM models. Vulnerability, poverty, net returns, and per capita income with and without climate change were computed by TOA-MD for current and future agricultural production systems. Representative agricultural pathways (RAPs) were formulated. Cost factors, future prices, household size, and farm size were formulated for the RAPs by expert opinion and results from the IMPACT model

Table 6.10 Climate change adaptation package for cotton-wheat cropping system in Punjab, Pakistan

Climate change adaptations		
Biophysical	Socioeconomic	Institutional and policy
• Virtual cultivars (heat- and drought-tolerant varieties)	• Construction of water storage	• Agricultural insurance/finance
• Plant population	• Participatory management approach	• Farm mechanization (mechanical cotton picker)
• Improved agricultural practices	• Increasing off-farm income opportunities	• Subsidies/taxation
• Efficient irrigation practices	• Population control measures	• Input/output price policies
• Changes in cropping patterns		• Trade, off-farm employment
• Soil reclamation projects		• Efficient input/output markets

Source: Modified and adopted from Ahmad et al.

will be decreased in future (Fig. 6.15). The findings of this study have been published as a book chapter and research paper as well (Ahmad et al. 2023b). This study also suggested adaptation package for this system (Table 6.10).

6.9 Conclusion

The major area under CWS is situated in Asia. Currently, productivity of CWS is low due to CLCV and stagnant yield. Among reasons for lower and stagnant yield, climate change is one of the key reasons. Due to insects, pests, diseases, and climate change, food and fiber security in developing countries is at risk. Scientists are finding solutions to control CLCV and quantify climate change influence on CWS phenology and productivity along with adapting climate-smart practices.

References

Adhikari P, Ale S, Bordovsky JP, Thorp KR, Modala NR, Rajan N, Barnes EM (2016) Simulating future climate change impacts on seed cotton yield in the Texas high plains using the CSM-CROPGRO-Cotton model. Agric Water Manag 164:317–330

Ahmad S, Abbas Q, Abbas G, Fatima Z, Atique-ur-Rehman NS, Younis H, Khan RJ, Nasim W, Rehman MH, Ahmad A, Rasul G, Khan MA, Hasanuzzaman M (2017) Quantification of climate warming and crop management impacts on cotton phenology. Plan Theory 6:7. https://doi.org/10.3390/plants6010007

Ahmad S, Abbas G, Ahmed M, Fatima Z, Anjum MA, Rasul G, Khan MA, Hoogenboom G (2019) Climate warming and management impact on the change of phenology of the rice-wheat cropping system in Punjab, Pakistan. Field Crop Res 230:46–61. https://doi.org/10.1016/j.fcr.2018.10.008

Ahmad S, Ahmad I, Ahmad B, Ahmad A et al (2023a) Regional integrated assessment of climate change impact on cotton production in a semi-arid environment. Clim Res 89:113–132. https://doi.org/10.3354/cr01710

Ahmad Z, Gui D, Ali S, Chen X, Qi Z (2023b) Evaluating climate change impacts on cotton phenology and yield under full and deficit irrigation conditions in an extremely arid Oasis. Int J Plant Prod 17:49–63

Ahmed M, Ijaz W, Ahmad S (2018) Adapting and evaluating APSIM-SoilP-Wheat model for response to phosphorus under rainfed conditions of Pakistan. J Plant Nutr 41(16):2069–2084. https://doi.org/10.1080/01904167.2018.1485933

Ahmed M, Stöckle CO, Nelson R, Higgins S, Ahmad S, Raza MA (2019) Novel multimodel ensemble approach to evaluate the sole effect of elevated CO_2 on winter wheat productivity. Sci Rep 9(1):7813. https://doi.org/10.1038/s41598-019-44251-x

Ahmed M, Hayat R, Ahmad M, ul-Hassan M, Kheir AMS, ul-Hassan F, ur-Rehman MH, Shaheen FA, Raza MA, Ahmad S (2022a) Impact of climate change on Dryland agricultural systems: a review of current status, potentials, and further work need. Int J Plant Prod. https://doi.org/10.1007/s42106-022-00197-1

Ahmed A, Hussain I, Nasrullah H, Khalil IA, Ali B, Hamid A, Imtiaz M (2022b) Impact of planting techniques on the productivity of the cotton-wheat rotation in Punjab, Pakistan. Int J Agric Biol 28:40–48

Amin A, Nasim W, Mubeen M, Ahmad A, Nadeem M, Urich P, Fahad S, Ahmad S, Wajid A, Tabassum F, Hammad HM (2018) Simulated CSM-CROPGRO-cotton yield under projected future climate by SimCLIM for southern Punjab, Pakistan. Agric Syst 167:213–222

Balota EL, Kanashiro M, Filho AC, Andrade DS, Dick RP (2004) Soil enzyme activities under long-term tillage and crop rotation systems in subtropical agro-ecosystems. Braz J Microbiol 35:300–306

Basal H, Karademir E, Goren HK, Sezener V, Dogan MN, Gencsoylu I, Erdogan O (2020) Cotton production in Turkey and Europe. In: Jabran K, Chauhan BS (eds) Cotton production. Wiley, Hoboken

Bokhari SAA, Ahmad B, Ali J, Ahmad S, Mushtaq H, Rasul G (2018) Future climate change projections of the Kabul River Basin using a multi-model ensemble of high-resolution statistically downscaled data. Earth Syst Environ 2(3):477–497

Brar BS, Dheri GS, Sekhon FS (2019) Productivity and profitability of cotton-wheat cropping system as influenced by complex fertilizers. Curr J Appl Sci Technol 38(6):1–7

Buttar GS, Sidhu HS, Singh V, Jat ML, Gupta R, Singh Y (2013) Relay planting of wheat in cotton: an innovative technology for enhancing productivity and profitability of wheat in cotton-wheat production system of South Asia. Exp Agric 49:19–30

Claessens L, Antle JM, Stoorvogel JJ, Valdivia RO, Thornton PK, Herrero M (2012) A method for evaluating climate change adaptation strategies for small-scale farmers using survey, experimental and modeled data. Agric Syst 111:85–95

Conrad C, Rahmann M, Machwitz M, Stulina G, Paeth H, Dech S (2013) Satellite based calculation of spatially distributed crop water requirements for cotton and wheat cultivation in Fergana Valley, Uzbekistan. Globe Planet Change 110:88–98

CRDC (2017) Australian cotton production manual, CRDC, Australia. Department of Agriculture and Fisheries 2018, Cotton growing

Daloz AS, Rydsaa JH, Hodnebrog O, Sillmann J, Oort BV, Mohr CW, Agrawal M, Emberson L, Stordal F, Zhang T (2021) Direct and indirect impacts of climate change on wheat yield in the Indo-Gangetic plain in India. J Agric Food Res 4:100132

Das TK, Bhattacharyya R, Sudhishri S, Sharma AR, Saharawat YS, Bandyopadhyay KK, Sepat S, Bana RS, Aggarwal P, Sharma RK, Bhatia A, Singh G, Datta SP, Kar A, Singh B, Singh P, Pathak H, Vyas AK, Jat ML (2014) Conservation agriculture in an irrigated cotton-wheat system of the western Indo-Gangetic Plains: crop and water productivity and economic profitability. Field Crop Res 158:24–33

Eckstein D, Hutfils ML, Winges M (2018) Global climate risk index 2019. In: Who suffers most from extreme weather events, vol 36

Ejaz M, Abbas G, Fatima Z, Iqbal P, Raza MA, Kheir AMS, Ahmed M, Kakar KM, Ahmad S (2022) Modelling climate uncertainty and adaptations for soybean-based cropping system. Int J Plant Prod. https://doi.org/10.1007/s42106-022-00190-8

FAO (2024) FAOSTAT database, Food and Agriculture Organization of the United Nations. Available at http://faostat.fao.org/ (Accessed 15 Feb 2024)

Farooqi AB, Khan AH, Mir H (2005) Climate change perspective in Pakistan. Pakistan J Meteorol 2(3)

Fatima Z, Ahmed M, Hussain M, Abbas G, Ul-Allah S, Ahmad S, Ahmed N, Ali MA, Sarwar G, Haque Eu, Iqbal P, Hussain S (2020) The fingerprints of climate warming on cereal crops phenology and adaptation options. Sci Rep 10:18013. https://doi.org/10.1038/s41598-020-74740-3

Fronzek S, Pirttioja N, Carter TR, Bindi M, Hoffmann H, Palosuo T, Ruiz-Ramos M, Tao F, Trnka M, Acutis M, Asseng S, Baranowski P, Basso B, Bodin P, Buis S, Cammarano D, Deligios P, Destain MF, Dumont B, Ewert F, Ferrise R, François L, Gaiser T, Hlavinka P, Jacquemin I, Kersebaum KC, Kollas C, Krzyszczak J, Lorite IJ, Minet J, Minguez MI, Montesino M, Moriondo M, Müller C, Nendel C, Ozturk I, Perego A, Rodriguez A, Ruane AC, Ruget F, Sanna M, Semenov MA, Slawinski C, Stratonovitch P, Supit I, Waha K, Wang E, Wu L, Zhao Z, Rotter RP (2018) Classifying multi-model wheat yield impact response surfaces showing sensitivity to temperature and precipitation change. Agric Syst 159:209–224

Funk CC, Brown ME (2009) Declining global per capita agricultural production and warming oceans threaten food security. Food Secur 1:271–289

Gérardeaux E, Sultan B, Palaï O, Guiziou C, Oettli P, Naudin K (2013) Positive effect of climate change on cotton in 2050 by CO_2 enrichment and conservation agriculture in Cameroon. Agron Sustain Dev 33:485–495

Gupta SC, Allmaras RR (1987) Models to assess the susceptibility of soils to excessive compaction. In: Stewart BA (ed) Advances in soil science. Springer, New York, pp 65–100

Haim D, Shechter M, Berliner P (2008) Assessing the impact of climate change on representative field crops in Israeli agriculture: a case study of wheat and cotton. Clim Chang 86:425–440

He L, Asseng S, Zhao G, Wu D, Yang X, Zhuang W, Jin N, Yu Q (2015) Impacts of recent climate warming, cultivar changes, and crop management on winter wheat phenology across the Loess Plateau of China. Agric For Meteorol 200:135–143

Hebbar KB, Venugopalan MV, Prakash AH, Aggarwal PK (2013) Simulating the impacts of climate change on cotton production in India. Clim Chang 118:701–713

Hillel D, Rosenzweig C (eds) (2015) Handbook of climate change and agroecosystems: the agricultural model intercomparison and improvement project (Agmip) integrated crop and economic assessments—joint publication with Asa, Cssa, and Sssa (In 2 Parts), vol 3. World scientific

Hoogenboom G (2021) The application of crop modeling for irrigation and water management. In: 6th Decennial National Irrigation Symposium, 6–8, December 2021, San Diego, California. American Society of Agricultural and Biological Engineers, p 1

Hu Q, Weiss A, Feng S, Baenziger PS (2005) Earlier winter wheat heading dates and warmer spring in the US Great Plains. Agric For Meteorol 135(1–4):284–290

IPCC (2023) Summary for policymakers. In: Core Writing Team, Lee H, Romero J (eds) Climate change 2023: synthesis report. A report of the intergovernmental panel on climate change. Contribution of working groups I, II and III to the sixth assessment report of the intergovernmental panel on climate change. IPCC, Geneva, Switzerland

Jabran K, Nawaz A, Uludag A, Ahmad S, Hussain M (2020) Cotton relay intercropping under continuous cotton-wheat cropping system. Cotton Prod Uses 16:311–320

Jalota SK, Buttar GS, Sood A, Chahal GBS, Ray SS, Panigrahy S (2008) Effects of sowing date, tillage and residue management on productivity of cotton (*Gossypium hirsutum* L.)-wheat (*Triticum aestivum* L.) system in northwest India. Soil Tillage Res 99:76–83

Javed MI, Sultan AA, Hassan S, Asghar A (2009) An efficiency analysis of Punjab's cotton-wheat system. Lahore J Econ 14:97–124

Khokhar Z, Hussain I, Khokhar B, Sohail M (2010) Effect of planting date on yield of wheat genotypes in Sindh. Pak J Agric Res 23(3–4)

Kidron GJ, Karnieli A, Benenson I (2010) Degradation of soil fertility following cycles of cotton-cereal cultivation in Mali, West Africa: a first approximation to the problem. Soil Tillage Res 106(2):254–262

Kienzler KM, Djanibekov N, Lamers JPA (2011) An agronomic, economic and behavioral analysis of N application to cotton and wheat in post-Soviet Uzbekistan. Agric Syst 104:411–418

Kilic H, Gursoy S (2010) Effect of seeding rate on yield and yield components of durum wheat cultivars in cotton-wheat cropping system. Sci Res Essays 5(15):2078–2084

Lalici B, Eitzinger J, Mihailovic DT, Thaler T, Jancici M (2013) Climate change impacts on winter wheat yield change—which climatic parameters are crucial in Pannonian lowland? J Agric Sci 151:757–774

Liu B, Martre P, Ewert F, Webber H, Waha K, Thorburn PJ, Ruane AC, Aggarwal PK, Ahmed M, Balkovič J, Basso B, Biernath C, Bindi M, Cammarano D, Cao W, Challinor AJ, De Sanctis G, Dumont B, Espadafor M, Eyshi Rezaei E, Fereres E, Ferrise R, Garcia-Vila M, Gayler S, Gao Y, Horan H, Hoogenboom G, Izaurralde RC, Jabloun M, Jones CD, Kassie BT, Kersebaum KC, Klein C, Koehler A-K, Maiorano A, Minoli S, Montesino San Martin M, Müller C, Naresh Kumar S, Nendel C, O'Leary GJ, Eivind Olesen J, Palosuo T, Porter JR, Priesack E, Ripoche D, Rötter RP, Semenov MA, Stöckle C, Stratonovitch P, Streck T, Supit I, Tao F, Van der Velde M, Wang E, Wolf J, Xiao L, Zhang Z, Zhao Z, Zhu Y, Asseng S (2023) AgMIP-Wheat multi-model simulations on climate change impact and adaptation for global wheat. Open Data J Agric Res 9:10–25. https://doi.org/10.18174/odjar.v9i0.18092

Long XX, Ju H, Wang JD, Gong SH, Li GY (2022) Impact of climate change on wheat yield and quality in the Yellow River Basin under RCP8.5 during 2020–2050. Adva Clim Chang Res 13:397–407

Ludwig F, Asseng S (2006) Climate change impacts on wheat production in a Mediterranean environment in Western Australia. Agric Syst 90:159–179

Munir Z, Shrestha S, Zaman M, Khan MI, Akram MM, Tahir MN (2022) Climate change impacts on wheat yield: a multi-modeling case study of central Punjab, Pakistan. Clim Res 87:13–37

Nasrullah HM, Zahid AR, Akhtar M, Majid A, Ali B, Rehman H, AlamJ AMI, Nawaz H (2017) Enhancement of wheat productivity as a relay crop in cotton-wheat cropping system. J Environ Agric Sci 13:25–30

Pavlova V, Shkolnik I, Pikaleva A, Efimo S, Karachenkova A, Kattsov V (2019) Future changes in spring wheat yield in the European Russia as inferred from a large ensemble of high-resolution climate projections. Environ Res Lett 14:2–12

Rahman MH, Ahmad A, Wang X, Wajid A, Nasim W, Hussain M, Ahmad B, Ahmad I, Ali Z, Ishaque W, Awais M, Shelia V, Ahmad S, Fahd S, Alam M, Ullah H, Hoogenboom G (2018) Multi-model projections of future climate and climate change impacts uncertainty assessment for cotton production in Pakistan. Agric For Meteorol 253–254:94–113

Rezaei E, Webber H, Gaiser T, Naab J, Ewert F (2015) Heat stress in cereals: mechanisms and modelling. Eur J Agron 64(0):98–113. https://doi.org/10.1016/j.eja.2014.10.003

Rosenzweig C, Jones JW, Hatfield JL, Ruane AC, Boote KJ, Thorburn P, Antle JM, Nelson GC, Porter C, Janssen S, Asseng S (2013) The agricultural model intercomparison and improvement project (AgMIP): protocols and pilot studies. Agric For Meteorol 170:166–182. https://doi.org/10.1016/j.agrformet.2012.09.011

Rosenzweig C, Antle JM, Ruane AC, Jones JW, Hatfield J, Boote KJ, Thorburn P, Valdivia RO, Descheemaeker K, Porter CH, Janssen S, Bartels W-L, Sulivan A, Mutter CZ (2016) Protocols for AgMIP Regional Integrated Assessments Version 7.0.. https://agmip.org/wp-content/uploads/2018/08/AgMIP-Protocols-for-Regional-Integrated-Assessment-v7-0-20180218-1-ilovepdf-compressed.pdf

Sajjad A, Anjum SA, Ahmad R, Waraich EA (2018) Relay cropping of wheat (Triticum aestivum L.) in cotton (Gossypium hirsutum L.) improves the profitability of cotton-wheat cropping system in Punjab, Pakistan. Environ Sci Pollut Res 25(1):782–789

Scott BJ, Eberbach PL, Evans J, Wade LJ (2010) Stubble retention in cropping systems in Southern Australia: benefits and challenges. EH Graham Centre Monograph No. 1

Semenov MA, Porter JR, Delecolle R (1993) Climatic change and the growth and development of wheat in the UK and France. Eur J Agron 2(4):293–304

Shabani F, Kotey B (2016) Climate change and agriculture research paper future distribution of cotton and wheat in Australia under potential climate change. J Agric Sci 154:175–185

Shah MA, Farooq M, Hussain M (2016) Productivity and profitability of cotton-wheat system as influenced by relay intercropping of insect resistant transgenic cotton in bed planted wheat. Eur J Agron 75:33–41

Sharma RK, Kumar S, Vatta K, Dhillon J, Reddy KN (2022) Impact of recent climate change on cotton and soybean yields in the southeastern United States. J Agric Food Res 9:100348

Singh Y, Kukal SS, Mangi LJ, Harminder SS (2014) Improving water productivity of wheat-based cropping systems in South Asia for sustained productivity. Adv Agron 127:157–258

Soule J, Carré D, Jackson W (1990) Ecological impact of modern agriculture. In: Agroecology. McGraw-Hill, New York, pp 165–188

Sultan MS, Kassaby ATE, Ghonema MH, Ogeaz AA, Allah AMA (2012) Relay intercropping wheat and cotton studies: II-Effect of sowing dates and ridge width on cotton. J Biol Sci 12(6):349–354

Tanveer M, Anjum SA, Hussain S, Cerdà A, Ashraf U (2017) Relay cropping as a sustainable approach: problems and opportunities for sustainable crop production. Environ Sci Pollut Res 24(8):6973–6988

Tariq M, Afzal MN, Ahmad M, Ahmed Z, Afzal MU et al (2022) Relay cropping of cotton in wheat improves productivity of cotton-wheat cropping system. PLoS One 17(4):e0266694. https://doi.org/10.1371/journal.pone.0266694

Traore B, Corbeels M, Wijk MTV, Rufino MC, Giller KE (2013) Effects of climate variability and climate change on crop production in southern Mali. Eur J Agron 49:115–125

Ullah S, You Q, Ullah W, Ali A, Xie W, Xie X (2019) Observed changes in temperature extremes over China–Pakistan economic corridor during 1980–2016. Int J Climatol 39:1457–1475. https://doi.org/10.1002/joc.5894

Vanli O, Ustundag BB, Ahmad I, Hernandez-Ochoa IM, Hoogenboom G (2019) Using crop modeling to evaluate the impacts of climate change on wheat in southeastern Turkey. Environ Sci Pollut Res 26(7):29397–29408

Voloudakis D, Karamanos A, Economoua G, Kalivas D, Vahamidis P, Kotoulas V, Kapsomenakis J, Zerefos C (2015) Prediction of climate change impacts on cotton yields in Greece under eight climatic models using the AquaCrop crop simulation model and discriminant function analysis. Agric Water Manag 147:116–128

Wang G, Feng L, Liu L, Zhang Y, Li A, Wang Z, Han Y, Li Y, Li C, Dong H (2021) Early relay intercropping of short-season cotton increases lint yield and earliness by improving the yield components and boll distribution under wheat-cotton double cropping. Agriculture 11:1294

Williams A, White N, Mushtaq S, Cockfield G, Power B, Kouadio L (2015) Quantifying the response of cotton production in eastern Australia to climate change. Clim Chang 129:183–196

Xiao D, Bai H, Liu DL (2018) Impact of future climate change on wheat production: a simulated case for China's wheat system. Sustainability 10(4):1–15

Yuan C, Zhen-yu L, Li H, Tambel LIM, Xiang Z, Yuan C, De-hua C (2021) Effects of plant density and mepiquat chloride application on cotton boll setting in wheat-cotton double cropping system. J Integr Agric 20(9):2372–2381

Chickpea-Mung Bean System

7

7.1 Introduction

Chickpea and mung bean cropping system is vital for ensuring food security under changing climatic conditions. As compared to other cropping systems, chickpea and mung bean cropping system requires less resources such as water and fertilizer. Gram (*Cicer arietinum* L.) is the world's third most significant pulse after dry beans and peas (Mitra et al. 2009; Grasso et al. 2022). Chickpeas are an important pulse crop with several nutritional and health advantages. Chickpeas are a high-nutritional-value grain legume. Gram is a very good basis of calories, protein, vitamins, minerals, fiber, as well as other phytochemicals that may be valuable to your health (Jukanti et al. 2012). It is grown on almost 11.52 million hectares, yielding ten million tonnes with a productivity of 863 kg/ha. Asia accounts for approximately 90% of global chickpea output. Every year, 14.24 million tonnes of chickpeas are produced worldwide. India is the world's greatest producer of chickpeas, producing 9.93 million tonnes annually. India alone produces more than 60% of the world's chickpeas. Turkey ranks second, producing 0.63 million tonnes/year. Russian Federation is the third greatest grower of chickpeas, producing 0.51 million tonnes/year. The USA ranks seventh with 0.28 million tonnes of production every year. Africa contributes 4.7% of worldwide gram area, with the majority of it located in East Africa. India is the world's leading producer of chickpeas. It covers 76% of Asia's gram-growing area. Other prominent chickpea-growing nations in the region include Pakistan and Iran. The mung bean (*Vigna radiata* L.) is a significant edible legume crop, with almost 6.0 million hectares produced globally (approximately 8.5% of the total pulse area) and used by the majority of Asian families. Mung bean is sometimes referred to as green gram, golden gram, as well as moong. It may be utilized as a green manure crop as well as animal fodder (Ganesan and Xu 2018) because of its drought tolerance, less input crop, and shorter crop duration. The worldwide mung bean area is approximately 7.31 Mha, and global production is around 5.3 M tonnes, with India and Myanmar generating about 30%, China 16%,

M. Ahmed et al., *Cropping Systems Modeling Under Changing Climate*,
https://doi.org/10.1007/978-981-97-0331-9_7

and Indonesia 5%. Chickpea and mung bean cropping system exists in Pakistan, in which chickpea crop is typically planted in Punjab, Pakistan, between the start of October and mid of November and reaped between earlier March and earlier April (Ejaz et al. 2022; Fatima et al. 2018, 2020, 2021; Nadeem et al. 2022; Naz et al. 2022). Mung bean crop is typically cultivated in Punjab, Pakistan, from the start of May until the end of June and harvested in August. Figure 7.1 shows the area and production of chickpea and mung bean in Pakistan from 1994 to 2018. Changes in

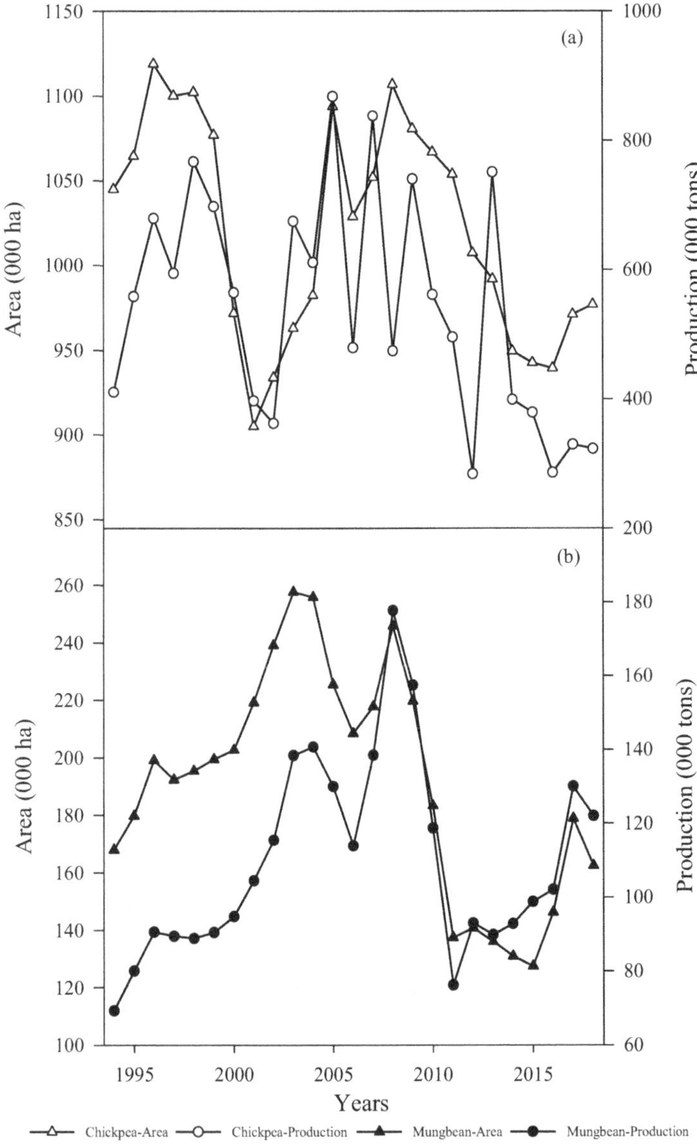

Fig. 7.1 Chickpea (**a**) and mung bean (**b**) area and production in Pakistan from 1994 to 2018. (Source: Fatima et al. 2021)

sowing dates, development of new cultivars requiring higher degree days, and cultivation of cultivars resistant to drought are some crucial techniques to lessen the threat posed by climatic unpredictability and enhance the ability of agricultural plants to adapt for the assurance of food security (Bahl 2015). We can assist plants in adjusting to their surroundings by using crop management techniques that can enhance crop development under a variety of environmental difficulties. In the chickpea and mung bean cropping system, it is crucial to pick the proper sowing timing, planting density, and irrigation techniques to resist weather challenges.

7.2 Issues of Chickpea-Mung Bean Cropping System

Increased productivity in the chickpea and mung bean cropping system is constrained by a number of reasons, including those relating to edaphic production, climate change, and technological concerns (Ali et al. 2013; Amin et al. 2018a, b, c). The main unfavorable environmental variables that restrict its production are the abiotic stressors, which include high temperatures and dryness at different growth stages during the productive season (Pratap et al. 2019). In fact, up to 50% and 20%, respectively, of the world's losses in chickpea output can be attributed to drought and high heat. Most of the area used for growing chickpeas is rainfed. Chickpea and mung bean production is highly concerning because of heat stress brought on by global warming. Many biotic and abiotic factors contribute to the low output of the chickpea and mung bean cropping system. Due to climate change, the production of mung beans and chickpeas is impeded by low soil fertility and erratic rainfall (Ahmed et al. 2016, 2018a, b, 2019, 2020, 2022; Ahmad and Hasanuzzaman 2012; Harris et al. 2018). Due to the unpredictable behavior of the monsoon season, which impacts sowing and planting in different parts of the world, chickpea and mung bean output is also constrained. The yield of mung beans and chickpeas is limited as a result of the significant geographical and temporal variation in rainfall. Production of chickpeas and mung beans is constrained as a result of soil moisture stress during critical developmental stages, notably seed filling. During critical stages of growth, high temperatures impact chickpea and mung bean output. The yield of mung beans and chickpeas is constrained due to biotic crop growth interferences. The output of soybeans is limited as a result of insufficient mechanization (Gull et al. 2020). Due to a lack of adoption of cutting-edge production technology with low-risk coverage capabilities, chickpea and mung bean output is constrained. The production of chickpeas and mung beans is constrained due to monoculture cultivation and insufficient varietal variety, which raises risk. Production of mung beans and chickpeas is limited by a dearth of timely access to high-quality inputs. The low production of soybeans is caused by a lack of technological expertise (Cable et al. 2008). Because planting time overlaps with the growing seasons of key crops, the growing area of the chickpea and mung bean cropping system is limited. Abiotic stresses, like heat, salinity, and drought, negatively affect morphology and physiology, along with the capacity for reproduction of the plants. In the end, this reduces their plasticity and capacity for adaptation to changing climates, which has a significant impact on the quality and quantity of the produce (Abbas et al. 2017, 2019, 2020a, b, c; Abbas and Ahmad 2018; Nair et al. 2019; Clark and Hobson 2022).

7.3 Decision Support System for Agrotechnology Transfer

7.3.1 Chickpea

To evaluate mutual impacts of cropping methods, soil managing, along with weather on agro-environmental pointers, and cropping system models are frequently employed as useful tools (Ahmad et al. 2012a, b, 2013, 2015a, b, c, 2016a, b, c, 2017a, b). They offer a wide variety of prediction data that may be used to develop and assess novel cropping systems (Ahmad et al. 2018, 2019, 2021, 2023a, b). To evaluate crop growth and yield benefits brought on by new technologies in various target conditions, plant growth simulation models may be utilized (Soltani and Sinclair 2011). As these models include parameters that describe the genetic characteristics of cultivars, these characteristics may be changed while staying within the observed range for genetic variability for determining if doing so would be advantageous for the target environment. The modeling of cropping systems is still being worked on, particularly for pulse-based intercropping systems. Evaluation of biophysical constraints on crop productivity can be aided by a strong crop model (Hajishabani et al. 2020). There aren't many models available right now that can be applied to a wide range of crops, soil categories, irrigation systems, and water managing techniques (Jones et al. 2003; Keating et al. 2003; Hoogenboom et al. 2019, 2021). The model has demonstrated its capacity to mimic a variety of crops and agricultural settings, making it one of the few models now accessible that have been created for such broad purposes. The present version would enable predictions real-time simulations of 20 fields, each with its own crop, irrigation system, irrigation plan, crop, and N-fertilizer. The model can simulate biomass, productivity, salinity leaching needs, nitrogen dynamics, nitrate leaching, temperature, water absorption, ET, groundwater level along with salinity, drainage, as well as soil moisture and salinity profiles (Carberry et al. 2009). The model can predict crop phenology, leaf growth in response to temperature, crop biomass accumulation in response to radiation intercepted, grain dry matter accumulation in response to time and temperature, and soil water balance (Rehman et al. 2015; Rahman et al. 2018, 2021; Sarwar et al. 2013, 2021; Shabbir et al. 2016). The sensitivity of phenology, leaf development and senescence, and biomass output to soil water content was developed (Mahmood et al. 2013; Mubeen et al. 2016; Nasim et al. 2018; Mubarak et al. 2022). The model makes use of freely accessible meteorological and soil data as well as a daily time step (Boote et al. 2002). Fatima et al. (2021) calibrated and evaluated CSM-CROPGRO model for chickpea and mung bean crops in Pakistan. Figure 7.2 displays the correlation between the anthesis and maturity dates of the years from 1980 to 1985 at the research locations as seen in the field and those predicted by the model. The CROPGRO-Chickpea and CROPGRO-Mung bean models performed well, based on the good agreement between field observations and model predictions, where slope = 0.97, R^2 = 0.88 for anthesis, and slope = 0.89, R^2 = 0.86 for maturity, for chickpea and slope = 0.81, R^2 = 0.71 for anthesis, and slope = 0.82, R^2 = 0.75 for maturity, for mung bean crop. Hajishabani et al. (2020) revealed from model calibration for various chickpea varieties that there were very good agreements between the phenological stages, leaf area index, biomass, and grain yield of

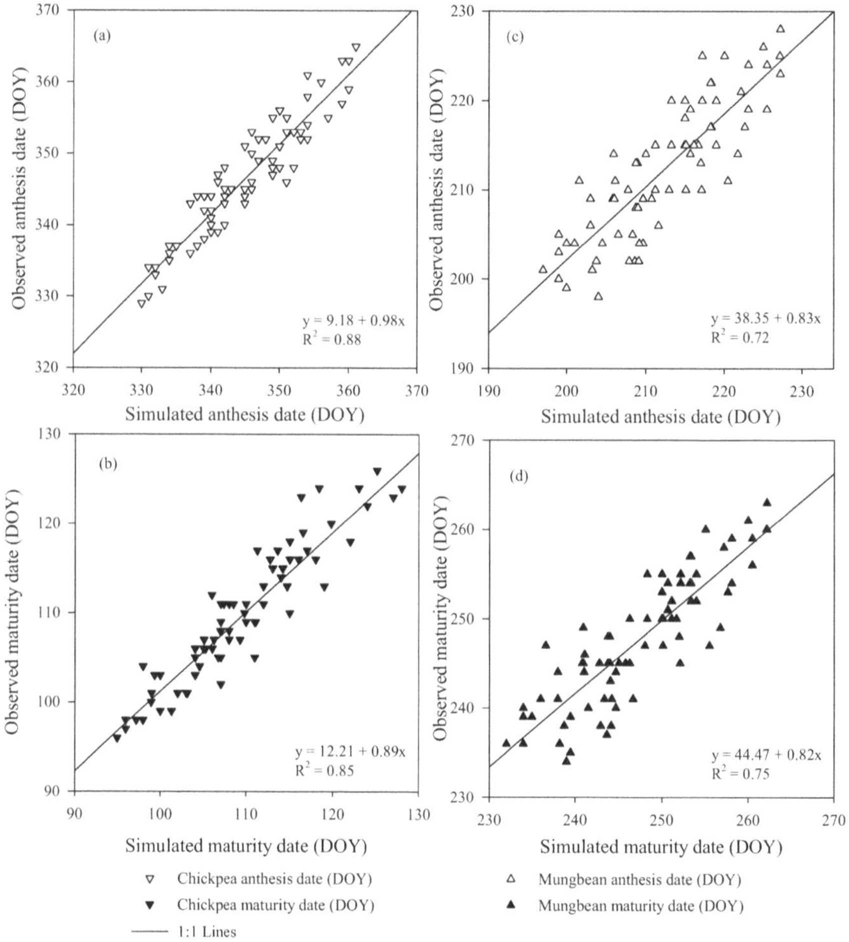

Fig. 7.2 Comparison of simulated and observed anthesis and maturity dates for chickpea (**a, b**; unfilled and filled down triangles) and mung bean (**c, d**; unfilled and filled up triangles) crops, respectively, for 15 locations in Punjab, Pakistan. (Source: Fatima et al. 2021)

observed as well as predicted values. Results revealed that CROPGRO-Chickpea model can simulate phenological stages and productivity for gram varieties. NRMSE values for Bivanj, Addel, Armaan, and ILC-482 of leaf area index were 26.12%, 27.91%, 28.37%, and 20.19%, respectively. *d*-Value for leaf area index ranged from 0.80 to 0.90. NRMSE average for evaluated varieties of biomass was 16.51%. *d*-value for biomass was 0.99. NRMSE mean for evaluated varieties of grain yield was 13.51%. *d*-value for grain yield ranged from 0.96 to 0.98. Patil and Patel (2017) calibrated and validated DSSAT-CROPGRO model with field trial data of two successive winter seasons in 2015 and 2016 at Anand, India. A rough estimate of the genetic coefficients for the chickpea cultivar GG-1 was made. Days to anthesis had a higher R^2 (0.97) between observed and simulated values, with a percent error (PE) of 8.08% and a D-index of 0.91. All of the parameters' PEs between observed and measured values were determined to be less than 10% off. Under various management scenarios, the model may be utilized to reliably estimate seed yield. Hence, the phenology and yield of chickpea may be simulated using the CROPGRO model. Chauhan et al. (2008) identified homoclimes (similar chickpea-growing environment) in the main chickpea-growing zones of the two countries Australia and India, using the APSIM chickpea model. The model was well evaluated and can be applied to determine appropriate cop management practices in homoclimes of chickpea. Silva et al. (2013) used SALTMED model for simulation of seed yield of chickpea. The findings of the SALTMED model's calibration and validation demonstrated that the model is capable of simulating soil moisture content, grain yield, and total dry biomass of several chickpea types in both wet and dry years with a high degree of accuracy. With more features and options than earlier iterations, the new SALTMED model offers academics and professionals an excellent tool for managing water, soil, and crops. Kherif et al. (2022) calibrated STICS model by using chickpea cultivated at various N rates throughout two cropping periods. We were able to model a variety of agronomic situations (temperature, N fertilization, and cropping system) thanks to this calibration in order to improve the management of intercrops (durum wheat-chickpea). The estimation of 37 parameters was done by the sequential optimization technique. According to our findings, STICS predicts the LAI, biomass, and N uptake of alone and intercropped cultivars with acceptable model performance. Also, the model accurately forecasted grain production with a negligible error (NRMSE 13%), particularly for the wheat crop (EF 0.50). Singh et al. (2014) used CROPGRO-Chickpea model. The observed productivity and predicted productivity throughout the test locations were strongly and significantly correlated, according to model calibration of the baseline chickpea cultivars. Also high was the *d*-value, a metric for model predictability. A substantial link between observed yields and simulated yields across the test sites was revealed by model validation using the independent 50% of data that were not utilized in calibration. These validation outcomes demonstrate the reliability of genetic coefficients for cultivars and applicability of the CROPGRO-Chickpea model for simulating chickpea growth and production under various soil-climate conditions and in response to climate change variables. Mohammed et al. (2017) used chickpea model in Northeastern Ethiopia. The model did a decent modeling of chickpea development, phenology, and yield under semiarid conditions in northern Ethiopia, as seen by the results of the model calibration and evaluation, which showed good agreement between observed and simulated data. The calibrated model was used to make crop management decisions and investigate how expected climate changes will affect chickpea. The genetic coefficients of chickpea crop in DSSAT and APSIM models are presented in Tables 7.1 and 7.2, respectively.

7.3.2 Mung Bean

Biswas et al. (2018) developed a MungGro model for calibration and evaluation purpose. The objective of the study was to develop a simple model for mung bean growth (MungGro), which was developed in Fortran Simulation Translator and

Table 7.1 Chickpea genetic coefficients and description in DSSAT model

Genetic coefficient	Description
EXPNO	Number of experiments used to estimate cultivar parameters
ECO#	Ecotype code of this cultivar, points to the ecotype in the ECO file
CSDL	Critical short-day length below which reproductive development progresses with day length effect (for long-day plants) (h)
PPSEN	Slope of the relative response of development to photoperiod with time (negative for long-day plants) (1/h)
EM-FL	Time between plant emergence and flower appearance (R1) (photothermal days)
FL-SH	Time between first flower and first pod (R3) (photothermal days)
FL-SD	Time between first flower and first seed (R5) (photothermal days)
SD-PM	Time between first seed (R5) and physiological maturity (R7) (photothermal days)
FL-LF	Time between first flower (R1) and end of leaf expansion (photothermal days)
LFMAX	Maximum leaf photosynthesis rate at 30 C, 350 vpm CO_2, and high light (mg CO_2/m^2-s)
SLAVR	Specific leaf area of cultivar under standard growth conditions (cm^2/g)
SIZLF	Maximum size of full leaf (three leaflets) (cm^2)
XFRT	Maximum fraction of daily growth that is partitioned to seed + shell
WTPSD	Maximum weight per seed (g)
SFDUR	Seed filling duration for pod cohort at standard growth conditions (photothermal days)
SDPDV	Average seed per pod under standard growing conditions (#/pod)
PODUR	Time required for cultivar to reach final pod load under optimal conditions (photothermal days)
THRSH	The maximum ratio of (seed/(seed + shell)) at maturity. Causes seed to stop growing as their dry weights increase until shells are filled in a cohort. (Threshing percentage)
SDPRO	Fraction of protein in seeds (g(protein)/g(seed))
SDLIP	Fraction of oil in seeds (g(oil)/g(seed))

Source: DSSAT v4.8; Hoogenboom et al. (2021)

followed the SUCROS2 guidelines (FST). A water balance subroutine was added to the G-E interactions, while the model was effectively calibrated and confirmed in Bangladeshi conditions. The growth and output of the test crop were precisely anticipated by MungGro. Chaki et al. (2019) used a reliable process for parameterizing, calibrating, and validating the model before conducting statistical analyses to assess the model's performance. For this, they used a variety of tillage, residue, N rates, and irrigation techniques in identical field settings. The confidence in the APSIM's ability to accurately represent various management techniques in rice-wheat systems comes from its stability in simulating system behavior without requiring model resets. Even though model performance in the full-N case was noticeably better than in the zero-N case, both are still regarded as satisfactory. In zero-N circumstances, biomass output tended to be over-simulated, especially in the first season. This increased in the second season, showing that the model was able to account for starting condition uncertainties, reduce model error, and simulate the system as intended. Timsina et al. (2000) used SWAGMAN Destiny simulation model; this

Table 7.2 Genetic coefficients, description, ranges, and guidelines for model calibration of chickpea in APSIM

Genetic coefficients	Description
x_pp_hi_incr	Photoperiod (h)
y_hi_incr	Rate of HI increase (1/d)
x_hi_max_pot_stress	Average stress at flowering
y_hi_max_pot	Maximum harvest index potential
cum_vernal_days	(d)
tt_emerg_to_endjuv	TT from emergence to the end of juvenile phase (°Cd)
est_days_emerg_to_init	Estimated days from emergence to floral initiation (d)
x_pp_endjuv_to_init	Photoperiod (h)
y_tt_endjuv_to_init	TT from the end of juvenile to floral initiation (°Cd)
x_pp_init_to_fower	Photoperiod (h)
Y_tt_init_to_fower	TT from initiation to flowering (°Cd)
x_pp_fower_to_start_grain	Photoperiod (h)
y_tt_fower_to_start_grain	TT from flowering to start grain fill (°Cd)
x_pp_start_to_end_grain	Photoperiod (h)
y_tt_start_to_end_grain	TT from start grain fill to end grain fill (°Cd)
tt_end_grain_to_maturity	TT from end grain fill to maturity (°Cd)
tt_maturity_to_ripe	TT from maturity to harvest ripe (°Cd)
x_stem_wt	Stem weight (g/plant)
y_height	Plant height (mm)

Source: Chauhan et al. (2008, 2017)

study investigated the possible advantages of shallow subsurface drainage on mung bean performance. The model was used to analyze several planting dates and water table scenarios for irrigated mung bean crops. The model was verified for grain and biomass yields for a location in northern Bangladesh. On all planting dates, yields were higher under a deep water table (DWT, 500 cm) than they were under a shallow water table (SWT, 50 cm) with drainage (at a depth of 50 cm). For both water table scenarios, the yields were higher in drained fields for planting on April 30, but higher in undrained fields for planting on March 15 of shallow subsurface drainage on mung bean performance. When shallow drainage was given for rainfed crops, yields were also higher. In every case, either a high water deficiency stress or a waterlogging stress was linked to smaller yields. The findings suggest that controlling water tables and choosing the best planting period might boost mung bean yield in rice-wheat systems in Bangladesh. Rahman et al. (2013) used both deterministic and ARIMA models, which have been best fitted for forecasting, and predictions have been created using best fitted deterministic and ARIMA models. According to the study, compared to mung bean productivity (CV of 52.87%), grass pea productivity was shown to be comparatively steady (CV of 44.38%). Throughout the research period, the growth rates of the production of mung bean and grass pea pulses varied from −21.55% to 20.26% and −19.51% to 14.44%, respectively. The production forecast demonstrated mung bean pulse's important potential in Bangladesh. Hence, policymakers, stakeholders, and receivers should act quickly to stop the drop in the region planted with sugarcane. The genetic coefficients of chickpea crop in APSIM models are presented in Table 7.3.

Table 7.3 Genetic coefficients, description, ranges, and guidelines for model calibration of mung bean in APSIM

Genetic coefficients	Description
x_pp_hi_incr	h
y_hi_incr	Rate of HI increase (optional) 1/day
x_hi_max_pot_stress	Maximum harvest index potential
tt_emergence	TT from emergence to the end of juvenile phase (°Cd)
est_days_emerg_to_init	Estimated days from emergence to floral initiation h
x_pp_end_of_juvenile	Photoperiod (d)
y_tt_end_of_juvenile	TT from the end of juvenile to floral initiation
x_pp_floral_initiation	(h)
y_tt_floral_initiation	TT from initiation to flowering (°Cd)
x_pp_flowering	(h)
y_tt_flowering	TT from flowering to start grain fill (°Cd)
x_pp_start_grain_fill	(h)
y_tt_start_grain_fill	(°Cd)
tt_end_grain_fill	(°Cd)
tt_maturity	TT from maturity to harvest ripe (°Cd)
x_stem_wt	Stem weight (g/plant)
y_height	Plant height (mm)

7.4 Climate Change and Temperature Trend

Temperature is increasing in chickpea-mung bean cropping system in most of the world. Koocheki et al. (2006) simulated climate change in chickpea-mung bean cropping system for Tabriz in Iran. Models from GFDL and GISS were used to forecast monthly average temperatures and rainfall under a CO_2 doubling scenario. A crop simulation model (OSBOL) was used to measure the influences of climate change on crop growth time, yield components, and water needs of various crops. Tabriz's average annual temperature upsurged by 4.6 and 4.3 °C, respectively, with a doubling of CO_2 concentration, according to GISS and GFDL estimations. According to the GISS findings, over the first 5 months of the year, doubling CO_2 will result in a rise in mean temperature of 3.7 °C and an increase in mean rainfall of 40%. Yet according to GFDL, over the first 5 months of the year, the mean temperature will rise by 4.7 °C, and the mean rainfall will plummet by 12%. Mubvuma (2018) indicated that a substantial rise in temperature ranged from 4.20 to 5.50 °C from 1950 to 2100 in Southern Africa. Fatima et al. (2021) studied the influence of climate change on chickpea and mung bean system in Pakistan. Increasing thermal trend was indicated in phases of phenology of chickpea and mung bean from 1980 to 2018 for entirely studied locations (Fig. 7.3). The temperature tendency ranged from 0.57 to 0.95 °C for sowing to anthesis phase and 0.71 to 0.98 for anthesis to maturity phase, whereas for S-M phase, it ranged from 0.82 to 1.15 °C/decade for gram (Fig. 7.4). Average chickpea values for S-A, A-M, and S-M phases were 0.78, 0.86, and 0.97 °C/decade, respectively. Similarly ranging values for temperature tendency for mung bean were from 0.44 to 0.78, 0.48 to 0.91, and 0.64 to 0.97 °C/decade for S-A, A-M, and S-M phases, respectively (Fig. 7.4). Averaged values for

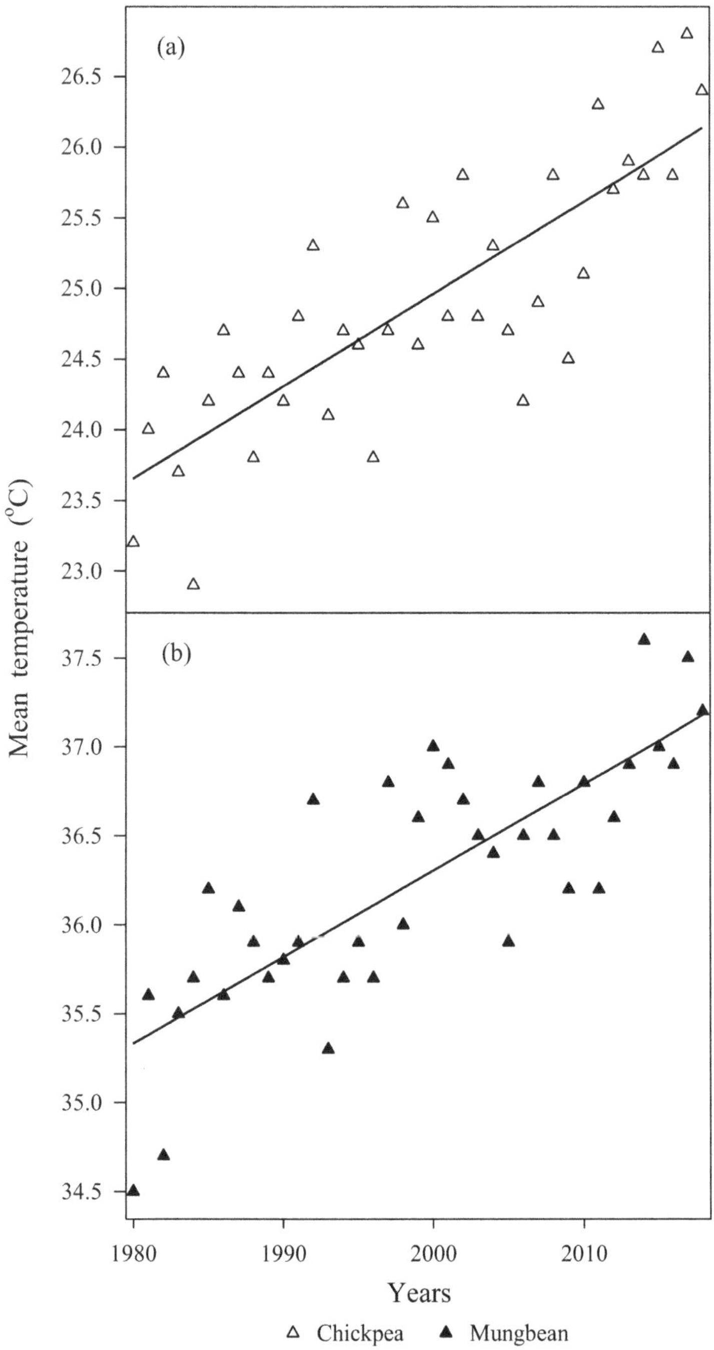

Fig. 7.3 Mean temperature trends in chickpea (**a**) and mung bean (**b**) crops from 1980 to 2018 in Punjab, Pakistan. (Source: Fatima et al. 2021)

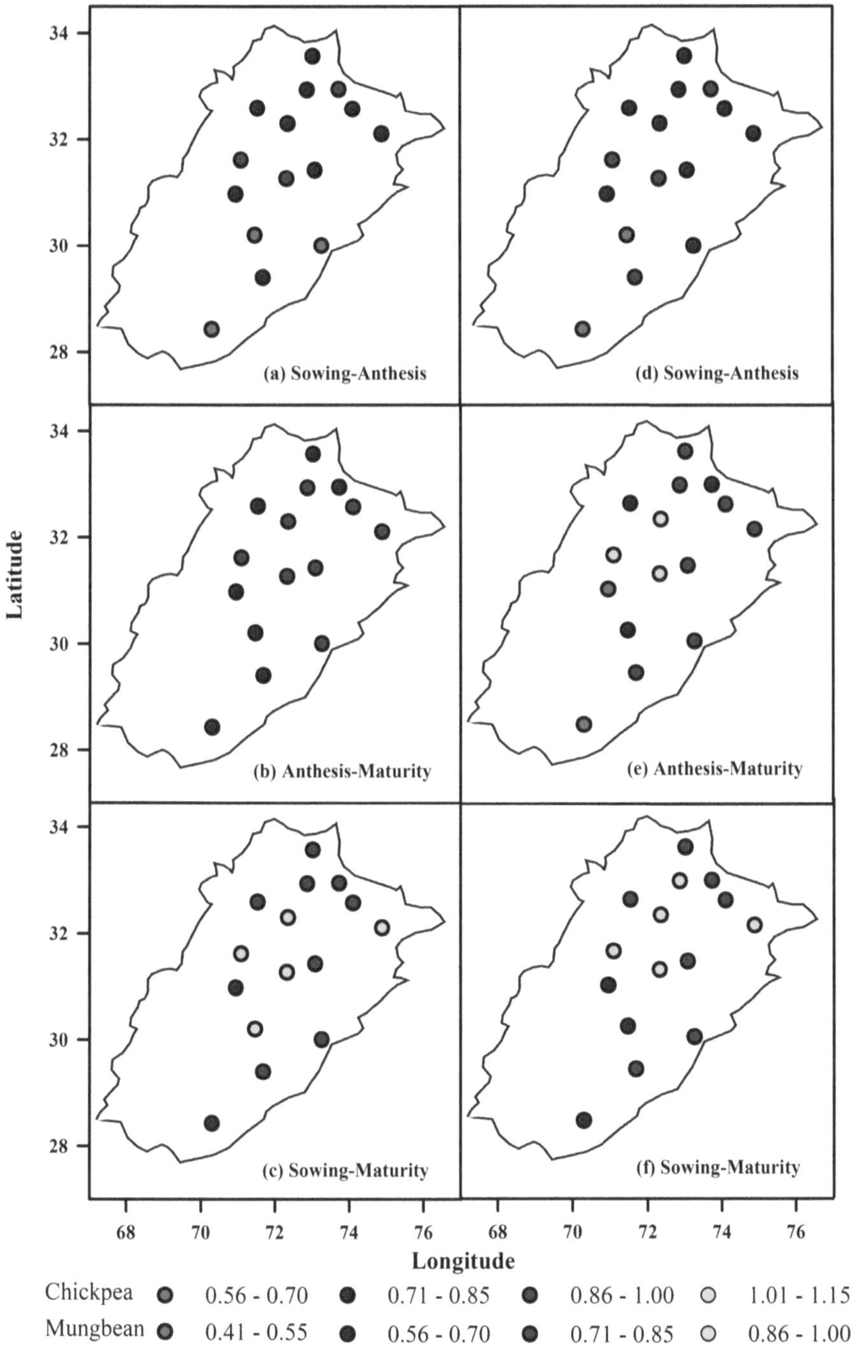

Fig. 7.4 Observed trends in mean temperature during the phenological phases {sowing-anthesis (**a, d**); anthesis-maturity (**b, e**); and sowing-maturity (**c, f**)} for chickpea (**a–c**; left) and mung bean (**d–f**; right) crops from 1980 to 2018 in Punjab, Pakistan. (Source: Fatima et al. 2021)

temperature tendency for mung bean for S-A, A-M, and S-M phases were 0.65, 0.73, and 0.81 °C/decade, respectively.

7.5 Climate Change Impact on Chickpea-Mung Bean Morphophysiological Responses

7.5.1 Chickpea

Climate change is predicted to alter the frequency, severity, and periods of climatic extremes. During the reproductive phase, heat stress has a negative impact on pollen viability, fertilization, pod set, and seed development. As a result, flowers and pods abscise, significantly reducing grain output. One of the most crucial elements influencing plant development is temperature (Liu et al. 2019). The reproductive phase of the plant is more susceptible than the vegetative phase, which frequently results in a drop in production, to extreme temperature changes, both low and high temperatures. Low temperatures cause cold stress, which alters several components of the photosynthetic apparatus and makes photosystems I (PSI) and II (PSII) highly temperature dependent, potentially resulting in photodamage. In plant tissues, cold stress led to increased membrane viscosity, slowed metabolism, delayed energy dissipation, radical production, and oxidative stress (Kumari et al. 2020). Heat stress from high temperatures limits photosynthesis in plants since it is the most thermosensitive element of plant activity, and PSII is more sensitive than PSI. As a result of osmotic adjustment, heat stress increases the accumulation of soluble carbohydrates in the leaves while decreasing the quantities of photosynthetic pigments. With no germination beyond 45 °C, decreased seedling development, and even seedling mortality, high temperatures have an impact on the germination of chickpea seeds (Devasirvatham et al. 2013). Genotypic diversity for high-temperature tolerance at seed germination has been found. Studies in controlled environments revealed that both tolerant and sensitive genotypes significantly increased their biomass at 35/25 °C, but exposure to 40/30 °C resulted in a loss in all genotypes' biomass at maturity, though more so in the sensitive genotypes. All phenological phases of chickpeas are restricted by heat stress; however, the reproductive stage is thought to be more vulnerable for temperature extremes as compared to vegetative stage (Kaushal et al. 2013). In general, heat stress in reproduction (1) decreases flowers, (2) increases flower abortion, (3) changes the numbers of anther locules, (4) results in pollen sterility and poor germination, (5) decreases fertilization along with stigma receptivity, (6) results in ovary abnormality, (7) decreases reallocation of photosynthates in seed, and (8) decreases seeds, seed weight, and productivity. By causing physiological anomalies, pre-anthesis exposure of gram to stress (35/20 °C) decreased anther growth and pollen output, along with fertility. High temperatures can cause structural abnormalities in anthers as well as pollens, like changes in anther locule quantity, thickening of the anther epidermal wall, and pollen sterility, which are major contributors to lowering chickpea output (Devi et al. 2022). In comparison to the female gametophyte, gram pollens are more susceptible toward heat stress. Chickpea stigma function and pollen fertility suffer at temperatures over 45 °C. Chlorophyll content and photosynthetic rate, along with membrane stability of leaf tissues, are some essential physiological characteristics that could be employed as markers of sensitivity (Ceylan et al. 2013). Gram is substantially more sensitive regarding membrane integrity along with photosystem II functions at higher temperatures of 50 °C for 48 h as compared to other legumes. By destroying the chloroplasts and harming the chlorophyll, stress causes leaf senescence in gram plants. Chlorosis results from heat stress, which decreases chlorophyll contents of gram leaf. This loss may happen. A heat-sensitive gram genotype suffered greater leaf damage under heat stress (>32/20 °C) than a heat-tolerant genotype (Zeitelhofer et al. 2022). This was because the heat-sensitive genotype experienced a greater decline in leaf water status (measured by RLWC), a potential failure in stomatal conductance, along with a restraint in root hydraulic conductivity. At rising temperatures, chickpea transpiration efficiency fell. Chickpea's photosystem II (PSI) activity was unaffected at 35 °C; however, at 46 °C, there was a notable decline that resulted in irreparable harm to the photosynthetic systems. Comparable substantial PS damage was seen in chickpea after 48 h at 50 °C. In the reproductive stage, temperatures exceeding 35 °C decreased photosynthesis, altered metabolic pathways, and reduced grain size (Kumar et al. 2013).

7.5.2 Mung Bean

Temperature stress changes fluidity of the mitochondrial, chloroplast, and plasmalemma membranes that could lead to the breakdown of the lipid bilayer and modify the shape of the protein and result in protein unfolding. Moreover, heat stress causes the creation of ROS, which harms other parts of the photosynthetic system and impairs metabolic function (HanumanthaRao et al. 2016). Compared to photosynthesis, respiration is more temperature sensitive. The cellular oxidizing capacity of gram plant during the vegetative stage significantly decreased at 45/35 °C, indicating inefficient respiration and energy production, potentially as a result of enzyme inactivation. In chickpea leaves during the reproductive phase, sucrose metabolism was impaired by the suppression of sucrose-synthesizing enzyme at high temperatures (Saleh et al. 2007). Because of this, sucrose supply to flowers in temperature-sensitive cultivars was significantly reduced, which had an impact on the functional and developmental characteristics of pollens and caused poor fertilization and pod set. During anthesis until maturity, higher temperature lowered starch deposit in gram seeds due to decreased ADP-glucose pyrophosphorylase along with starch synthase activity, which in turn caused a decrease in grain weight. Around the period of blooming, the chickpea is subjected to temperature stress, which has a negative impact on development and a number of metabolic activities and changes redox cell state (Khattak et al. 2006). ROS formation results in oxidative loss to essential biological constituents like lipids, nucleic acids, proteins, and pigments, along with enzymes at high temperatures. Alkoxyl radicals, hydroxyl radicals, superoxide, and other free radicals as well as non-radicals such as hydrogen peroxide and singlet oxygen make up the ROS-induced oxidative damage (Kareem et al. 2022).

7.6 Impact of Climate Change on Phenology

At higher temperatures, phenological parameters like days to first flowering, days to 50% flowering, and days to crop maturity have a significant effect. Plants under stress are driven to mature by fleeing the heat. Hence, phenological features can be used to detect earliness. Under heat stress, a significant fluctuation in chickpea phenology was seen (Browning et al. 2019). Fatima et al. (2021) studied the influence of climate change on phenology on gram-mung bean system in Pakistan. Due to the need for thermal time (increasing degree days °C) to reach any developmental stage, days to 50% blooming in particular were postponed and days to crop maturity were accelerated. At the local, regional, and global levels, climate change is having an effect on the phenology, growth, and productivity of many agricultural systems. In order to develop solutions for the chickpea-mung bean system (CMS) to lessen the negative effects of climate change, long-term observed phenological variations in the CMS were utilized to determine the link between crop practices, climate warming, and phenology. For chickpeas and mung beans, respectively, the observed thermal tendency from planting to maturity ranged from 0.82 to 1.15 °C/decade between 1980 and 2018. Figures 7.5 and 7.6 indicate that observed gram phenological stages were earlier by average 7.04 for sowing, 6.76 for emergence, 4.31 for anthesis, and 2.15 for maturity days/decade, while gram phases were reduced by 2.73 for sowing to anthesis, 2.16 for anthesis to maturity, and 4.89 for sowing to maturity days per decade. For mung bean, sowing 6.24, emergence 5.97, flowering 3.76, and maturity 2.01 days/decade happened earlier. Duration of mung bean phenological phases was shortened with an average 2.45 for sowing to anthesis, 1.76 for sowing to maturity, and 4.23 for anthesis to maturity days per decade, respectively (Fig. 7.7). For mung bean, regression coefficient for observed stages and phases with temperature is characterized in Table 7.4. Chickpea and mung bean cultivars had increased thermal time requirements for phenological phases, which ranged from 72 to 104 and 85 to 124 °C days/decade at entire locations (Fig. 7.8). Phenology stages as well as phases for both crops, gram and mung bean, associated negatively with increasing

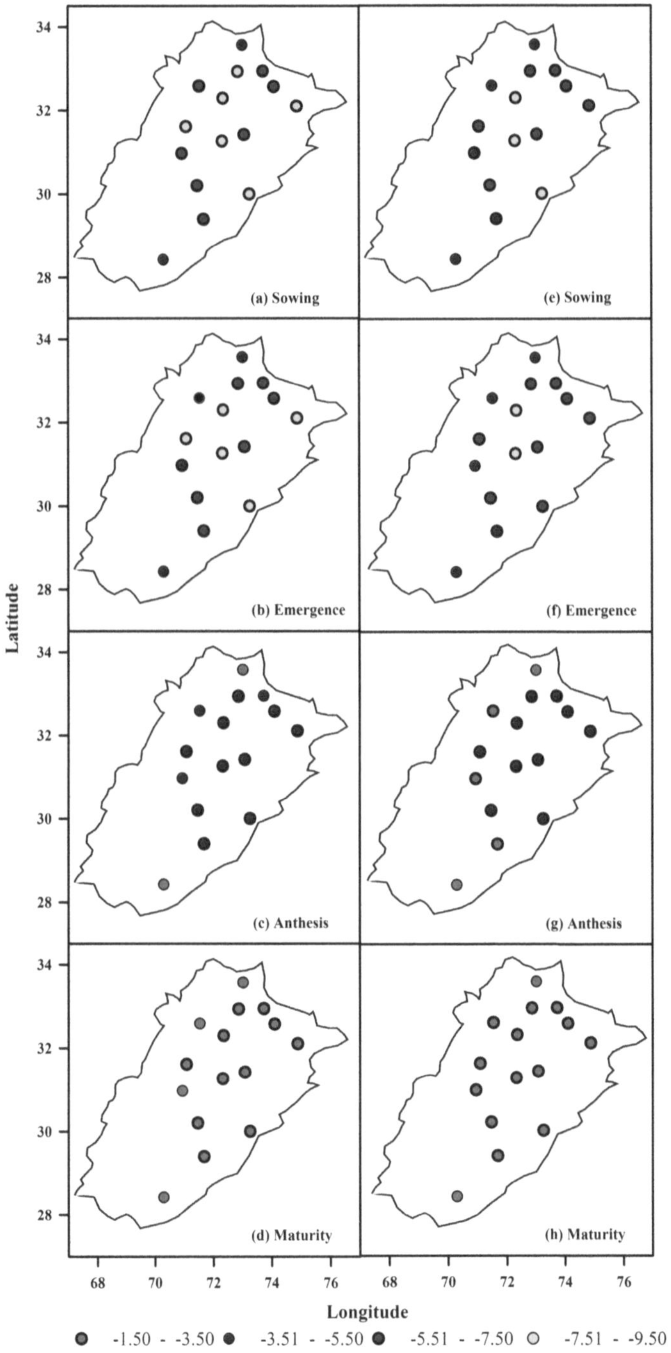

Fig. 7.5 Observed trends in phenology of chickpea (**a–d**) and mung bean (**e–h**) [sowing, emergence, anthesis, and maturity]; circles with black borderline indicate statistically significant trends at $p = 0.05$ levels from 1980 to 2018 in Punjab, Pakistan. (Source: Fatima et al. 2021)

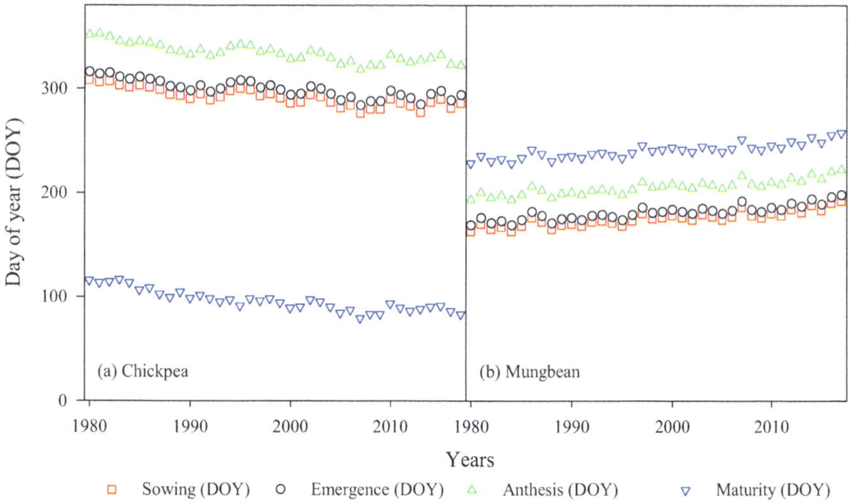

Fig. 7.6 Time series plots of farmer field observed dates of sowing, emergence, anthesis, and maturity of chickpea (**a**) and mung bean (**b**) crops in Punjab, Pakistan. (Source: Fatima et al. 2021)

temperature at the whole locations studied (Fig. 7.9). CROPGRO model-projected phenology was more sensitive in comparison to observed for gram and mung bean (Table 7.5). CROPGRO-Chickpea and CROPGRO-Legume models for common gram along with mung bean cultivars over a period of 38 years at the locations showed that the simulated phenology stages were accelerated with a greater thermal trend than the stages actually observed. Research demonstrated that, over the past few decades, planting newly developed pulse cultivars with higher thermal time requirements largely compensated changes in chickpea (33%) and mung bean (20%) phenology caused by rising temperature. Hence, newly developed cultivars of chickpea and mung bean that require more degree days and have better temperature tolerance must be acquainted in order to mitigate the effects of global warming (Tariq et al. 2017, 2018a, b; Tarriq et al. 2021; Hussain et al. 2018; Khan et al. 2019, 2020; Zahoor et al. 2022).

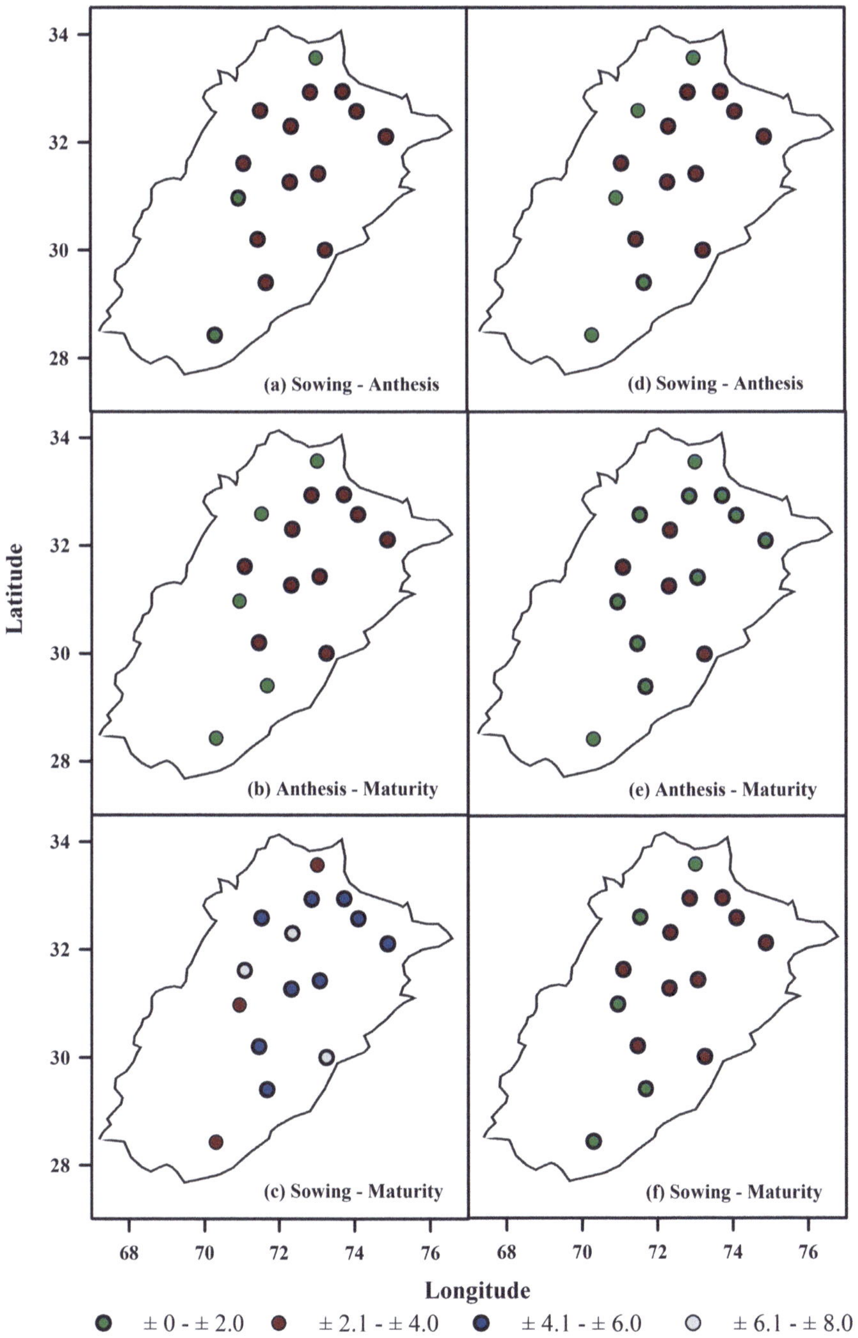

Fig. 7.7 Observed trends in phenology {(**a, d**) sowing-anthesis; (**b, e**) anthesis-maturity; and (**c, f**) sowing-maturity; circles with black borderline indicate statistically significant trends at $p = 0.05$ levels} of chickpea (**a–c**; left) and mung bean (**d–f**; right) from 1980 to 2018 in Punjab, Pakistan. (Source: Fatima et al. 2021)

Table 7.4 Average observed phenology of chickpea and mung bean in Punjab, Pakistan, during the period of 1980–2018

Site name	Sowing	Emergence	Anthesis[a]	Maturity[b]	S-A	A-M	S-M
Chickpea							
Rahim Yar Khan	290 ± 2.1	297 ± 3.5	329 ± 5.1	82 ± 2.9	39 ± 5.2	118 ± 9.3	157 ± 6.8
Bahawalpur	297 ± 5.8	304 ± 4.9	333 ± 4.9	75 ± 5.3	36 ± 6.3	107 ± 6.4	143 ± 7.1
Bahawalnagar	284 ± 4.5	291 ± 4.1	315 ± 8.3	76 ± 3.7	31 ± 5.9	126 ± 3.7	157 ± 5.3
Multan	293 ± 8.1	300 ± 7.2	326 ± 7.9	84 ± 8.5	33 ± 4.2	123 ± 2.5	156 ± 4.9
Layyah	291 ± 6.5	298 ± 6.9	326 ± 6.3	83 ± 4.9	35 ± 8.5	122 ± 3.8	157 ± 7.3
Jhang	304 ± 7.2	311 ± 5.8	335 ± 2.7	72 ± 5.2	31 ± 6.1	102 ± 5.1	133 ± 5.8
Faisalabad	281 ± 3.7	288 ± 2.9	324 ± 6.2	81 ± 5.9	43 ± 9.0	122 ± 5.8	165 ± 6.4
Bhakkar	285 ± 4.3	292 ± 3.6	318 ± 9.3	80 ± 6.4	33 ± 4.2	127 ± 4.6	160 ± 5.2
Narowal	282 ± 9.2	289 ± 8.5	320 ± 8.4	87 ± 3.3	38 ± 5.8	132 ± 4.1	170 ± 7.3
Khushab	298 ± 2.6	305 ± 3.1	333 ± 6.5	81 ± 4.5	35 ± 5.2	113 ± 5.2	148 ± 6.2
Gujrat	307 ± 3.1	314 ± 2.7	336 ± 4.9	69 ± 7.6	49 ± 2.8	088 ± 7.2	127 ± 7.3
Mianwali	304 ± 5.9	311 ± 4.3	337 ± 2.7	79 ± 8.3	33 ± 4.1	107 ± 6.2	140 ± 5.2
Chakwal	278 ± 8.5	285 ± 8.1	320 ± 4.1	85 ± 6.5	42 ± 5.9	130 ± 4.9	172 ± 9.4
Jhelum	287 ± 4.2	294 ± 3.9	323 ± 7.3	89 ± 3.8	36 ± 3.1	131 ± 5.2	167 ± 8.2
Rawalpindi	300 ± 6.7	307 ± 5.2	331 ± 4.2	76 ± 4.1	31 ± 2.9	110 ± 8.2	141 ± 5.8
Mung bean							
Rahim Yar Khan	178 ± 3.2	184 ± 2.9	212 ± 6.2	241 ± 5.1	34 ± 4.2	29 ± 8.3	63 ± 4.1
Bahawalpur	185 ± 4.8	191 ± 4.1	216 ± 3.5	241 ± 4.7	31 ± 2.5	25 ± 7.4	56 ± 3.6
Bahawalnagar	172 ± 8.1	177 ± 7.8	198 ± 4.9	229 ± 3.5	26 ± 5.4	31 ± 9.2	57 ± 7.4
Multan	181 ± 6.1	188 ± 5.7	209 ± 2.5	246 ± 8.3	28 ± 6.7	37 ± 7.4	65 ± 5.3
Layyah	179 ± 7.3	185 ± 6.9	209 ± 6.1	243 ± 4.9	30 ± 4.0	34 ± 4.2	64 ± 3.5
Jhang	192 ± 3.6	197 ± 2.8	218 ± 7.2	245 ± 3.7	26 ± 3.2	27 ± 3.6	53 ± 8.2
Faisalabad	169 ± 4.1	165 ± 3.6	207 ± 4.6	231 ± 6.2	38 ± 5.7	24 ± 6.1	62 ± 5.1
Bhakkar	173 ± 2.9	178 ± 2.5	201 ± 3.9	234 ± 2.9	28 ± 3.8	33 ± 5.8	61 ± 6.9
Narowal	170 ± 6.4	173 ± 5.2	203 ± 6.4	238 ± 4.1	33 ± 7.4	35 ± 4.1	68 ± 4.2

Khushab	186 ± 5.1	191 ± 4.7	216 ± 3.8	248 ± 7.3	30 ± 5.9	32 ± 3.5	62 ± 7.6
Gujrat	195 ± 8.5	199 ± 6.9	219 ± 7.4	250 ± 6.9	24 ± 9.1	31 ± 4.8	55 ± 6.2
Mianwali	192 ± 4.9	197 ± 3.5	220 ± 9.3	252 ± 5.2	28 ± 4.7	32 ± 8.1	60 ± 5.8
Chakwal	166 ± 5.3	172 ± 5.6	203 ± 6.1	232 ± 3.1	37 ± 3.1	29 ± 7.4	66 ± 9.3
Jhelum	175 ± 2.8	181 ± 4.1	206 ± 5.4	245 ± 6.3	31 ± 5.3	39 ± 5.3	70 ± 4.7
Rawalpindi	188 ± 5.7	193 ± 3.8	214 ± 2.5	245 ± 4.7	26 ± 5.9	31 ± 7.2	57 ± 5.3

Source: Fatima et al. (2021)

[a]50% Anthesis

[b]50% Physiological maturity

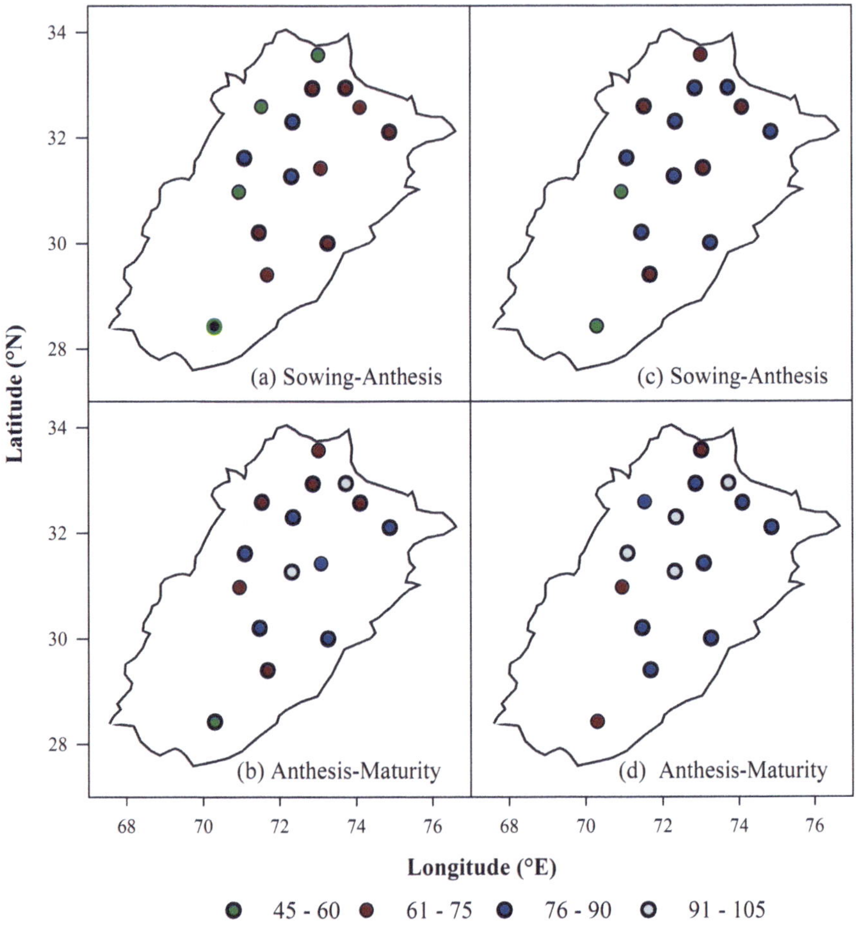

Fig. 7.8 Observed trends in thermal time required for chickpea (**a, b**; left) and mung bean (**c, d**; right) for sowing-anthesis (**a**) and anthesis-maturity (**b**); circles with black border indicate statistically significant trends at $p = 0.05$ levels in Punjab, Pakistan. (Source: Fatima et al. 2021)

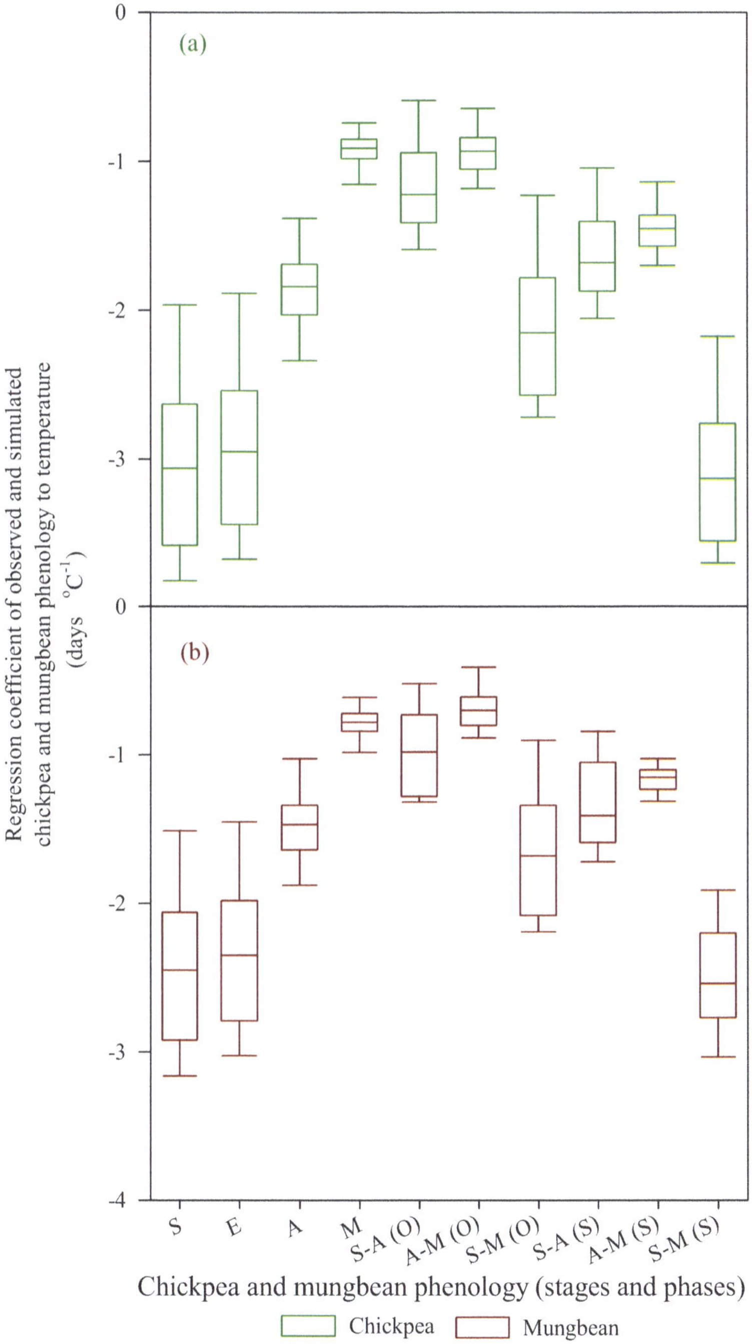

Fig. 7.9 Observed phenology (stages and phases) and simulated phenology (phases) versus temperature trends of chickpea (**a**) and mung bean (**b**) crop cultivars at each location in Punjab, Pakistan, from 1980 to 2018. (*S* sowing, *E* emergence, *A* anthesis, *M* maturity, *S-A* sowing-anthesis, *A-M* anthesis-maturity, *S-M* sowing-maturity, *O* observed, *S* simulated). The central line represents median, box limits are 25th and 75th percentiles, and whiskers represent the minimum and maximum values. (Source: Fatima et al. 2021)

Table 7.5 Summary of observed phenology response to temperature for chickpea and mung bean in Punjab, Pakistan, for 1980–2018

Phenology	No. of neg.[a]	No. of pos.[b]	No. of sig. neg.[c]	No. of sig. pos.[d]	Reg. mean[e] (days $^{\circ}C^{-1}$)
Chickpea observed stages and phases					
Sowing	15	0	14	0	−3.03
Emergence	15	0	13	0	−2.91
Anthesis	15	0	11	0	−1.85
Maturity	15	0	12	0	−0.92
Sowing-anthesis	15	0	14	0	−1.17
Anthesis-maturity	15	0	10	0	−0.93
Sowing-maturity	15	0	12	0	−2.10
Chickpea simulated phases					
Sowing-anthesis	15	0	12	0	−1.63
Anthesis-maturity	15	0	14	0	−1.44
Sowing-maturity	15	0	15	0	−3.08
Mung bean observed stages and phases					
Sowing	15	0	13	0	−2.43
Emergence	15	0	12	0	−2.33
Anthesis	15	0	14	0	−1.47
Maturity	15	0	14	0	−0.78
Sowing-anthesis	15	0	11	0	−0.96
Anthesis-maturity	15	0	13	0	−0.68
Sowing-maturity	15	0	14	0	1.65
Mung bean simulated phases					
Sowing-anthesis	15	0	14	0	−1.33
Anthesis-maturity	15	0	12	0	−1.16
Sowing-maturity	15	0	13	0	−2.50

Source: Fatima et al. (2021)
[a]Number of locations with negative regression coefficients
[b]Number of locations with positive regression coefficients
[c]Number of locations with significant negative regression coefficients
[d]Number of locations with significant positive regression coefficients
[e]Mean of regression coefficients

7.7 Impact of Climate Change on Current and Future Scenarios of Crop Production

7.7.1 Chickpea

Chickpea production is extremely sensitive due to climate change (Mar et al. 2018). Chickpea production is under danger as a result of climate change in food insecure areas. Table 7.6 indicates that chickpea-mung bean cropping system is affected due to climate change across the world. Yield of chickpea and mung bean is/can be reduced as a result of increasing climate warming in current and future scenarios. Kadiyala et al. (2016) studied the impact of climate change on chickpea in India, in which the sequence analysis tool of DSSAT model was used to research how future climate change may affect chickpea productivity. The findings showed that, in comparison to the baseline environment, the expected climatic change by 2069 (during the mid-century period) will reduce chickpea output by 4.3–18.6% across the several areas evaluated. According to the simulation study's results on yield gains from several adaptation methods, moving forward the sowing window by one fortnight and applying a critical irrigation at 60 days after sowing were both shown to be helpful in raising chickpea yields under a climate change scenario. Kadiyala et al. (2016), in their study in India, revealed that in comparison to the current environment, the predicted climatic change by 2069 (the mid-century timeframe) will result in a 4.3–18.6% reduction in chickpea production throughout the investigated

Table 7.6 Comparison of the responses of chickpea and mung bean phenology with average temperature using the observed and simulated data in Punjab, Pakistan, during 1980–2018

Phenology	Regression coefficient (days °C^{-1})[a] Obs. data	Sim. data	Difference between obs. and sim. data (days °C^{-1})[b]	t-Test (p-value)
Chickpea				
Sowing-anthesis	−1.17	−1.63	0.46	0.091**
Sowing-maturity	−0.93	−1.44	0.52	0.0075**
Anthesis-maturity	−2.10	−3.08	0.97	0.0008**
Mung bean				
Sowing-anthesis	−0.96	−1.33	0.37	0.0024**
Anthesis-maturity	−0.68	−1.16	0.48	0.0037**
Sowing-maturity	−1.65	−2.50	0.85	0.0019**

Source: Fatima et al. (2021)
**Significant at the 0.01 probability level
[a]Mean of regression coefficients
[b]Duration of sowing to maturity

regions. Singh et al. (2014) used CSM-CROPGRO-Chickpea model. They studied the impact of climatic changes on the production of chickpea at certain locations in South Asia along with East Africa. They also explored possible reimbursements of including heat tolerance characters in gram crop with the help of gram model besides virtual cultivar way. Climate change by 2050 (including CO_2) raised chickpea yields by 17% at Hisar and Indore, 18% at Zaloke, 25% at Debre Zeit, and 18% at Kabete compared to the baseline meteorological data, while yields declined by 16% at Nandhyal and 7% at Ukiriguru. When compared to yields under the current atmospheric CO_2 concentration, the yield gain from increased CO_2 by 2050 ranged from 7% to 20% across locations; however, variations in temperature as well as precipitation had an opposite effect on yield across the locations. Koocheki et al. (2006) studied under simulated climate change in Tabriz, Iran. The OSBOL model under GDFL projections found that yield, biomass, and water requirements for irrigated sunflower increased, and the time from planting to maturity and harvest index decreased. With rainfed sunflower, the biomass increased, but the harvest index and days to maturity decreased. For irrigated chickpea, doubling carbon dioxide resulted in a decrease in days to maturity, biomass, yield, and water needs. Climate change has decreased the number of days needed for rainfed chickpeas to reach maturity while increasing harvest index, biomass, and yield. Mohammed et al. (2017) evaluated the effects of predicted climate change in Northeastern Ethiopia for the time periods of 2030 and 2050 under all of the RCPs, both with and without CO_2 fertilization. According to the results of the climate change effect analysis on chickpea, under CO_2 fertilization conditions across all the RCPs, grain yield is expected to dramatically rise both by 2030 and 2050 compared to baseline (1961–1990) grain yield. Grain yield, however, would not considerably rise between 2030 and 2050 across all scenarios, according to simulations conducted without CO_2. It may be assumed, in general, that chickpea will profit from the anticipated climatic changes in northern Ethiopia based on the projection results. Meghdadi and Hajarpoor (2015) assessed the impact of climate change in Iran. The results show that, despite the average yield increase by 33.4%, the yield stability over years decreased (an increase of 21.2% in variation coefficient). The increase in yield appears to have been caused by the good impact of doubling carbon dioxide on photosynthesis and the ensuing reduction in transpiration. The findings further demonstrated that rising temperatures were the primary driver of changes across years. The county with the largest reaction to the future climate was Zanjan, where grain yield increased by 46.5%, while yield stability decreased by 37.7%.

7.7.2 Mung Bean

Higher temperature along with waterlogging decreased development and productivity of mung bean. Elevated CO_2 rates minimized temperature-associated productivity decrease rate (Basu et al. 2016). Biswas et al. (2018) indicated that the grain production of mung bean was significantly reduced by 12% in normal soil moisture conditions due to an increase in temperature under the same CO_2 level. If CO_2 levels climb to 650 ppm, the 1 °C temperature increase-related drop in grain yield is predicted to be made up. Fu et al. (2016) used the CSM-CROPGRO model to analyze the relationship between changes in yield and the combined effects of CO_2, temperature, and precipitation. They found that grain yield was predicted to decrease under the climate combination that included the extremely high temperature of +7.4 °C, and yield increased as a result of elevated CO_2 and precipitation. Ahmad et al. (2022) applied the validated CSM-CROPGRO model to simulate the growth and yield of peanut cultivars under temporal variation. They suggested that Peanut sowing at mid-April can be recommended for the farmers to avoid environmental stress and appropriate use of available resources for sustainable yield. Results depicted the potential of model for selection of appropriate genotype, planting dates and other peanut management practices in the region. The quantification of the influence of predicted changes in climatic parameters, such as atmospheric CO_2, temperature, and rainfall, on mung bean crop yield was investigated. The influence of climate change on grain yield ranged from 7.5% to 21.3% depending on the treatment. Cultivar I3VS2 was predicted to have the biggest yield drop and I1VS2 to have the lowest, with a mean yield reduction of 7.5% due to climate warming condition. Lateff et al. (2018) revealed that raising the minimum and maximum temperatures by 1 °C resulted in a 10.2% and 9.8% per plant and per hectare decrease in

Table 7.7 Seed yield reduction/increase in chickpea-mung bean cropping systems across various regions in the world

Country/ continent	Model	Climatic scenarios	Seed yield reduction/ increase (%)	References
India	DSSAT	2039–2069	19	Kadiyala et al. (2016)
India	CROPGRO	2015–2050	17	Singh et al. (2014)
Myanmar	CROPGRO	2015–2050	18	Singh et al. (2014)
Ethiopia	CROPGRO	2015–2050	25	Singh et al. (2014)
Kenya	CROPGRO	2015–2050	18	Singh et al. (2014)
Tanzania	CROPGRO	2015–2050	7	Singh et al. (2014)
Iran	OSBOL	2065–2095	13	Koocheki et al. (2006)
Ethiopia	DSSAT	2039–2069	45	Mohammed et al. (2017)
South Africa	AquaCrop	1950–2100	16	Mubvuma (2018)
Iran	APSIM	2015–2050	47	Meghdadi and Hajarpoor (2015)
Bangladesh	MungGro	2001–2017	16	Biswas et al. (2018)
India	CROPGRO	2071–2100	21	Karande et al. (2018)
Egypt	APSIM	2020–2050	10	Lateff et al. (2018)

seed production, respectively. According to this study, Egypt's mung bean growing requires a specific heat regime. The mung bean plant has demonstrated a high sensitivity to changes in minimum and maximum temperatures. The heat regime

(22–32 °C) is the one that is most beneficial for the cultivation of mung beans in Egypt (Table 7.7).

7.8 Adaption Strategies

Changes in sowing dates, cultivation of cultivars resistant to drought, and development of new cultivars demanding higher degree days in chickpea and mung bean cropping system are some crucial techniques to reduce the threat posed by climatic unpredictability and improve agricultural plants' ability to adapt for the assurance of food security (Bahl 2015). With the use of crop management strategies that can improve crop development under a variety of environmental challenges, we can also help plants adapt to their environment. To combat weather pressures, it is important to choose the right sowing time, planting density, and watering procedures in chickpea and mung bean cropping system. In cropping systems or growing environments when the crop is subjected to heat stress in the reproductive period, gram and mung bean cultivars with improved heat tolerance will reduce yield losses. Kadiyala et al. (2016) made adaptation strategies for chickpea in India. According to the simulation study's results on yield gains from several adaptation methods, moving forward the sowing window by one fortnight and applying a critical irrigation at 60 days after sowing were both shown to be helpful in raising chickpea yields under a climate change scenario. Singh et al. (2014) revealed that up to 22% of sites' yields benefited from drought resistance under historic along with climate change situations. In baseline environment, heat tolerance improved chickpea output by 9% at Hisar and Indore and 13% at Hisar, Indore, Nandhyal, and Ukiriguru. Mohammed et al. (2017) revealed that rendering to the simulation consequences, given the current and projected climatic conditions, the use of two supplemental irrigation, at flower initiation and pod setting phases, and early seeding greatly increases grain productivity of chickpea in Northeastern Ethiopia. To boost chickpea productivity in the current and future climatic conditions, it is crucial to choose suitable cultivars based on the local agroecology. Khatun et al. (2021) indicated that among the main adaptation options to climate change, crop production can be increased by adoption of higher fertilizer application rates, crop diversification, changes in the amount of land used for pulse farming, land fragmentation, relay cropping, and seed treatment.

7.9 Conclusion

Climate change and the chickpea-mung bean cropping system are intimately connected in several ways, as climate change is the main reason for heat stress that harms the chickpea-mung bean cropping system. Climate change affects the land and its chickpea-mung bean cropping system in a variety of ways, including variabilities in yearly precipitation; mean temperature; heat waves; shifts in weeds, pests, or microbes; global changes in air carbon dioxide or ozone level; as well as sea-level variations. The prospect of variable global climate has piqued scientists' interest, as these changes are having a detrimental influence on global agricultural productivity and jeopardizing world food security. Depending on some forecasted projections, the chickpea-mung bean cropping system is an endangered activity that is being harmed by climate change. Food security and ecosystem resilience are currently the most pressing global concerns. Climate-smart chickpea-mung bean cropping systems are the only method to reduce the damaging effect of climatic fluctuations on crop adaptability before they have a significant influence on global crop yields.

References

Abbas Q, Ahmad S (2018) Effect of different sowing times and cultivars on cotton fiber quality under stable cotton-wheat cropping system in southern Punjab, Pakistan. Pak J Life Soc Sci 16(2):77–84

Abbas G, Ahmad S, Ahmad A, Nasim W, Fatima Z, Hussain S, Rehman MH, Khan MA, Hasanuzzaman M, Fahad S, Boote KJ, Hoogenboom G (2017) Quantification the impacts of climate change and crop management on phenology of maize-based cropping system in Punjab, Pakistan. Agric For Meteorol 247:42–55

Abbas G, Fatima Z, Riaz F, Iqbal P, Hussain S, Zakir I, Iqbal M, Haque E, Ahmad S (2019) Study of sowing dates and hybrids effect in maize-based cropping system under arid conditions of Southern Punjab, Pakistan. Pak J Life Soc Sci 17(1):24–31

Abbas G, Ahmad S, Hussain M, Fatima Z, Hussain S, Iqbal P, Ahmed M, Farooq M (2020a) Sowing date and hybrid choice matters production of maize-maize system. Int J Plant Prod 14:583–595

Abbas G, Fatima Z, Hussain M, Hussain S, Atique-ur-Rehman, Sarwar N, Ahmed M, Ahmad S (2020b) Nitrogen rate and hybrid selection matters productivity of maize-maize cropping system under irrigated arid environment of southern Punjab, Pakistan. Int J Plant Prod 14:309–320

Abbas Q, Atique-ur-Rehman SN, Hussain S, Abbas G, Khan MN, Fatima Z, Naz S, Younis H, Ahmed M, Ullah H, Iqbal P, Iqbal M, Kan M, Ahmad S (2020c) Effect of different sowing times and cultivars on wheat grain quality under cotton-wheat cropping system in Southern Punjab, Pakistan. Pak J Life Soc Sci 18(1):41–50

Ahmad S, Hasanuzzaman M (2012) Integrated effect of plant density, N rates and irrigation regimes on the biomass production, N content, PAR use efficiencies and water productivity of rice under irrigated semiarid environment. Notulae Botanicae Horti Agrobotanici Cluj-Napoca 40(1):201–211

Ahmad S, Ahmad A, Soler CMT, Ali H, Zia-Ul-Haq M, Anothai J, Hussain A, Hoogenboom G, Hasanuzzaman G (2012a) Application of the CSM-CERES-Rice model for evaluation of plant density and nitrogen management of fine transplanted rice for an irrigated semiarid environment. Precis Agric 13(2):200–218

Ahmad S, Ali H, Ismail M, Shahzad MI, Nadeem M, Anjum MA, Zia-Ul-Haq M, Firdous N, Khan MA (2012b) Radiation and nitrogen use efficiencies of C_3 winter cereals to nitrogen split application. Pak J Bot 44(1):139–149

Ahmad S, Ahmad A, Ali H, Hussain A, Garcia y Garcia A, Khan MA, Zia-Ul-Haq M, Hasanuzzaman M, Hoogenboom G (2013) Application of the CSM-CERES-Rice model for evaluation of plant density and irrigation management of transplanted rice for an irrigated semiarid environment. Irrig Sci 31(3):491–506

Ahmad A, Ashfaq M, Rasul G, Wajid SA, Khaliq T, Rasul F, Saeed U, Rahman MH, Hussain J, Baig IA, Naqvi SAA, Bokhari SAA, Ahmad S, Naseem W, Hoogenboom G, Valdivia RO (2015a) Impact of climate change on the rice-wheat cropping system of Pakistan. In: Hillel D,

Rosenzweig C (eds) Handbook of climate change and agro-ecosystems, vol 3. Imperial College Press and the American Society of Agronomy, London, pp 219–258

Ahmad S, Ali H, Atique-ur-Rehman, Khan RJ, Ahmad W, Fatima Z, Abbas G, Irfan M, Ali H, Khan MA, Hasanuzzaman M (2015b) Measuring leaf area of winter cereals by different techniques: a comparison. Pak J Life Soc Sci 13(2):117–125

Ahmad S, Raza I, Muhammad D, Ali H, Hussain S, Dogan H, Zia-Ul-Haq M (2015c) Radiation, water and nitrogen use efficiencies of *Gossypium hirsutum* L. Turk J Agric For 39(5):825–837

Ahmad S, Ali H, Farooq U, Khan SU, Atique-ur-Rehman, Sarwar N, Shahzad AN, Doğan H, Hussain S, Sultan MT, Waheed A, Zia-ul-Haq M, Hussain K, Khan MA (2016a) Improving nitrogen and radiation-use-efficiencies of C_4 summer cereals by split nitrogen applications under irrigated arid environment. Turk J Agric For 40(2):280–289

Ahmad S, Hussain S, Fatima Z, Abbas G, Atique-ur-Rehman, Khan MR, Younis H, Naz S, Sohail M, Ajmal M, Abbas N, Akhtar M, Rauf A, Khan M, Ali Z, Hassan M, Rizwan M, Bajwa RS, Ajmal A, Naz S, Ali H, Khan AA, Ali M, Sarwar G, Khan MA, Hasanuzzaman M (2016b) Application of DSSAT model for sowing date management of C_4 summer cereals for fodder and grain crops under irrigated arid environment. Pak J Life Soc Sci 14(2):104–114

Ahmad S, Nadeem M, Abbas G, Fatima Z, Khan RJ, Ahmed M, Ahmad A, Rasul G, Khan MA (2016c) Quantification of the effects of climate warming and crop management on sugarcane phenology. Clim Res 71(1):47–61

Ahmad S, Abbas G, Fatima Z, Khan RJ, Anjum MA, Ahmed M, Khan MA, Porter CH, Hoogenboom G (2017a) Quantification of the impacts of climate warming and crop management on canola phenology in Punjab, Pakistan. J Agron Crop Sci 203(5):442–452

Ahmad S, Abbas Q, Abbas G, Fatima Z, Atique-ur-Rehman, Naz S, Younis H, Khan RJ, Nasim W, Rehman MH, Ahmad A, Rasul G, Khan MA, Hasanuzzaman M (2017b) Quantification of climate warming and crop management impacts on cotton phenology. Plan Theory 6:1–16

Ahmad S, Atique-ur-Rehman, Ejaz M, Fatima Z, Kan M, Ahmad M (2018) Agricultural land-use change of major field crops in Pakistan (1961–2014). Sci Technol Dev 37(3):113–121

Ahmad A, Abbas G, Ahmed M, Fatima Z, Anjum MA, Rasul G, Khan MA, Hoogenboom G (2019) Climate warming and management impact on the change of phenology of the rice-wheat cropping system in Punjab, Pakistan. Field Crop Res 230:46–61

Ahmad A, Ashfaq M, Rasul G, Wajid SA, Ahmad I, Khaliq T, Nasir J, Rasul F, Riaz F, Ahmad B, Ahmad S, Baig IA, Valdivia RO, Hoogenboom G (2021) Development of climate change adaptation strategies for cotton–wheat cropping system of Punjab Pakistan. In: Rosenzweig C, Mutter CZ, Contreras EM (eds) Handbook of climate change and agroecosystems: climate change and farming system planning in Africa and South Asia: AgMIP stakeholder-driven research. World Scientific Publishing Europe Ltd, London, pp 277–327

Ahmad A, Wajid A, Saleem MH, Shehzadi M, Rehan M, Hussaan M, Faheem M, Javed MT, El-Sheikh MA, Alshaya H, Ali S (2022) Simulating the growth and yield of peanut cultivars under temporal variation using Cropgro-peanut (DSSAT) model. Pak J Bot 54(3). https://doi.org/10.30848/PJB2022-3(40)

Ahmad S, Ahmad I, Ahmad B, Ahmad A, Wajid A, Khaliq T, Abbas G, Wilkerson CJ, Hoogenboom G (2023a) Regional integrated assessment of climate change impact on cotton production in a semi-arid environment. Clim Res 89:113–132

Ahmad S, Raza MA, Hussain S, Abbas G, Fatima Z, Ahmed M, Goheer MA, Garcia y Garcia A, Wilkerson CJ, Hoogenboom G (2023b) Identification of weak links in the production technology of farmer for bridging canola yield-gap. J Agric Sci 161(2):241–253. https://doi.org/10.1017/S0021859623000187

Ahmed M, Akram MN, Asim M, Aslam M, F-u Hassan Higgins S, Stockle CO, Hoogenboom G (2016) Calibration and validation of APSIM-wheat and CERES-wheat for spring wheat under rainfed conditions: models evaluation and application. Comput Electron Agric 123:384–401

Ahmed I, Rahman MH, Ahmad A, Ahmad S, Hussain J, Ullah A, Judge J (2018a) Assessing the impact of climate variability on maize using simulation modeling under semi-arid environment of Punjab, Pakistan. Environ Sci Pollut Res 25(23):28413–28430

Ahmed M, Ijaz W, Ahmad S (2018b) Adapting and evaluating APSIM-SoilP-Wheat model for response to phosphorus under rainfed conditions of Pakistan. J Plant Nutr 41(16):2069–2084

Ahmed M, Stockle CO, Nelson R, Higgins S, Ahmad S, Raza M (2019) Novel multimodal ensemble approach to evaluate the sole effect of elevated CO_2 on winter wheat productivity. Sci Rep 9:7813

Ahmed M, Ahmad S, Tariq M, Fatima Z, Aslam Z, Raza MA, Iqbal N, Akmal M, Fayyaz-ul-Hassan, Abbasi NA, Hayat R (2020) Wastes to be the source of nutrients and energy to mitigate climate change and ensure future sustainability: options and strategies. J Plant Nutr 43(6):896–920

Ahmed M, Hayat R, Ahmad M, Mahmood-ul-Hassan, Kheir AMS, Fayyaz-ul-Hassan, Rehman MH, Shaheen FA, Raza MA, Ahmad S (2022) Impact of climate change on dryland agricultural systems: a review of current status, potentials, and further work need. Int J Plant Prod 16:341–363

Ali H, Iqbal N, Ahmad S, Shahzad AN, Sarwar N (2013) Performance of late sown wheat crop under different planting geometries and irrigation regimes in arid climate. Soil Tillage Res 130:109–119

Amin A, Nasim W, Fahad S, Ali S, Ahmad S, Rasool A, Saleem N, Hammad HM, Sultana SR, Mubeen M, Bakhat HF, Ahmad N, Shah GM, Adnan M, Noor M, Basir A, Saud S, Rahman MH, Paz JO (2018a) Evaluation and analysis of temperature for historical (1996–2015) and projected (2030–2060) climates in Pakistan using SimCLIM climate model: ensemble application. Atmos Res 213:422–436

Amin A, Nasim W, Mubeen M, Ahmad A, Nadeem M, Urich P, Fahad S, Ahmad S, Wajid A, Tabassum F, Hammad HM, Sultana SR, Anwar S, Baloch SK, Wahid A, Wilkerson CJ, Hoogenboom G (2018b) Simulated CSM-CROPGRO-cotton yield under projected future climate by SimCLIM for Southern Punjab, Pakistan. Agric Syst 167:213–222

Amin A, Nasim W, Mubeen M, Sarwar S, Urich P, Ahmad A, Wajid A, Khaliq T, Rasul F, Hammad HM, Rehmani MIA, Mubarak H, Mirza N, Wahid A, Ahmad S, Fahad S, Ullah A, Khan MN, Ameen A, Amanullah, Shahzad B, Saud S, Alharby H, Ata-Ul-Karim ST, Adnan M, Islam F, Ali QS (2018c) Regional climate assessment for temperature and precipitation in Southern Punjab (Pakistan) using SimCLIM climate model for different temporal scales. Theor Appl Climatol 131:121–131

Bahl PN (2015) Climate change and pulses: approaches to combat its impact. Agric Res 4:103–108

Basu PS, Singh UMMED, Kumar ANIL, Praharaj CS, Shivran RK (2016) Climate change and its mitigation strategies in pulses production. Indian J Agron 61:S71–S82

Biswas JC, Kalra N, Maniruzzaman M, Choudhury AK, Jahan MAHS, Hossain MB, Ishtiaque S, Haque MM, Kabir W (2018) Development of mungbean model (MungGro) and its application for climate change impact analysis in Bangladesh. Ecol Model 384:1–9

Boote KJ, Mínguez MI, Sau F (2002) Adapting the CROPGRO legume model to simulate growth of faba bean. Agron J 94(4):743–756

Browning DM, Snyder KA, Herrick JE (2019) Plant phenology: taking the pulse of rangelands. Rangelands 41(3):129–134

Cable JM, Ogle K, Williams DG, Weltzin JF, Huxman TE (2008) Soil texture drives responses of soil respiration to precipitation pulses in the Sonoran Desert: implications for climate change. Ecosystems 11:961–979

Carberry PS, Hochman Z, Hunt JR, Dalgliesh NP, McCown RL, Whish JPM, Robertson MJ, Foale MA, Poulton PL, Van Rees H (2009) Re-inventing model-based decision support with Australian dryland farmers. 3. Relevance of APSIM to commercial crops. Crop Pasture Sci 60(11):1044–1056

Ceylan HA, Türkan I, Sekmen AH (2013) Effect of coronatine on antioxidant enzyme response of chickpea roots to combination of PEG-induced osmotic stress and heat stress. J Plant Growth Regul 32:72–82

Chaki AK, Gaydon DS, Dalal RC, Bellotti WB, Menzies NW (2019) November. Evaluation of the APSIM model in conservation agriculture practices of the Rice-Wheat-Mungbean System in the Eastern Gangetic Plains (EGP) of Rajshahi, Bangladesh. In ASA, CSSA and SSSA International Annual Meetings, 2019. ASA-CSSA-SSSA

Chauhan Y, Wright G, Rachaputi N, McCosker K (2008) Identifying chickpea homoclimes using the APSIM chickpea model. Aust J Agric Res 59:260–269. https://doi.org/10.1071/AR07380

Chauhan Y, Allard S, Williams R, Williams B, Mundree S, Chenu K, Rachaputi NC (2017) Characterisation of chickpea cropping systems in Australia for major abiotic production constraints. Field Crop Res 204:120–134. https://doi.org/10.1016/j.fcr.2017.01.008

Clark R, Hobson K (2022) Climate change: aerial insectivores struggle to keep pace with earlier pulses of nutritious aquatic foods. Curr Biol 32(6):R267–R269

Devasirvatham V, Gaur PM, Mallikarjuna N, Raju TN, Trethowan RM, Tan DK (2013) Reproductive biology of chickpea response to heat stress in the field is associated with the performance in controlled environments. Field Crop Res 142:9–19

Devi P, Jha UC, Prakash V, Kumar S, Parida SK, Paul PJ, Prasad PV, Sharma KD, Siddique KH, Nayyar H (2022) Response of physiological, reproductive function and yield traits in cultivated chickpea (Cicer arietinum L.) under heat stress. Front Plant Sci 13:880519. https://doi.org/10.3389/fpls.2022.880519

Ejaz M, Abbas G, Fatima Z, Iqbal P, Raza MA, Kheir AMS, Ahmed M, Kakar KM, Ahmad S (2022) Modelling climate uncertainty and adaptations for soybean-based cropping system. Int J Plant Prod 16:235–250

Fatima Z, Abbas Q, Khan A, Hussain S, Ali MA, Abbas G, Younis H, Naz S, Ismail M, Shahzad MI, Nadeem M, Farooq U, Khan SU, Javed K, Khan AA, Ahmed M, Khan MA, Ahmad S (2018) Resource-use-efficiencies of C_3 and C_4 cereals under split nitrogen regimes. Agronomy 8(5):69

Fatima Z, Ahmad M, Hussain M, Abbas G, Ul-Allah S, Ahmad S, Ahmed N, Ali MA, Sarwar G, Haque E, Iqbal P, Hussain S (2020) The fingerprints of climate warming on cereal crops phenology and adaptation options. Sci Rep 10:18013

Fatima Z, Abbas G, Iqbal P, Zakir I, Khan MA, Kamal GM, Ahmed M, Ahmad S (2021) Quantification of climate warming and crop management impacts on phenology of pulses-based cropping systems. Int J Plant Prod 15:107–123

Fu T, Ha B, Ko J (2016) Simulation of CO_2 enrichment and climate change impacts on soybean production. Int Agrophy 30:25–37

Ganesan K, Xu B (2018) A critical review on phytochemical profile and health promoting effects of mung bean (*Vigna radiata*). Food Sci Human Wellness 7(1):11–33

Grasso N, Lynch NL, Arendt EK, O'Mahony JA (2022) Chickpea protein ingredients: a review of composition, functionality, and applications. Compr Rev Food Sci Food Saf 21(1):435–452

Gull R, Bhat TA, Sheikh TA, Wani OA, Fayaz S, Nazir A, Saad AA, Jan S, Nazir I (2020) Climate change impact on pulse in India - a review. J Pharmacogn Phytochem 9(4):3159–3166

Hajishabani H, Mondani F, Bagheri A (2020) Simulation effects of sowing date on growth and yield of rainfed chickpea (*Cicer arietinum* L.) by CROPGRO-Chickpea model. Iranian J Field Crops Res 18(2):197–212

HanumanthaRao B, Nair RM, Nayyar H (2016) Salinity and high temperature tolerance in mungbean [Vigna radiata (L.) Wilczek] from a physiological perspective. Frontiers in. Plant Sci 7:957

Harris RM, Beaumont LJ, Vance TR, Tozer CR, Remenyi TA, Perkins-Kirkpatrick SE, Mitchell PJ, Nicotra AB, McGregor S, Andrew NR, Letnic M (2018) Biological responses to the press and pulse of climate trends and extreme events. Nat Clim Chang 8(7):579–587

Hoogenboom G, Porter CH, Boote KJ, Shelia V, Wilkens PW, Singh U, White JW, Asseng S, Lizaso JI, Moreno LP, Pavan W, Ogoshi R, Hunt LA, Tsuji GY, Jones JW (2019) The DSSAT crop modeling ecosystem. In: Boote KJ (ed) Advances in crop modeling for a sustainable agriculture. Burleigh Dodds Science Publishing, Cambridge, pp 173–216. https://doi.org/10.19103/AS.2019.0061.10

Hoogenboom G, Porter CH, Shelia V, Boote KJ, Singh U, White JW, Pavan W, Oliveira FAA, Moreno-Cadena LP, Lizaso JI, Asseng S, Pequeno DNL, Kimball BA, Alderman PD, Thorp KR, Jones MR, Cuadra SV, Vianna MS, Villalobos FJ, Ferreira TB, Batchelor WD, Koo J, Hunt LA, Jones JW (2021) Decision Support System for Agrotechnology Transfer (DSSAT) Version 4.8 (DSSAT.net). DSSAT Foundation, Gainesville, FL

Hussain M, Ahmad S, Hussain S, Lal R, Ul-Allah S, Nawaz A (2018) Rice in saline soils: physiology, biochemistry, genetics and management. Adv Agron 148:231–287

Jones JW, Hoogenboom G, Porter CH, Boote KJ, Batchelor WD, Hunt LA, Wilkens PW, Singh U, Gijsman AJ, Ritchie JT (2003) The DSSAT cropping system model. Eur J Agron 18(3–4):235–265

Jukanti AK, Gaur PM, Gowda CLL, Chibbar RN (2012) Nutritional quality and health benefits of chickpea (*Cicer arietinum* L.): a review. Br J Nutr 108(S1):S11–S26

Kadiyala MDM, Charyulu DK, Nedumaran S, Shyam DM, Gumma MK, Bantilan MCS (2016) Agronomic management options for sustaining chickpea yield under climate change scenario. J Agrometeorol 18(1):41–47

Karande BI, Patel HR, Yadav SB, Vasani MJ, Patil DD (2018) Impact of projected climate change on summer mungbean in Gujarat, India. Int J Curr Microbiol App Sci 7(8):4178–4189

Kareem HA, Saleem MF, Saleem S, Rather SA, Wani SH, Siddiqui MH, Alamri S, Kumar R, Gaikwad NB, Guo Z, Niu J (2022) Zinc oxide nanoparticles interplay with physiological and biochemical attributes in terminal heat stress alleviation in mungbean (*Vigna radiata* L.). Front Plant Sci 13:101

Kaushal N, Awasthi R, Gupta K, Gaur P, Siddique KH, Nayyar H (2013) Heat-stress-induced reproductive failures in chickpea (*Cicer arietinum*) are associated with impaired sucrose metabolism in leaves and anthers. Funct Plant Biol 40(12):1334–1349

Keating BA, Carberry PS, Hammer GL, Probert ME, Robertson MJ, Holzworth D, Huth NI, Hargreaves JNG, Meinke H, Hochman Z, McLean G, Verburg K, Snow V, Dimes JP, Silburn M, Wang E, Brown S, Bristow KL, Asseng S, Chapman S, McCown RL, Freebairn DM, Smith CJ (2003) An overview of APSIM: a model designed for farming systems simulation. Eur J Agron 18:267–288

Khan A, Fahad S, Khan A, Saud S, Adnan M, Wahid F, Noor M, Nasim W, Hammad HM, Bakhat HF, Ahmad S, Rehman MH, Wang D, Sönmez O (2019) Managing tillage operations and manure to restore soil carbon stocks in wheat-maize cropping system. Agron J 111:1–10

Khan MN, Hussain M, Abbas G, Fatima Z, Iqbal P, Khan A, Zakir I, Ahmed M, Mateen A, Khan MA, Ahmad S (2020) Improving resource use efficiencies of sugarcane at farmer field under arid environment. Int J Agric Biol 24(5):1279–1285

Khattak GSS, Saeed I, Muhammad T (2006) Breeding for heat tolerance in mungbean (*Vigna radiata* (L.) Wilczek). Pak J Bot 38(5):1539–1550

Khatun M, Begum MEA, Rashid MA, Miah MAM, Hasan MK (2021) Effect of climate change adaptation strategies on production efficiency of chickpea and lentil in Rajshahi District. Bangladesh J Agric Econ 42:55–72

Kherif O, Seghouani M, Justes E, Plaza-Bonilla D, Bouhenache A, Zemmouri B, Dokukin P, Latati M (2022) The first calibration and evaluation of the STICS soil-crop model on chickpea-based intercropping system under Mediterranean conditions. Eur J Agron 133:126449

Koocheki A, Nassiri M, Soltani A, Sharifi H, Ghorbani R (2006) Effects of climate change on growth criteria and yield of sunflower and chickpea crops in Iran. Clim Res 30(3):247–253

Kumar S, Thakur P, Kaushal N, Malik JA, Gaur P, Nayyar H (2013) Effect of varying high temperatures during reproductive growth on reproductive function, oxidative stress and seed yield in chickpea genotypes differing in heat sensitivity. Arch Agron Soil Sci 59(6):823–843

Kumari P, Rastogi A, Yadav S (2020) Effects of heat stress and molecular mitigation approaches in orphan legume, chickpea. Mol Biol Rep 47:4659–4670

Lateff EA, El-Salam MA, Selim MM, Tawfik MM, El-Kramany MF, Farrag AA (2018) Effect of climate change on mungbean growth and productivity under Egyptian conditions. Int J Agric For Life Sci 2(2):16–23

Liu Y, Li J, Zhu Y, Jones A, Rose RJ, Song Y (2019) Heat stress in legume seed setting: effects, causes, and future prospects. Front Plant Sci 10:938

Mahmood A, Ullah H, Shahzad AN, Ali H, Ahmad S, Zia-Ul-Haq M, Honermeier B, Hasanuzzaman M (2013) Dry matter yield and chemical composition of sorghum cultivars with varying planting density and sowing date. Sains Malaysiana 42(10):1529–1538

Mar S, Nomura H, Takahashi Y, Ogata K, Yabe M (2018) Impact of erratic rainfall from climate change on pulse production efficiency in lower Myanmar. Sustainability 10(2):402

Meghdadi N, Hajarpoor A (2015) Simulating climate change impacts on production of chickpea in Zanjan province. J Crop Prod 7(4):1–22

Mitra A, Banerjee K, Sengupta K, Gangopadhyay A (2009) Pulse of climate change in Indian Sundarbans: a myth or reality? Natl Acad Sci Lett 32(1):19

Mohammed A, Tana T, Singh P, Molla A, Seid A (2017) Identifying best crop management practices for chickpea (*Cicer arietinum* L.) in Northeastern Ethiopia under climate change condition. Agric Water Manag 194:68–77

Mubarak MK, Hussain K, Abbas G, Altaf MT, Baloch FS, Ahmad S (2022) Productivity of sorghum (*Sorghum bicolor* L.) at diverse irrigation regimes and sowing dates in semi-arid and arid environment. Turk J Agric For 46(1):1–18

Mubeen M, Ahmad A, Wajid A, Khaliq T, Hammad HM, Sultana SR, Ahmad S, Fahad S, Nasim W (2016) Application of CSM-CERES-Maize model in optimizing irrigated conditions. Outlook Agric 45(3):173–184

Mubvuma MT (2018) Planting date as an adaptive strategy to improve yield of Chickpea (*Cicer arietinum*) under climate change condition in Southern Africa. Doctoral dissertation

Nadeem M, Khan MN, Abbas G, Fatima Z, Iqbal P, Ahmed M, Raza MA, Rehman A, Haq EU, Hayat A, Ali M, Ahmad S (2022) Application of CSM-CANEGRO model for climate change impact assessment and adaptation for sugarcane in semi-arid environment of Southern Punjab, Pakistan. Int J Plant Prod 16:443–466

Nair RM, Pandey AK, War AR, Hanumantharao B, Shwe T, Alam AKMM, Pratap A, Malik SR, Karimi R, Mbeyagala EK, Douglas CA (2019) Biotic and abiotic constraints in mungbean production—progress in genetic improvement. Front Plant Sci 10:1340

Nasim W, Amin A, Fahad S, Awais M, Khan N, Mubeen M, Wahid A, Turan V, Rehman MH, Ihsan MZ, Ahmad S, Hussain S, Mian IA, Khan B, Jamal Y (2018) Future risk assessment by estimating historical heat wave trends with projected heat accumulation using SimCLIM climate model in Pakistan. Atmos Res 205:118–133

Naz S, Ahmad S, Abbas G, Fatima Z, Hussain S, Khan MA, Ahmed M, Khan A, Fahad S, Nasim W, Ercisli S, Wilkerson CJ, Hoogenboom G (2022) Modeling the impact of climate warming on potato phenology. Eur J Agron 132:126404

Patil DD, Patel HR (2017) Calibration and validation of CROPGRO (DSSAT 4.6) model for chickpea under middle GUJARAT agroclimatic region. Int J Agric Sci 9(27):4342–4344

Pratap A, Gupta S, Basu PS, Tomar R, Dubey S, Rathore M, Prajapati US, Singh P, Kumari G (2019) Towards development of climate smart mungbean: challenges and opportunities. In: Kole C (ed) Genomic designing of climate-smart pulse crops. Springer, Cham, pp 235–264. https://doi.org/10.1007/978-3-319-96932-9_5

Rahman NMF, Aziz MA, Rahman MM, Mohammad N (2013) Modeling on grass pea and mung bean pulse production in Bangladesh using ARIMA Model. IOSR J Agric Vet Sci 6(1):20–31

Rahman MH, Ahmad A, Wang X, Wajid A, Nasim W, Hussain M, Ahmad B, Ahmad I, Ali Z, Ishaque W, Awais M, Muddasir M, Shelia V, Ahmad S, Fahad S, Alam M, Ullah H, Hoogenboom G (2018) Multi-model projections of future climate and climate change impacts uncertainty assessment for cotton production in Pakistan. Agric For Meteorol 253–254:94–113

Rahman MH, Ahmad I, Wang D, Fahad S, Afzal M, Ghaffar A, Saddique Q, Khan MA, Saud S, Hassan S, Fahad M, Din M, Ahmad M, Yue Z, Shah RA, Sönmez O, Nasim W (2021) Influence of semi-arid environment on radiation use efficiency and other growth attributes of lentil crop. Environ Sci Pollut Res 28:13697–13711

Rehman A, Ashfaq M, Naqvi SAA, Adil SA, Bashir K, Ahmad A, Ahmad S, Ali A, Imran A (2015) Is climate change worsening the poverty of maize growers? Evidence from Punjab province of Pakistan. Ciencia e Tecnica Vitivinicola 30(1):105–116

Saleh AA, Abdel-Kader DZ, El Elish AM (2007) Role of heat shock and salicylic acid in antioxidant homeostasis in mungbean (*Vigna radiata* L.) plant subjected to heat stress. Am J Plant Physiol 2:344–355

Sarwar N, Ali H, Maqsood M, Ullah E, Shahzad AN, Shahzad M, Mubeen K, Shahid MA, Ahmad S (2013) Phenological response of rice plants to different micronutrients application under water saving paddy fields on calcareous soil. Turk J Field Crop 18(1):52–57

Sarwar N, Atique-ur-Rehman, Farooq O, Wasaya A, Hussain M, El-Shehawi AM, Ahmad S, Brestic M, Mahmoud SF, Zivcak M, Farooq S (2021) Integrated nitrogen management improves productivity and economic returns of wheat-maize cropping system. J King Saud Univ Sci 33:101475

Shabbir RN, Waraich EA, Ali H, Nawaz F, Ashraf MY, Ahmad R, Awan MI, Ahmad S, Irfan M, Hussain S, Ahmad Z (2016) Supplemental exogenous NPK application alters biochemical processes to improve yield and drought tolerance in wheat (*Triticum aestivum* L.). Environ Sci Pollut Res 23(3):2651–2662

Silva LL, Ragab R, Duarte I, Lourenço E, Simões N, Chaves MM (2013) Calibration and validation of SALTMED model under dry and wet year conditions using chickpea field data from Southern Portugal. Irrig Sci 31:651–659

Singh P, Nedumaran S, Boote KJ, Gaur PM, Srinivas K, Bantilan MCS (2014) Climate change impacts and potential benefits of drought and heat tolerance in chickpea in South Asia and East Africa. Eur J Agron 52:123–137

Soltani A, Sinclair TR (2011) A simple model for chickpea development, growth and yield. Field Crop Res 124(2):252–260

Tariq M, Yasmeen A, Ahmad S, Hussain N, Afzal MN, Hasanuzzaman M (2017) Shedding of fruiting structures in cotton: factors, compensation and prevention. Trop Subtrop Agroecosyst 20(2):251–262

Tariq M, Afzal MN, Muhammad D, Ahmad S, Shahzad AN, Kiran A, Wakeel A (2018a) Relationship of tissue potassium content with yield and fiber quality components of Bt cotton as influenced by potassium application methods. Field Crop Res 229:37–43

Tariq M, Ahmad S, Fahad S, Abbas G, Hussain S, Fatima Z, Wajid N, Mubeen M, Rehman M, Khan MA, Adnan M, Wilkerson CJ, Hoogenboom G (2018b) The impact of climate warming and crop management on phenology of sunflower-based cropping systems in Punjab, Pakistan. Agric For Meteorol 256–257:270–282

Tarriq M, Fatima Z, Iqbal P, Nahar K, Ahmad S, Hasanuzzaman M (2021) Sowing dates and cultivars mediated changes in phenology and yield traits of cotton-sunflower cropping system in arid environment. Int J Plant Prod 15:291–302

Timsina J, Godwin D, Connor DJ (2000) Water management for mungbean within rice-wheat cropping systems of Bangladesh: a simulation study. In: Proceedings—The Third International Symposium on Systems Approaches for Agricultural Development Griffith, NSW 2680, Australia

Zahoor A, Ahmed M, Fayyaz-ul-Hassan SG, Ahmad S (2022) Ontogeny growth and radiation use efficiency of canola (*Brassica napus* L.) under various nitrogen management strategies and contrasting environments. Int J Plant Prod 16(1):195–208

Zeitelhofer M, Zhou R, Ottosen CO (2022) Physiological responses of chickpea genotypes to cold and heat stress in flowering stage. Agronomy 12(11):2755

Soybean-Soybean System 8

8.1 Introduction

Soybean-soybean cropping system is important for edible oil and poultry feed in various regions of the world. Soybean is grown in both spring and fall seasons in Pakistan, India, and other countries in the world (Ejaz et al. 2022; Fatima et al. 2018, 2020, 2021; Nadeem et al. 2022; Naz et al. 2022). Climate warming is seriously influencing this cropping system. Soybean is a very important crop across the globe (Ntiamoah et al. 2022; Soratto et al. 2022). Soybean-soybean cropping system is playing an important role in soybean production in all over the world. Soybean is grown in two growing times of kharif season and rabi season. During the kharif season, utmost usual time of planting is beginning of the monsoon while spring sowing is done between mid-January and end of February. The legume family includes soybean, sometimes known as "golden beans." A large legume crop that offers a number of essential nutritional components is the soybean. Soybean is one of the world's important grain and oil crops, and any change in soybean production in soybean-soybean cropping systems under future climatic scenarios would affect grain as well as edible oil security on a regional and worldwide scale (Singh et al. 2020; Sharma et al. 2022). Soybean (*G. max* (L.) Merr.), which was historically first cultivated in China, is a significant protein source for people and also makes excellent animal feed. It has an 18–22% oil content and 40–42% excellent-grade protein. About two-thirds of the world's protein concentrate for cattle feeding comes from it, and it produces 25% of the world's total edible oil production. Soybean meal is an important component of fish and poultry diets. One of the most important sources of dietary energy for people is soybean (Fernandes et al. 2022). It is cultivated on roughly 6% of the world's arable land and is valued economically for its high-quality oil and protein. Being the most significant legume crop in the world, soybean produces industrial by-products, human food, and animal feed because of its plentiful seeds that are high in protein and vegetable oil (Tariq et al. 2017, 2018a, b; Tarriq et al. 2021; Hussain et al. 2018; Khan et al. 2019, 2020; Zahoor et al. 2022).

M. Ahmed et al., *Cropping Systems Modeling Under Changing Climate*,
https://doi.org/10.1007/978-981-97-0331-9_8

Table 8.1 Area and production of soybean-soybean systems across various regions in the world

Countries	Winter soybean		Summer soybean		Soybean-soybean system		
	Area (Mha)	Production (MMT)	Area (Mha)	Production (MMT)	Area (Mha)	Production (MMT)	Percent of world production (percent)
Brazil	28.4	121.6	6.3	31.4	34.7	153.0	39
USA	4.2	10.0	31.4	106.3	35.6	116.3	30
Argentina	14.3	39.8	2.0	5.7	16.3	45.5	12
China	1.1	1.8	6.8	18.5	7.9	20.3	5
India	2.1	3.3	9.3	8.7	11.4	12.0	3
Paraguay	2.6	8.4	0.9	1.6	3.5	10.0	3
Canada	0.2	0.7	2.5	5.8	2.7	6.5	2
Russia	0.4	0.2	2.1	5.3	2.5	5.5	1

Source: USDA (2022)
Mha million hectare, *MMT* million metric tonne

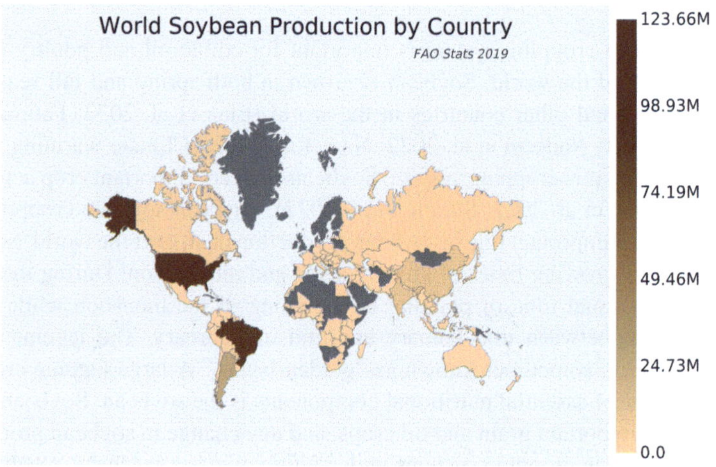

Fig. 8.1 World soybean production by country. (Source: FAO 2024)

Soybean seeds, being members of the Leguminosae (Fabaceae) family, are high in protein and vital fatty acids. It is now also respected as a potential facility for the manufacture of biodiesel (Cabanos et al. 2021; Blanco et al. 2022). Furthermore, the obtainability of vital nutritional supplement in the soybean and increasing feeding have caused an increase in demand for the soybean grown. Soybean was primarily domesticated in China, where there are around 23,000 varieties, and after that was imported into the USA and Brazil. At the global level, soybean production is projected to be 311.1 million metric tonnes during 2023 and 371.3 million metric tonnes during 2030. Nonetheless, the quantity level in 2030 is expected to be almost 1.7 times that of 2005–2007. The top three primary soybean-producing countries are the USA, Brazil, and Argentina (Table 8.1 and Fig. 8.1). These three nations dominate worldwide soybean production, accounting for 80% of global supply.

8.2 Issues of Soybean-Soybean Cropping System

There are a number of factors that limit increased productivity in soybean-soybean cropping system, including those related to climate warming, edaphic, production, and technology issues (Abbas et al. 2017, 2019, 2020a, b, c; Abbas and Ahmad 2018; Ahmad et al. 2012a, b, 2013, 2015a, b, c, 2016a, b, c, 2017a, b, 2018, 2019, 2021, 2023a, b). The majority of the soybean farmland is rainfed (de Araújo et al. 2019; Karges et al. 2022). Heat stress caused by global warming is a major concern in global soybean production. Soybean is a short-day plant that is extremely sensitive to the duration of the day. As a result, individual soybean cultivars have limited adaptation across latitudes and planting dates (Allen and Boote 2000; Zhao et al. 2022). Low production in the soybean-soybean cropping system is caused by a variety of biotic and abiotic stressors. Production of soybean is hampered through low soil productivity and unpredictable precipitation owing to climate change (Ahmed et al. 2016, 2018a, b, 2019, 2020, 2022; Ahmad and Hasanuzzaman 2012; Ali et al. 2013; Amin et al. 2018a, b, c). Soybean yield is also limited due to the irregular behavior of the monsoon season, which affects sowing and planting in various areas of the world (Junior 2022). Because of the considerable regional and temporal fluctuation in rainfall, soybean output is constrained. Soybean output is limited owing to soil moisture stress throughout important development phases, particularly seed filling. Soybean production is hampered by high-temperature stress at crucial development phases. Soybean output is limited owing to biotic crop growth interferences. Due to inadequate automation, soybean production is constrained. Soybean yield is limited owing to a lack of adoption of modern production technology-low-risk coverage capabilities (Umburanas et al. 2022). Soybean output is limited owing to monoculture farming and inadequate varietal diversification, which increases risk. Soybean production is hampered by a lack of timely availability of high-quality inputs. Soybean productivity is limited owing to a lack of technological knowledge. Growing area of soyabean-soybean cropping system is low because soybean sowing time is overlapped with maize, sunflower, and other crops (Tariq et al. 2017, 2018a, b; Tarriq et al. 2021; Hussain et al. 2018; Khan et al. 2019, 2020; Zahoor et al. 2022; Lopez et al. 2022).

8.3 Decision Support System for Agrotechnology Transfer

Soybean growth prediction models play an important role in decision-making in the agro-industry and policymakers regarding risks and possibility of climate change acclimatization strategies, as well as directing plant scientists and breeders toward the required traits in improved varieties and soybean-soybean cropping system management practices to alleviate worldwide climate warming impact (Ruíz-Nogueira et al. 2001; Bhatia et al. 2008; Jones et al. 2003; Keating et al. 2003; Hoogenboom et al. 2019, 2021). Soybean crop model intercomparison studies have revealed significant errors and biases, and regrettably, they do not always integrate current understanding of soybean plant responses to high atmospheric pressure

(Bebeley et al. 2022). Soybean crop models are intended to determine soybean crop production in soybean-soybean cropping system (and other relevant soil-plant system parameters) as a function of weather and soil conditions, plant-specific features, and a variety of agricultural management strategies (Fodor et al. 2017). Soybean yield prediction models have been developed and are being used as tools to evaluate various management alternatives for a specific site, such as fertilizer application rates, plant population, sowing times, and land usage. Using soil knowledge and long-term meteorological data, the DSSAT allows for extrapolation of outcomes of short-period tests to further years besides places. Many researchers from across the world have thoroughly tested and assessed the DSSAT, APSIM, CropSyst, AquaCrop, and other models for soybean, finding strong correlations between actual and predicted values for an extensive variety of experimental methods versus field data along with climatic variables in soybean-based cropping systems. These soybean models simulate inorganic fertilizer management, crop managing strategies for extension systems, study of variety, sowing time skills, and soybean cultivar performance in diverse agroecologies. Bebeley et al. (2022) used CSM-CROPGRO-Soybean model. The phenology, development, and productivity characteristics of soybean cultivars were simulated, and actual values were in good agreement, according to the model calibration and validation assessment findings. This indicated that DSSAT-CROPGRO-Soybean model was successful in simulating soybean development along with productivity in Nigeria. The study also demonstrated the model's potential as a tool for calculating the ideal soybean planting window in savannas. Regarding productivity, simulation results also reveal a variety of agroecologies and soybean genotypes. Kothari et al. (2022) used soybean crop model of assessment of climate change. Reliable soybean productivity predictions at climate change situations are critical for quantifying our capacity to provide for an increasing population and developing adaptation strategies to fulfill forthcoming food needs. They selected ten famous models for predicting soybean productivity at altering temperature and carbon dioxide concentrations in this first soybean multimodel study to evaluate uncertainty for soybean productivity predictions in response under conditions (Fig. 1.1). Soybean simulation models were initially calibrated using good-quality observed datasets of five different contexts. In simulated yield reactions to rising temperature and carbon dioxide, they discovered significant diversity across models. For instance, with a +3 °C temperature increase in the coldest site in Argentina, certain soybean models predicted a 24% decrease in seed yield, while others predicted a 29% gain. Models estimated a productivity drop ranging from 38% decline at +3 °C temperature increase to no influence on productivity in hottest areas in Brazil. Depending on the location and treatment, measured soybean seed production ranged from 3297 to 6179 kg/ha (Fig. 8.2). When compared to the blind calibration, which was based solely on phenological observations, the variability and normalized root mean square error (nRMSE) for the simulated yield across the ten models were reduced after full calibration using extensive in-season observations (Fig. 8.3). The responses of the models to changes from the baseline temperature of −3, 0, +3, +6, and 9 °C varied significantly (Fig. 8.4). After blind calibration and full calibration, individual models simulated yield across all sites with nRMSE ranging from 9% to 31% and 20% to 51%, respectively. Models

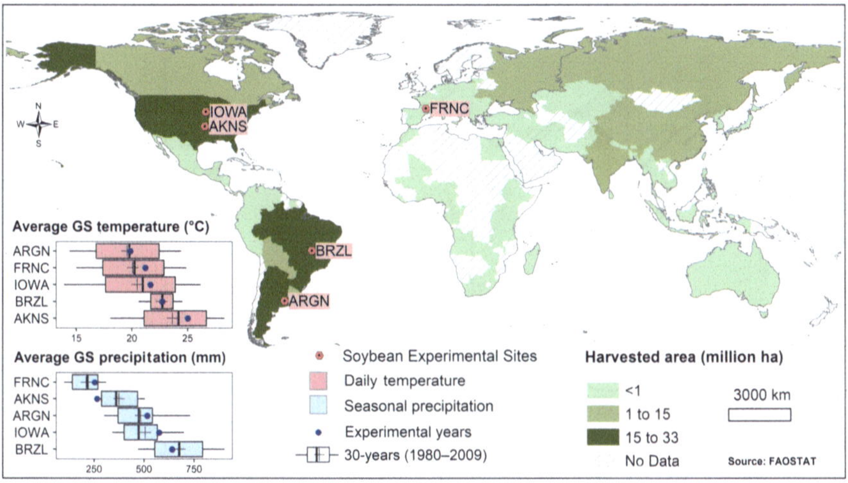

Fig. 8.2 Locations of experimental sites used for model calibration under baseline conditions and soybean-harvested area from 2010 to 2019 (FAO 2024). Azul, Argentina (ARGN); Brasilia, Brazil (BRZL); Auzeville, France (FRNC); and Ames, IA (IOWA), and Fayetteville, Arkansas (AKNS), USA. Figure inserts show average daily growing season (GS) temperature and total precipitation during the experimental years (blue points) and during 30 years (1980–2009) used for sensitivity analysis (boxplots). Boxplot bounds show 25th and 75th percentiles, whiskers are 10th and 90th percentiles, the median is shown as a vertical bold line, and the mean as a plus symbol. (Source: Modified and adopted from Kothari et al. 2022)

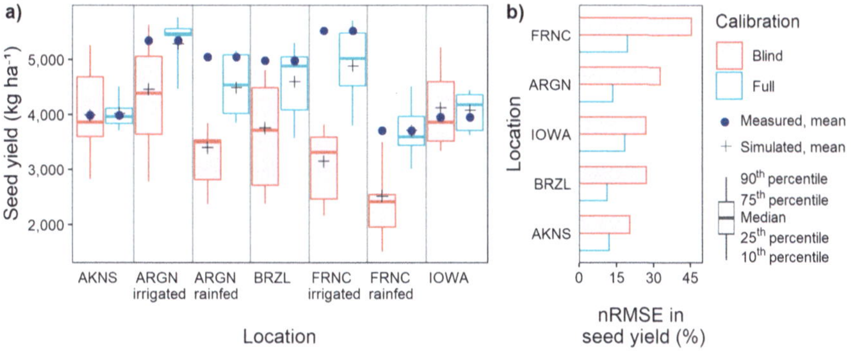

Fig. 8.3 Mean observed and simulated yield and normalized root mean square error (nRMSE) for simulation of yield at each location after *blind* and *full* calibration. (**a**) Measured (blue symbol) and simulated seed yield across ten models (boxplots) with data averaged across treatments within a location and irrigation system (rainfed or irrigated), and (**b**) average nRMSE across models for simulation of seed yield by location after *blind* and *full* calibration. (Source: Modified and adopted from Kothari et al. 2022)

predicted a productivity increase ranging from 6% to 31% when concentration of carbon dioxide was raised from 360 to 540 ppm. Although model calibration had an unanticipated impact on changing yield responses to temperature for some of the models, it did not lower variability across models. The significant level of

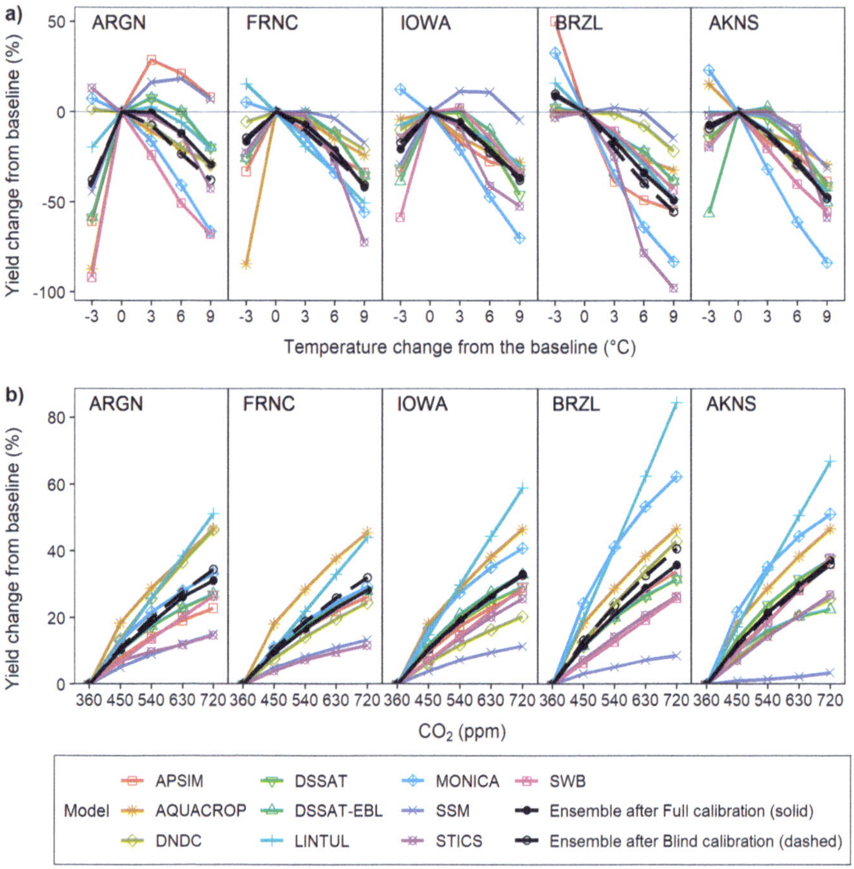

Fig. 8.4 Simulated yield change (%) in response to (**a**) temperature and (**b**) [CO_2] variation from the baseline at each site. Results are shown for each model after *full* calibration, and for the model ensemble after *blind* and *full* calibration. The locations are sorted from cold (Argentina) to hot (Arkansas), left to right, based on mean growing season temperature. Data averaged across 30 years of simulations under no water limitations (automatic irrigation). (Source: Modified and adopted from Kothari et al. 2022)

uncertainty in model responses shows that individual models have limited application for predicting the future of food due to climatic changes. The significant level of uncertainty in simulations of soybean yields connected to individual models and their parametrization, however, may be effectively reduced by the ensemble mean of simulations across models. Ensemble average productivity response to carbon dioxide as well as temperature was comparable to those documented in literature. This analysis underscores the need for additional soybean model development with trials under higher carbon dioxide and temperature to lower the uncertainty from the individual models and is the first to show the advantages of employing an ensemble for legume models for climate change forecasts. Using the CROPGRO-Soybean model, Jagtap and Jones (2002) estimated soybean yields under different cropping management combinations of three cultivars and three planting dates. The majority of soybean models have been created for field soybeans for animal feed and oil production. Because soybean sprout has distinct agronomic properties than field soybeans, simulation models created for field soybeans have limited ability to estimate yields of soybean sprout genotypes. Ejaz et al. (2022) calibrated and evaluated CSM-CROPGRO-Soybean model in soybean-soybean cropping system under semiarid conditions in Pakistan. Evaluation of soybean calibrated model was satisfactory as a result of satisfactory values of statistical indices (Table 8.2) (Mahmood et al. 2013; Mubeen et al. 2016; Nasim et al. 2018; Mubarak et al. 2022; Rehman et al. 2015; Rahman et al. 2018, 2021; Sarwar et al. 2013, 2021; Shabbir et al. 2016). The genetic coefficients of soybean crop in DSSAT model are presented in Table 8.3.

Table 8.2 Summary of observed and simulated results during model calibration with data recorded on the sowing date of 1st February in spring soybean season and 1st August in autumn soybean season during 2019

			Spring season			Autumn season		
Variable	Unit	Cultivars	Sim.	Obs.	Error (%)	Sim.	Obs.	Error (%)
Anthesis	day	'Williams-82'	58	58	0.00	35	34	2.94
		'NARC-1'	58	57	1.75	37	35	5.71
		'Ajmeri'	58	57	1.75	34	34	0.00
Maturity	day	'Williams-82'	112	111	0.90	80	79	1.27
		'NARC-1'	109	108	0.93	78	77	1.30
		'Ajmeri'	95	94	1.06	70	69	1.45
Max. LAI		'Williams-82'	3.93	3.497	15.69	6.09	5.541	9.91
		'NARC-1'	3.58	3.186	16.01	6.13	5.478	11.90
		'Ajmeri'	2.87	2.554	12.37	4.61	4.195	9.89
Grain yield	Kg ha^{-1}	'Williams-82'	3716	3207	11.69	2858	2343	21.98
		'NARC-1'	3080	2705	13.86	2500	2025	23.46
		'Ajmeri'	2401	2036	17.93	1917	1506	27.29
Total dry matter	Kg ha^{-1}	'Williams-82'	6747	6104	10.53	5456	5019	8.71
		'NARC-1'	6637	5906	12.38	5492	5152	6.60
		'Ajmeri'	4940	4496	9.88	4781	4351	9.88

Source: Modified and adopted from Ejaz et al. (2022)

Table 8.3 Genetic coefficients, description, ranges, and guidelines for model calibration of soybean in DSSAT

Genetic coefficient	Description
VAR#	Identification code or number for a specific cultivar
VAR-NAME	Name of cultivar
EXPNO	Number of experiments used to estimate cultivar parameters
ECO#	Ecotype code of this cultivar, points to the Ecotype in the ECO file
CSDL	Critical short-day length below which reproductive development progresses with no day length effect (for short-day plants) (h)

(continued)

Table 8.3 (continued)

Genetic coefficient	Description
PPSEN	Slope of the relative response of development to photoperiod with time (positive for short-day plants) (1/h)
EM-FL	Time between plant emergence and flower appearance (R1) (photothermal days)
FL-SH	Time between first flower and first pod (R3) (photothermal days)
FL-SD	Time between first flower and first seed (R5) (photothermal days)
SD-PM	Time between first seed (R5) and physiological maturity (R7) (photothermal days)
FL-LF	Time between starburst (R1) and end of leaf expansion (photothermal days)
LFMAX	Maximum leaf photosynthesis rate at 30 °C, 350 vpm CO_2, and high light (mg CO_2/m^2-s)
SLAVR	Specific leaf area of cultivar under standard growth conditions (cm^2/g)
SIZLF	Maximum size of full leaf (three leaflets) (cm^2)
XFRT	Maximum fraction of daily growth that is partitioned to thalamus, seed + shell
WTPSD	Maximum weight per seed (g)
SFDUR	Seed filling duration for pod cohort at standard growth conditions (photothermal days)
SDPDV	Average seed per pod under standard growing conditions (#/pod)
PODUR	Time required for cultivar to reach the final pod load under optimal conditions (photothermal days)
THRSH	Threshing percentage. The maximum ratio of (seed/(seed + shell)) at maturity. Causes seeds to stop growing as their dry weight increases until shells are filled in a cohort
SDPRO	Fraction of protein in seeds (g(protein)/g(seed))
SDLIP	Fraction of oil in seeds (g(oil)/g(seed))

Source: DSSAT v4.8; Hoogenboom et al. (2021)

8.4 Climate Change and Temperature Trend

Climate change is a serious threat to soybean-soybean cropping system all over the world. Thermal trend is increasing due to climate warming in soybean-soybean cropping system. Mall et al. (2004) studied that in Indian soybean-soybean cropping system, the expected temperature increase is very seasonal and spatially dependent. The temperature rise across South India during the monsoon season is expected to be less than 1.5 °C by the 2050s. Tan et al. (2021) found that during the soybean growth periods from 1992 until 2018, mean temperature of 0.34 ± 0.09 °C/decade along with accumulated rainfall (6.66 ± 0.93 mm/decade) increased; however, total overall sunshine hours (33.98 ± 1.05 h/decade) declined. Goldblum (2009) revealed that in the USA, CRCM-modeled average monthly temperatures indicate generally consistent hotter circumstances when compared to the 1961–1990 normal period, with mean annual temperature 3.75 °C higher in the 2050s and 2060s. The most dramatic warming is expected during the winter months from December to February, when mean monthly temperatures are expected to be from 5 to 6 °C higher than the 1961–1990 average period. Temperatures are expected to be from 2 to 3 °C higher than typical for much of the spring and fall months, while temperatures in the summer months of July and August are expected to be 4.5 °C higher than 1961–1990 climatic normal. Yuzbashkandi and Khalilian (2020) indicated that in the soybean-soybean copping system in Iran, the estimated increase in lowest temperature, highest temperature, as well as rainfall during 2021–2040 was around 1.40 °C, 1.20 °C, and 13 cm, respectively. In the soybean-soybean copping system, the lowest and highest temperature standard deviations would reduce. The rainfall standard deviation increased in comparison to historical weather data from 1980 to 2010, revealing increasing variability in the chronological spreading of rainfall in upcoming decades in the studied region in Iran. The climate of Gorleston province in the soybean-soybean cropping system will become wetter as well as warmer in the early twenty-first century. Air temperature will be more severe in the future scenario in soybean-growing areas.

8.5 Climate Change Impact on Soybean Morphophysiological Responses

Climate warming had a big impact on how well soybean seeds germinated, which decreased their economic worth and made them more susceptible to disease assault. Heat stress due to climate change lowers the ability of plants to absorb carbon, damages membranes and stomata, and inhibits the action of many enzymes (Huang et al. 2019). Due to rapid Chl breakdown, HT stress causes a decrease in the chlorophyll concentration of leaves, which triggers the senescence of leaves. Climate warming affects the metabolic processes, which changes how plants develop. The development and growth of soybeans are adversely affected by climate warming (Chebrolu et al. 2016). Researchers tested heat-tolerant and -sensitive genotypes of soybean under heat stress. The heat-sensitive genotype's seed germination was considerably impacted by heat stress (42 °C), but the heat-tolerant genotype's germination was unaffected by similar factors. When soybean plants were subjected to 42 °C, climate warming had an impact on the protein content of the seeds and the protein storage vacuoles. The structure and integrity of the membrane in vacuoles, which store protein, are disturbed by heat stress (Cohen et al. 2021). Similar to this, a different study found that after 5–10 days of exposure, acute heat stress reduced the amount of salicylic acid (SA) in soybean. Activity of several metabolites and amino acids in soybean was likewise impacted by climate warming. When plants were subjected to intense heat stress, their concentration of secondary metabolites, lipids, proteins, and carbohydrates decreased (Couëdel et al. 2021). Climate warming had a severe impact on soybean genotypes' ability to produce seeds, which eventually reduced output. These results suggested that heat stress might impact any stage of plant development and reduce the overall soybean yield per plant. Plants subjected to heat stress over 24 h suffered leaf damage and had lower proline and soluble sugar concentrations. The structure and development of soybean roots are impacted by heat stress. Another study found that heat stress decreased the chlorophyll content of soybean plants. Additionally, climate warming decreased the assimilates' accessibility to soybean grains and the rate at which young leaves were converted to sink grains (Ortiz et al. 2022). Climate warming caused oxidative damage, decreased stomatal conductance, and lowered shoot and dry root weight. Climate warming has an impact on the growth of shoots and roots, germination of seeds, production of reactive oxygen species, activity of antioxidant enzymes, and total protein content. An important climatic condition that adversely impacts plant growth and development and significantly reduces soybean crop output globally is the rising ambient temperature (Narayanan et al. 2020). As a result, high temperatures due to climate warming cause decreased seed germination, which promotes pathogenic infection and reduces the soybean crop's economic output. High temperatures can decrease the effectiveness of photosynthesis and transpiration in plants, which negatively affects physio-biochemical processes in the plant system and ultimately degrades crop output and quality (Feng et al. 2019). On a physio-biochemical level, climate warming tempts protein denaturation, boosts lipid fluidity in cell membranes, and overproduces reactive oxygen species (ROS), which eventually impairs the function of the photosynthetic machinery (Song et al. 2017).

8.6 Impact of Climate Change on Soybean Phenology

Soybean phenological stages and phases are a critical indicator of climate change worldwide. Spatial and temporal features of soybean crop phenology must be revealed in order to mitigate the negative consequences of climate change and guide national agricultural production. Climate as well as agronomic management practices including planting time and cultivar characteristics affect soybean crop phenology. High temperature has a significant impact on soybean-soybean cropping system production when it happens throughout pivotal phases of crop development. Temperature increases shorten soybean crop life span by triggering earlier anthesis and decreasing grain filling phase (St-Marseille et al. 2019). Understanding the interaction impacts of climate change along with crop managing practices for soybean phenology could be utilized to develop climate change adaption strategies. Temperature, rainfall, as well as solar irradiance are all major influences of soybean crop growth and production. Climate change is described by rising temperatures, changing rainfall pattern, and an increase in the occurrence of severe weather events. Rising temperature hastens soybean crop phenological development and shortens the growing season, possibly reducing soybean crop production in soybean-soybean copping system. Furthermore, soybean development is jeopardized by drought caused by decreasing precipitation. A deeper knowledge of how soybean crop reacts to climate change can give a scientific foundation for adapting to and mitigating the effects of climate change in soybean-soybean copping system. He et al. (2020) revealed that between 1981 and 2010, changes in soybean phenology were seen in the key soybean region of eastern China. Sowing, emergence, flowering, and maturity dates were observed to be delayed by a mean value of 1.78, 0.83, 0.19, as well as 0.62 days/decade, respectively. Furthermore, durations of vegetative development phase and soybean planting season were reduced by 0.62 and 1.16 days/decade, respectively. In contrast, the reproductive time was extended by a mean of 0.43 days each/decade. Crop management strategies had a bigger impact than climatic change on planting, emergence, and time to maturity. The magnitude of change in phenology trends caused by management and climate warming was opposing. Climate change had a greater impact on flowering dates, vegetative and reproductive phases, and duration of the growing period than crop management practices. The foremost climatic element impacting most soybean phenological stages and phases was the mean temperature. Tan et al. (2021) revealed that on a nationwide scale, soybean phenological sages such as planting, E, trifoliate, A, and podding dates were delayed, but M dates advanced. Vegetative duration (0.52 ± 0.24 days/decade) and total growth phase (1.32 ± 0.30 days/decade) for soybean were both reduced, whereas reproductive duration (0.05 ± 0.26 days/decade) was somewhat enhanced. Regional tendencies for soybean phenology phases and crucial growth periods differed. Soybean phenology stages were delayed in the Huang soybean-growing region but accelerated in the southern soybean region. Furthermore, the important growing phases in the north soybean-growing region were significantly decreased. The vulnerability of important phases of growth of soybean to temperatures was generally negative, although the sensitivity to rainfall and sunlight hours varied by area. Numerous phenological phases, in particular, were adversely sensitive to sunlight hours. Liu and Dai (2020) stated that from 1992 to 2011 in China, consequences revealed that North region spring season growing soybean's S-A, A-M, and overall S-M phases were decreased, but summer season growing soybean's phenological phases were lengthened. Southern region growing spring soybean S-A, A-M, and overall S-M phases were also extended. Anthropogenic global warming independently increased the sowing to anthesis durations for northern spring, summer, as well as southern spring by 0.3, 0.52, and 0.3 days/decade, respectively, while decreasing total growth durations by 1.43, 0.74, and 6.3 days/decade. Rising temperatures accelerated soybean phenology and reduced crucial growth phases.

8.7 Impact of Climate Change on Current and Future Scenarios of Soybean Production

Changes in climate have posed significant obstacles to the agriculture sector. Crop yields, notably soybean yields, are expected to fall owing to climatic stress factors such as extreme temperature changes, drought, floods, as well as excessive saltiness across the world. Soybean-soybean cropping system may profit from rising carbon dioxide concentrations in the atmosphere, but these benefits are likely to be offset by heat and drought conditions triggered by climate change (Osei et al. 2023). According to the World Bank statement, changing climate might result in a 5% yield loss from present food production system at the end of 2030 and a 30% decline at the end of 2080. Kucharik and Serbin (2008) suggested that in the USA, the yield of soybean in soybean-soybean cropping system might possibly fall by 16% for every degree (°C) if summer-time temperatures rise in the future; however, if summer precipitation totals were modestly raised by 50 mm, yields might improve by 5–10%, offsetting some of the negative impacts of rising temperatures. While northern US Belt regions like Wisconsin might benefit from a warmer climate regime and management changes that lengthen the crop-growing season in the spring and fall, mid- to high-latitude soybean crop productivity may be threatened by increased summer season warming unless adaptive actions are taken. Mall et al. (2004) predicted in India that reduced soybean yields are expected for practically all areas in all the GCM-projected climate warming scenarios (at the time of a doubling of carbon dioxide levels). In diverse circumstances, the average soybean yield drop varied by location, from 14% in Pune (western India) to 23% in Gwalior (Central India). In comparison to other regions of India, western and southern India show less of a decline in soybean output. Under climatic scenarios produced by the UKMO model for the present CO_2 atmosphere and the doubling carbon dioxide atmosphere, the average production was shown to be considerably influenced. Guo et al. (2022) revealed that in China from 1990 to 2017, though the rate of rise in soybean yield reduced after 2010, more than 90% of the soybean-growing areas still exhibited an increase of soybean production. Climate change lowered soybean production by 1%, but adaptation strategies such as improvements in crop management methods and genetic factors improved soybean yield by 16% and 6%, respectively. The amount of pesticide applied and the overall number of agricultural equipment were mostly responsible for the beneficial effects of agricultural management methods. do Rio et al. (2016) revealed a reduction in soybean production by approximately 35% for the region of southern Brazil. According to Justino et al. (2013), the Midwest and northern Brazilian areas had a rise in soybean production up to 60%. After 2010, soybean yield exhibited a flat or even a decline with only improvements to agricultural management methods. In 49% of the areas, soybean yield would decline in the future if technology remained at its 2010s' level. Gong et al. (2022) indicated that concerning the effects of climate changes on soybean crop yield, China's climatic potential productivity grew from 1961 to 2010 and showed a rising trend from south to north and west to east. Under the climatic scenarios, the climate potential production of soybean will drop in the USA and southern Brazil, and a later planting date is advised to prevent probable thermal and water scarcity stress. Battisti et al. (2018) assessed climate change impact on soybean production. Three climate scenarios were assessed in the soybean-growing region: the baseline (from 1961 to 2014) along with two projected climate situations for mid-century (from 2041 to 2070) with lowest (+2.2 °C, A1BLs) and highest (+3.2 °C, A1BHs) deltas for temperature along with carbon dioxide level of 600 ppm. For that region, soybean production might be lowered by up to 28%. Goldblum (2009) revealed that in the USA, soybean production is inversely connected to mean monthly temperature in central and southern Illinois throughout the summer and favorably correlated with precipitation in the same regions between July and August. da Silva et al. (2021) determined the influence of climate change on soybean yield. They stated that in the average future climatic scenario, soybean yields across 16 regions varied from +1% to +32% from present yields. However, they revealed a 5% increase in the soybean yield production risk for RCP 8.5. The primary cause of these outcomes was attributed to the good impact of rising carbon dioxide levels on soybean crop production, which outweighs the detrimental impacts of rising temperatures and water shortage stress on rainfed Brazilian soybeans. Araji et al. (2018) projected in Iran that during the end of 2030s, all studied parameters caused a rise in soybean production and biomass that was positively correlated with meteorological factors. For practically all parameters, the most notable association between soybean yield and biomass was represented by the maximum temperature. Finally, soybean may eventually reach their ideal threshold temperature, which would result in higher yield starting in the year 2030. Yoon et al. (2021) considered the impact of climate change on soybean cultivars in South Korea. In comparison to historic yield, the maximum production drop of 27% was recorded in 'Soy-15' cultivated under the SSP585 scenario from 2040 to 2059. 'Soy-15' yield was 2.21 Mg/ha in the historical period and 1.61 Mg/ha in the future period from 2040 to 2059. The largest yield decrease of 14% was likewise seen for the 'Soy-6' cultivar in the future scenario from 2040 to 2059 under the SSP585 circumstance. 'Soy-0' had the lowest yield loss in projected warming scenarios, with yield dropping by approximately 3–8% in the soybean-growing region. Mohanty et al. (2017) predicted increase in temperature trend and thus caused lessened soybean production. Increasing the thermal trend while decreasing rainfall by 10% reduced soybean production by 10%. However, increases in temperature and rainfall did not result in favorable soybean growth. Lowering the temperature from the base by 1 °C and increasing precipitation by more than 10% increased soybean yield, but raising the temperature by 1 °C with no changes in precipitation resulted in a 10–15% decrease in soybean production in soybean-growing regions in India. Using the CSM-CROPGRO-Soybean model, Lal et al. (1999) predicted a 50% increase in soybean production in Central India assuming a doubling of carbon dioxide concentration. Nevertheless, a 3 °C increase in surface atmospheric temperature nearly negates the benefits of a doubling of concentrations of carbon dioxide by causing advancement in anthesis stage and decreasing the grain filling phase, which reduces the entire length of the crop and, consequently, production. It has been shown that soybean crop in Central India is more susceptible to upsurge in Tmax than decreases in Tmin. Grain yield is limited to roughly 32% by a 10% reduction in daily precipitation (Table 8.4).

Table 8.4 Soybean yield reduction/increase in soybean-soybean cropping systems across various regions in the world

Country/ continent	Model	Climatic scenarios	Grain yield reduction/ increase (%)	References
Pakistan	CROPGRO	2039–2069	−36	Ejaz et al. (2022)
USA	APSIM	1976–2006	−16	Kucharik and Serbin (2008)
India	CROPGRO	2010–2050	−23	Mall et al. (2004)
China	CROPGRO	2021–2060	−49	Guo et al. (2022)
Brazil	MONICA	2041–2070	−28	Battisti et al. (2018)
USA	CRCM	2039–2063	−31	Goldblum (2009)
Iran	Econometric	2021–2040	−26	Yuzbashkandi and Khalilian (2020)
USA	CROPGRO	2040–2070	−34	Mera et al. (2006)
Brazil	DSSAT	2040–2069	+32	da Silva et al. (2021)
Iran	AquaCrop	2020–2039	+19	Araji et al. (2018)
South Korea	ALMANAC	2040–2059	−27	Yoon et al. (2021)
India	APSIM	1980–2010	−15	Mohanty et al. (2017)
America	HERMES	2083–2099	−26	Ma et al. (2021)
India	DSSAT	1980–2013	−32	Ramteke et al. (2015)

8.8 Soybean Case Study in Pakistan

Ejaz et al. (2022) assessed the influence of change in climatic conditions on soybean in the soybean-soybean cropping system in Pakistan. During 2019 and 2020, field trials were carried out for spring and fall period at grower's farm in Multan. In the spring growing period experiment, soybean cultivars comprising 'Williams 82', 'NARC 1', and 'Ajmeri' were kept in the main plots for all growing periods, whereas planting dates included 10th January, 20th January, 1st February, 10th February, and 20th February in the spring growing season and 1st July, 15th July, 1st August, 15th August, and 30th August in the autumn growing season and were allotted to the subplots. All experiment observed data were used for calibration along with evaluation of soybean model. After that, the model was applied for climate change assessment in soybean-soybean cropping system in Pakistan. Results showed that model calibration and evaluation were good in the prediction of time series datasets for LAI as well as biomass of soybean cultivars for spring and fall growing seasons (Figs. 8.5, 8.6, 8.7, and 8.8). Post-calibration, model evaluation was good because of satisfactory ranges for errors and root mean square errors. Regarding sensitivity test, soybean growth simulation model was responsive to CO_2, air temperatures, as well as precipitation. Soybean production was lessened as a result of climate warming in all growing seasons in Pakistan. Climate warming minimized more soybean production in the fall period as compared to spring period (Fig. 8.9). Without adaptation approaches, grain productivity will be lessened by 22.7% for spring along with 35.9% for fall period for mid-century (2039–2069) at whole GCM at RCP 8.5

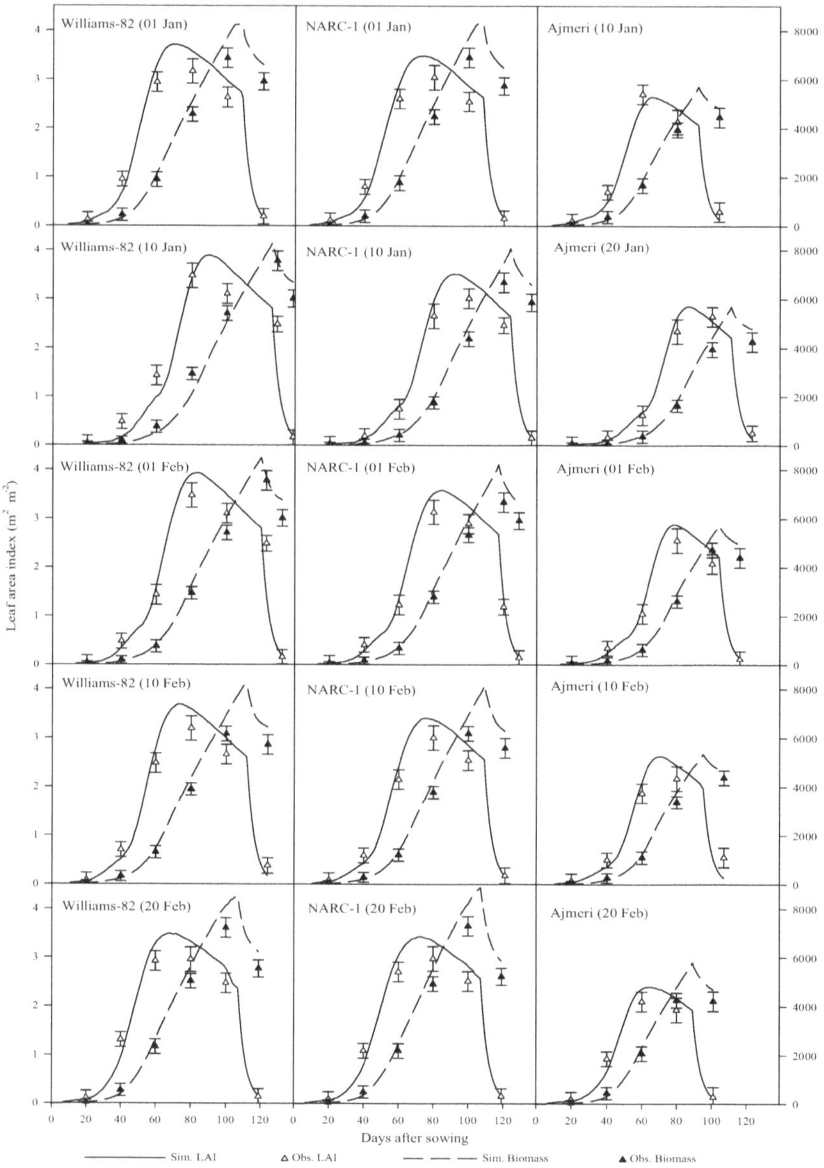

Fig. 8.5 Effect of cultivars and sowing dates on leaf area index and biomass of spring soybean during 2019 at the study site. (Source: Modified and adopted from Ejaz et al. 2022)

(Fig. 8.10). Implementing adaptative approaches of 7 days' early planting for spring along with 10 days' delayed planting for fall, in addition to sowing of climate-resilient genotypes, could diminish the adverse influence of climatic changes.

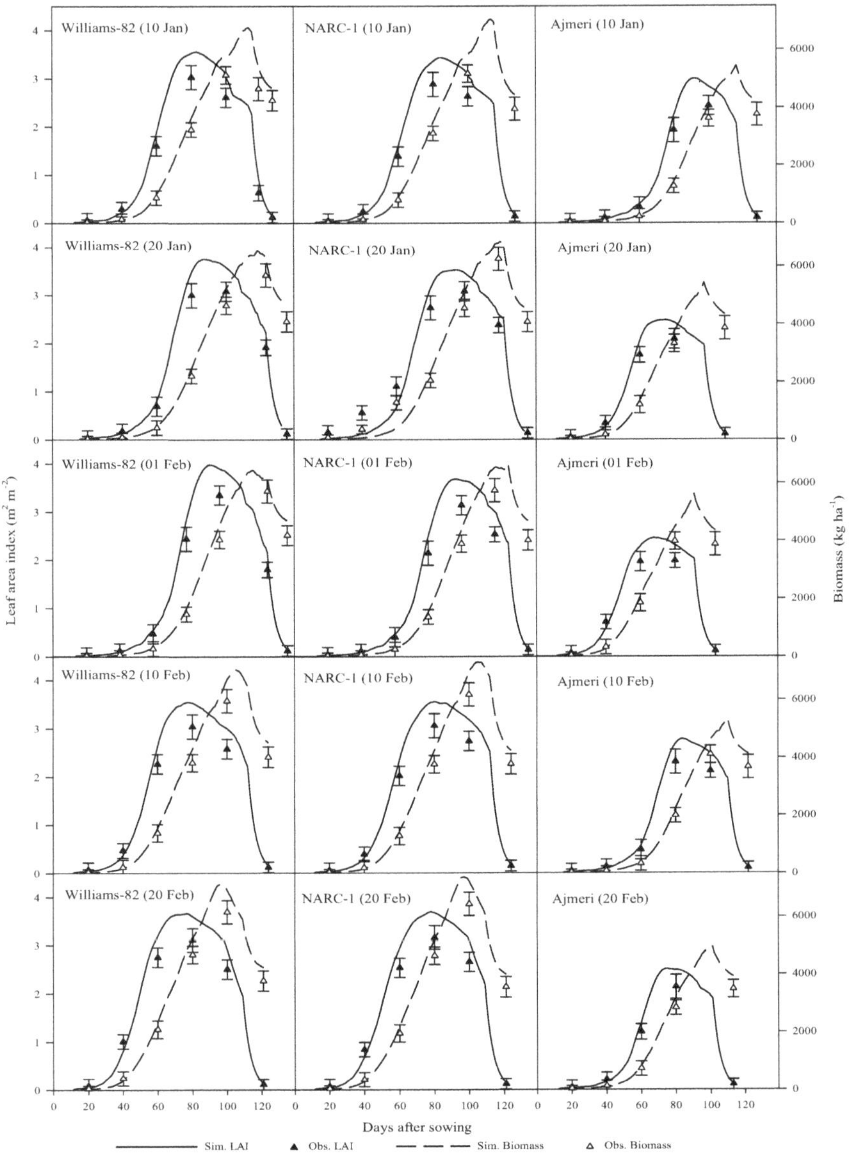

Fig. 8.6 Effect of cultivars and sowing dates on leaf area index and biomass of spring soybean during 2020 at the study site. (Source: Modified and adopted from Ejaz et al. 2022)

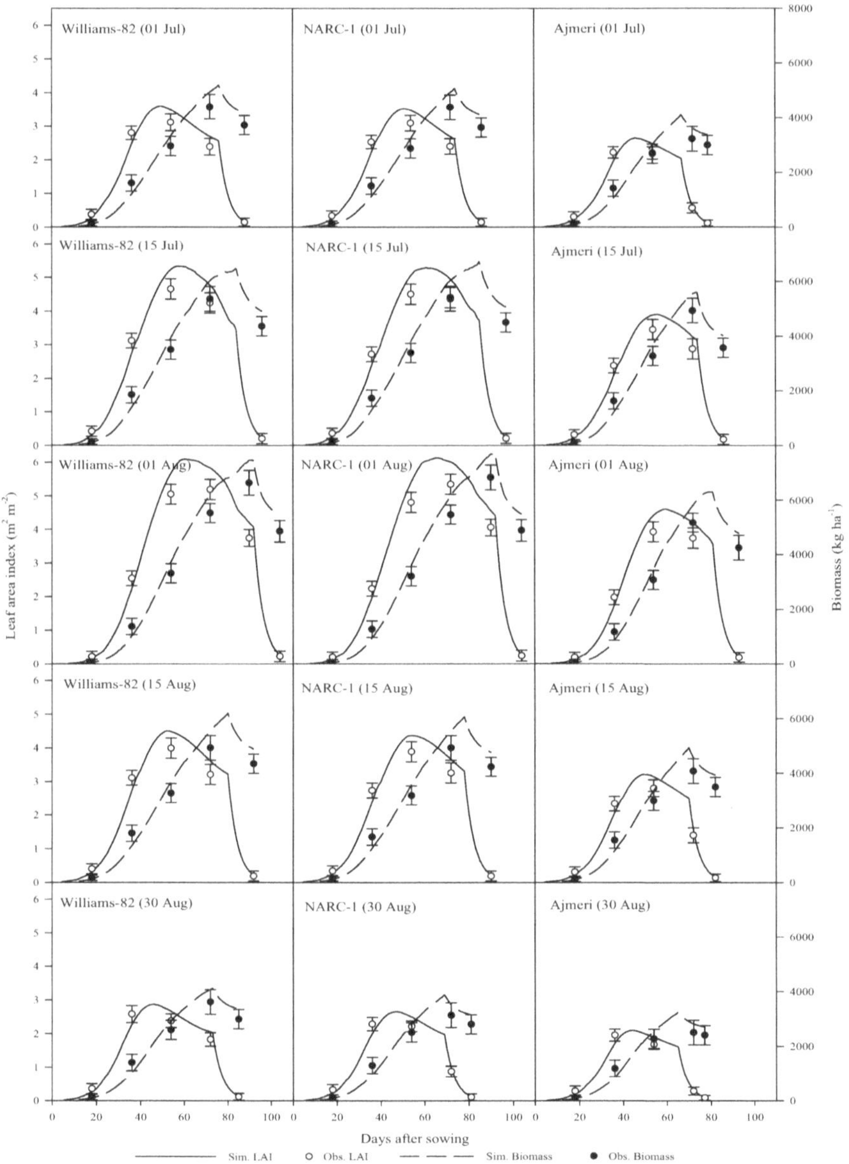

Fig. 8.7 Effect of cultivars and sowing dates on leaf area index and biomass of autumn soybean during 2019 at the study site. (Source: Modified and adopted from Ejaz et al. 2022)

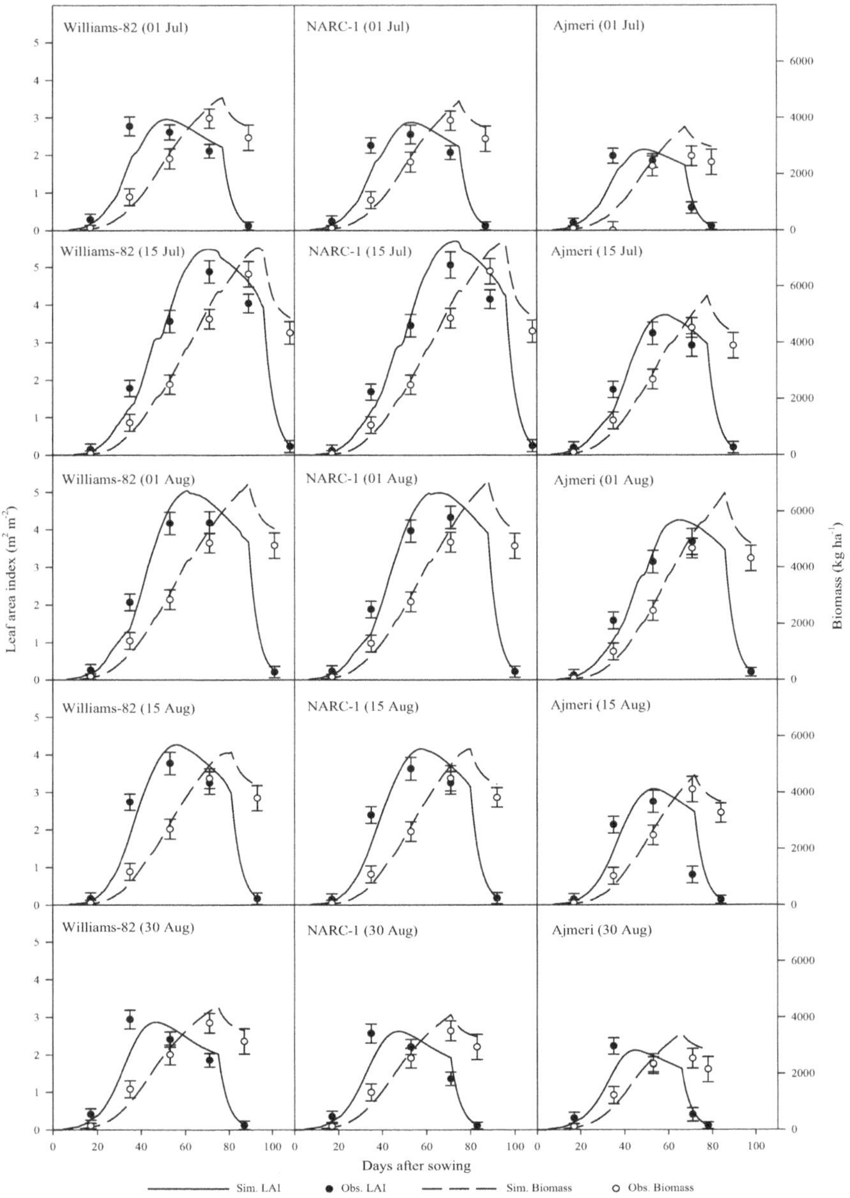

Fig. 8.8 Effect of cultivars and sowing dates on leaf area index and biomass of autumn soybean during 2020 at the study site. (Source: Modified and adopted from Ejaz et al. 2022)

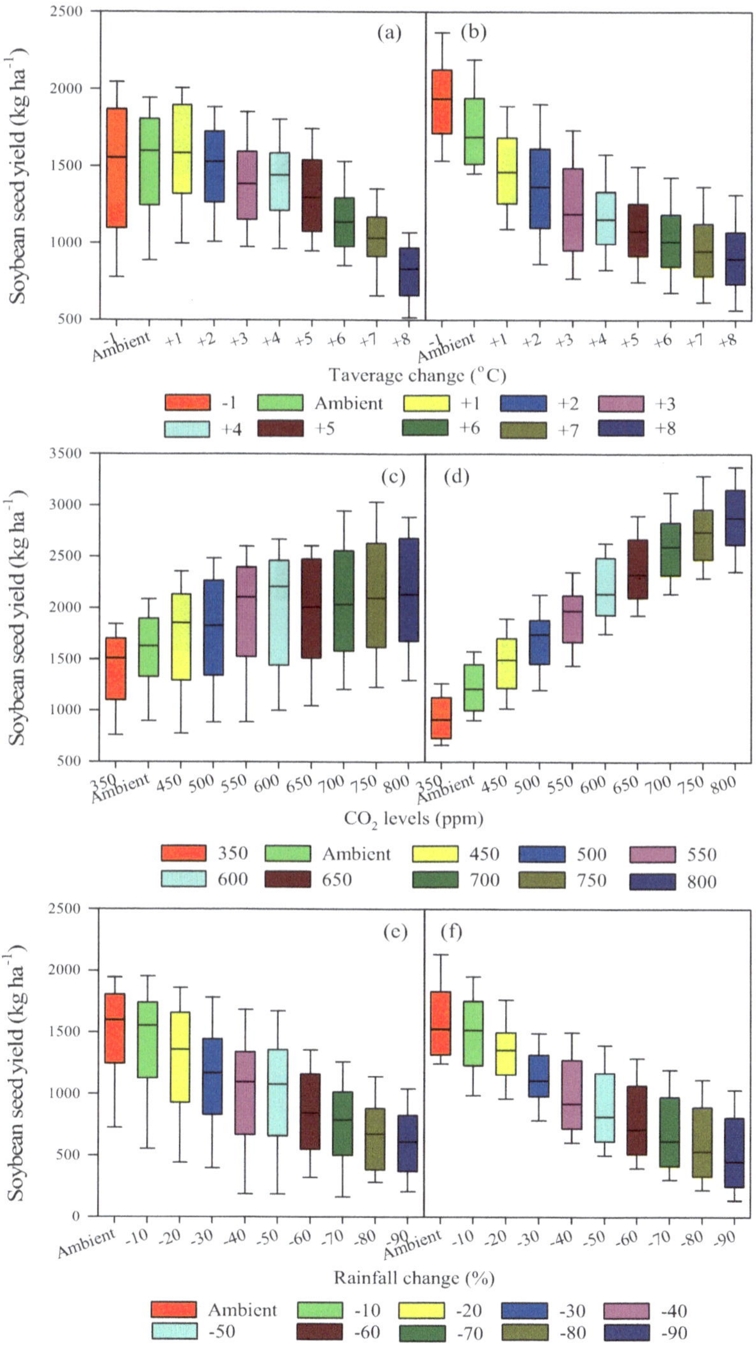

Fig. 8.9 Model sensitivity analysis at different Tave (**a, b**), CO_2 (**c, d**), and rainfall (**e, f**), for spring soybean (**a, c, e**) and autumn soybean (**b, d, f**) at the study site. (Source: Modified and adopted from Ejaz et al. 2022)

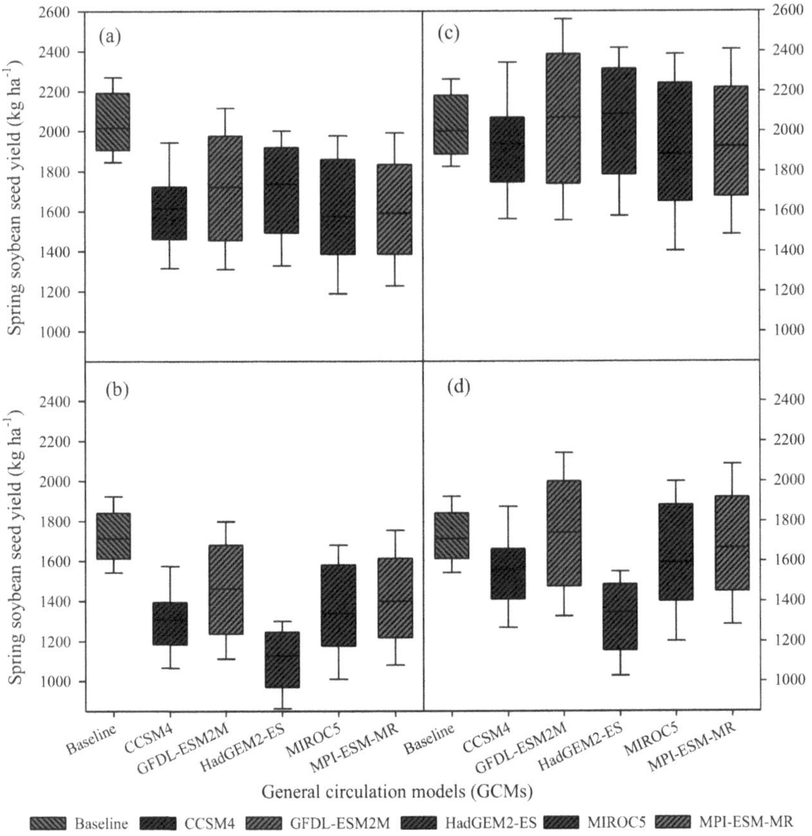

Fig. 8.10 Integrated climate change impact assessment on seed yield of spring soybean without adaptation (**a**) and with adaptation (**b**) and autumn soybean without adaptation (**c**) and with adaptation strategies (**d**) for different GCMs at the study site. (Source: Modified and adopted from Ejaz et al. 2022)

8.9 Adaption Strategies

In the present and upcoming decades, it is essential to adopt climate-smart soybean system (soybean-based climate-smart innovation, considering the interaction among changes in climatic conditions, genotypes, and crop husbandry practices for example, optimizes planting window along with plant population) to enhance soybean production (Bao et al. 2015). The most crucial tool for increasing soybean yield and for adopting climate change resilience is appropriate crop management practices in soybean-soybean cropping system. Adopting the ideal planting window could be a smart strategy for combating climate warming condition (Feng et al. 2021; Yoon et al. 2021). To maintain and boost soybean plant productivity, heat stress-tolerant genotype development and agronomic approaches that minimize the negative impacts of high temperatures are necessary. Farmers have historically employed longer duration cultivars and postponed sowing dates as management adjustments to climate change, and they can continue to do so. Soybean crop phenology is influenced by a variety of agronomic management techniques, including sowing dates and cultivar traits, as well as environmental factors (primarily thermal time needed). Previous research has demonstrated that altering the sowing dates and cultivars utilized will mitigate phenology fluctuations brought on by the climate (He et al. 2020). Selection of soybean crop cultivars for adaptability against climate warming depends on irrigation management as well as maturity group. In comparison to rainfed soybeans, irrigated soybean would improve soybean production. Enhancement in soybean plant populations would also lessen the impacts of high air temperatures. Changing the planting date and choosing soybean cultivars with higher growing degree day needs, higher temperature resistance, and lengthy development periods are efficient adaptations to minimize the negative impact of climate change in soybean-soybean cropping system regions (Liu and Dai 2020). A number of antioxidants such as proline (Pro) and glycine-betaine (GB) function as friendly solutes or osmo-protectants that can be employed to lessen the negative effects of heat stress due to climate warming. Various osmo-protectants can affect soybean plant growth in a number of ways, including through foliar sprays, pre-sowing seed treatments, and rooting media. According to various reports, using Pro lessens the negative impacts of environmental pressures. Applications of GB improve soybean plants' ability to tolerate challenging conditions. Stay-green (SG) cultivars, also known as delayed senescence varieties, are a key source of germplasm for the genetic modification of soybean plants to reduce heat stress due to climate warming. Climate-resilient cultivars preserve their green leaves and photosynthetic potential for a longer period of time following anthesis than normal cultivars (MacCarthy et al. 2022). Climate-smart soybean plants exhibit delayed leaf senescence under stress and enhanced soybean crop yield, demonstrating their tolerance to biotic and abiotic challenges. Through several methods, the use of antioxidants, osmo-protectants, and phytohormones may increase the heat tolerance in soybean plants. As a result, the antioxidative defense system and activity of a number of compounds, including antioxidants, compatible solutes, and hormones, against higher temperature stress are strong and can significantly alleviate the problem of seasonal climate warming heat stress (Ramteke et al. 2015). Adjusting the cropping schedule and crop rotation, creating and promoting the adoption of high-yielding cultivars, and employing sustainable technological solutions are some potential adaptation strategies for sustained soybean productivity. The best method for reducing the thermal effects of climate change should be to delay the sowing date or make it earlier for the soybean crop in all regions of the world (Gong et al. 2022).

8.10 Conclusion

Soybean-soybean cropping system may profit from rising carbon dioxide concentrations in the atmosphere, but these benefits are likely to be offset by heat and drought conditions triggered by climate change. High temperature has a significant impact on soybean-soybean cropping system production when it happens throughout pivotal phases of crop development. Temperature increases shorten soybean crop life span by triggering earlier anthesis and decreasing grain filling phase. The shortened crop period resulted in reduced yield per plant; consequently, an increase in temperatures resulted in a decrease in soybean productivity under a warmer environmental condition. Across the world, reports of heat-stressed soybean crop have increased in frequency in recent years. Even irrigated soybean crop experiences heat stress and considerable evaporation losses. In these circumstances, photosynthesis decreases, soybean plant shifts from a growth route to a survival one, and yields are therefore reduced. In order to assess the effect of climatic inconsistency and change on soybean crop production, identify adaptation strategies and suitable management practices, and develop mitigation measures to reduce the adverse effect of climatic changes, comprising extremity of events, on soybean production, it is essential to have a strong understanding of the association between climatic capriciousness, crop management practices, and soybean yield. Farmers should adopt adaptation strategies to increase soybean production under climate change circumstances.

References

Abbas Q, Ahmad S (2018) Effect of different sowing times and cultivars on cotton fiber quality under stable cotton-wheat cropping system in southern Punjab, Pakistan. Pak J Life Soc Sci 16(2):77–84

Abbas G, Ahmad S, Ahmad A, Nasim W, Fatima Z, Hussain S, Rehman MH, Khan MA, Hasanuzzaman M, Fahad S, Boote KJ, Hoogenboom G (2017) Quantification the impacts of climate change and crop management on phenology of maize-based cropping system in Punjab, Pakistan. Agric For Meteorol 247:42–55

Abbas G, Fatima Z, Riaz F, Iqbal P, Hussain S, Zakir I, Iqbal M, Haque E, Ahmad S (2019) Study of sowing dates and hybrids effect in maize-based cropping system under arid conditions of Southern Punjab, Pakistan. Pak J Life Soc Sci 17(1):24–31

Abbas G, Ahmad S, Hussain M, Fatima Z, Hussain S, Iqbal P, Ahmed M, Farooq M (2020a) Sowing date and hybrid choice matters production of maize-maize system. Int J Plant Prod 14:583–595

Abbas G, Fatima Z, Hussain M, Hussain S, Atique-ur-Rehman, Sarwar N, Ahmed M, Ahmad S (2020b) Nitrogen rate and hybrid selection matters productivity of maize-maize cropping system under irrigated arid environment of southern Punjab, Pakistan. Int J Plant Prod 14:309–320

Abbas Q, Atique-ur-Rehman, Sarwar N, Hussain S, Abbas G, Khan MN, Fatima Z, Naz S, Younis H, Ahmed M, Ullah H, Iqbal P, Iqbal M, Kan M, Ahmad S (2020c) Effect of different sowing times and cultivars on wheat grain quality under cotton-wheat cropping system in Southern Punjab, Pakistan. Pak J Life Soc Sci 18(1):41–50

Ahmad S, Hasanuzzaman M (2012) Integrated effect of plant density, N rates and irrigation regimes on the biomass production, N content, PAR use efficiencies and water productivity of rice under irrigated semiarid environment. Notulae Botanicae Horti Agrobotanici Cluj-Napoca 40(1):201–211

Ahmad S, Ahmad A, Soler CMT, Ali H, Zia-Ul-Haq M, Anothai J, Hussain A, Hoogenboom G, Hasanuzzaman G (2012a) Application of the CSM-CERES-Rice model for evaluation of plant density and nitrogen management of fine transplanted rice for an irrigated semiarid environment. Precis Agric 13(2):200–218

Ahmad S, Ali H, Ismail M, Shahzad MI, Nadeem M, Anjum MA, Zia-Ul-Haq M, Firdous N, Khan MA (2012b) Radiation and nitrogen use efficiencies of C_3 winter cereals to nitrogen split application. Pak J Bot 44(1):139–149

Ahmad S, Ahmad A, Ali H, Hussain A, Garcia y Garcia A, Khan MA, Zia-Ul-Haq M, Hasanuzzaman M, Hoogenboom G (2013) Application of the CSM-CERES-Rice model for evaluation of plant density and irrigation management of transplanted rice for an irrigated semiarid environment. Irrig Sci 31(3):491–506

Ahmad A, Ashfaq M, Rasul G, Wajid SA, Khaliq T, Rasul F, Saeed U, Rahman MH, Hussain J, Baig IA, Naqvi SAA, Bokhari SAA, Ahmad S, Naseem W, Hoogenboom G, Valdivia RO (2015a) Impact of climate change on the rice-wheat cropping system of Pakistan. In: Hillel D, Rosenzweig C (eds) Handbook of climate change and agro-ecosystems, vol 3. Imperial College Press and the American Society of Agronomy, London, pp 219–258

Ahmad S, Ali H, Atique-ur-Rehman, Khan RJ, Ahmad W, Fatima Z, Abbas G, Irfan M, Ali H, Khan MA, Hasanuzzaman M (2015b) Measuring leaf area of winter cereals by different techniques: a comparison. Pak J Life Soc Sci 13(2):117–125

Ahmad S, Raza I, Muhammad D, Ali H, Hussain S, Dogan H, Zia-Ul-Haq M (2015c) Radiation, water and nitrogen use efficiencies of *Gossypium hirsutum* L. Turk J Agric For 39(5):825–837

Ahmad S, Ali H, Farooq U, Khan SU, Atique-ur-Rehman, Sarwar N, Shahzad AN, Doğan H, Hussain S, Sultan MT, Waheed A, Zia-ul-Haq M, Hussain K, Khan MA (2016a) Improving nitrogen and radiation-use-efficiencies of C_4 summer cereals by split nitrogen applications under irrigated arid environment. Turk J Agric For 40(2):280–289

Ahmad S, Hussain S, Fatima Z, Abbas G, Atique-ur-Rehman, Khan MR, Younis H, Naz S, Sohail M, Ajmal M, Abbas N, Akhtar M, Rauf A, Khan M, Ali Z, Hassan M, Rizwan M, Bajwa RS, Ajmal A, Naz S, Ali H, Khan AA, Ali M, Sarwar G, Khan MA, Hasanuzzaman M (2016b) Application of DSSAT model for sowing date management of C_4 summer cereals for fodder and grain crops under irrigated arid environment. Pak J Life Soc Sci 14(2):104–114

Ahmad S, Nadeem M, Abbas G, Fatima Z, Khan RJ, Ahmed M, Ahmad A, Rasul G, Khan MA (2016c) Quantification of the effects of climate warming and crop management on sugarcane phenology. Clim Res 71(1):47–61

Ahmad S, Abbas G, Fatima Z, Khan RJ, Anjum MA, Ahmed M, Khan MA, Porter CH, Hoogenboom G (2017a) Quantification of the impacts of climate warming and crop management on canola phenology in Punjab, Pakistan. J Agron Crop Sci 203(5):442–452

Ahmad S, Abbas Q, Abbas G, Fatima Z, Atique-ur-Rehman, Naz S, Younis H, Khan RJ, Nasim W, Rehman MH, Ahmad A, Rasul G, Khan MA, Hasanuzzaman M (2017b) Quantification of climate warming and crop management impacts on cotton phenology. Plan Theory 6:1–16

Ahmad S, Atique-ur-Rehman, Ejaz M, Fatima Z, Kan M, Ahmad M (2018) Agricultural land-use change of major field crops in Pakistan (1961–2014). Sci Technol Dev 37(3):113–121

Ahmad A, Abbas G, Ahmed M, Fatima Z, Anjum MA, Rasul G, Khan MA, Hoogenboom G (2019) Climate warming and management impact on the change of phenology of the rice-wheat cropping system in Punjab, Pakistan. Field Crop Res 230:46–61

Ahmad A, Ashfaq M, Rasul G, Wajid SA, Ahmad I, Khaliq T, Nasir J, Rasul F, Riaz F, Ahmad B, Ahmad S, Baig IA, Valdivia RO, Hoogenboom G (2021) Development of climate change adaptation strategies for cotton–wheat cropping system of Punjab, Pakistan. In: Rosenzweig C, Mutter CZ, Contreras EM (eds) Handbook of climate change and agroecosystems: climate change and farming system planning in Africa and South Asia: AgMIP stakeholder-driven research. World Scientific Publishing Europe Ltd, London, pp 277–327

Ahmad S, Ahmad I, Ahmad B, Ahmad A, Wajid A, Khaliq T, Abbas G, Wilkerson CJ, Hoogenboom G (2023a) Regional integrated assessment of climate change impact on cotton production in a semi-arid environment. Clim Res 89:113–132

Ahmad S, Raza MA, Hussain S, Abbas G, Fatima Z, Ahmed M, Goheer MA, Garcia y Garcia A, Wilkerson CJ, Hoogenboom G (2023b) Identification of weak links in the production technology of farmer for bridging canola yield-gap. J Agric Sci 161:241–253. https://doi.org/10.1017/S0021859623000187

Ahmed M, Akram MN, Asim M, Aslam M, F-u Hassan Higgins S, Stockle CO, Hoogenboom G (2016) Calibration and validation of APSIM-wheat and CERES-wheat for spring wheat under rainfed conditions: models evaluation and application. Comput Electron Agric 123:384–401

Ahmed I, Rahman MH, Ahmad A, Ahmad S, Hussain J, Ullah A, Judge J (2018a) Assessing the impact of climate variability on maize using simulation modeling under semi-arid environment of Punjab, Pakistan. Environ Sci Pollut Res 25(23):28413–28430

Ahmed M, Ijaz W, Ahmad S (2018b) Adapting and evaluating APSIM-SoilP-Wheat model for response to phosphorus under rainfed conditions of Pakistan. J Plant Nutr 41(16):2069–2084

Ahmed M, Stockle CO, Nelson R, Higgins S, Ahmad S, Raza M (2019) Novel multimodal ensemble approach to evaluate the sole effect of elevated CO_2 on winter wheat productivity. Sci Rep 9:7813

Ahmed M, Ahmad S, Tariq M, Fatima Z, Aslam Z, Raza MA, Iqbal N, Akmal M, Fayyaz-ul-Hassan, Abbasi NA, Hayat R (2020) Wastes to be the source of nutrients and energy to mitigate climate change and ensure future sustainability: options and strategies. J Plant Nutr 43(6):896–920

Ahmed M, Hayat R, Ahmad M, Mahmood-ul-Hassan, Kheir AMS, Fayyaz-ul-Hassan, Rehman MH, Shaheen FA, Raza MA, Ahmad S (2022) Impact of climate change on dryland agricultural systems: a review of current status, potentials, and further work need. Int J Plant Prod 16:341–363

Ali H, Iqbal N, Ahmad S, Shahzad AN, Sarwar N (2013) Performance of late sown wheat crop under different planting geometries and irrigation regimes in arid climate. Soil Tillage Res 130:109–119

Allen LJ, Boote KJ (2000) Crop ecosystem responses to climatic change: soybean. In: Reddy KR, Hodges HF (eds) Climate change and global crop productivity. CABI, Oxon, pp 133–160

Amin A, Nasim W, Fahad S, Ali S, Ahmad S, Rasool A, Saleem N, Hammad HM, Sultana SR, Mubeen M, Bakhat HF, Ahmad N, Shah GM, Adnan M, Noor M, Basir A, Saud S, Rahman MH, Paz JO (2018a) Evaluation and analysis of temperature for historical (1996–2015) and projected (2030–2060) climates in Pakistan using SimCLIM climate model: ensemble application. Atmos Res 213:422–436

Amin A, Nasim W, Mubeen M, Ahmad A, Nadeem M, Urich P, Fahad S, Ahmad S, Wajid A, Tabassum F, Hammad HM, Sultana SR, Anwar S, Baloch SK, Wahid A, Wilkerson CJ, Hoogenboom G (2018b) Simulated CSM-CROPGRO-cotton yield under projected future climate by SimCLIM for Southern Punjab, Pakistan. Agric Syst 167:213–222

Amin A, Nasim W, Mubeen M, Sarwar S, Urich P, Ahmad A, Wajid A, Khaliq T, Rasul F, Hammad HM, Rehmani MIA, Mubarak H, Mirza N, Wahid A, Ahmad S, Fahad S, Ullah A, Khan MN, Ameen A, Amanullah, Shahzad B, Saud S, Alharby H, Ata-Ul-Karim ST, Adnan M, Islam F, Ali QS (2018c) Regional climate assessment for temperature and precipitation in Southern Punjab (Pakistan) using SimCLIM climate model for different temporal scales. Theor Appl Climatol 131:121–131

Araji HA, Wayayok A, Bavani AM, Amiri E, Abdullah AF, Daneshian J, Teh CBS (2018) Impacts of climate change on soybean production under different treatments of field experiments considering the uncertainty of general circulation models. Agric Water Manag 205:63–71

Bao Y, Hoogenboom G, McClendon RW, Paz JO (2015) Potential adaptation strategies for rainfed soybean production in the south-eastern USA under climate change based on the CSM-CROPGRO-Soybean model. J Agric Sci 153(5):798–824

Battisti R, Sentelhas PC, Parker PS, Nendel C, Gil MDS, Farias JR, Basso CJ (2018) Assessment of crop-management strategies to improve soybean resilience to climate change in Southern Brazil. Crop Pasture Sci 69(2):154–162

Bebeley JF, Kamara AY, Jibrin JM, Akinseye FM, Tofa AI, Adam AM, Kamai N, Solomon R (2022) Evaluation and application of the CROPGRO-soybean model for determining optimum sowing windows of soybean in the Nigeria savannas. Sci Rep 12(1):6747

Bhatia VS, Singh P, Wani SP, Chauhan GS, Rao AK, Mishra AK, Srinivas K (2008) Analysis of potential yields and yield gaps of rainfed soybean in India using CROPGRO-Soybean model. Agric For Meteorol 148(8–9):1252–1265

Blanco A, Högy P, Zikeli S, Pignata ML, Rodriguez JH (2022) Assessment of elevated CO_2 concentrations and heat stress episodes in soybean cultivars growing in heavy metal polluted soils: crop nutritional quality and food safety. Environ Pollut 303:119123

Cabanos C, Matsuoka Y, Maruyama N (2021) Soybean proteins/peptides: a review on their importance, biosynthesis, vacuolar sorting, and accumulation in seeds. Peptides 143:170598

Chebrolu KK, Fritschi FB, Ye S, Krishnan HB, Smith JR, Gillman JD (2016) Impact of heat stress during seed development on soybean seed metabolome. Metabolomics 12:1–14

Cohen I, Zandalinas SI, Fritschi FB, Sengupta S, Fichman Y, Azad RK, Mittler R (2021) The impact of water deficit and heat stress combination on the molecular response, physiology, and seed production of soybean. Physiol Plant 172(1):41–52

Couëdel A, Edreira JIR, Lollato RP, Archontoulis S, Sadras V, Grassini P (2021) Assessing environment types for maize, soybean, and wheat in the United States as determined by spatio-temporal variation in drought and heat stress. Agric For Meteorol 307:108513

da Silva EHFM, Antolin LAS, Zanon AJ, Junior ASA, de Souza HA, dos Santos Carvalho K, Junior NAV, Marin FR (2021) Impact assessment of soybean yield and water productivity in Brazil due to climate change. Eur J Agron 129:126329

de Araújo MLS, Sano EE, Bolfe ÉL, Santos JRN, dos Santos JS, Silva FB (2019) Spatiotemporal dynamics of soybean crop in the Matopiba region, Brazil (1990–2015). Land Use Policy 80:57–67

do Rio A, Sentelhas PC, JRB F, RNR S, Ferreira RC (2016) Alternative sowing dates as a mitigation measure to reduce climate change impacts on soybean yields in southern Brazil. Int J Climatol 36(11):3664–3672

Ejaz M, Abbas G, Fatima Z, Iqbal P, Raza MA, Kheir AM, Ahmed M, Kakar KM, Ahmad S (2022) Modelling climate uncertainty and adaptations for soybean-based cropping system. Int J Plant Prod 16(2):235–250

FAO (2024) FAOSTAT database, Food and Agriculture Organization of the United Nations. Available at http://faostat.fao.org/ (Accessed 15 Feb 2024

Fatima Z, Abbas Q, Khan A, Hussain S, Ali MA, Abbas G, Younis H, Naz S, Ismail M, Shahzad MI, Nadeem M, Farooq U, Khan SU, Javed K, Khan AA, Ahmed M, Khan MA, Ahmad S (2018) Resource-use-efficiencies of C_3 and C_4 cereals under split nitrogen regimes. Agronomy 8(5):69

Fatima Z, Ahmad M, Hussain M, Abbas G, Ul-Allah S, Ahmad S, Ahmed N, Ali MA, Sarwar G, Haque E, Iqbal P, Hussain S (2020) The fingerprints of climate warming on cereal crops phenology and adaptation options. Sci Rep 10:18013

Fatima Z, Atique-ur-Rehman, Abbas G, Iqbal P, Zakir I, Khan MA, Kamal GM, Ahmed M, Ahmad S (2021) Quantification of climate warming and crop management impacts on phenology of pulses-based cropping systems. Int J Plant Prod 15:107–123

Feng ZJ, Liu N, Zhang GW, Niu FG, Xu SC, Gong YM (2019) Investigation of the AQP family in soybean and the promoter activity of TIP2; 6 in heat stress and hormone responses. Int J Mol Sci 20(2):262

Feng L, Wang H, Ma X, Peng H, Shan J (2021) Modeling the current land suitability and future dynamics of global soybean cultivation under climate change scenarios. Field Crop Res 263:108069

Fernandes RDM, de Melo DM, Elli EF, Battisti R (2022) Climate change impacts on rainfed and irrigated soybean yield in Brazil's new agricultural frontier. Theor Appl Climatol 147(1–2):803–816

Fodor N, Challinor A, Droutsas I, Ramirez-Villegas J, Zabel F, Koehler AK, Foyer CH (2017) Integrating plant science and crop modeling: assessment of the impact of climate change on soybean and maize production. Plant Cell Physiol 58(11):1833–1847

Goldblum D (2009) Sensitivity of corn and soybean yield in Illinois to air temperature and precipitation: the potential impact of future climate change. Phys Geogr 30(1):27–42

Gong L, Liu D, Jiang L, Li X, Lv J (2022) Distribution characteristics of climate potential productivity of soybean in frigid region and its response to climate change. Environ Sci Pollut Res 29(5):7452–7464

Guo S, Zhang Z, Guo E, Fu Z, Gong J, Yang X (2022) Historical and projected impacts of climate change and technology on soybean yield in China. Agric Syst 203:103522

He L, Jin N, Yu Q (2020) Impacts of climate change and crop management practices on soybean phenology changes in China. Sci Total Environ 707:135638

Hoogenboom G, Porter CH, Boote KJ, Shelia V, Wilkens PW, Singh U, White JW, Asseng S, Lizaso JI, Moreno LP, Pavan W, Ogoshi R, Hunt LA, Tsuji GY, Jones JW (2019) The DSSAT crop modeling ecosystem. In: Boote KJ (ed) Advances in crop modeling for a sustainable agriculture. Burleigh Dodds Science Publishing, Cambridge, pp 173–216. https://doi.org/10.19103/AS.2019.0061.10

Hoogenboom G, Porter CH, Shelia V, Boote KJ, Singh U, White JW, Pavan W, Oliveira FAA, Moreno-Cadena LP, Lizaso JI, Asseng S, Pequeno DNL, Kimball BA, Alderman PD, Thorp KR, Jones MR, Cuadra SV, Vianna MS, Villalobos FJ, Ferreira TB, Batchelor WD, Koo J, Hunt LA, Jones JW (2021) Decision Support System for Agrotechnology Transfer (DSSAT) Version 4.8 (DSSAT.net). DSSAT Foundation, Gainesville

Huang Y, Xuan H, Yang C, Guo N, Wang H, Zhao J, Xing H (2019) GmHsp90A2 is involved in soybean heat stress as a positive regulator. Plant Sci 285:26–33

Hussain M, Ahmad S, Hussain S, Lal R, Ul-Allah S, Nawaz A (2018) Rice in saline soils: physiology, biochemistry, genetics and management. Adv Agron 148:231–287

Jagtap SS, Jones JW (2002) Adaptation and evaluation of the CROPGRO-soybean model to predict regional yield and production. Agric Ecosyst Environ 93:73–85

Jones JW, Hoogenboom G, Porter CH, Boote KJ, Batchelor WD, Hunt LA, Wilkens PW, Singh U, Gijsman AJ, Ritchie JT (2003) The DSSAT cropping system model. Eur J Agron 18(3–4):235–265

Junior VJ (2022) Soybean production in Paraguay: agribusiness, economic change and agrarian transformations. J Agrar Chang 22(2):317–340

Justino F, Oliveira E, Rodrigues R, Gonçalves P, Souza P, Stordal F, Marengo J, Silva T, Delgado R, Lindemann D, Costa L (2013) Lindeman Mean and interannual variability of maize and soybean in Brazil under global warming conditions. Am J Clim Change 2:237–253

Karges K, Bellingrath-Kimura SD, Watson CA, Stoddard FL, Halwani M, Reckling M (2022) Agro-economic prospects for expanding soybean production beyond its current northerly limit in Europe. Eur J Agron 133:126415

Keating BA, Carberry PS, Hammer GL, Probert ME, Robertson MJ, Holzworth D, Huth NI, Hargreaves JNG, Meinke H, Hochman Z, McLean G, Verburg K, Snow V, Dimes JP, Silburn M, Wang E, Brown S, Bristow KL, Asseng S, Chapman S, McCown RL, Freebairn DM, Smith CJ (2003) An overview of APSIM: a model designed for farming systems simulation. Eur J Agron 18:267–288

Khan A, Fahad S, Khan A, Saud S, Adnan M, Wahid F, Noor M, Nasim W, Hammad HM, Bakhat HF, Ahmad S, Rehman MH, Wang D, Sönmez O (2019) Managing tillage operations and manure to restore soil carbon stocks in wheat-maize cropping system. Agron J 111:1–10

Khan MN, Hussain M, Abbas G, Fatima Z, Iqbal P, Khan A, Zakir I, Ahmed M, Mateen A, Khan MA, Ahmad S (2020) Improving resource use efficiencies of sugarcane at farmer field under arid environment. Int J Agric Biol 24(5):1279–1285

Kothari K, Battisti R, Boote KJ, Archontoulis SV, Confalone A, Constantin J, Cuadra SV, Debaeke P, Faye B, Grant B, Hoogenboom G (2022) Are soybean models ready for climate change food impact assessments? Eur J Agron 135:126482

Kucharik CJ, Serbin SP (2008) Impacts of recent climate change on Wisconsin corn and soybean yield trends. Environ Res Lett 3(3):034003

Lal M, Singh KK, Srinivasan G, Rathore LS, Naidu D, Tripathi CN (1999) Growth and yield response of soybean in Madhya Pradesh, India to climate variability and change. Agric For Meteorol 93:53–70

Liu Y, Dai L (2020) Modelling the impacts of climate change and crop management measures on soybean phenology in China. J Clean Prod 262:121271

Lopez JR, Winter JM, Elliott J, Ruane AC, Porter CH, Hoogenboom G, Anderson M, Hain C (2022) Sustainable use of groundwater may dramatically reduce irrigated production of maize, soybean, and wheat. Earth's Future 10(1):2021EF002018

Ma L, Fang QX, Sima MW, Burkey KO, Harmel RD (2021) Simulated climate change effects on soybean production using two crop modules in RZWQM2. Agron J 113(2):1349–1365

MacCarthy DS, Traore PS, Freduah BS, Adiku SG, Dodor DE, Kumahor SK (2022) Productivity of soybean under projected climate change in a semi-arid region of West Africa: sensitivity of current production system. Agronomy 12(11):2614

Mahmood A, Ullah H, Shahzad AN, Ali H, Ahmad S, Zia-Ul-Haq M, Honermeier B, Hasanuzzaman M (2013) Dry matter yield and chemical composition of sorghum cultivars with varying planting density and sowing date. Sains Malaysiana 42(10):1529–1538
Mall RK, Lal M, Bhatia VS, Rathore LS, Singh R (2004) Mitigating climate change impact on soybean productivity in India: a simulation study. Agric For Meteorol 121(1–2):113–125
Mera RJ, Niyogi D, Buol GS, Wilkerson GG, Semazzi FH (2006) Potential individual versus simultaneous climate change effects on soybean (C3) and maize (C4) crops: an agrotechnology model based study. Glob Planet Chang 54(1–2):163–182
Mohanty M, Sinha NK, McDermid SP, Chaudhary RS, Reddy KS, Hati KM, Somasundaram J, Lenka S, Patidar RK, Prabhakar M, Cherukumalli SR (2017) Climate change impacts vis-a-vis productivity of soybean in vertisol of Madhya Pradesh. J Agrometeorol 19(1):10–16
Mubarak MK, Hussain K, Abbas G, Altaf MT, Baloch FS, Ahmad S (2022) Productivity of sorghum (*Sorghum bicolor* L.) at diverse irrigation regimes and sowing dates in semi-arid and arid environment. Turk J Agric For 46(1):1–18
Mubeen M, Ahmad A, Wajid A, Khaliq T, Hammad HM, Sultana SR, Ahmad S, Fahad S, Nasim W (2016) Application of CSM-CERES-Maize model in optimizing irrigated conditions. Outlook Agric 45(3):173–184
Nadeem M, Khan MN, Abbas G, Fatima Z, Iqbal P, Ahmed M, Raza MA, Rehman A, Haq EU, Hayat A, Ali M, Ahmad S (2022) Application of CSM-CANEGRO model for climate change impact assessment and adaptation for sugarcane in semi-arid environment of Southern Punjab, Pakistan. Int J Plant Prod 16:443–466
Narayanan S, Zoong-Lwe ZS, Gandhi N, Welti R, Fallen B, Smith JR, Rustgi S (2020) Comparative lipidomic analysis reveals heat stress responses of two soybean genotypes differing in temperature sensitivity. Plan Theory 9(4):457
Nasim W, Amin A, Fahad S, Awais M, Khan N, Mubeen M, Wahid A, Turan V, Rehman MH, Ihsan MZ, Ahmad S, Hussain S, Mian IA, Khan B, Jamal Y (2018) Future risk assessment by estimating historical heat wave trends with projected heat accumulation using SimCLIM climate model in Pakistan. Atmos Res 205:118–133
Naz S, Ahmad S, Abbas G, Fatima Z, Hussain S, Khan MA, Ahmed M, Khan A, Fahad S, Nasim W, Ercisli S, Wilkerson CJ, Hoogenboom G (2022) Modeling the impact of climate warming on potato phenology. Eur J Agron 132:126404
Ntiamoah EB, Li D, Appiah-Otoo I, Twumasi MA, Yeboah EN (2022) Towards a sustainable food production: modelling the impacts of climate change on maize and soybean production in Ghana. Environ Sci Pollut Res 29(48):72777–72796
Ortiz AC, De Smet I, Sozzani R, Locke AM (2022) Field-grown soybean shows genotypic variation in physiological and seed composition responses to heat stress during seed development. Environ Exp Bot 195:104768
Osei E, Jafri SH, Saleh A, Gassman PW, Gallego O (2023) Simulated climate change impacts on corn and soybean yields in Buchanan County, Iowa. Agriculture 13(2):268
Rahman MH, Ahmad A, Wang X, Wajid A, Nasim W, Hussain M, Ahmad B, Ahmad I, Ali Z, Ishaque W, Awais M, Muddasir M, Shelia V, Ahmad S, Fahad S, Alam M, Ullah H, Hoogenboom G (2018) Multi-model projections of future climate and climate change impacts uncertainty assessment for cotton production in Pakistan. Agric For Meteorol 253-254:94–113
Rahman MH, Ahmad I, Wang D, Fahad S, Afzal M, Ghaffar A, Saddique Q, Khan MA, Saud S, Hassan S, Fahad M, Din M, Ahmad M, Yue Z, Shah RA, Sönmez O, Nasim W (2021) Influence of semi-arid environment on radiation use efficiency and other growth attributes of lentil crop. Environ Sci Pollut Res 28:13697–13711
Ramteke R, Gupta GK, Singh DV (2015) Growth and yield responses of soybean to climate change. Agric Res 4(3):319–323
Rehman A, Ashfaq M, Naqvi SAA, Adil SA, Bashir K, Ahmad A, Ahmad S, Ali A, Imran A (2015) Is climate change worsening the poverty of maize growers? Evidence from Punjab province of Pakistan. Ciencia e Tecnica Vitivinicola 30(1):105–116
Ruíz-Nogueira B, Boote KJ, Sau F (2001) Calibration and use of CROPGRO-soybean model for improving soybean management under rainfed conditions. Agric Syst 68(2):151–173

Sarwar N, Ali H, Maqsood M, Ullah E, Shahzad AN, Shahzad M, Mubeen K, Shahid MA, Ahmad S (2013) Phenological response of rice plants to different micronutrients application under water saving paddy fields on calcareous soil. Turk J Field Crop 18(1):52–57

Sarwar N, Atique-ur-Rehman, Farooq O, Wasaya A, Hussain M, El-Shehawi AM, Ahmad S, Brestic M, Mahmoud SF, Zivcak M, Farooq S (2021) Integrated nitrogen management improves productivity and economic returns of wheat-maize cropping system. J King Saud Univ Sci 33:101475

Shabbir RN, Waraich EA, Ali H, Nawaz F, Ashraf MY, Ahmad R, Awan MI, Ahmad S, Irfan M, Hussain S, Ahmad Z (2016) Supplemental exogenous NPK application alters biochemical processes to improve yield and drought tolerance in wheat (*Triticum aestivum* L.). Environ Sci Pollut Res 23(3):2651–2662

Sharma RK, Kumar S, Vatta K, Dhillon J, Reddy KN (2022) Impact of recent climate change on cotton and soybean yields in the southeastern United States. J Agric Food Res 9:100348

Singh G, Dukariya G, Kumar A (2020) Distribution, importance and diseases of soybean and common bean: a review. Biotechnol J Int 24(6):86–98

Song K, Yim WC, Lee BM (2017) Expression of heat shock proteins by heat stress in Soybean. Plant Breed Biotechnol 5(4):344–353

Soratto RP, Guidorizzi FV, Sousa WS, Gilabel AP, Job AL, Calonego JC (2022) Effects of previous fall–winter crop on spring–summer soybean nutrition and seed yield under no-till system. Agronomy 12(12):2974

St-Marseille AFG, Bourgeois G, Brodeur J, Mimee B (2019) Simulating the impacts of climate change on soybean cyst nematode and the distribution of soybean. Agric For Meteorol 264:178–187

Tan Q, Liu Y, Dai L, Pan T (2021) Shortened key growth periods of soybean observed in China under climate change. Sci Rep 11(1):1–12

Tariq M, Yasmeen A, Ahmad S, Hussain N, Afzal MN, Hasanuzzaman M (2017) Shedding of fruiting structures in cotton: factors, compensation and prevention. Trop Subtrop Agroecosyst 20(2):251–262

Tariq M, Afzal MN, Muhammad D, Ahmad S, Shahzad AN, Kiran A, Wakeel A (2018a) Relationship of tissue potassium content with yield and fiber quality components of Bt cotton as influenced by potassium application methods. Field Crop Res 229:37–43

Tariq M, Ahmad S, Fahad S, Abbas G, Hussain S, Fatima Z, Wajid N, Mubeen M, Rehman M, Khan MA, Adnan M, Wilkerson CJ, Hoogenboom G (2018b) The impact of climate warming and crop management on phenology of sunflower-based cropping systems in Punjab, Pakistan. Agric For Meteorol 256–257:270–282

Tariq M, Fatima Z, Iqbal P, Nahar K, Ahmad S, Hasanuzzaman M (2021) Sowing dates and cultivars mediated changes in phenology and yield traits of cotton-sunflower cropping system in arid environment. Int J Plant Prod 15:291–302

Umburanas RC, Kawakami J, Ainsworth EA, Favarin JL, Anderle LZ, Dourado-Neto D, Reichardt K (2022) Changes in soybean cultivars released over the past 50 years in southern Brazil. Sci Rep 12(1):508

USDA National Agricultural Statistics Service (2022) Census of Agriculture. Complete data available at www.nass.usda.gov/AgCensus

Yoon CY, Kim S, Cho J, Kim S (2021) Modeling the impacts of climate change on yields of various Korean soybean sprout cultivars. Agronomy 11(8):1590

Yuzbashkandi SS, Khalilian S (2020) On projecting climate change impacts on soybean yield in Iran: an econometric approach. Environ Process 7(1):73–87

Zahoor A, Ahmed M, Fayyaz-ul-Hassan SG, Ahmad S (2022) Ontogeny growth and radiation use efficiency of canola (*Brassica napus* L.) under various nitrogen management strategies and contrasting environments. Int J Plant Prod 16(1):195–208

Zhao J, Han T, Wang C, Shi X, Wang K, Zhao M, Chen F, Chu Q (2022) Assessing variation and driving factors of the county-scale water footprint for soybean production in China. Agric Water Manag 263:107469

Sugarcane System 9

9.1 Introduction

Sugarcane (*Saccharum officinarum* L.) is a large perennial grass that belongs to the Poaceae family. It is mostly cultivated in the subtropical and tropical regions of the world and is a crop of agro-industry that is cultivated on a large scale (Vandana et al. 2020). It is a C4 plant and is cultivated as a cash crop (Moore and Frederik 2014) in many countries. In terms of world sugar production, sugarcane provides 75% of sugar (Kumar et al. 2019). When there is a lack of energy in the body, sugar is a cheap energy source that can help with nutritional issues (FAO 1997). According to estimates, sugarcane production of 1 tonne needs 402 m^3 of water; thus, sugarcane crop consumes a lot of water. The primary material source for Egypt's sugar and molasses industries is thought to be sugarcane. Egyptians also drink it as fresh juice, and their by-products are used in the production of yeast, ethyl alcohol, acetic and citric acid, and pulp and paper (El-Kholi 2008). Due to global warming, there is a predicted increase in average temperature (4.3–4.9 °C) in 2080 in Pakistan (Akram and Hamid 2015; Abbas et al. 2017, 2019, 2020a, b, c; Abbas and Ahmad 2018; Ahmad et al. 2012a, b, 2013, 2015a, b, c, 2016a, b, c, 2017a, b, 2018, 2019, 2021, 2023a, b). A combination of long-term changes in global weather (climate change), brought on by both man-made and natural causes, may lead to significant environmental and ecological problems that have already had an impact on and will continue to have an impact on agriculture (Ahmed et al. 2016, 2018a, b, 2019, 2020, 2022; Ahmad and Hasanuzzaman 2012; Ali et al. 2013; Amin et al. 2018a, b, c). From the middle of the eighteenth century, the amount of carbon dioxide (CO_2) in the atmosphere has risen by around 30%, which increased deforestation, fossil fuel consumption, and industrial activities (Houghton et al. 2001). In many parts of the world, extreme temperatures and drought have been the two main factors affecting agricultural output and economic consequences (Ejaz et al. 2022; Fatima et al. 2018, 2020, 2021; Nadeem et al. 2022; Naz et al. 2022; Jones et al. 2003; Keating et al. 2003; Hoogenboom et al. 2019, 2021). The agricultural industry has difficulties in

M. Ahmed et al., *Cropping Systems Modeling Under Changing Climate*,
https://doi.org/10.1007/978-981-97-0331-9_9

ensuring food security for a growing global population while safeguarding the environment and the health of its ecosystems under circumstances of climate change (Rosenzweig et al. 2007). At the end of the twenty-first century, atmospheric concentration of CO_2 might be doubled (from 500 to 800 ppm) from present levels in a high-emission scenario (Mahmood et al. 2013; Mubeen et al. 2016; Nasim et al. 2018; Mubarak et al. 2022; Rehman et al. 2015; Rahman et al. 2018, 2021; Sarwar et al. 2013, 2021; Shabbir et al. 2016). Rising atmospheric CO_2 concentration and greenhouse gases are directly related to global warming (Tariq et al. 2017, 2018a, b; Tarriq et al. 2021; Hussain et al. 2018; Khan et al. 2019, 2020; Zahoor et al. 2022). GHG emission levels and other factors that raise global surface mean temperatures are projected to rise by 1.1–2.9 °C (low emission) or 2.0–5.4 °C (high emission) by 2100 compared to 1980–1999 geographical areas (IPCC 2014). Extreme weather conditions impact sugarcane growth and production worldwide (Zhao and Li 2015). To fulfill sugar demand, approximately 630 MT of sugarcane production is required with 11.5% of sugar recovery (currently 10.5%) (DES 2016). Sugarcane crop grows in more than 100 countries (Farooq and Gheewala 2020) including Brazil (Silva et al. 2019a, b), Egypt (Abdrabbo et al. 2021), Pakistan (Farooq and Gheewala 2020), India (Verma et al. 2023), Mexico (Baez-Gonzalez et al. 2018), Thailand (Pipitpukdee et al. 2020), China (Ruan et al. 2018), Indonesia (Riajaya 2020), South Africa (Jones et al. 2015), etc. Sugarcane production was recorded to be high in the continent America, then Asia, Africa, and Oceania (Table 9.1). Among all these countries, Brazil accounted for the highest production of sugarcane which was 41% of the world followed by India 19.8%, China 5.8%, Pakistan 4.32%, Thailand 4.0%,

Table 9.1 Production of sugarcane continent-wise in 2020

Continents	Production (thousand tonnes)
Americas	1,009,249
Asia	732,434
Africa	95,667
Oceania	32,365
Europe	–
World	1,869,715

Table 9.2 Production of sugarcane in 2020 country-wise

Countries	Production (thousand tonnes)
Brazil	757,117
India	370,500
China	108,652
Pakistan	81,009
Thailand	74,968
Mexico	53,953
Australia	30,283
Indonesia	28,914

Table 9.3 Production of raw sugar (2015–2019) in thousand tonnes

Countries	2015	2016	2017	2018	2019	Mean
Brazil	35,194	40,534	36,745	28,001	27,732	33,641
India	30,460	27,385	22,200	34,309	34,300	29,730
China	11,758	9664	10,245	11,440	11,811	10,983
USA	7711	8289	8172	8505	7374	8010
Mexico	5985	6945	7043	6226	6710	6581
Russian Federation	5743	6045	6665	6273	7310	6407
Pakistan	5115	7048	6621	5462	4881	5825
France	4692	4310	5648	5801	4897	5069
Australia	4816	5058	4939	4605	4517	4787

etc. (Table 9.2), while the highest raw sugar production was accounted by Brazil, India, China, etc. (Table 9.3).

9.2 Issues of Sugarcane Crop

In the sugarcane system, variation in the production of sugarcane in different areas is due to changes in temperature and rainfall patterns, natural hazards, low capacity of cultivars against vulnerability, no proper forecasting, and poor strategies (Gawander 2007). High labor costs have also played a significant role in the low profitability of sugarcane, due to low prices and the unavailability of mechanized operation (field management to harvesting) in hilly places (Zhao and Li 2015). In the sugarcane system, climate change (increased temperature) has increased evapotranspiration (Abdrabbo et al. 2021); hence, to fulfill the irrigation need, diesel pumping irrigation method is used (Henggeler et al. 2004) that increases the cost of production and emission of CO_2 from agricultural machinery (Naohiro and Takahide 2015). Availability issue of land preparation instruments, lack of irrigation water, and insect and disease problems affect the production of sugarcane (Khan et al. 2016). Financial challenges, such as a lack of capital, high input costs, low output costs, and late payments, among others, are tied to economic issues and limits. All farmers agreed that high input costs pose a serious obstacle to the efficient cultivation of sugarcane. A big problem with sugarcane is its procurement faced by the farmers. Insufficient resources and a minimum cost for the crop are the major issues for growers of sugarcane. Technical barriers to production include a lack of scientific understanding, poor soil preparation, seed shortage, inadequate insecticides and pesticides, insufficient irrigation, and natural disasters, among other things. The majority of farmers said that villagers were used to chop the sugarcane tops in order to use them as livestock feed. Sugarcane theft and top cutting are also social issues (Nazir et al. 2013).

9.3 Sugarcane Modeling

Different crop models were used to study the crop development and production of sugarcane systems for the future like CROPWAT, CERES, CropSyst, SWAP/WOFOST, EPIC, DSSAT, AquaCrop (Hunink and Droogers 2011), INFOCROP (Aggarwal et al. 2006), APSIM, and QCANE (Verma et al. 2023). While QCANE is a less frequently used but well-known model, CANEGRO and APSIM are being used in the world's key cane-producing countries. Australian models include APSIM and QCANE, whereas South African models include CANEGRO (O'Leary 1999), but the main model used for sugarcane in Asia, America, and Africa is the DSSAT-CANEGRO model (O'Leary 2000). Models for crop simulation are software tools that are computer based that are utilized to assess how climatic, agronomic, edaphic, and hydrologic factors affect crop development and production. These models, which conceptually describe a streamlined crop system, are quantified tools founded on scientific knowledge. Understanding interactions of the underlying environmental elements, plant growth mechanisms, and management methods that influence plant growth and yield has allowed for the development of crop simulation models (Battisti et al. 2018). Crop growth and development of sugarcane (tillering, canopy development, root growth, water stress, phenology, and biomass portioning and accumulation) were simulated by CANEGRO model (Singels and Bezuidenhout 2002). The genetic coefficients of sugarcane crop in DSSAT and APSIM models are presented in Tables 9.4 and 9.5, respectively. A dynamic crop development simulation model called CANEGRO-Sugarcane can simulate the effects of many climatic conditions and help develop adaptation plans for dealing with climate change in irrigated arid and semiarid environments (Jones et al. 2003; Hoogenboom et al.

Table 9.4 Sugarcane crop genetic coefficients and their description in DSSAT model

Genetic coefficient	Description
MaxPARCE	Maximum (no stress) radiation conversion efficiency expressed as assimilate produced before respiration, per unit PAR (g/MJ)
APFMX	Maximum fraction of dry mass increments that can be allocated to aerial dry mass (t/t)
STKPFMAX	Fraction of daily aerial dry mass increments partitioned to stalk at high temperatures in a mature crop (t/t on a dry mass basis)
SUCA	Sucrose partitioning parameter: maximum sucrose contents in the base of the stalk (t/t)
TBFT	Sucrose partitioning: temperature at which partitioning of unstressed stalk mass increments to sucrose is 50% of the maximum value
LFMAX	Maximum number of green leaves a healthy, adequately watered plant will have after it is old enough to lose some leaves
MXLFAREA	Maximum leaf area assigned to all leaves above leaf number MXLFARNO (cm^2)
MXLFARNO	Leaf number above which leaf area is limited to MXLFAREA

(continued)

Table 9.4 (continued)

Genetic coefficient	Description
LER0	Maximum leaf elongation rate, per °Cd
PI1	Phyllocron interval 1
PI2	Phyllocron interval 2
PSWITCH	Leaf number at which the phyllocron changes
TDELAY	Delay between primary shoot appearance and appearance of first secondary shoot
TAR0	Maximum tiller appearance rate, per primary shoot, per °Cd
POPTT16	Stalk population at/after 1600 degree days (/m^2)
TTPLNTEM	Thermal time to emergence for a plant crop
TTRATNEM	Thermal time to emergence for a ratoon crop
CHUPIBASE	Thermal time from emergence to the start of stalk growth
SER0	Maximum stalk elongation rate, per °Cd
TT_POPGROWTH	Thermal time to peak tiller population
LG_AMBASE	Aerial mass (fresh mass of stalks, leaves, and water attached to them) at which lodging starts; t/ha
AQP_UP5	Not used yet. Available soil water depletion fraction at which photosynthesis water stress starts

Source: DSSAT v4.8; Hoogenboom et al. (2021)

Table 9.5 Sugarcane crop genetic coefficients and their description in APSIM model

Genetic coefficients	Description
leaf_size	Size of each leaf serving as inflection points for leaf size curve (mm^2)
leaf_size_no	Leaf number from top leaf (leaf number 1) (leaf)
cane_fraction	Fraction of accumulated biomass partitioned to cane (g^{-1})
sucrose_fraction_stalk	Fraction of accumulated biomass partitioned to sucrose (g^{-1})
stress_factor_stalk	Stress factor for sucrose accumulation (°Cd)
min_sstem_sucrose	Minimum cane stalk biomass before sugar accumulation begins ($g\ m^{-2}$)
min_sstem_sucrose_redn	Minimum sugarcane stem accumulation reduced under stress ($g\ m^{-2}$)
tt_emerg_to_begcane	Degree days required from emergence to the start of stalk growth (°Cd)
tt_begcane_to_flowering	Degree days required from the start of stalk growth to the start of flowering (°Cd)
tt_flowering_to_crop_end	Degree days from flowering to crop death/end (°Cd)
green_leaf_no	Green leaf number (leaves)
tillerf_leaf_size	Expansion factor applied to leaf size due to tillering
tillerf_leaf_size_no	Leaf number from top leaf (leaf number 1) (leaf)

Source: Dias and Sentelhas (2017), Peng et al. (2020); APSIM model

2019, 2021). Process-based models will be helpful in measuring the influence of drought and heat on crop output, even though growers and mill specialists may not be aware of the effect in terms of lower biomass production (Siebert et al. 2014). For a global evaluation of climate change for a future scenario, the crop growth CANEGRO-Sugarcane model is employed. Task-based models depend on sugarcane development and growth, including phenology, tillering, canopy development, accumulation and partitioning of biomass, water stress, root development, and lodging, that simulates sugarcane growth by the use of climate and water inputs. It is created to replicate the entire plant, root and stalk biomass, plant phenology, sucrose concentration, and other factors using a daily time step (Dias and Inman-Bamber 2020). Singels et al. (2013) used the CANEGRO model to predict the future sugarcane production at constant (360 ppm) atmospheric carbon dioxide concentration from the baseline period (1980–2010), in which yield decreased by 4.15% and 4.65% in rainfed sugarcane crop of Brazil and irrigated sugarcane crop of Australia, but in South Africa, the predicted yield increased by 2.58%. In Brazilian sugarcane, the CANEGRO model showed satisfactory results. Model predictions were checked by the LAI, sucrose content, dry mass, and soil water content (Nassif et al. 2012). Nadeem et al. (2022) used the CANEGRO-Sugarcane model for the sugarcane cropping system in Pakistan, in which model performance was good for both spring and autumn sugarcane. The model performed well when it came to forecasting spring sugarcane growth, sugarcane production, and total dry matter. For spring sugarcane, calibrated results revealed that the model's predicted error values were 2.47%, 3.23%, and 3.54%, respectively, and showed maximum LAI of cultivars (Figs. 9.1, 9.2, and 9.3). The model performed better when it came to forecasting autumn sugarcane growth, sugarcane production, and total dry matter. For autumn sugarcane, calibrated results revealed that the model-predicted error values were 1.82%, 2.35%, and 2.95%, respectively (Figs. 9.4 and 9.5). In terms of ponda sugarcane, throughout the calibration procedure for a variety of ponda sugarcanes, simulation process of growth, sugarcane yield, and total dry matter performed better as per the model. According to calibration data, the model simulated a higher maximum LAI than what was actually measured. Between the simulated and observed highest LAI, there was a 6.04% error. The difference between the total TDM that was simulated and the observed value, 26.50 t ha^{-1}, was 3.61% (Figs. 9.6, 9.7, and 9.8).

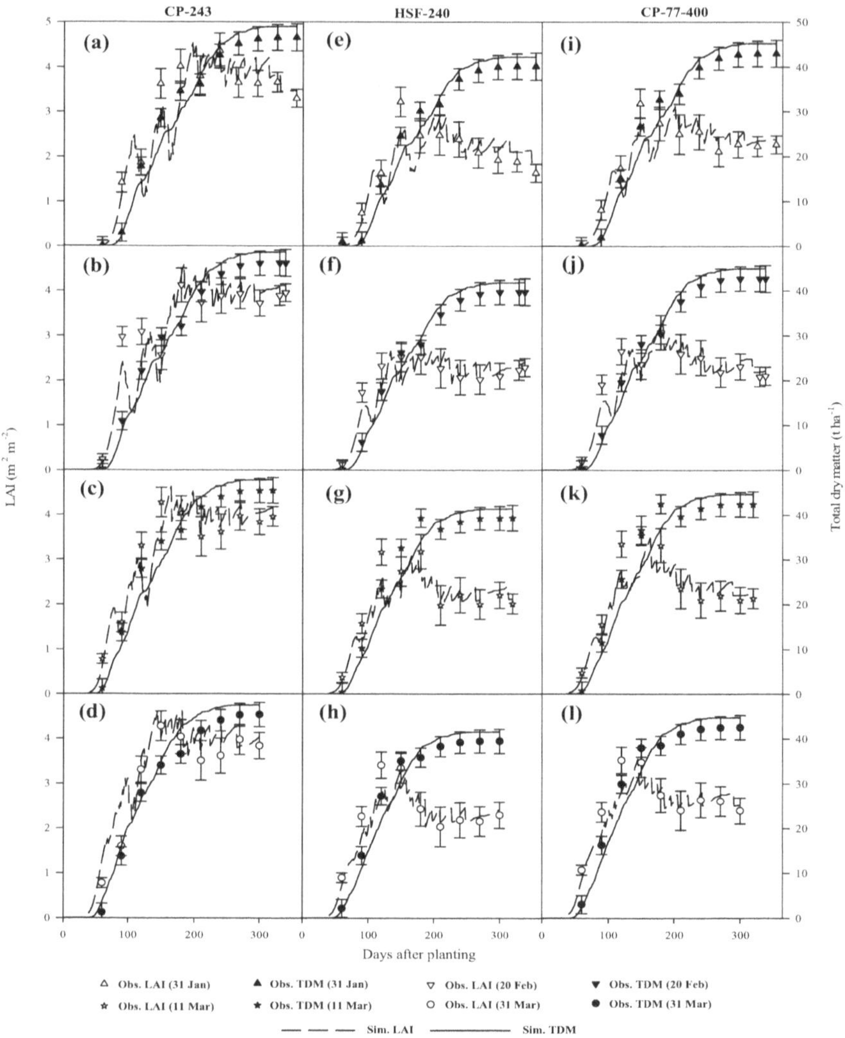

Fig. 9.1 Effect of cultivars and sowing dates on leaf area index and total dry matter of spring sugarcane during 2013–2014 at the study site (Source: Adopted from Nadeem et al. 2022)

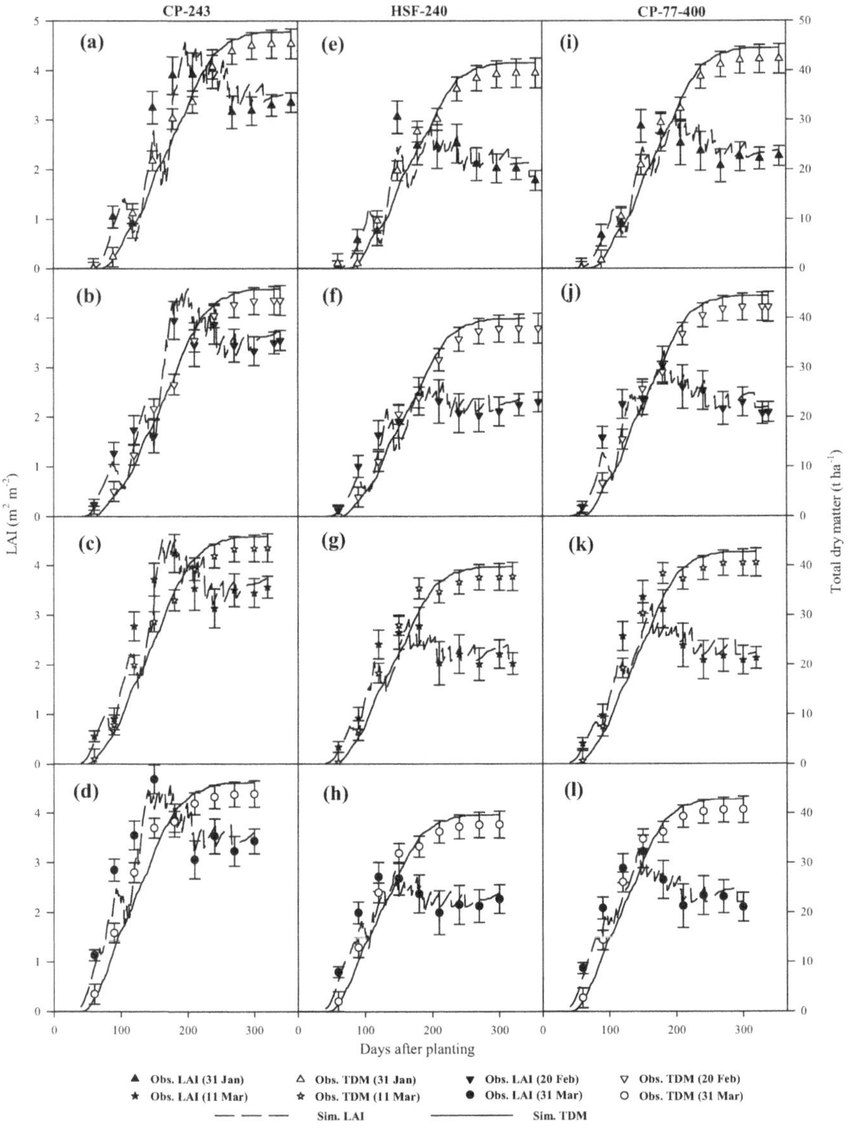

Fig. 9.2 Effect of cultivars and sowing dates on leaf area index and total dry matter of spring sugarcane during 2014–2015 at the study site (Source: Adopted from Nadeem et al. 2022)

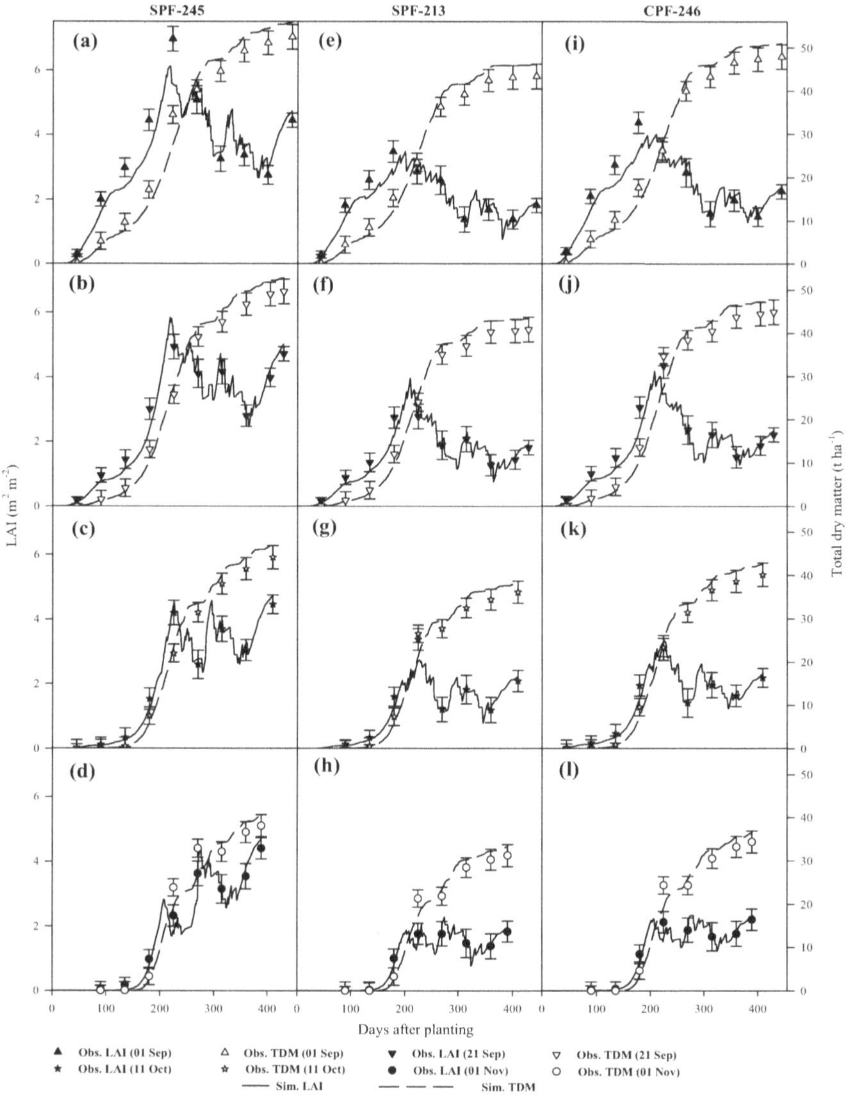

Fig. 9.3 Effect of cultivars and sowing dates on leaf area index and total dry matter of autumn sugarcane during 2013–2014 at the study site (Source: Adopted from Nadeem et al. 2022)

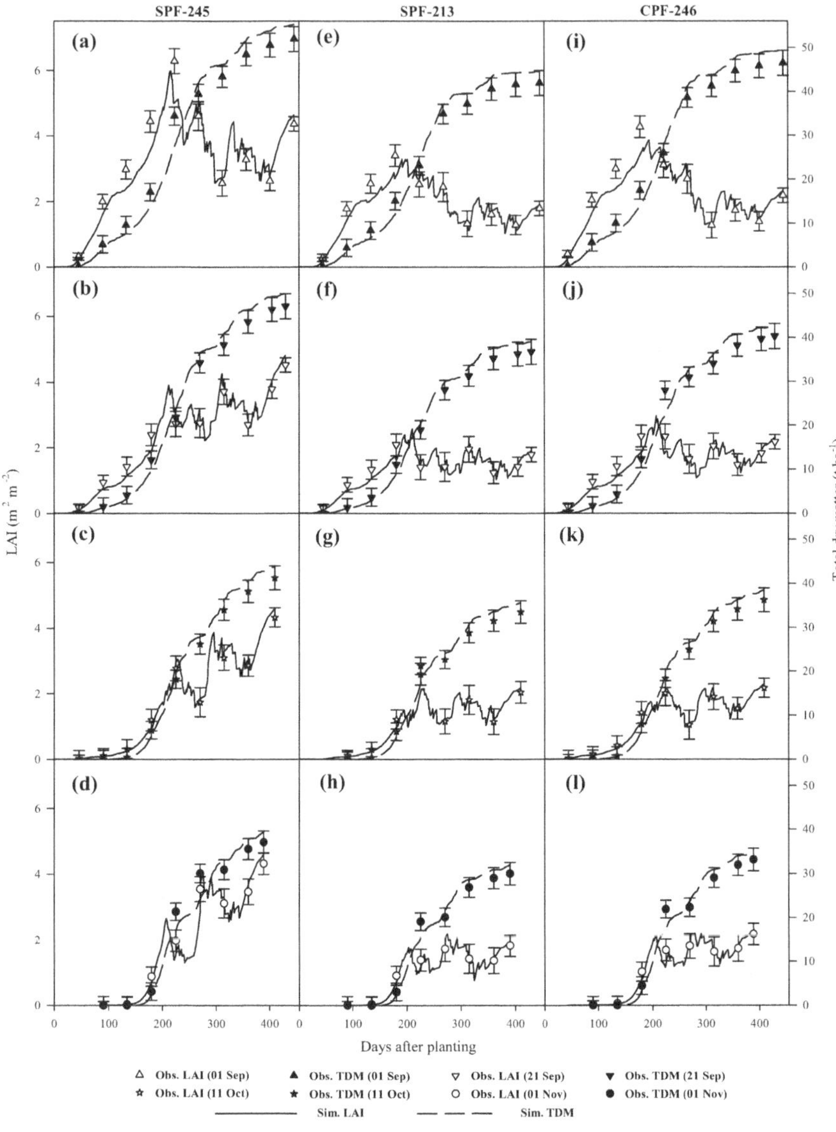

Fig. 9.4 Effect of cultivars and sowing dates on leaf area index and total dry matter of autumn sugarcane during 2014–2015 at the study site (Source: Adopted from Nadeem et al. 2022)

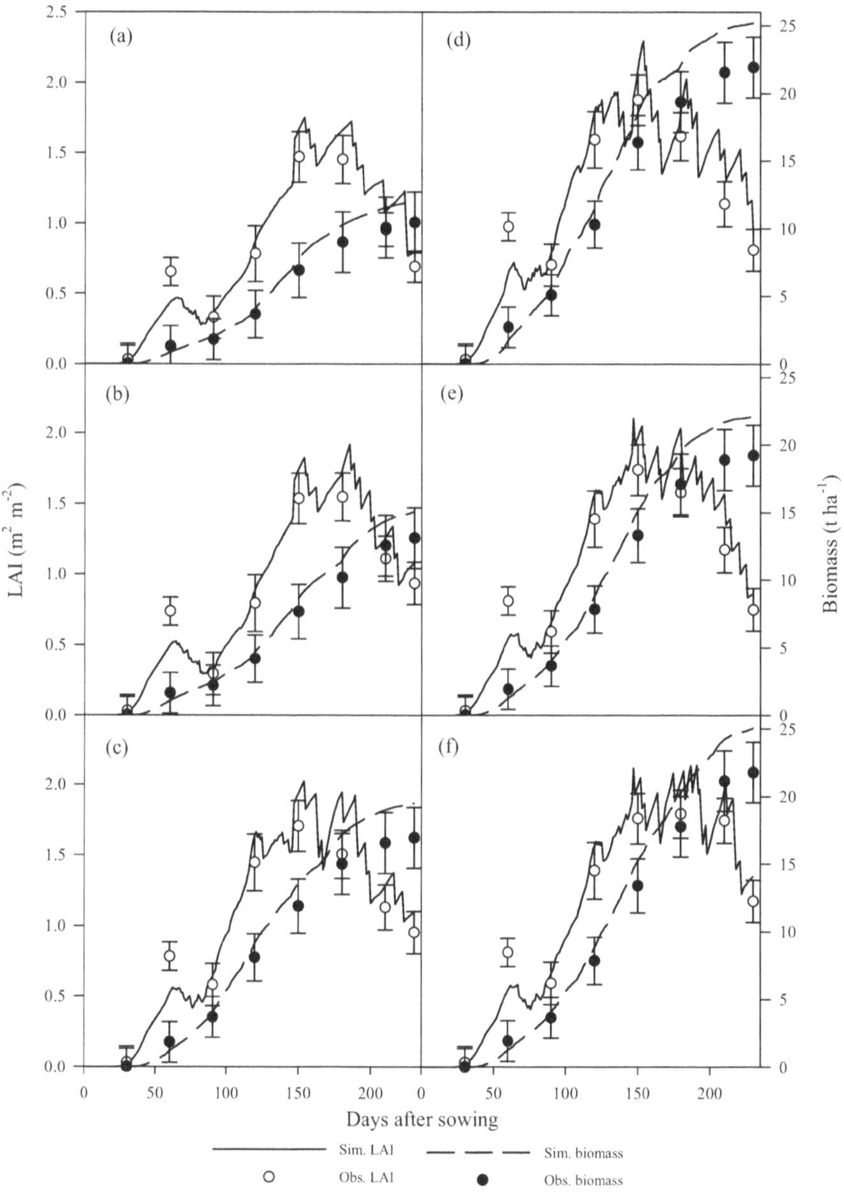

Fig. 9.5 Effect of irrigation regimes on leaf area index and biomass of *ponda* sugarcane during 2017 at the study site (Source: Adopted from Nadeem et al. 2022)

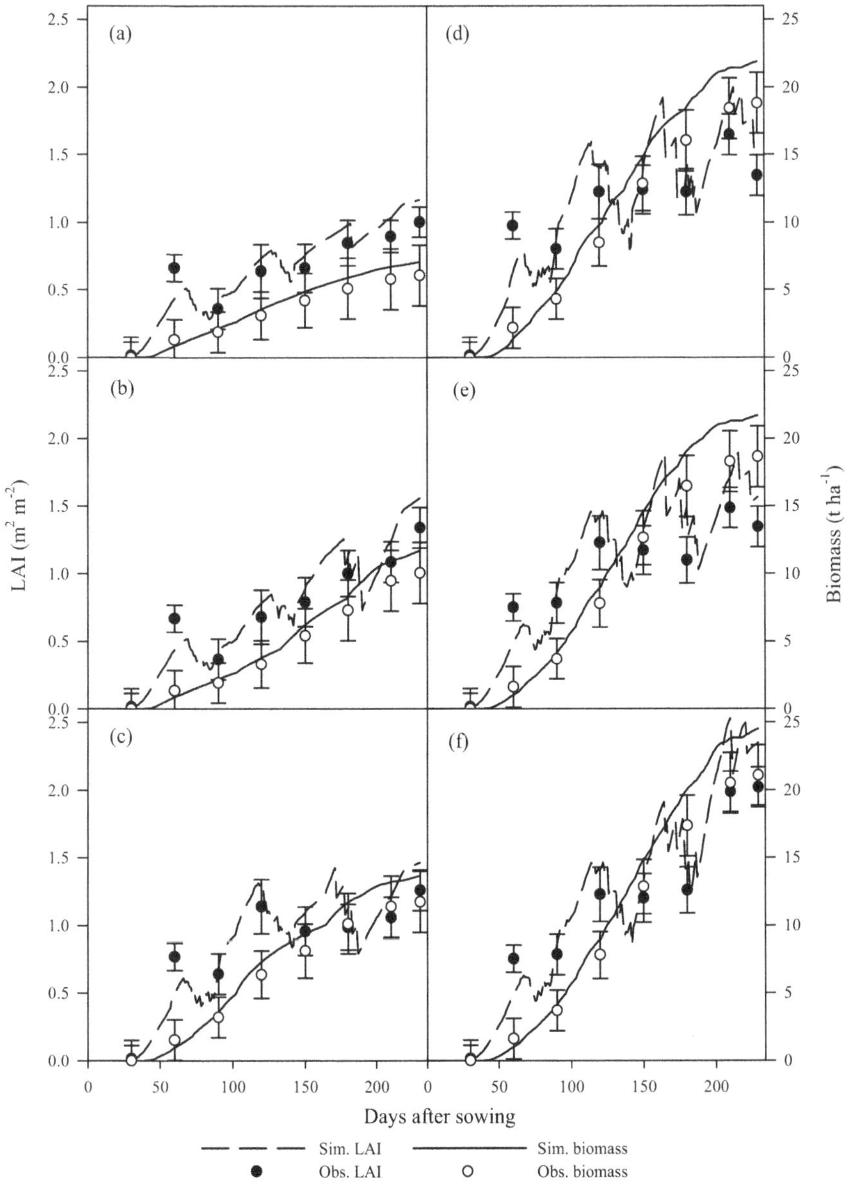

Fig. 9.6 Effect of irrigation regimes on leaf area index and biomass of *ponda* sugarcane during 2018 at the study site (Source: Adopted from Nadeem et al. 2022)

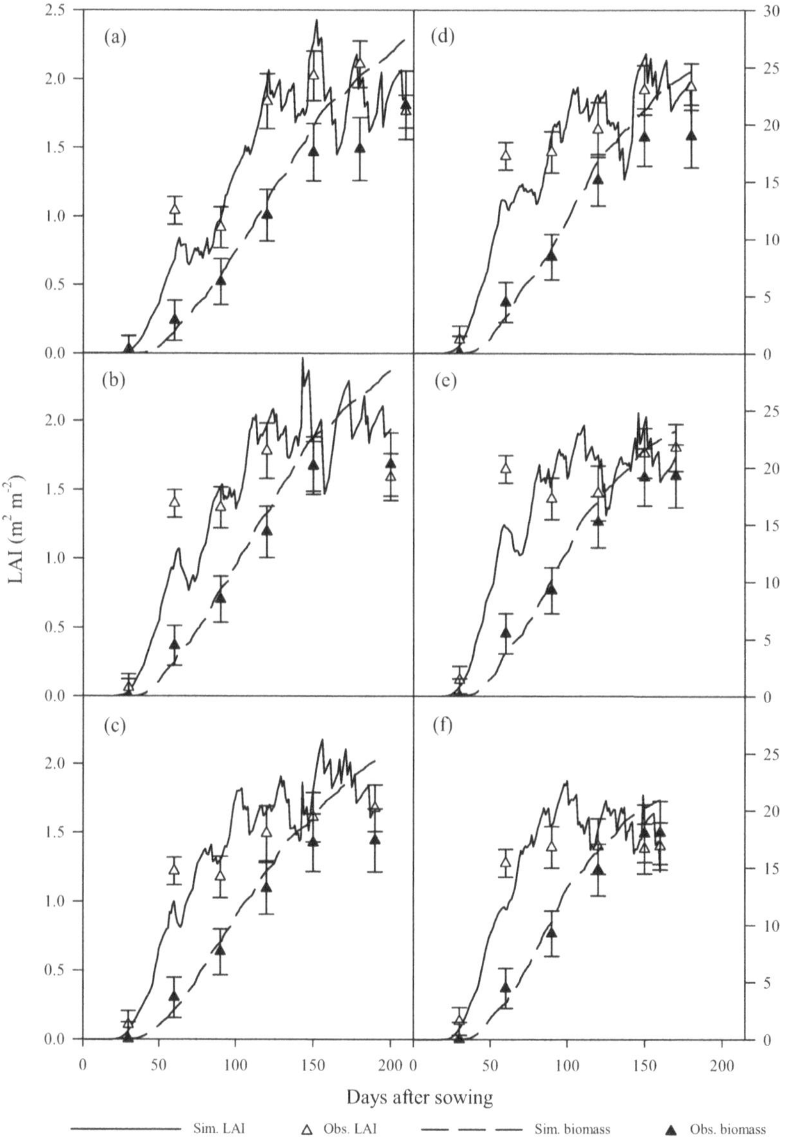

Fig. 9.7 Effect of sowing dates on leaf area index and biomass of *ponda* sugarcane during 2017 at the study site (Source: Adopted from Nadeem et al. 2022)

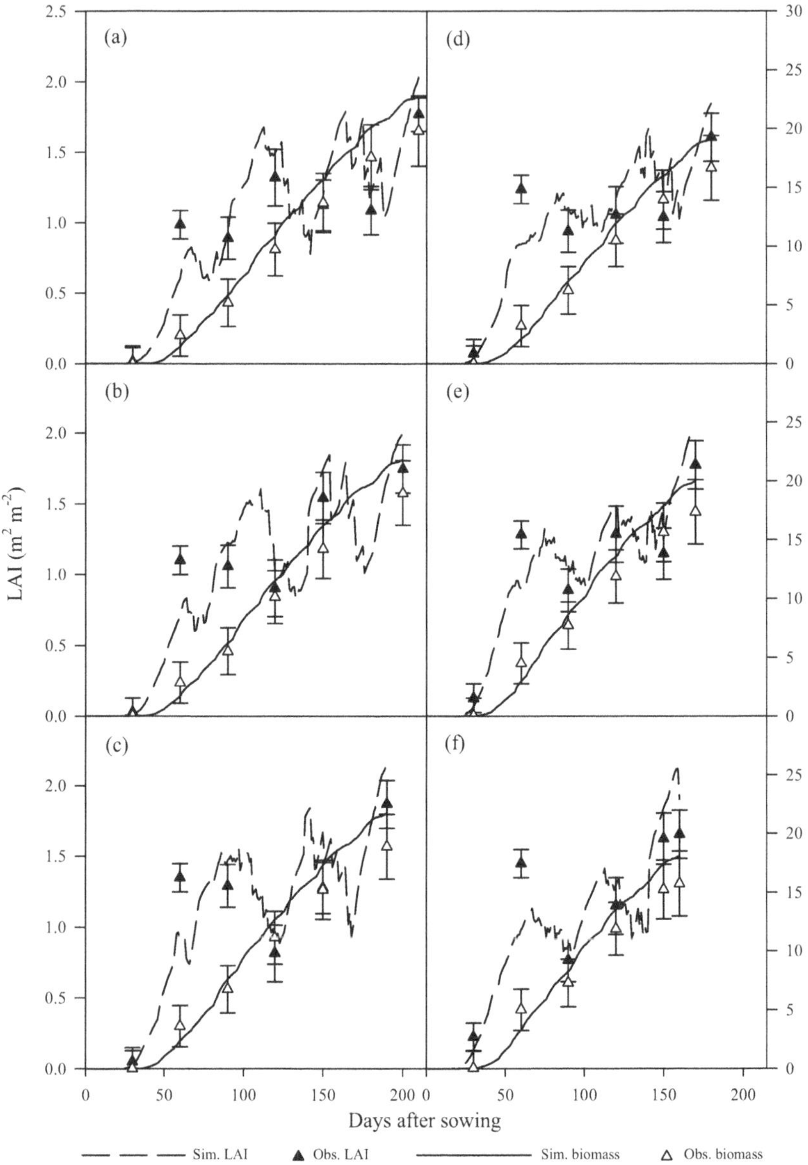

Fig. 9.8 Effect of sowing dates on leaf area index and biomass of *ponda* sugarcane during 2018 at the study site (Source: Adopted from Nadeem et al. 2022)

9.4 Climate Change Impact on Sugarcane Phenology

The overview of sugarcane phenological stages and phases is presented in Fig. 9.9. Seasonal changes affect the life cycle of sugarcane (rainy, winter, and summer). Being commonly used to describe, air temperature, solar radiation, and availability of moisture are thought to be the primary climatic factors influencing this crop's development, production, and quality (Marengo et al. 2014). Due to climate change, an increase in precipitation and temperature affected the sugarcane system in the future (Deressa et al. 2005) and phenological stages of sugarcane affected by temperature stress and crop damage increased when it reached threshold level (Folberth et al. 2014). Rainfall and temperature fluctuation can have both direct and indirect effects on sugarcane phenology and yield (Glasziou et al. 1965; Mali et al. 2014). Sugarcane grows well in 28–32 °C temperatures, whereas crop tillering and lateral growth are reduced at a temperature above 45 °C, but its growth also slows down when the temperature is below 20 °C. Sugarcane growth requires an ideal 70–85% of relative humidity and for ripening phase 55–75% relative humidity. Sugarcane cultivation is not recommended when the relative humidity is less than 50% during the growing season (Shukla et al. 2017). Sugarcane sett germination is adversely affected by rainfall at sowing time (Binbol et al. 2006), and production decreases when rainfall occurs at the germination phase (Kumar 1984).

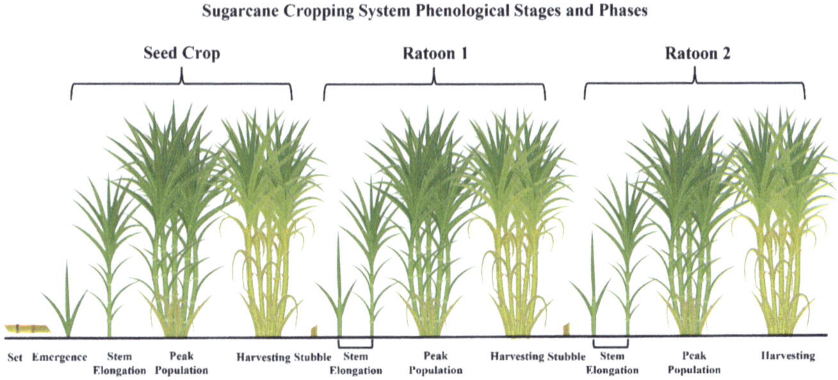

Fig. 9.9 Overview of sugarcane phenology

9.5 Impact of Climate Change on Phenology: A Case Study in Pakistan

Ahmad et al. (2016a, b, c) studied the impact of climate change on sugarcane phenology and productivity in the sugarcane system in Pakistan. The CSM-CANEGRO-Sugarcane model was used to assess the variability of the phenology of sugarcane cropping system (Fig. 9.10), agronomic practices, and cultivar alternations in

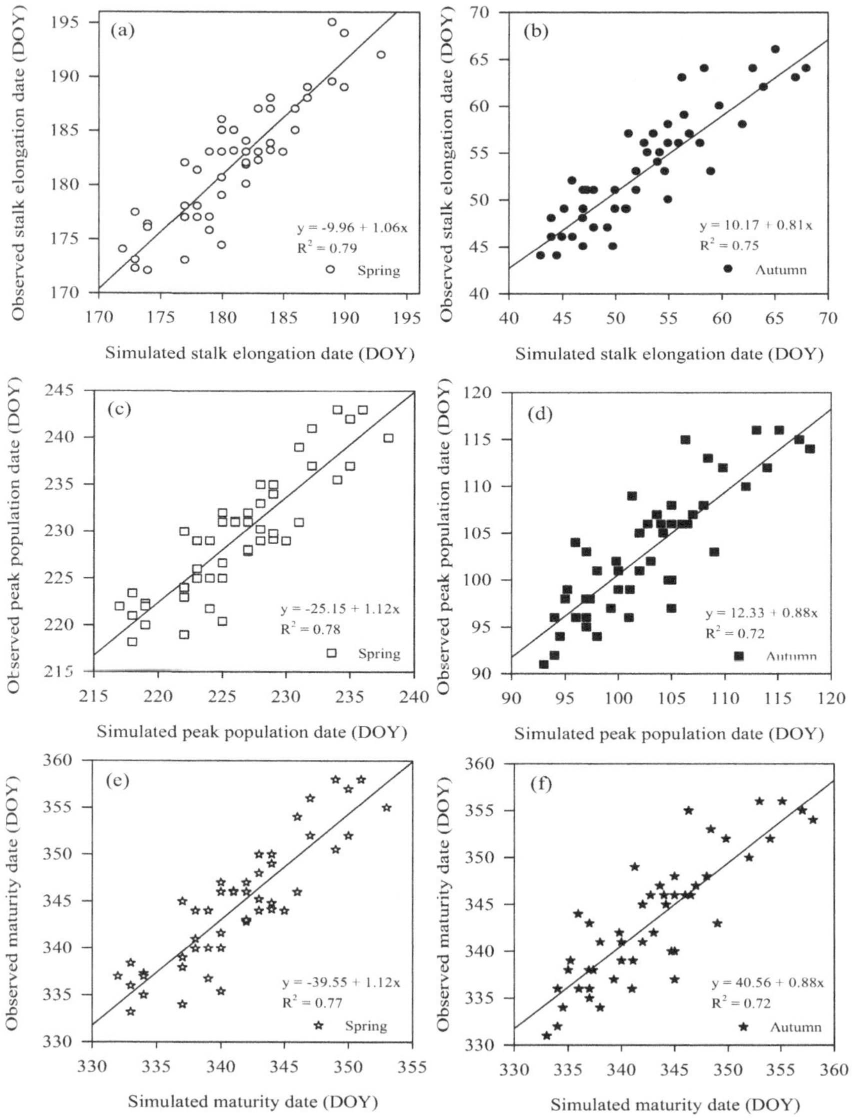

Fig. 9.10 Comparison of simulated and observed stalk elongation dates, peak population dates, and maturity dates for spring (**a, c, e**; unfilled symbols) and autumn (**b, d, f**; filled symbols) sugarcane season planted crop, respectively, during 1983–1985 at ten stations in Punjab, Pakistan (Source: Adopted from Ahmad et al. 2016c)

Punjab, Pakistan. Between 1980 and 2018, variations in spring and autumn sugarcane phenology were detected in Punjab, Pakistan (Fig. 9.11). In Table 9.6, spring sugarcane sowing was earlier by an average of 2.87 days/decade, whereas autumn sugarcane sowing was delayed by an average of 6.59 days/decade. In spring sugarcane, stalk elongation was earlier by 2.9–5.6 days/decades, but in autumn sugarcane, it was 2.6–6.4 days/decade. Similarly, for spring and autumn sugarcane in the sugarcane-sugarcane cropping system, the planting to stalk elongation, stalk elongation to peak population, and peak population to maturity phase were on average reduced by 1.60, 0.54, and 1.40 days/decade and 2.21, 1.25, and 0.96 days/decade (Fig. 9.11). The variation in phenological phases of spring and autumn sugarcane was related to temperature increase between 1980 and 2014 (Fig. 9.11). An increase in temperature affects the temporal and spatial variation of spring and autumn sugarcane phenological phases in Punjab, Pakistan (Table 9.7). The temperature response of phenological phases in field-recorded data was lower than that in model-simulated phenological data (Table 9.8). The CANEGRO-Sugarcane model with standard cultivars for all sites and years revealed that model-predicted phenology has accelerated more than field-observed phenology with climate change (Fig. 9.11). These results indicated that throughout the period of 1980–2014, shifts in cultivars needing high total growing degree days and earlier and late planting may have partially offset the detrimental influences of climate change on the phenology of spring and autumn growing sugarcane (Fig. 9.12).

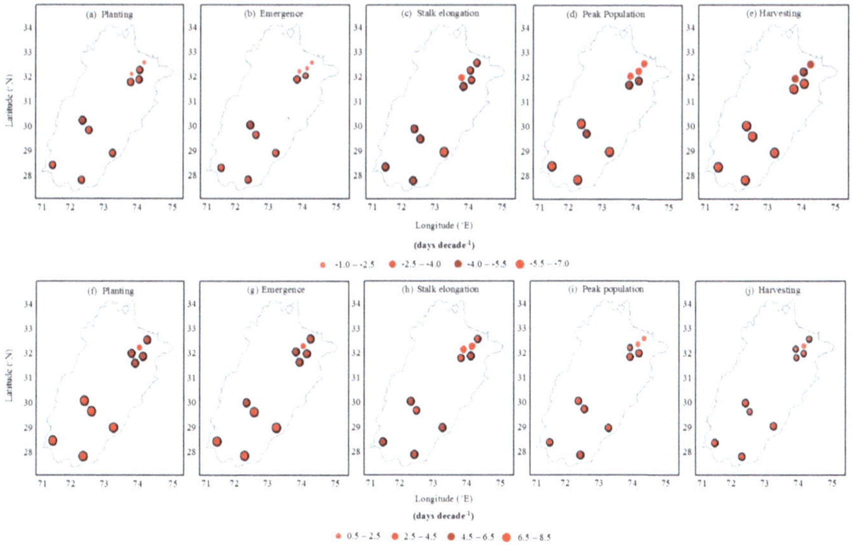

Fig. 9.11 Observed trends in phenology stages of spring and autumn season planted sugarcane crop {planting (**a, f**); emergence (**b, g**); stalk elongation (**c, h**); peak population (**d, i**); and harvesting (**e, j**), respectively; circles with black borderline indicate statistically significant trends at $p = 0.05$ levels} from 1980 to 2014 in Punjab, Pakistan (Source: Adopted from Ahmad et al. 2016c)

Table 9.6 Average observed phenology of spring- and autumn-planted sugarcane during 1980–2014 in Punjab, Pakistan (Source: Adopted from Ahmad et al. 2016c)

Site name	Planting	Emergence	Stalk elongation[a]	Peak population	Harvesting
Spring sugarcane					
Sialkot	34 ± 4.6	45 ± 5.8	174 ± 5.1	227 ± 6.3	357 ± 7.5
Gujranwala	40 ± 5.3	54 ± 6.1	178 ± 7.3	225 ± 6.1	354 ± 6.8
Hafizabad	38 ± 5.1	46 ± 4.8	179 ± 6.1	229 ± 7.4	361 ± 7.2
Sheikhupura	44 ± 6.4	52 ± 5.6	183 ± 7.5	236 ± 5.8	349 ± 6.6
Nankana Sahib	36 ± 4.2	47 ± 4.0	170 ± 6.1	242 ± 8.3	358 ± 9.6
Multan	47 ± 5.5	58 ± 5.3	172 ± 6.8	227 ± 6.2	362 ± 8.4
Lodhran	54 ± 6.2	67 ± 5.7	188 ± 6.1	236 ± 7.6	351 ± 7.3
Bahawalpur	48 ± 5.4	61 ± 5.1	193 ± 6.6	230 ± 6.1	347 ± 8.2
Bahawalnagar	52 ± 6.1	63 ± 5.6	192 ± 7.8	238 ± 8.1	346 ± 9.6
Rahim Yar Khan	50 ± 5.2	62 ± 5.1	186 ± 6.9	243 ± 9.4	352 ± 10.3
Autumn sugarcane					
Sialkot	252 ± 4.5	261 ± 4.8	50 ± 6.4	94 ± 5.8	350 ± 8.3
Gujranwala	246 ± 5.3	258 ± 5.1	56 ± 6.1	101 ± 6.8	353 ± 7.2
Hafizabad	260 ± 6.2	271 ± 6.1	48 ± 7.3	98 ± 7.9	348 ± 8.8
Sheikhupura	255 ± 6.7	267 ± 7.3	41 ± 5.8	103 ± 7.2	351 ± 8.2
Nankana Sahib	262 ± 5.8	270 ± 6.4	63 ± 6.7	92 ± 8.3	340 ± 9.1
Multan	259 ± 6.2	268 ± 5.7	66 ± 7.2	104 ± 9.6	357 ± 10.3
Lodhran	266 ± 5.2	279 ± 6.1	59 ± 8.4	110 ± 9.3	344 ± 8.9
Bahawalpur	261 ± 7.1	275 ± 8.2	54 ± 9.3	98 ± 8.1	349 ± 10.5
Bahawalnagar	256 ± 6.6	267 ± 7.1	65 ± 8.2	97 ± 9.3	342 ± 9.2
Rahim Yar Khan	250 ± 5.8	263 ± 6.4	68 ± 8.8	102 ± 7.5	353 ± 6.2

Source: Ahmad et al. (2016c)
[a]50% Stalk elongation, peak population

Table 9.7 Summary of observed and simulated phenology responses to temperature for spring- and autumn-planted sugarcane from 1980 to 2014 in Punjab, Pakistan (Source: Adopted from Ahmad et al. 2016c)

Phenology	No. of Neg.[a]	No. of Pos.[b]	No. of Sig. Neg.[c]	No. of Sig. Pos.[d]	Reg. Mean[e] (days °C^{-1})
Spring sugarcane stages and phases (observed)					
Planting	10	0	8	0	−1.29
Emergence	10	0	7	0	−1.18
Stalk elongation	10	0	9	0	−2.01
Peak population	10	0	7	0	−2.25
Harvesting	10	0	8	0	−2.88
Planting-stalk elongation	10	0	9	0	−0.72
Stalk elongation-peak population	10	0	9	0	−0.57
Peak population-harvest	10	0	8	0	−0.63
Planting-harvest	10	0	9	0	−1.59
Spring sugarcane phases (simulated)					
Planting-stalk elongation	10	0	8	0	−1.18
Stalk elongation-peak population	10	0	9	0	−0.82
Peak population-harvest	10	0	8	0	−1.00
Planting-harvest	10	0	9	0	−1.98

(continued)

Table 9.7 (continued)

Phenology	No. of Neg.[a]	No. of Pos.[b]	No. of Sig. Neg.[c]	No. of Sig. Pos.[d]	Reg. Mean[e] (days °C^{-1})
Autumn sugarcane stages and phases (observed)					
Planting	10	0	0	9	2.97
Emergence	10	0	0	9	2.80
Stalk elongation	10	0	0	8	1.96
Peak population	10	0	0	8	1.41
Harvesting	10	0	0	9	0.98
Planting-stalk elongation	10	0	8	0	−1.05
Stalk elongation-peak population	10	0	7	0	−0.56
Peak population-harvest	10	0	9	0	−0.43
Planting-harvest	10	0	8	0	−2.01
Autumn sugarcane phases (simulated)					
Planting-stalk elongation	10	0	9	0	−1.49
Stalk elongation-peak population	10	0	8	0	−0.81
Peak population-harvest	10	0	8	0	−0.79
Planting-harvest	10	0	10	0	−2.27

Source: Ahmad et al. (2016c)
[a]Number of locations with negative regression coefficients
[b]Number of locations with positive regression coefficients
[c]Number of locations with significant negative regression coefficients
[d]Number of locations with significant positive regression coefficients
[e]Mean of regression coefficients

Table 9.8 Comparison of the responses of spring- and autumn-planted sugarcane phenology with average temperature using the observed and simulated data during 1980–2014 in Punjab, Pakistan (Source: Adopted from Ahmad et al. 2016c)

Phenology	Regression coefficient using observed data (days °C^{-1}) Reg mean[a]		Regression coefficient using simulated data (days °C^{-1}) Reg mean[b]		Difference between obs. and sim. data (days °C^{-1})		t-Test (*p*-value)	
Phases/season	Spring	Autumn	Spring	Autumn	Spring	Autumn	Spring	Autumn
Planting-stalk elongation	−0.72	−1.05	−1.18	−1.49	0.46	0.44	0.0012*	0.0025*
Stalk elongation-peak population	−0.57	−0.56	−0.82	−0.81	0.25	0.25	0.0014*	0.0022*
Peak population-harvest	−0.63	−0.43	−1.00	−0.79	0.37	0.36	0.0017*	0.0026*
Planting-harvest	−1.59	−2.01	−1.98	−2.27	0.39	0.26	0.0015*	0.012*

Source: Ahmad et al. (2016c)
**Significant at the 0.01 probability level
[a]Mean of regression coefficients
[b]Duration of sowing to maturity

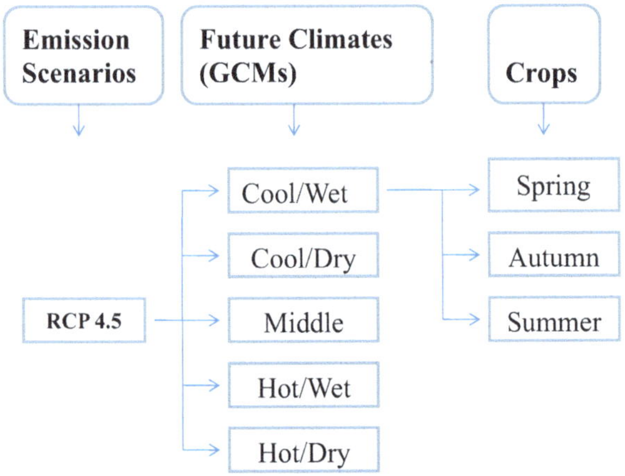

Fig. 9.12 Schematic methodology of climate change impact simulations for sugarcane at study sites (Source: Adopted from Nadeem et al. 2022)

9.6 CTWN Model Sensitivity for Sugarcane

For spring and autumn sugarcane crop 1980–2014 baseline weather data and for ponda crop 1980–2019 weather data were run under the CANEGRO sugarcane model to study the analysis of model sensitivity. Each degree of CTW analysis for spring, autumn, and ponda sugarcane crops had an impact on fresh cane production. It was discovered that the model was responsive to different CO_2 levels. Although the fresh yield of cane remained nearly constant at some levels between 650 and 700 ppm, the model was responsive to gradually increasing the CO_2 level from 350 to 600 ppm. While fresh cane yield decreased with an increase in temperature during spring and autumn, chewing sugarcane production increased with lower average temperatures than the baseline ambient temperature. The model responded to changes in rainfall at every level. The methodology of climate change impact assessment on sugarcane productivity has been presented in Fig. 9.12. In sugarcane (spring, autumn, and ponda), production of fresh cane declined with the low rainfall level as shown in Fig. 9.13.

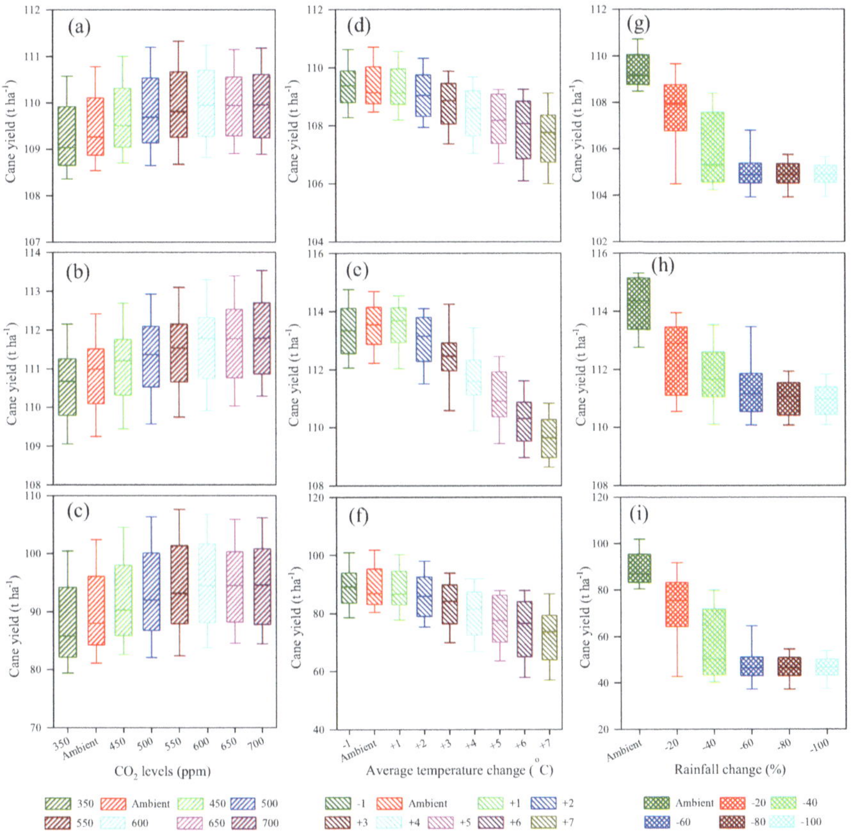

Fig. 9.13 Carbon dioxide (**a–c**), temperature (**d–f**), and rainfall (**g–i**) sensitivities for spring (**a, d, g**), autumn (**b, e, h**), and *ponda* (**c, f, i**) for sugarcane crops at study sites (Source: Adopted from Nadeem et al. 2022)

9.7 Climate Change Impact on Sugarcane Production Systems

In Table 9.9, global warming's effects on temperature, precipitation, and solar radiation trends have both favorable and unfavorable effects on sugarcane production (Srivastava and Rai 2012). The production of sugarcane and the occurrences of disease and pests may be influenced by the rising cloud cover, lack of direct sun radiation, and drought risk (Vishwakarma et al. 2013). Pokkah boeng disease of sugarcane increased due to more cloudy weather, and increasing humidity (70–80%) under ideal temperature affects sugarcane production (Sharma et al. 2013). In some places in Brazil, climate change boosted sugarcane water usage efficiency and sugarcane productivity. They anticipated that sugarcane production in 2050 might be 15–59% greater than today's average (Marin et al. 2012). In Fiji, sugarcane production declined due to drought conditions in 1983, 1997, 1998, and 2003, but in 1997,

Table 9.9 Impact of climate change on sugarcane across various countries

Sr. no.	Country	Model	Climatic scenarios and study periods	Yield increase/ decrease	References
1	Swaziland	CANEGRO-Sugarcane model	GCM (HadCM3), SRES (A2 and B2), (2020s, 2050s, and 2080s)	11–14% increase in sucrose yield	Knox et al. (2010)
2	Thailand	Existing Model IV and Spatial Regression	RCP (4.5 and 8.5) (1989–2016, 2046–2055)	25–35% decrease	Pipitpukdee et al. (2020)
4	Pakistan	CROPWAT 8.0 and AquaCrop 4.0	RCPs (4.5 and 8.5) and 3 RCMs (ACCESS1.0, CNRM-CM5, MPI-ESM-LR) (1981–2005, 2020, 2050, 2080)	Increased by 0.3–4.6%	Farooq and Gheewala (2020)
5	China and Brazil	Soil-Plant-Atmosphere model	MERRA2 6 SSPs of CMIP6 (ssp119, ssp126, ssp245, ssp370, ssp434, ssp585) (1980–2018, 2015–2100)	Decreased up to 4–12%	Flack-Prain et al. (2021)
6	Mexico	ALMANAC model	A2 scenario, (2008–2010, 2021–2030, 2031–2040, 2041–2050)	Increased up to 1–13%	Baez-Gonzalez et al. (2018)
7	Northern India	CANEGRO-Sugarcane model	(1971–2000, 2040–2060) RCP (4.5, 6.0, and 8.5) and RCM (RegCM4)	Stalk fresh mass and sucrose content decline by 3–25%	Sonkar et al. (2019)
8	Egypt		RCPs (3, 4.5, 6, 8.5) (1971–2000, 2011–2040, 2041–2070, 2071–2100)	Increase in irrigation requirement	Abdrabbo et al. (2021)
9	Pakistan	CANEGRO-Sugarcane model	5 GCMs (CCSM4, GFDL ESM2M, HadGEM2-ES, MIROC5, and MPI-ESM-MR) and RCP 8.5 (1980–2018, 2039–2069)	Yield decreased by 15 22%	Nadeem et al. (2022)
10	South Africa	CANEGRO-Sugarcane model	3 GCMs (MIROC3 2 MEDRES, MPI ECHAM5, UKMO HADCM3) (1980–2010, 2070–2100)	Yield increased by 11%	Jones et al. (2015)
11	India	CANEGRO-Sugarcane model	A2 and B2 scenarios, CO_2 Conc. (720 and 500 ppm)	Yield increased by 4–9.75%	Verma et al. (2023)

there was maximum reduction in yield (16–46%) faced by tropic cyclones (Gawander 2007; FAO 2014; Factfish 2015). In Guangxi China, in 2008, production of sugar, sugarcane, and yield was 9.41 Mt., 77.1 Mt., and 83.8 t ha^{-1} in favorable climatic conditions, but in 2009, temperature rose from −4 to −6 °C, and in 2010, drought occurred in early growing season, and in December, extreme freezing temperature declined the production of sugar, sugarcane, and yield to 6.75 Mt., 55.7 Mt. and 56.3 t ha^{-1} (Li and Yang 2014; Li et al. 2011). Due to drought conditions, sugarcane yield declined to 18% in 2003–2004 (Li and Wei 2006). The Existing Model IV and spatial regression model were used to know the impact of climate change on the sugarcane system under the climate scenario RCPs 4.5 and 8.5 from the base period (1989–2016) and in the future (2046–2055) in Thailand. Climate model projection showed the rise of temperature in the growing season, and maximum extreme temperature affects sugarcane growth and production. Sugarcane growing season temperature was increased under RCPs 4.5 and 8.5 ranging from 1.08 to 1.22 and 1.48 to 1.68 °C, with maximum increase in extreme temperature ranging from 1.21 to 1.55 and 1.61 to 1.86 °C. Due to climate change (increase in temperature) in the future (2046–2055), sugarcane harvested area, yield, and production will decline to 1.29–2.49%, 23.95–33.26%, and 24.94–34.93%, respectively (Pipitpukdee et al. 2020). Climate change had an adverse effect on the requirements of the sugarcane system in Egypt. In Egypt, the availability of water is a serious issue, and the increased temperature raises the evapotranspiration during present and future conditions that were studied under climate scenario RCPs 3.0, 4.5, 6.0, and 8.5 from 1971 to 2000, 2011 to 2040, 2041 to 2070, and 2071 to 2100. To fulfill water requirements, fuel consumption rises to 5–12% and cost of irrigation rises to 12–21% in all scenarios (Abdrabbo et al. 2021). Sugarcane affected by climate change was predicted by the global climate model (HadCM3) with SRES (A2 and B2) in Swaziland for the future (2020s, 2050s, and 2080s) time period. With the effect of climate change, irrigation requirements will increase to 9% in the future and sucrose yield is estimated to increase by 11–14% (Knox et al. 2010). Climate change is a critical worldwide issue since it can create a demanding condition for plant growth. This makes it difficult to produce different crops sustainably around the world. This study evaluates the influence of climate change on Pakistan's future irrigation requirements and sugarcane production for the 1981–2005 baseline and future timeframes of the 2020s, 2050s, and 2080s. With the use of crop model, three regional climatic models (ACCESS1.0, CNRM-CM5, MPI-ESM-LR) were used under RCPs 4.5 and 8.5. Because the expected increase in rainfall under forecasted weather circumstances cannot make up for the rising evaporative demand, the results showed a consequent increase in irrigation requirement for crop growth induced by the projected temperature increase. Findings indicated that, given the predicted climate, sugarcane production was generally increasing, in which the yield range increased (0.3–4.6%) from 0.37 to 6.48 t ha^{-1} (Farooq and Gheewala

2020). In Pakistan, sugarcane production might be enhanced from 3493.4 thousand tonnes by the rise in temperature to 1 °C and production increased from 17.22 thousand tonnes by the increase of 1 mm precipitation (Magsi et al. 2018), but with an increase of 1–2 °C in 2030, yield will be reduced to 13.56–40.09%, respectively (Siddiqui et al. 2012). In Southern China, sugarcane yield showed a positive response to climate change, in which yield increased by 44.2% and 23.5% under RCPs 4.5 and 8.5 for the future time period (2030s, 2060s, and 2090s) (Ruan et al. 2018). In Vietnam, sugarcane predicted yield increased from 1.73% to 8.26% in the future (2020s, 2055s, 2090s) under RCPs 4.5 and 8.5 (Lee and Dang 2018). In Brazil, in 2050, the yield of sugarcane increased to 15–59% from the current average yield (Marin et al. 2012). In Paraiba, Brazil, climate change showed positive and negative impacts on sugarcane production. Increased temperature declined production, whereas production increased due to rainfall (Silva et al. 2019a, b), while Marin et al. (2012) reported that rainfed sugarcane production in 2050 would be increased by 24% in Brazil. Lobell et al. (2008) predicted how climate change will affect sugarcane production; it was found that while sugarcane would be resistant to temperature increases, it would be subject to variations in rainfall. By 2030, East Africa's sugarcane production is expected to be down by less than 5% from 1998 to 2002 levels. According to the Agricultural Land Management Alternatives model, sugarcane production will increase by 1–13% under the A2 scenario from baseline (2008–2010) as a result of future climate change scenarios (Baez-Gonzalez et al. 2018). In India, climate projection with model predictions indicated that sugarcane production will be increased by 4–9.75% under the CO_2 concentration 500 and 720 ppm under the A2 and B2 scenarios (Verma et al. 2023). Due to heat stress, productivity of sugarcane would decline by 23% in Brazil from 2041 to 2070 (de Carvalho et al. 2015). For the mid-century evaluation of the effects of climate change, five GCMs (GFDL-ESM 2M, CCSM4, HadGEM2-ES, MPI-ESM-MR, and MIROC5) at RCP 8.5 were utilized (2039–2069). These five GCMs for site-specific RCP 8.5 were used to run model simulations of the CANEGRO-Sugarcane at the study site (Fig. 9.14). When comparing baseline weather data (1980–2018), average temperature climbed by nearly 0.94 °C, whereas other climate forecasts predict that average temperature would rise by 3–5 °C by the middle of the century. Crops grown in the spring, fall, and ponda sugarcane will produce less fresh cane in these conditions. According to the results, spring, autumn, and ponda sugarcane produced the highest 110.92, 113.59, and 113.59 t ha^{-1} fresh cane yield at baseline scenarios. During the mid-century, production of fresh cane will be decreased in all GCMs in comparison to baseline. In ponda and spring sugarcane, maximum production of fresh cane was decreased under MPI-ESM-MR but in autumn sugarcane under CCSM4. At various GCMs, the yield of fresh cane in future scenarios will decline between 15.31% and 22.57% in the mid-century (Fig. 9.14).

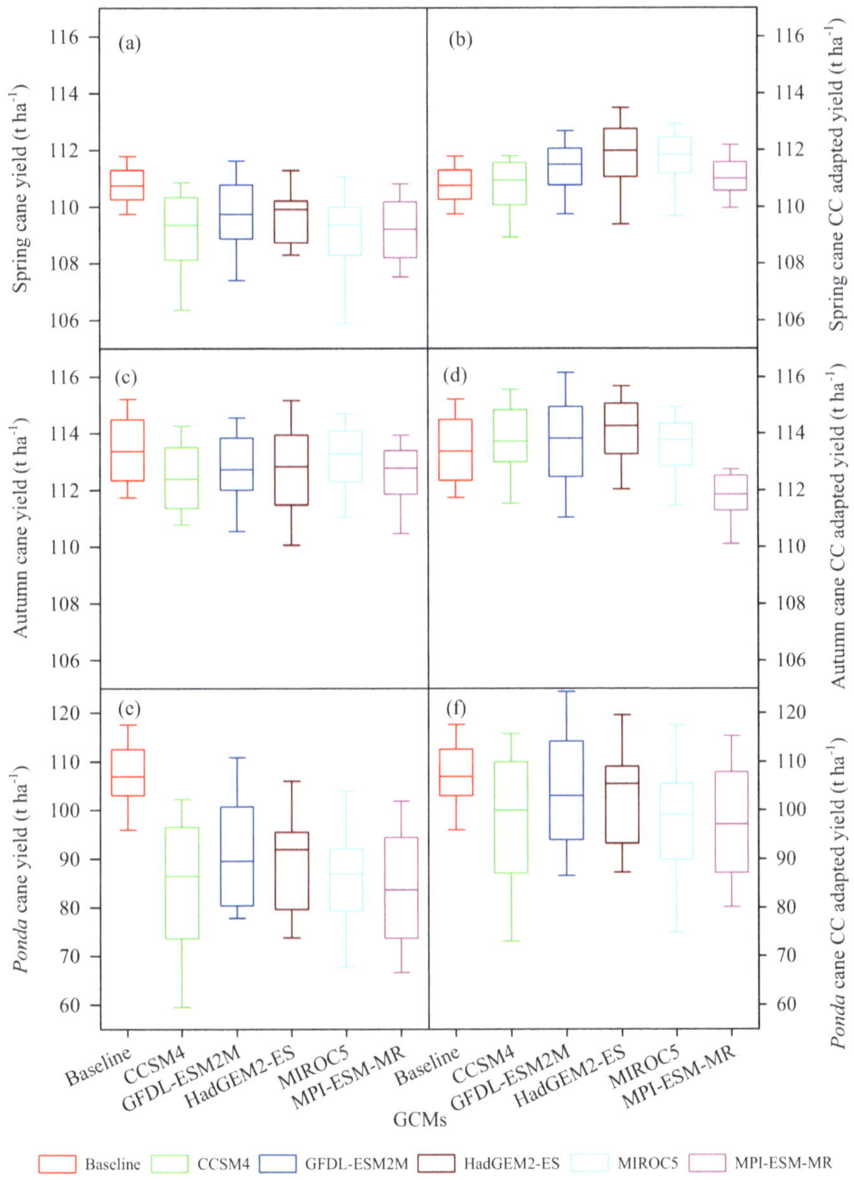

Fig. 9.14 Integrated climate impact assessment for spring (**a**), autumn (**c**), and *ponda* (**e**) without adaptation and with adaptation yields for spring (**b**), autumn (**d**), and *ponda* (**f**) sugarcane crops at study sites (Source: Adopted from Nadeem et al. 2022)

9.8 Climate Change's Impact on Sugarcane Morphophysiology

Raised carbon dioxide enhanced sugarcane water-use efficiency, photosynthesis, productivity, and biomass in a controlled environment. Stomatal conductance declined under higher CO_2 and is primarily responsible for sugarcane's enhanced water usage efficiency (Vu and Allen Jr 2009a, b; Souza et al. 2008). As the temperature rises quickly rather than moderately, high-temperature stress in C4 plant limits photosynthesis because of enzyme inhibition, which limits biomass formation (Crafts-Brandner and Salvucci 2002). Sugarcane production would be increased by 13% under 2070-year scenarios in Australia and Brazil (Da Silva et al. 2008), but according to Marin et al. (2012), in future, dryland sugarcane yield in Brazil and South Africa will increase by 9% and 20%, respectively.

9.9 Climate Change's Impact on Sugarcane Quality

Sugar accumulation in sugarcane is reduced because photorespiration increases and the conversion of sugar content into glucose and fructose is enhanced because abiotic stress increases due to high temperatures (Binbol et al. 2006; Gawander 2007). Over 26.6 °C, it has been observed that the accumulation of sucrose decreases (Clements 1980). Changes in climatic conditions (drought, cloud cover, lack of solar radiation, etc.) lower the sugarcane yield and quality (Vishwakarma et al. 2013). In sugarcane system, high temperature and doubled atmospheric CO_2 showed positive response to sugarcane properties, and there was increase in stem juice volume (124%), sucrose content (29%), stalk fresh mass (24%), stem fresh weight (55%), stem biomass (60%), and sugarcane yield (15–59%) (Zhao and Li 2015). In future (2100), due to climate change, sugarcane yield will increase to 4–20%, but yield of sucrose will vary from −33 to +13% in Brazil, Australia, and Africa (Singels et al. 2014), but Jones et al. (2015) reported that in irrigated areas, yield increased by 14% and 15% in rainfed areas. Sugar accumulation in sugarcane will decline due to increasing temperature that affects the fraction of sucrose contents into fructose and glucose and photorespiration (Binbol et al. 2006; Gawander 2007). Sonkar et al. (2019) used regional climate model with different CO_2 concentrations under RCP scenarios to predict the impact of climate change on sugarcane. Sugarcane stalk fresh mass (SFM) and sucrose mass (SM) will decline by 3–25% due to increase of temperature by 1–4% in future. Under elevated CO_2 levels, the observed SFM and SM increased by 2–14%. SFM showed a positive effect on the combined response of increased temperature and elevated CO_2 concentration, but SM showed decline. In future (2040–2060), irrigated and rainfed sugarcane projected SFM value increased by 7–47% and 3–39%, but projected SM value decreased by 6–37% and 9–69%, respectively.

9.10 Climate Change's Impact of Pest on Sugarcane

In the sugarcane system, climate change may affect sugarcane through rust infections, hurricanes, and heavy storms, which can promote the *Xanthomonas albilineans*-caused leaf scald that affects the yield and production of the crop (Everingham et al. 2002). The prevalence of smut illness, which is caused by *Sporisorium scitamineum* (Syd.), is probably going to rise in sugarcane as a result of the warm weather. The indications of ratoon stunting disease are made worse by the frequently dry weather (Matthieson 2007). In Australia, loss in the sugar industry was 150–210 million Aus$ in yield due to a new pathogen of orange rust that declines the performance of better cultivars (Braithwaite et al. 2009).

9.11 Adaptation Strategies

Sugarcane scientists and other researchers can generate a computerized searchable database to plan hybridization (across or within species) for specific breeding requirements and use morphological and physiological features to evaluate superior clones against abiotic and biotic challenges (Inman-Bamber et al. 2012). According to Edme et al. (2005), with the use of mitigation strategies, sugar yield and sucrose content were enhanced by 47.0% and 26% in the organic soil of Florida, which is the comparison with the yield and sucrose content of 1968–2000. In Fig. 9.14, the findings show that, in the absence of adaptation methods, sugarcane production for the spring, autumn, and ponda crops will decrease at all GCMs in the future scenario (2039–2069). Sugarcane adaptation solutions for spring, autumn, and ponda crops can help to reduce the negative effects of climate change in a future scenario. Various management options for adaptations like irrigation regimes and planting dates were used. To reduce the negative impact of climate in the future, optimum planting time will be a very impressive adaptation in the sugarcane system. In the spring crop, 15–20 days' early planting will increase yield from 2.5% to 5.1%; in the autumn crop, cane yield will increase from 1.9% to 4.5%, while cane yield will increase from 2.1% to 4.4% because planting will be earlier by 18–25 days. Under different GCMs, cane yield increased from 0.8% to 2.1% with the increasing efficiency of irrigation frequency and quantity by 15% to reduce the climate effect. In model strategies, cane yield increased from 3.2% to 5.7% by developing heat-tolerant cultivars. Results revealed that cane yield was enhanced with the influence of adaptation measures in comparison to without adaptation measures in all GCMs when all of the abovementioned adaptation choices were applied in the CANEGRO model.

9.12 Conclusion

The input resources are very costly; therefore, economic resources and welfare of the farming community are the need of the hour. But the higher prices of the inputs have made agriculture a very expensive business. Climate change and climate variability are the major challenges for successful crop production. The adaptation of climate-smart sugarcane production is the only solution to harvest higher yields along with profits for farming families in the world.

References

Abbas Q, Ahmad S (2018) Effect of different sowing times and cultivars on cotton fiber quality under stable cotton-wheat cropping system in southern Punjab, Pakistan. Pak J Life Soc Sci 16(2):77–84

Abbas G, Ahmad S, Ahmad A, Nasim W, Fatima Z, Hussain S, Rehman MH, Khan MA, Hasanuzzaman M, Fahad S, Boote KJ, Hoogenboom G (2017) Quantification the impacts of climate change and crop management on phenology of maize-based cropping system in Punjab, Pakistan. Agric For Meteorol 247:42–55

Abbas G, Fatima Z, Riaz F, Iqbal P, Hussain S, Zakir I, Iqbal M, Haque E, Ahmad S (2019) Study of sowing dates and hybrids effect in maize-based cropping system under arid conditions of Southern Punjab, Pakistan. Pak J Life Soc Sci 17(1):24–31

Abbas G, Ahmad S, Hussain M, Fatima Z, Hussain S, Iqbal P, Ahmed M, Farooq M (2020a) Sowing date and hybrid choice matters production of maize-maize system. Int J Plant Prod 14:583–595

Abbas G, Fatima Z, Hussain M, Hussain S, Atique-ur-Rehman, Sarwar N, Ahmed M, Ahmad S (2020b) Nitrogen rate and hybrid selection matters productivity of maize-maize cropping system under irrigated arid environment of southern Punjab, Pakistan. Int J Plant Prod 14:309–320

Abbas Q, Atique-ur-Rehman, Sarwar N, Hussain S, Abbas G, Khan MN, Fatima Z, Naz S, Younis H, Ahmed M, Ullah H, Iqbal P, Iqbal M, Kan M, Ahmad S (2020c) Effect of different sowing times and cultivars on wheat grain quality under cotton-wheat cropping system in Southern Punjab, Pakistan. Pak J Life Soc Sci 18(1):41–50

Abdrabbo MAA, Farag AAHA, Radwan MAM, Heggi HM, Aboelsoud, Singla C, Sharda R (2021) Climate change impact on economic and irrigation requirements for sugarcane crop in Egypt. Fut Food J Food Agric Soc 9(2):1–13

Aggarwal PK, Kalra N, Chander S, Pathak H (2006) InfoCrop: a dynamic simulation model for the assessment of crop yields, losses due to pests, and environmental impact of agro-ecosystems in tropical environments. Agric Syst 89(1):1–25

Ahmad S, Hasanuzzaman M (2012) Integrated effect of plant density, N rates and irrigation regimes on the biomass production, N content, PAR use efficiencies and water productivity of rice under irrigated semiarid environment. Notulae Botanicae Horti Agrobotanici Cluj-Napoca 40(1):201–211

Ahmad S, Ahmad A, Soler CMT, Ali H, Zia-Ul-Haq M, Anothai J, Hussain A, Hoogenboom G, Hasanuzzaman G (2012a) Application of the CSM-CERES-Rice model for evaluation of plant density and nitrogen management of fine transplanted rice for an irrigated semiarid environment. Precis Agric 13(2):200–218

Ahmad S, Ali H, Ismail M, Shahzad MI, Nadeem M, Anjum MA, Zia-Ul-Haq M, Firdous N, Khan MA (2012b) Radiation and nitrogen use efficiencies of C_3 winter cereals to nitrogen split application. Pak J Bot 44(1):139–149

Ahmad S, Ahmad A, Ali H, Hussain A, Garcia y Garcia A, Khan MA, Zia-Ul-Haq M, Hasanuzzaman M, Hoogenboom G (2013) Application of the CSM-CERES-Rice model for evaluation of plant density and irrigation management of transplanted rice for an irrigated semiarid environment. Irrig Sci 31(3):491–506

Ahmad A, Ashfaq M, Rasul G, Wajid SA, Khaliq T, Rasul F, Saeed U, Rahman MH, Hussain J, Baig IA, Naqvi SAA, Bokhari SAA, Ahmad S, Naseem W, Hoogenboom G, Valdivia RO (2015a) Impact of climate change on the rice-wheat cropping system of Pakistan. In: Hillel D, Rosenzweig C (eds) Handbook of climate change and agro-ecosystems, vol 3. Imperial College Press and the American Society of Agronomy, London, pp 219–258

Ahmad S, Ali H, Atique-ur-Rehman, Khan RJ, Ahmad W, Fatima Z, Abbas G, Irfan M, Ali H, Khan MA, Hasanuzzaman M (2015b) Measuring leaf area of winter cereals by different techniques: a comparison. Pak J Life Soc Sci 13(2):117–125

Ahmad S, Raza I, Muhammad D, Ali H, Hussain S, Dogan H, Zia-Ul-Haq M (2015c) Radiation, water and nitrogen use efficiencies of *Gossypium hirsutum* L. Turk J Agric For 39(5):825–837

Ahmad S, Ali H, Farooq U, Khan SU, Atique-ur-Rehman, Sarwar N, Shahzad AN, Doğan H, Hussain S, Sultan MT, Waheed A, Zia-ul-Haq M, Hussain K, Khan MA (2016a) Improving nitrogen and radiation-use-efficiencies of C_4 summer cereals by split nitrogen applications under irrigated arid environment. Turk J Agric For 40(2):280–289

Ahmad S, Hussain S, Fatima Z, Abbas G, Atique-ur-Rehman, Khan MR, Younis H, Naz S, Sohail M, Ajmal M, Abbas N, Akhtar M, Rauf A, Khan M, Ali Z, Hassan M, Rizwan M, Bajwa RS, Ajmal A, Naz S, Ali H, Khan AA, Ali M, Sarwar G, Khan MA, Hasanuzzaman M (2016b) Application of DSSAT model for sowing date management of C_4 summer cereals for fodder and grain crops under irrigated arid environment. Pak J Life Soc Sci 14(2):104–114

Ahmad S, Nadeem M, Abbas G, Fatima Z, Khan RJ, Ahmed M, Ahmad A, Rasul G, Khan MA (2016c) Quantification of the effects of climate warming and crop management on sugarcane phenology. Clim Res 71(1):47–61

Ahmad S, Abbas G, Fatima Z, Khan RJ, Anjum MA, Ahmed M, Khan MA, Porter CH, Hoogenboom G (2017a) Quantification of the impacts of climate warming and crop management on canola phenology in Punjab, Pakistan. J Agron Crop Sci 203(5):442–452

Ahmad S, Abbas Q, Abbas G, Fatima Z, Atique-ur-Rehman, Naz S, Younis H, Khan RJ, Nasim W, Rehman MH, Ahmad A, Rasul G, Khan MA, Hasanuzzaman M (2017b) Quantification of climate warming and crop management impacts on cotton phenology. Plan Theory 6:1–16

Ahmad S, Atique-ur-Rehman, Ejaz M, Fatima Z, Kan M, Ahmad M (2018) Agricultural land-use change of major field crops in Pakistan (1961–2014). Sci Technol Dev 37(3):113–121

Ahmad A, Abbas G, Ahmed M, Fatima Z, Anjum MA, Rasul G, Khan MA, Hoogenboom G (2019) Climate warming and management impact on the change of phenology of the rice-wheat cropping system in Punjab, Pakistan. Field Crop Res 230:46–61

Ahmad A, Ashfaq M, Rasul G, Wajid SA, Ahmad I, Khaliq T, Nasir J, Rasul F, Riaz F, Ahmad B, Ahmad S, Baig IA, Valdivia RO, Hoogenboom G (2021) Development of climate change adaptation strategies for cotton–wheat cropping system of Punjab Pakistan. In: Rosenzweig C, Mutter CZ, Contreras EM (eds) Handbook of climate change and agroecosystems: climate change and farming system planning in Africa and South Asia: AgMIP stakeholder-driven research. World Scientific Publishing Europe Ltd, London, pp 277–327

Ahmad S, Ahmad I, Ahmad B, Ahmad A, Wajid A, Khaliq T, Abbas G, Wilkerson CJ, Hoogenboom G (2023a) Regional integrated assessment of climate change impact on cotton production in a semi-arid environment. Clim Res 89:113–132

Ahmad S, Raza MA, Hussain S, Abbas G, Fatima Z, Ahmed M, Goheer MA, Garcia y Garcia A, Wilkerson CJ, Hoogenboom G (2023b) Identification of weak links in the production technology of farmer for bridging canola yield-gap. J Agric Sci 161(2):241–253. https://doi.org/10.1017/S0021859623000187

Ahmed M, Akram MN, Asim M, Aslam M, F-u Hassan Higgins S, Stockle CO, Hoogenboom G (2016) Calibration and validation of APSIM-wheat and CERES-wheat for spring wheat under rainfed conditions: models evaluation and application. Comput Electron Agric 123:384–401

Ahmed I, Rahman MH, Ahmad A, Ahmad S, Hussain J, Ullah A, Judge J (2018a) Assessing the impact of climate variability on maize using simulation modeling under semi-arid environment of Punjab-Pakistan. Environ Sci Pollut Res 25(23):28413–28430

Ahmed M, Ijaz W, Ahmad S (2018b) Adapting and evaluating APSIM-SoilP-Wheat model for response to phosphorus under rainfed conditions of Pakistan. J Plant Nutr 41(16):2069–2084

Ahmed M, Stockle CO, Nelson R, Higgins S, Ahmad S, Raza M (2019) Novel multimodal ensemble approach to evaluate the sole effect of elevated CO_2 on winter wheat productivity. Sci Rep 9:7813

Ahmed M, Ahmad S, Tariq M, Fatima Z, Aslam Z, Raza MA, Iqbal N, Akmal M, Fayyaz-ul-Hassan, Abbasi NA, Hayat R (2020) Wastes to be the source of nutrients and energy to mitigate climate change and ensure future sustainability: options and strategies. J Plant Nutr 43(6):896–920

Ahmed M, Hayat R, Ahmad M, Mahmood-ul-Hassan, AMS K, Fayyaz-ul-Hassan, Rehman MH, Shaheen FA, Raza MA, Ahmad S (2022) Impact of climate change on dryland agricultural systems: a review of current status, potentials, and further work need. Int J Plant Prod 16:341–363

Akram N, Hamid AG (2015) Climate change: a threat to the economic growth of Pakistan. Prog Dev Stud 15:73–86

Ali H, Iqbal N, Ahmad S, Shahzad AN, Sarwar N (2013) Performance of late sown wheat crop under different planting geometries and irrigation regimes in arid climate. Soil Tillage Res 130:109–119

Amin A, Nasim W, Fahad S, Ali S, Ahmad S, Rasool A, Saleem N, Hammad HM, Sultana SR, Mubeen M, Bakhat HF, Ahmad N, Shah GM, Adnan M, Noor M, Basir A, Saud S, Rahman MH, Paz JO (2018a) Evaluation and analysis of temperature for historical (1996–2015) and projected (2030–2060) climates in Pakistan using SimCLIM climate model: ensemble application. Atmos Res 213:422–436

Amin A, Nasim W, Mubeen M, Ahmad A, Nadeem M, Urich P, Fahad S, Ahmad S, Wajid A, Tabassum F, Hammad HM, Sultana SR, Anwar S, Baloch SK, Wahid A, Wilkerson CJ, Hoogenboom G (2018b) Simulated CSM-CROPGRO-cotton yield under projected future climate by SimCLIM for southern Punjab, Pakistan. Agric Syst 167:213–222

Amin A, Nasim W, Mubeen M, Sarwar S, Urich P, Ahmad A, Wajid A, Khaliq T, Rasul F, Hammad HM, Rehmani MIA, Mubarak H, Mirza N, Wahid A, Ahmad S, Fahad S, Ullah A, Khan MN, Ameen A, Amanullah SB, Saud S, Alharby H, Ata-Ul-Karim ST, Adnan M, Islam F, Ali QS (2018c) Regional climate assessment for temperature and precipitation in Southern Punjab (Pakistan) using SimCLIM climate model for different temporal scales. Theor Appl Climatol 131:121–131

Baez-Gonzalez AD, Kiniry JR, Meki MN, Williams JR, Cilva MA, Gonzalez JLR, Estala AM (2018) Potential impact of future climate change on sugarcane under dryland conditions in Mexico. J Agron Crop Sci 204:515–528

Battisti R, Sentelhas PC, Boote KJ (2018) Sensitivity and requirement of improvements of four soybean crop simulation models for climate change studies in Southern Brazil. Int J Biometeorol 62(5):823–832

Binbol NL, Adebayo AA, Kwon-Ndung (2006) Influence of climate factors on the growth and yield of sugarcane at Numan, Nigeria. Clim Res 32:247–252

Braithwaite KS, Croft BJ, Magarey RC, Scharaschkin T (2009) Phylogenetic placement of the sugarcane orange rust pathogen *Puccinia kuehnii* in a historical and regional context. Australas Plant Pathol 38(4):380–388

Clements HF (1980) Sugarcane crop logging and crop control. University Press, Hawaii

Crafts-Brandner SJ, Salvucci ME (2002) Sensitivity of photosynthesis in a C4 plant, maize, to heat stress. Plant Physiol 129:1773–1780

Da Silva FC, Diaz-Ambrona CGH, Buckeridge MS, Souza A, Barbieri V, Dourado ND (2008) Sugarcane and climate change effects of CO_2 on potential growth and development. Acta Hortic 802:331–336

de Carvalho AL, Menezes RSC, Nobrega RS, de Pinto A, Siqueira JPHB, Ometto C, von Randow C, Giarolla A (2015) Impact of climate changes on potential sugarcane yield in Pernambuco, northeastern region of Brazil. Renew Energy 78:26–34

Deressa T, Hassan R, Poonyth D (2005) Measuring the impact of climate change on South African agriculture: the case of sugarcane growing regions. Agrekon 44:524–542

DES (2016) Agricultural statistics at a glance 2016. Department of Agriculture Cooperation and Framers Welfare, Ministry of Agriculture and farmers Welfare, Directorate of Economics and Statistics, Government of India, New Delhi

Dias HB, Inman-Bamber G (2020) Sugarcane: contribution of process-based models for understanding and mitigating impacts of climate variability and change on production. In: Ahmed M (ed) Systems modeling. Springer, Singapore, pp 217–260

Dias HB, Sentelhas PC (2017) Evaluation of three sugarcane simulation models and their ensemble for yield estimation in commercially managed fields. Field Crop Res 213:174–185

Edme SJ, Miller JD, Glaz B, Tai PYP, Comstock JC (2005) Genetic contribution to yield gains in the Florida sugarcane industry across 33 years. Crop Sci 45(1):92–97

Ejaz M, Abbas G, Fatima Z, Iqbal P, Raza MA, Kheir AMS, Ahmed M, Kakar KM, Ahmad S (2022) Modelling climate uncertainty and adaptations for soybean-based cropping system. Int J Plant Prod 16:235–250

El-Kholi MMA (2008) Sugar Crops Research Institute, Giza (Egypt): a profile. Sugar Technol 10(3):189–196

Everingham YL, Muchow RC, Stone RC, InmanBamber NG, Singels A, Bezuidenhout CN (2002) Enhanced risk management and decision-making capability across the sugarcane industry value chain based on seasonal climate forecasts. Agric Syst 74(3):459–477

Factfish (2015) Sugar cane, production quantity (tons)—for all countries. http://www.factfish.com/statistic/sugar+cane,+production+quantity

FAO (1997) Is sugar pure, white and deadly? In: Proceedings of the Fiji/FAO Asia Pacific Sugar Conference, Suva, Fiji, 29–31 Oct 1997, pp 41–44

FAO (2014) Report, Food and Agricultural Organization, United Nations: Economic and Social Department, The Statistical Division, FAO 2013 FAOSTAT, 2014

Farooq N, Gheewala SH (2020) Assessing the impact of climate change on sugarcane and adaptation actions in Pakistan. Acta Geophys 68:1489–1503

Fatima Z, Abbas Q, Khan A, Hussain S, Ali MA, Abbas G, Younis H, Naz S, Ismail M, Shahzad MI, Nadeem M, Farooq U, Khan SU, Javed K, Khan AA, Ahmed M, Khan MA, Ahmad S (2018) Resource-use-efficiencies of C_3 and C_4 cereals under split nitrogen regimes. Agronomy 8(5):69

Fatima Z, Ahmad M, Hussain M, Abbas G, Ul-Allah S, Ahmad S, Ahmed N, Ali MA, Sarwar G, Haque E, Iqbal P, Hussain S (2020) The fingerprints of climate warming on cereal crops phenology and adaptation options. Sci Rep 10:18013

Fatima Z, Atique-ur-Rehman, Abbas G, Iqbal P, Zakir I, Khan MA, Kamal GM, Ahmed M, Ahmad S (2021) Quantification of climate warming and crop management impacts on phenology of pulses-based cropping systems. Int J Plant Prod 15:107–123

Flack-Prain S, Shi L, Zhu P, Rocha HRD, Cabral O, Hu S, Williams M (2021) The impact of climate change and climate extremes on sugarcane production. GCB Bioenergy 13:408–424

Folberth C, Yang H, Gaiser T, Liu J, Wang X, Williams J, Schulin R (2014) Effects of ecological and conventional agricultural intensification practices on maize yields in sub-Saharan Africa under potential climate change. Environ Res Lett 9:044004

Gawander J (2007) Impact of climate change on sugar-cane production in Fiji. WMO Bull 56(1):34–39

Glasziou KT, Bull TA, Hatch MD, Whiteman PC (1965) Effects of temperature, photoperiod duration, and diurnal and seasonal temperature changes on growth and ripening. Aust J Biol Sci 18:53–66

Henggeler J, Henry D, Kenny NC (2004) Measuring diesel fuel consumption to estimate engine efficiency, University of Missouri. https://www.researchgate.net/profile/Arvind_Singh56/post/How_to_measure_the_efficiency_of_diesel_engine/attachment/59d6407179197b807799ca7c/AS%3A431123860856833%401479799346198/download/1.pdf

Hoogenboom G, Porter CH, Boote KJ, Shelia V, Wilkens PW, Singh U, White JW, Asseng S, Lizaso JI, Moreno LP, Pavan W, Ogoshi R, Hunt LA, Tsuji GY, Jones JW (2019) The DSSAT crop modeling ecosystem. In: Boote KJ (ed) Advances in crop modeling for a sustainable agriculture. Burleigh Dodds Science Publishing, Cambridge, pp 173–216. https://doi.org/10.19103/AS.2019.0061.10

Hoogenboom G, Porter CH, Shelia V, Boote KJ, Singh U, White JW, Pavan W, Oliveira FAA, Moreno-Cadena LP, Lizaso JI, Asseng S, Pequeno DNL, Kimball BA, Alderman PD, Thorp KR, Jones MR, Cuadra SV, Vianna MS, Villalobos FJ, Ferreira TB, Batchelor WD, Koo J, Hunt

LA, Jones JW (2021) Decision Support System for Agrotechnology Transfer (DSSAT) Version 4.8 (DSSAT.net). DSSAT Foundation, Gainesville, FL

Houghton RA, Lawrence KT, Hackler JL, Brown S (2001) The spatial distribution of forest biomass in the Brazilian Amazon: a comparison of estimates. Glob Chang Biol 7(7):731–746

Hunink JE, Droogers P (2011) Climate change impact assessment on crop production in Uzbekistan. World Bank study on reducing vulnerability to climate change in Europe and Central Asia (ECA) Agricultural Systems, Wageningen

Hussain M, Ahmad S, Hussain S, Lal R, Ul-Allah S, Nawaz A (2018) Rice in saline soils: physiology, biochemistry, genetics and management. Adv Agron 148:231–287

Inman-Bamber NG, Lakshmanan P, Park S (2012) Sugarcane for water-limited environments: theoretical assessment of suitable traits. Field Crop Res 134:95–104

IPCC (2014) Summary for policymakers. In: Climate change 2014: impacts, adaptation, and vulnerability. Part A: global and sectoral aspects. Contribution of Working Group II to the Fifth Assessment Report of the Intergovernmental Panel on Climate Change (Field CB, Barros VR, Dokken DJ et al., eds). Cambridge University Press, Cambridge, pp 1–32

Jones JW, Hoogenboom G, Porter CH, Boote KJ, Batchelor WD, Hunt LA, Wilkens PW, Singh U, Gijsman AJ, Ritchie JT (2003) The DSSAT cropping system model. Eur J Agron 18(3–4):235–265

Jones MR, Singels A, Ruane AC (2015) Simulated impacts of climate change on water use and yield of irrigated sugarcane in South Africa. Agric Syst 139:260–270

Keating BA, Carberry PS, Hammer GL, Probert ME, Robertson MJ, Holzworth D, Huth NI, Hargreaves JNG, Meinke H, Hochman Z, McLean G, Verburg K, Snow V, Dimes JP, Silburn M, Wang E, Brown S, Bristow KL, Asseng S, Chapman S, McCown RL, Freebairn DM, Smith CJ (2003) An overview of APSIM: a model designed for farming systems simulation. Eur J Agron 18:267–288

Khan AQ, Idrees M, Aldosari F (2016) An assessment of main problems faced by farming community in sugarcane production of district Peshawar. Int J Agric Ext Rural Dev 3(1):149–155

Khan A, Fahad S, Khan A, Saud S, Adnan M, Wahid F, Noor M, Nasim W, Hammad HM, Bakhat HF, Ahmad S, Rehman MH, Wang D, Sönmez O (2019) Managing tillage operations and manure to restore soil carbon stocks in wheat-maize cropping system. Agron J 111:1–10

Khan MN, Hussain M, Abbas G, Fatima Z, Iqbal P, Khan A, Zakir I, Ahmed M, Mateen A, Khan MA, Ahmad S (2020) Improving resource use efficiencies of sugarcane at farmer field under arid environment. Int J Agric Biol 24(5):1279–1285

Knox JW, Rodríguez Díaz JA, Nixon DJ, Mkhwanazi M (2010) A preliminary assessment of climate change impacts on sugarcane in Swaziland. Agric Syst 103(2):63–72

Kumar RK (1984) Yield response of sugarcane to weather variations in North-East Andhra Pradesh, India. Arch Meteorol Geophys Bioclimatol Ser B 35(3):265–276

Kumar D, Malik N, Sengar (2019) Physio-biochemical insights into sugarcane genotypes under water stress. Biol Rhythm Res 52(1):92–115

Lee SK, Dang TA (2018) Application of AquaCrop model to predict sugarcane yield under climate change impact: a case study of Son Hoa district, Phu Yen province in Vietnam. Res Crops 19(2):310–314

Li YR, Wei YA (2006) Sugar industry in China: R & D and policy initiatives to meet sugar and biofuel demand of future. Sugar Technol 8(4):203–216

Li YR, Yang LT (2014) Chinese sugarcane and sugar industry in recent 10 years. In: Li Y-R, Rao GP, Solomon S (eds) Proceedings of the 5th International Association of Professionals in Sugar and Integrated Technologies: Green Technologies for Sustainable Growth of Sugar & Integrated Industries in Developing Countries, pp 1–10

Li YR, Fang FX, Yang LT (2011) Long duration cold stress and frost induced losses to sugarcane production in Guangxi (China) during 2010–11 and its control measures. In: Proceedings of the 4th IAPSIT International Sugar Conference (IS '11) & Expo (Balancing Sugar and Energy Production in Developing Countries: Sustainable Technologies and Marketing Strategies), Army Printing Press, IBDC Publishers, Lucknow, pp 353–358

Lobell DB, Burke M, Tebaldi C, Mastrandrea MD, Falcon WP, Naylor RL (2008) Prioritizing climate change adaptation needs for food security in 2030. Science 319:607–610

Magsi H, Rashid MK, Khushk GM, Khatyan MA, Babar MN (2018) Impact of climate change on major crops of Pakistan. SALU Commerce Econ Rev 4(1):70–77

Mahmood A, Ullah H, Shahzad AN, Ali H, Ahmad S, Zia-Ul-Haq M, Honermeier B, Hasanuzzaman M (2013) Dry matter yield and chemical composition of sorghum cultivars with varying planting density and sowing date. Sains Malaysiana 42(10):1529–1538

Mali SC, Shrivastava PK, Thakare HS (2014) Impact of weather changes on sugarcane production. Res Environ Life Sci 7(4):243–246

Marengo JA, Chou SC, Torres RR, Giarolla A, Alves LM, Lyra A (2014) Climate change in Central and South America: recent trends, future projections, and impacts on regional agriculture. CCAFS Working Paper no. 73. CGIAR Research Program on Climate Change, Agriculture and Food Security (CCAFS). Copenhagen. http://www.ccafs.cgiar.org

Marin FR, Jones JW, Singles A, Royce F, Assad ED, Pellegrino GQ, Justino F (2012) Climate change impacts on sugarcane attainable yield in southern Brazil. Clim Chang 117:227–239

Matthieson L (2007) Climate change and the Australian Sugarcane Industry: impacts, adaptation and R&D opportunities, SRDC technical report. Sugar Research and Development, Brisbane

Moore PH, Frederik CS (2014) Sugarcane: physiology, biochemistry and functional biology. Wiley-Blackwell, Hoboken

Mubarak MK, Hussain K, Abbas G, Altaf MT, Baloch FS, Ahmad S (2022) Productivity of sorghum (*Sorghum bicolor* L.) at diverse irrigation regimes and sowing dates in semi-arid and arid environment. Turk J Agric For 46(1):1–18

Mubeen M, Ahmad A, Wajid A, Khaliq T, Hammad HM, Sultana SR, Ahmad S, Fahad S, Nasim W (2016) Application of CSM-CERES-Maize model in optimizing irrigated conditions. Outlook Agric 45(3):173–184

Nadeem M, Khan MN, Abbas G, Fatima Z, Iqbal P, Ahmed M, Raza MA, Rehman A, Haq EU, Hayat A, Ali M, Ahmad S (2022) Application of CSM-CANEGRO model for climate change impact assessment and adaptation for sugarcane in semi-arid environment of Southern Punjab, Pakistan. Int J Plant Prod 16:443–466

Naohiro A, Takahide K (2015) Development of high efficiency water pump. Honda R D Tech Rev 27(1):142

Nasim W, Amin A, Fahad S, Awais M, Khan N, Mubeen M, Wahid A, Turan V, Rehman MH, Ihsan MZ, Ahmad S, Hussain S, Mian IA, Khan B, Jamal Y (2018) Future risk assessment by estimating historical heat wave trends with projected heat accumulation using SimCLIM climate model in Pakistan. Atmos Res 205:118–133

Nassif DSP, Marin FR, Filho WJP, Resende RS, Pellegrino GQ (2012) Parameterization and evaluation of the DSSAT/CANEGRO model for Brazilian sugarcane varieties. Pesqui Agropecuaria Bras 47(3):311–318

Naz S, Ahmad S, Abbas G, Fatima Z, Hussain S, Khan MA, Ahmed M, Khan A, Fahad S, Nasim W, Ercisli S, Wilkerson CJ, Hoogenboom G (2022) Modeling the impact of climate warming on potato phenology. Eur J Agron 132:126404

Nazir A, Jariko GA, Junejo MA (2013) Factors Affecting Sugarcane Production in Pakistan. Pak J Commer Soc Sci 7(1):128–140

O'Leary GJ (1999) A review of three sugarcane simulation models in their prediction of sucrose yield. Proc South Afr Sugar Technol Assoc 73:33–34

O'Leary GJ (2000) A review of three sugarcane simulation models with respect to their prediction of sucrose yield. Field Crop Res 68(2):97–111

Peng T, Fu J, Jiang D, Du J (2020) Simulation of the growth potential of sugarcane as an energy crop based on the APSIM model. Energy 13:2173

Pipitpukdee S, Attavanich W, Bejranonda S (2020) Climate change impacts on sugarcane production in Thailand. Atmos 11:408

Rahman MH, Ahmad A, Wang X, Wajid A, Nasim W, Hussain M, Ahmad B, Ahmad I, Ali Z, Ishaque W, Awais M, Muddasir M, Shelia V, Ahmad S, Fahad S, Alam M, Ullah H, Hoogenboom

G (2018) Multi-model projections of future climate and climate change impacts uncertainty assessment for cotton production in Pakistan. Agric For Meteorol 253–254:94–113

Rahman MH, Ahmad I, Wang D, Fahad S, Afzal M, Ghaffar A, Saddique Q, Khan MA, Saud S, Hassan S, Fahad M, Din M, Ahmad M, Yue Z, Shah RA, Sönmez O, Nasim W (2021) Influence of semi-arid environment on radiation use efficiency and other growth attributes of lentil crop. Environ Sci Pollut Res 28:13697–13711

Rehman A, Ashfaq M, Naqvi SAA, Adil SA, Bashir K, Ahmad A, Ahmad S, Ali A, Imran A (2015) Is climate change worsening the poverty of maize growers? Evidence from Punjab province of Pakistan. Ciencia e Tecnica Vitivinicola 30(1):105–116

Riajaya PD (2020) Rainy season period and climate classification in sugarcane plantation regions in Indonesia, 1st International Conference on Sustainable Plantation 2019. IOP Conf Ser Earth Environ Sci 418:012058

Rosenzweig C, Casassa G, Karoly DJ (2007) Assessment of observed changes and responses in natural and managed systems. In: Climate change 2007: impacts, adaptation and vulnerability. Contribution of Working Group II to the Fourth Assessment Report of the Intergovernmental Panel on Climate Change (Parry ML, Canziani OF, Palutikof JP, van der Linden PJ, Hanson CE, eds). Cambridge University Press, Cambridge, pp 79–131

Ruan H, Feng P, Wang B, Xing H, Leary GJO, Huang Z, Guo H, Liu DL (2018) Future climate change projects positive impacts on sugarcane productivity in southern China. Eur J Agron 96:108–119

Sarwar N, Ali H, Maqsood M, Ullah E, Shahzad AN, Shahzad M, Mubeen K, Shahid MA, Ahmad S (2013) Phenological response of rice plants to different micronutrients application under water saving paddy fields on calcareous soil. Turk J Field Crops 18(1):52–57

Sarwar N, Atique-ur-Rehman, Farooq O, Wasaya A, Hussain M, El-Shehawi AM, Ahmad S, Brestic M, Mahmoud SF, Zivcak M, Farooq S (2021) Integrated nitrogen management improves productivity and economic returns of wheat-maize cropping system. J King Saud Univ Sci 33:101475

Shabbir RN, Waraich EA, Ali H, Nawaz F, Ashraf MY, Ahmad R, Awan MI, Ahmad S, Irfan M, Hussain S, Ahmad Z (2016) Supplemental exogenous NPK application alters biochemical processes to improve yield and drought tolerance in wheat (*Triticum aestivum* L.). Environ Sci Pollut Res 23(3):2651–2662

Sharma J, Upadhyay AK, Adsule PG, Sawant SD, Sharma AK, Satisha J, Yadav DS, Ramteke SD (2013) Effect of climate change on grape and its value added products. In: Singh HCP, Rao NKS, Shivashankar KS (eds) Climate-resilient horticulture: adaptation and mitigation strategies. Springer, New Delhi, pp 67–80

Shukla SK, Sharma L, Awasthi SK, Pathak AD (2017) AICRP (S) Technical Bulletin No. 1 ICAR-All India coordinated research project on Sugarcane. Sugarcane in India (Package of practices for different agro-climatic zones). ICAR-Indian Institute of Sugarcane Research, Lucknow, 2017

Siddiqui R, Samad G, Nasir M, Jalil HH (2012) The impact of climate change on major agricultural crops: evidence from Punjab, Pakistan. Pak Dev Rev 51(4):261–276

Siebert S, Ewert F, Rezaei EE, Kage H, GraB R (2014) Impact of heat stress on crop yield-on the importance of considering canopy temperature. Environ Res Lett 9:044012

Silva WKDM, De Freitas GP, Junior LMC, Pinto PALDA, Abrahao R (2019a) Effects of climate change on sugarcane production in the state of Paraíba (Brazil): A panel data approach (1990–2015). Clim Chang 154:195–209

Silva WKDM, Freitas GPD, Coelho Jn LM, Pinto PALDA, Abrahao R (2019b) Effects of climate change on sugarcane production in the state of Paraíba (Brazil): a panel data approach (1990–2015). Climate Change 154:195–209

Singels A, Bezuidenhout CN (2002) A new method of simulating dry matter partitioning in the CANEGRO sugarcane model. Field Crop Res 78(2):151–164

Singels A, Jones M, Marin F, Ruane A, Thorburn P (2013) Predicting climate change impacts on sugarcane production at sites in Australia, Brazil and South Africa using the Canegro Model. Sugar Technol 16:347–355

Singels A, Jones M, Marin F, Ruane A, Thorburn P (2014) Predicting climate change impacts on sugarcane production at sites in Australia Brazil and South Africa using the Canegro model. Sugar Technol 164:347–355

Sonkar G, Singh N, Mall RK, Singh KK, Gupta A (2019) Simulating the impacts of climate change on sugarcane in diverse agro-climatic zones of Northern India using CANEGRO-Sugarcane model. Sugar Technol 22:460–272

Souza APD, Gaspar M, Silva EAD, Ulian EC, Waclawovsky AJ, Nishiyama MY Jr, Dos Santos RV, Teixeira MM, Souza GM, Buckeridge MS (2008) Elevated CO_2 increases photosynthesis, biomass and productivity, and modifies gene expression in sugarcane. Plant Cell Environ 31(8):1116–1127

Srivastava AK, Rai MK (2012) Review: sugarcane production: impact of climate change and its mitigation. Biodiversitas 13(4):214–227

Tariq M, Yasmeen A, Ahmad S, Hussain N, Afzal MN, Hasanuzzaman M (2017) Shedding of fruiting structures in cotton: factors, compensation and prevention. Trop Subtrop Agroecosyst 20(2):251–262

Tariq M, Afzal MN, Muhammad D, Ahmad S, Shahzad AN, Kiran A, Wakeel A (2018a) Relationship of tissue potassium content with yield and fiber quality components of Bt cotton as influenced by potassium application methods. Field Crop Res 229:37–43

Tariq M, Ahmad S, Fahad S, Abbas G, Hussain S, Fatima Z, Wajid N, Mubeen M, Rehman M, Khan MA, Adnan M, Wilkerson CJ, Hoogenboom G (2018b) The impact of climate warming and crop management on phenology of sunflower-based cropping systems in Punjab, Pakistan. Agric For Meteorol 256–257:270–282

Tarriq M, Fatima Z, Iqbal P, Nahar K, Ahmad S, Hasanuzzaman M (2021) Sowing dates and cultivars mediated changes in phenology and yield traits of cotton-sunflower cropping system in arid environment. Int J Plant Prod 15:291–302

Vandana P, Singh D, Srivastava S, Guru GDR (2020) Effect of climate change on sugarcane crop: a review. J Pharmacogn Phytochem 6:255–261

Verma AK, Garg PK, Prasad KSH, Dadhwal VK (2023) Variety-specific sugarcane yield simulations and climate change impacts on sugarcane yield using DSSAT-CSM-CANEGRO model. Agric Water Manag 275:108034

Vishwakarma SK, Kumar P, Nigam A, Singh A, Kumar A (2013) Pokkah Boeng: an emerging disease of sugarcane. J Plant Pathol Microbiol 4(3):170–174

Vu JCV, Allen LH Jr (2009a) Growth at elevated CO_2 delays the adverse effects of drought stress on leaf photosynthesis of the C4 sugarcane. J Plant Physiol 166(2):107–116

Vu JCV, Allen LH Jr (2009b) Stem juice production of the C4 sugarcane (Saccharum officinarum) is enhanced by growth at double-ambient CO_2 and high temperature. J Plant Physiol 166(11):1141–1151

Zahoor A, Ahmed M, Fayyaz-ul-Hassan, Shabbir G, Ahmad S (2022) Ontogeny growth and radiation use efficiency of canola (*Brassica napus* L.) under various nitrogen management strategies and contrasting environments. Int J Plant Prod 16(1):195–208

Zhao D, Li YR (2015) Climate change and sugarcane production, potential impact and mitigation strategies. Int J Agron 2015:547386

Potato-Potato System 10

10.1 Introduction

The potato, also known as *Solanum tuberosum* L., is an essential crop for human nutrition grown in various countries worldwide (Gustavsen 2021). It is the fourth most significant food crop after wheat, rice, and maize. It has become the primary source of income as well as food in countries that are still developing (Raymundo et al. 2017). These nations are experiencing a higher population growth rate than other regions in the world (Lutz and Samir 2010). More than a billion people around the world rely on potatoes as a staple food, making this crop an increasingly important part of future efforts to ensure global food security. There are approximately 20 million hectares of land that are used for planting globally. The average yield per hectare is 17 tonnes, resulting in 370 million tonnes worth approximately US $50 billion each year. Potatoes are produced in the largest quantities in China, and India is second in potato production worldwide (Table 10.1). The crop of potatoes is grown in regions with a high incidence of malnutrition, food security, and poverty; these include the Indo-Genetic basin in South Asia and the tropical highlands of Africa (Bruinsma 2003; Thiele et al. 2010). According to Mishra et al. (2020), because it contains a variety of phytonutrients, vitamins, minerals, proteins, carbohydrates, and dietary fibers, there is a possibility that it could play a role in the enhancement of human health. As a result, it is possible to consider it a significant food crop to ensure the nutritional security of the global population (Abbas et al. 2017, 2019, 2020a, b, c; Abbas and Ahmad 2018; Ahmad et al. 2012a, b, 2013, 2015a, b, c, 2016a, b, c, 2017a, b, 2018, 2019, 2021, 2023a, b). The potato is both a food crop and a vegetable that can be used in various preparations. Compared to wheat and rice, the dry matter yield of potato is up to 74.5% higher, and the amount of energy it produces per unit area is up to 58% higher. Because of their resistance to low fertility and drought conditions, potatoes can be grown successfully in cool or warm environments. Potato crops, on the other hand, cannot withstand high temperatures or humidity (Haas et al. 2009; Chakraborty et al. 2010). Because of its

M. Ahmed et al., *Cropping Systems Modeling Under Changing Climate*,
https://doi.org/10.1007/978-981-97-0331-9_10

Table 10.1 Area and production of potato-potato system across various regions in the world

Countries	Area (Mha)	Production (MMT)	Yield (kg ha^{-1})
China	4.81	90.32	18,764.2
India	2.15	48.52	22,561.1
Ukraine	1.32	22.50	17,049.8
Russian Federation	1.31	22.39	17,049.9
USA	0.41	20.60	49,762.4
Bangladesh	0.47	9.74	20,410.6
Germany	0.25	8.92	35,371.9
France	0.19	7.87	39,377.3

ancestry, the potato thrives in moderately cool climates that do not experience frost. On the other hand, it does not do as well in regions that experience high temperatures (Haverkort and Verhagen 2008; Ahmed et al. 2016, 2018a, b, 2019a, b, 2020, 2022; Ahmad and Hasanuzzaman 2012; Ali et al. 2013; Amin et al. 2018a, b, c).

10.2 Background

Peruvian farmers are responsible for developing four recognized potato species and 5000 potato varieties that are still grown today. The first potatoes were cultivated approximately 8000 years ago in the Central Andes of Peru. Conquistadors from Spain introduced potatoes to Europe in the sixteenth century. At first, the Europeans thought that the tubers were unfit for human consumption and should be reserved for animals or the poor. However, in the eighteenth century, early-maturing potato varieties were developed through selective breeding. They flourished in the harsh growing conditions of the north and arrived at a critical moment, just as a significant portion of continental Europe was being wiped out by famine. Suddenly, the importance of the potato as a crop for ensuring a steady food supply was recognized, and European farmers began growing it. Because potatoes can be grown quickly and at a low cost, they saved entire populations from starvation, provided food for the newly emerging urban working class, and freed up people living in rural areas to find work in factories during the nineteenth century. However, the tubers cloned and grown in multiple locations across Europe and North America belonged to a small number of genetically related varieties. In 1845, a disease known as late blight wiped out Ireland's potato crop, which led to a famine that claimed the lives of at least one million people and forced another million to leave the country. Following the Irish disaster, efforts were made to develop varieties that were more productive and disease resistant. Many modern potato varieties were developed by breeders in Europe and North America, who used potato germplasm originating in Chile as a source.

The potato was first cultivated on the Indo-Pakistan subcontinent by the Portuguese. In the nineteenth century, with the British's encouragement, it began to be cultivated. After being recognized for the first time as a vegetable, the potato quickly became one of the most significant vegetables. In recent years, this has also become of great importance, mainly because it is referred to as the "king of vegetables" in Pakistan (Qasim et al. 2013).

10.3 Potato Cropping System and Global Distribution

The cultivation of potatoes around the world utilizes a variety of different methods. Due to the country's agro-geographic conditions, it can be cultivated in Pakistan during spring, summer, and autumn. Potatoes are typically grown in Pakistan using a cropping method known as the potato-potato cropping system. Tubers are able to be grown in Brazil during the months of both spring and fall (Bisognin et al. 2008a). The average environmental temperature, photoperiod, and irradiation all rise during the spring growing season, going from 15 °C, 12 h, and 4.11 MJ/m^2/day to 22 °C, 15 h, and 24 MJ/m^2/day, respectively (Bisognin et al. 2008b). The average environmental temperature, photoperiod, and irradiation all drop during the autumn growing season, going from 28 °C, 13 h, and 19.9 MJ/m^2/day to 12 °C, 11 h, and 4.63 MJ/m^2/day, respectively (Bisognin et al. 2008b). These contrasting environmental conditions during spring and autumn growing seasons can potentially result in different processing qualities of potato tubers when stored at varying temperatures. During the springtime development of the plant, an increase in photoperiod, temperature, and irradiance can lead to a rise in the amount of carbohydrates accumulated in the tubers (Bisognin et al. 2008b; Ejaz et al. 2022; Fatima et al. 2018, 2020, 2021; Naz et al. 2022; Jones et al. 2003; Keating et al. 2003; Hoogenboom et al. 2019, 2021).

During plant and tuber development in the autumn cultivation, day length and temperature were decreasing, whereas in the spring, these parameters were increasing along with plant development (Alexopoulos et al. 2006). Larkin (2003) found more specific soil microbial changes and characteristics associated with potato-potato cropping systems. Then he explained how these modifications would affect crop productivity, health, and disease management. This strategy's goal was to implement crop management practices that optimized the soil's microbial environment to reduce disease, increase productivity, and ensure the strategy's long-term viability. The genetic makeup of the potato tubers' dry matter and the potato chips' color is determined by environmental conditions during the growing season and the temperature at which they are stored (Kawchuk et al. 2008). Potatoes grown in the spring have more dry matter and a shorter dormancy than tubers grown in the fall (De Freitas et al. 2006; Bisognin et al. 2008a).

Spring potatoes account for over 70% of total potato production through their cultivation. Although spring potatoes are very important, their productivity is low (only 26 tonnes per hectare) compared to potato yield in the USA or Europe (greater than 40 tonnes/hectare). The short growing season (less than 100 days) is the primary cause of this low yield. The planting of early spring potatoes is held up by frost, which occurs in early spring. In addition, harvesting process takes place before the onset of the early summer monsoons, which typically occur in late June, to protect the tubers from being damaged by waterlogging. This takes place in the early summer (Mahmood et al. 2013; Mubeen et al. 2016; Nasim et al. 2018;

Mubarak et al. 2022; Rehman et al. 2015; Rahman et al. 2018, 2021; Sarwar et al. 2013, 2021; Shabbir et al. 2016).

The crop is becoming more crucial for preserving food stability and security, particularly in areas of developing nations with extensive agricultural production (Hijmans and Spooner 2001; DeFauw et al. 2012). China dominates global potato industry because it has the most potato production area (30%) and production (24%) (Jansky et al. 2009). Potatoes are primarily cultivated in China's north (49% of the total area), southwest (39%), south (7%), and central plains (5%), each of which is classified as an agroecological zone. Potatoes were grown on an area of approximately 2.13 million hectares in India in 2016, and the country's annual production of potatoes came in at almost 44 million tonnes, with an average yield of 20.5 tonnes per hectare. Due to the fact that the United Nations anticipates a 19% increase in the country's population by the year 2050, India is facing a significant obstacle in its efforts to boost the production of all food crops, including potatoes, in order to satisfy anticipated demand.

Recent data on potato production and its consumption showed that the amount consumed in Europe and North America has decreased. In the developing world, where it is necessary to ensure food security, potatoes proved very effective (Wijesinha-Bettoni and Mouillé 2019). Potato production is becoming more vital in food grain-based cropping systems due to its many benefits, including the capability to produce more dry matter in a shorter time and the adaptability to be incorporated into various cropping systems. These benefits are driving the growth of the potato industry. Potato-based cropping systems are subject to a great deal of pressure as a result of various factors including an expanding human population, shifting weather patterns, and diminishing natural resources (Tariq et al. 2017, 2018a, b, c; Tarriq et al. 2021; Hussain et al. 2018; Khan et al. 2019, 2020; Zahoor et al. 2022).

10.4 Decision Support System for Agrotechnology Transfer for Potato System

In order to maximize the crop's potential and maintain production levels, farmers need to implement more diverse farming techniques and exercise greater discretion in the management of their available resources. These days, several different tools and methods are utilized to maximize the effectiveness of resource utilization in crop production. The use of simulation models as decision-supporting tools is an important part of coming to logical conclusions about how to most effectively and precisely utilize available resources. According to Singh et al. (2005), an alternative method that has the potential to increase the effectiveness of agronomic experimentation is the utilization of crop models. The relationship between plant growth and yield and the variables associated with both concepts have frequently been determined using older statistical models that corroborated various independent variables with dependent variables of interest (Singh et al. 2008). The applicability of these regression-based statistical yield models is constrained because they express yield without accounting for primary biological and physical processes that are involved in plant systems. Dynamic crop models with a higher extrapolative value can be used to quantify relationship between yield and the factors affecting it.

The analysis of a crop production system can be done more systematically with the help of crop simulation models, which represent a simplified version of a crop production system using nonlinear mathematical equations. The crop simulation models have reached a level of accuracy that is sufficient and have gained widespread acceptance (Easterling et al. 1996). Semiarid and arid tropical crop models can predict crop development, growth, and yield. These models consider genotype selection and management options for agricultural sustainability (Wolday and Hruy 2015). Crop simulation models can determine yield potential, crop management alternatives, climate change and variable effects on yield, and multi-environment testing efficiency. These models also improve multi-environment testing (Singh et al. 1999). It can be utilized as a decision support system for training and education, research and policy creation, interdisciplinary collaboration, and statistical weather-yield connections for economic modeling (Rivington and Koo 2010).

The International Benchmark Sites Network for Agrotechnology Transfer (IBSNAT) is responsible for the creation of the crop model known as Decision Support System for Agrotechnology Transfer (DSSAT), which is currently considered to be one of the most popular crop models (Hoogenboom et al. 2021). IBSNAT's project on crop modeling software resulted in the development of 16 different crop models, one of which was the DSSAT SUBSTOR Potato model. This model simulates daily growth and development while accounting for the effects of soil water and nitrogen balance. It is similar to CERES models, including water and nitrogen balance. There have been many crop models developed for potatoes. In the literature, there are over 30 different potato models. Each of these models has a unique structure. Most potato models have never been validated with field measurements, despite some of these models being used in studies on the effects of climate change, nitrogen fertilizer management, and irrigation management. It has been determined whether or not the potato models CROPSYSTVB-CSPOTATO, EXpert-N-SPASS, and LINTUL-NPOTATO are accurate representations of the nitrogen dynamics in the crop. Two different potato models, LPOTCO and AQUACROP, have been evaluated to see how they perform in environments with high levels of atmospheric CO_2. No models have been tested with high-temperature or heat-stress data. Versions of LINTUL and SUBSTOR-Potato are among the potato models that have been tested and implemented the most (Raymundo et al. 2018).

Crop models have the potential to play a pivotal role in the implementation of time-bound management strategies. Scientists created a program called BAsic CROp growth Simulator (BACROS), and this model served as a guide for the creation of other models and as the foundation for the creation of other models. With a focus on the tropics and subtropics, the International Benchmark Sites Network for Agrotechnology Transfer (IBSNAT) was founded in 1982. Its main output was a Decision Support System for Agrotechnology Transfer. Keeping in mind the idea of a decision-supporting system was a driving force behind the establishment of this network as its primary objective. The DSSAT package comprised a strategy evaluation program to assess various crop production options for higher yield and net returns as well as a database management system for soil, weather, genetic coefficients, management inputs, crop simulation models, and a series of weather generation programs. To simulate crop growth, the program included agricultural simulation models. This package includes several crop growth and yield models, each created on a common framework with a comparable topology (Ritchie 1985). The DSSAT group of crop models' structural design made it possible to use them for long-term simulations of crop rotations and successions (Bowen et al. 1998). In order to forecast crop growth throughout the course of a single growing season, the DSSAT group of crop models was created. Models that can mimic the growth of 16 distinct types of crops are included in the DSSAT version 3.5. According to Bowen et al. (1998), the DSSAT model bases its calculations on temperature, photoperiod, amount of radiation that is intercepted, as well as amount of soil water and nitrogen supply, to simulate the accumulation and partitioning of biomass as well as the phenological development of potato crops on a daily basis. As a result, the model enables users to simulate the results of growth in the absence of constraints (potential production), growth in the presence of a moisture deficit, and growth that is influenced by the interaction between the limited availability of nutrients (nitrogen and phosphorus) and water.

On the other hand, the LINTUL2 model was used to simulate growth in water-limited situations by adding the water balance of crop and soil into the LINTUL1 model. This was done in order to compare the two models' results (van Haren et al. 1998). In order to determine whether or not the LINTUL2 model is accurate, this was carried out. Van Keulen and Stol (1995) used this model to select the genotypes best suited to a variety of climatic circumstances and to assess the potential yield for about 1000 distinct locations. The WOFOST model is an additional one designed to estimate the growth of an annual crop in a given set of particular soil and meteorological conditions (World Food Studies). The potential production as well as two different levels of the constrained output can both be calculated with the help of this model. Scientists examined the SUBSTOR (Simulation of Underground Bulking STorage ORgans)-Potato model in order to estimate the yield advantage of employing various nitrogen management strategies. This was done in order to simulate underground bulk storage organs. Jégo et al. (2008) did a study in which they used the STICS soil-crop model to estimate the influence that various farming practices have on the quantity of nitrate that is leached away from the root zone of potatoes. The genetic coefficients of potato in DSSAT model are presented in Table 10.2.

In a similar vein, the CropSystVB-CS Potato model was used to analyze the fate and transport of nitrogen below the potato rhizosphere under varied degrees of nitrogen and irrigation management practices. This was done to better understand how nitrogen and irrigation management practices affect the fate and transport of nitrogen in the rhizosphere of potatoes (Alva et al. 2010). To better understand how nitrogen and water management practices affect potato growth and development, Gayler et al. (2002) combined the Soil–Plant–Atmosphere System Simulation (SPASS) with the model Expert-N to examine effects of different nitrogen fertilizers on potato cultivar growth and nitrogen absorption. The INFOCROP-POTATO model, which was built from the generic INFOCROP model, was utilized to

Table 10.2 Genetic coefficients, description, ranges, and guidelines for model calibration of potato in DSSAT

Genetic coefficient	Description
VAR#	Identification code or number for a specific cultivar
VAR-NAME	Name of cultivar
EXPNO	Number of experiments used to estimate cultivar parameters
ECO#	Ecotype code of this cultivar, points to the ecotype in the ECO file
G2	Leaf area expansion rate after tuber initiation (cm^2/m^2 d)
G3	Potential tuber growth rate (g/m^2 d)
PD	Index that suppresses tuber growth during the period that immediately follows tuber induction
P2	Tuber initiation sensitivity to long photoperiods
TC	Upper critical temperature for tuber initiation (C)

Source: DSSAT v4.8; Hoogenboom et al. (2021)

optimize the date of planting, conduct tests with nitrogen and water stress, and forecast the pre-harvest yield (Singh et al. 2005; Singh et al. 2008). The authors of the study (Manorama et al. 2015) optimized the planting and harvesting dates for several potato-growing regions in the Nilgiri hills of Tamil Nadu, India, by making use of the INFOCROP-POTATO model. These regions are located in India. Therefore, the information that was gained from the simulation can be interpreted in such a way as to make it easier to make precise decisions regarding the many different aspects of potato production.

10.5 Impact of Climate Change on Potato Phenology

Changes in phenology have been noted in many different groups of wild and ornamental plants (herbs, grasses, and trees) and in agricultural crops. These observations have been reported from a variety of latitudinal and climatic zones (Parmesan and Yohe 2003; Richardson et al. 2013). These observations have been made in both wild plants and ornamental plants. The influence of climate change on phenology has been the subject of multiple research papers (Schwartz 1999). Research in the field of phenology can provide information about how plants adapt to shifting weather patterns.

Numerous global, regional, and local studies have examined how climate change affects crop phenology and agricultural output (Ahmad et al. 2019; Fatima et al. 2021). The warming trend shortened spring and fall potato crop S-TI and TI-M phases, shortening all stages of growth (Adavi et al. 2018). There is mounting evidence that different agricultural rotations will experience earlier anthesis and maturity as a result of climate change (Hu et al. 2005; Li et al. 2014; Xiao and Tao 2016). Previous experimental research on a wide range of crops and dynamic models of diverse cropping systems (Ahmad et al. 2019; Fatima et al. 2021) have shown that the duration of the phenological phases is getting shorter over the course of time.

According to Kleinwechter et al. (2016), the temperature is a more important factor to consider when growing potatoes during the summer. The higher temperature will hasten the process of the plant growing, which in turn will hasten the flowering and ripening of the fruit. As a result, climate change and global warming are anticipated to shorten the growing season's length (Daccache et al. 2011; Eyshi Rezaei et al. 2015). There is a good chance that a shorter growth period will result in a lower potato yield, and this is because there will be less time available to make use of the resources needed for production. It was also reported by Moradi et al. that the anthesis stage of maize would advance less under the conditions of future climate change (2013, 2014). As was mentioned earlier, an increase in temperature will hasten a plant's growth and development stages and, as a result, shorten the time it takes for the plant to mature (Parry et al. 1999; Parry et al. 2004). This may result in a reduction in the amount of potatoes that can be harvested.

Warmer temperatures could make it possible to plant potatoes earlier in the year, which would be desirable because it would promote earlier harvesting (Bakhtiyor 2015) and reduce the risk of summer drought (Supit et al. 2012). However, the harvesting of potatoes at an earlier stage in the year is likely to be accompanied by an increased risk of frost damage across the majority of northern Europe. The models for the future suggested that a warmer environment and earlier planting could alter the time of harvest up to 1 month sooner than it is now projected to be.

10.6 Potato-Potato System: A Case Study of Pakistan

Using a modeling approach, Naz et al. (2022) investigated the impacts of the thermal trend to determine whether or not there is a correlation between the observed phenology of potatoes and the increasing temperatures. The research was carried out at 12 different locations in Punjab, Pakistan, using phenological observations for both the spring and autumn potato crops between 1980 and 2018. The CSM-SUBSTOR-Potato model was calibrated using the cultivar grown most frequently at each site between 1980 and 1982. The model was then tested using phenology data from 1983 through 1985 (Fig. 10.1). The warming trend was noticeably more pronounced in the spring than it was in the fall. According to the observations made of the weather in the surrounding area between 1980 and 2018, the average temperature of the air had risen by 0.8 °C over the previous decade. From 1980 to 2018, differences in spring and autumn phenology had a statistically significant inverse correlation with temperature rise. This correlation was found to exist between 1980 and 2018 (Figs. 10.2 and 10.3). The stages that were observed during the spring had an average advance of 6.2 days/decade for sowing, 6.0 days/decade for emergence, 3.8 days/decade for tuber start, and 2.0 days/decade for maturity (Fig. 10.4; Table 10.3). When looking at the stages that were observed throughout the fall, however, there was a delay of 5.2 days/decade for sowing, 5.1 days for emergence, 3.3 days for the commencement of tuber development, and 2.3 days for maturity. This delay was detected across all four stages. It took an average of 2.4 fewer days/decade for sowing to be followed by tuber initiation, 1.8 fewer days/decade for

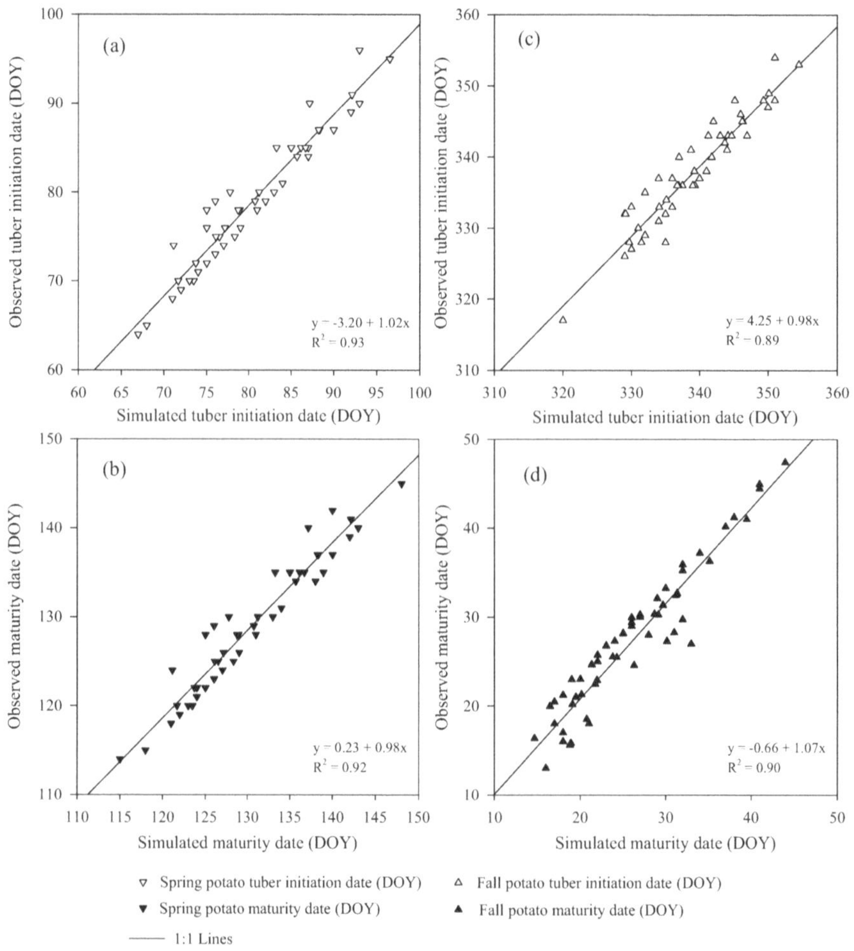

Fig. 10.1 Comparison of simulated and observed tuber initiation and maturity dates for spring (**a, b**; unfilled and filled down triangles) and fall season (**c, d**; unfilled and filled up triangles) potato crop, respectively, during 1980–1985 for ten locations in Punjab, Pakistan (Source: Adopted from Naz et al. 2022)

tuber initiation to be followed by maturity for spring, and 4.2 fewer days/decade for sowing to be followed by maturity. The length of the fall phase shrunk by an average of 2.9 days/decade for sowing to maturity, 1.9 days/decade for tuber start to maturity, and 1.0 days/decade for sowing to maturity, respectively. These reductions were seen across all three stages (Figs. 10.5, 10.6, and 10.7). The adoption of gradually new evolving cultivars caused the thermal time requirement of spring and autumn potatoes to be at its lowest point in 1980 and its highest point in 2018, respectively (Fig. 10.8). The temperature did not positively correlate with the planting dates for either spring or autumn potatoes (Fig. 10.9; Tables 10.4 and 10.5).

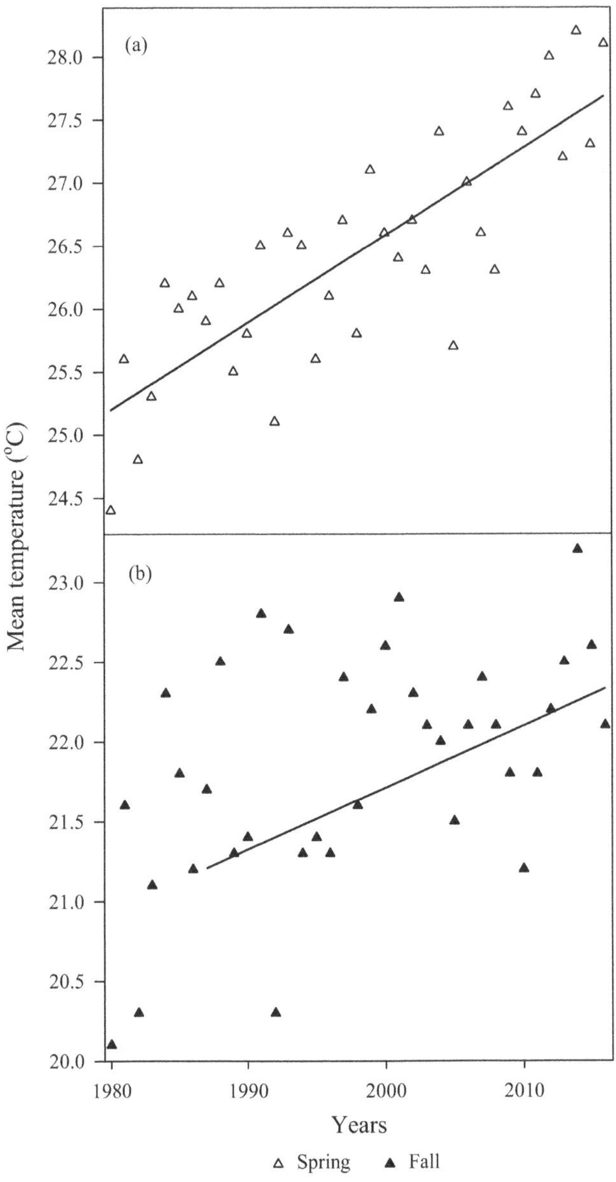

Fig. 10.2 Mean temperature trends in spring (**a**) and fall (**b**) season potato crop from 1980 to 2016 in Punjab, Pakistan (Source: Adopted from Naz et al. 2022)

The shift in the timing of management is not catching up in Germany to the changing phenology of the plants as quickly as it is elsewhere (Menzel et al. 2006). The timing of spring sowing is contingent not only on the type of soil, but also on the weather conditions that have come before, those that are currently occurring,

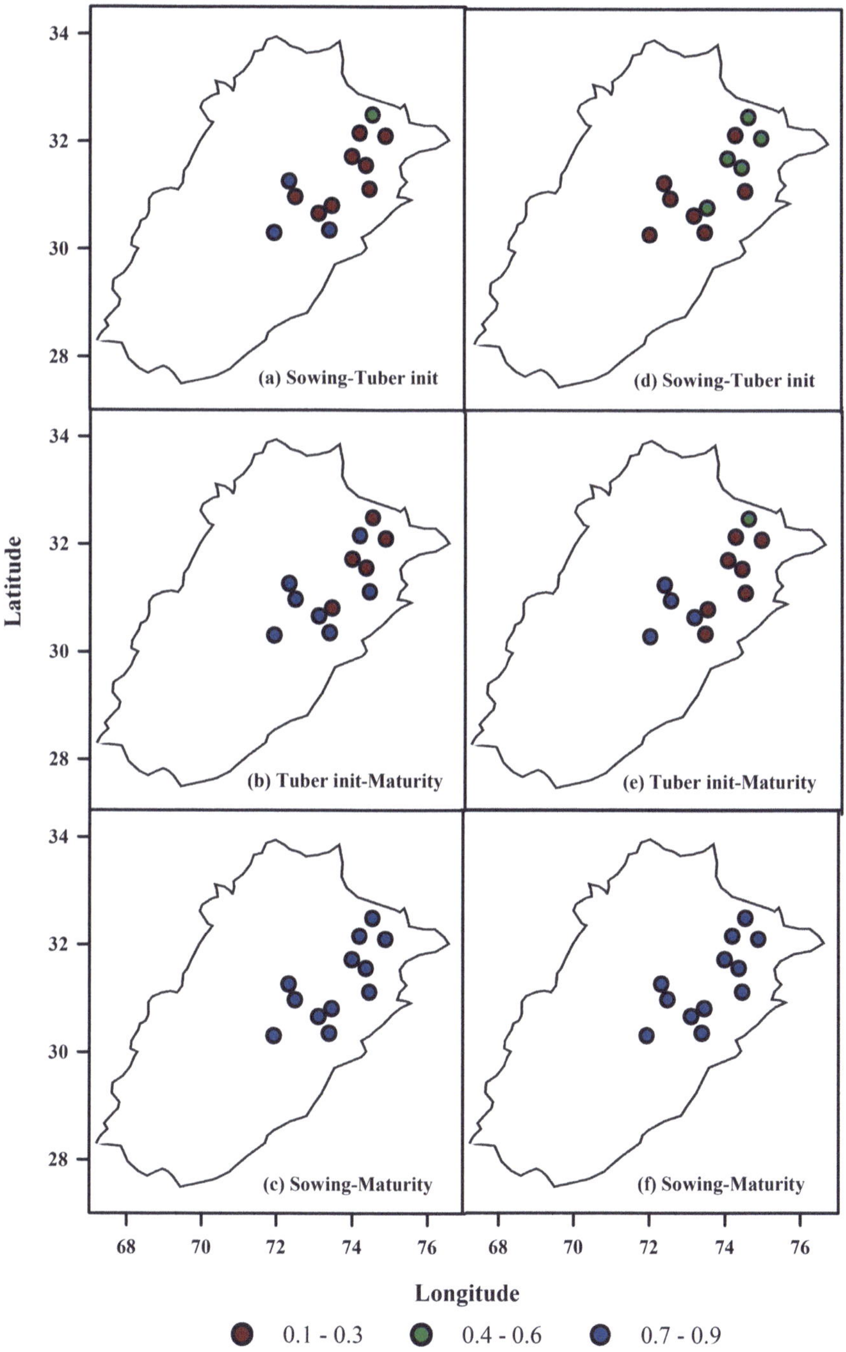

Fig. 10.3 Observed trends in mean temperature during the phenological phases {sowing-tuber initiation (**a, d**); tuber initiation-maturity (**b, e**); and sowing-maturity (**c, f**)} for spring (**a–c**; left) and fall (**d–f**; right) season's potato crop from 1980 to 2016 in Punjab, Pakistan (Source: Adopted from Naz et al. 2022)

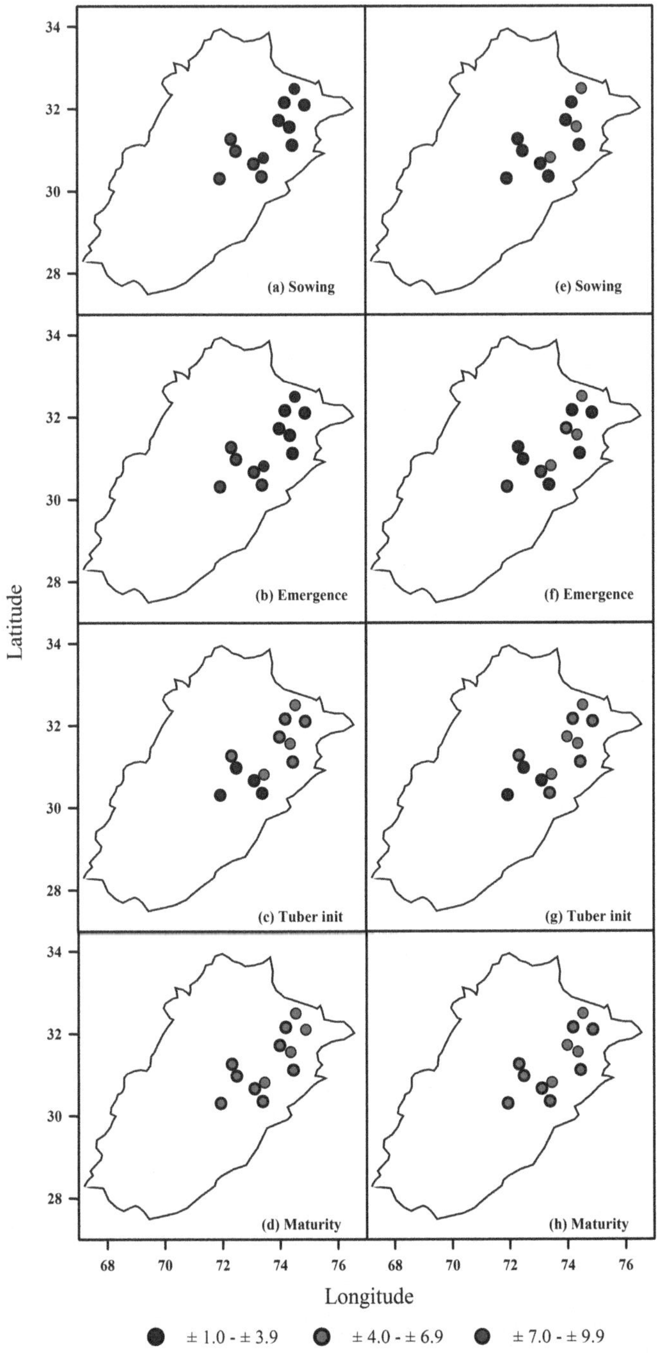

Fig. 10.4 Observed trends in the phenology of spring (**a–d**) and fall (**e–h**) season potato [sowing, emergence, tuber initiation, and maturity]; circles with black borderline indicate statistically significant trends at $p = 0.05$ levels from 1980 to 2016 in Punjab, Pakistan (Source: Adopted from Naz et al. 2022)

Table 10.3 Average observed phenology of spring and fall potato in Punjab, Pakistan, during the period of 1980–2016

Locations	Sowing (DOY)	Emergence (DOY)	Tuber initiation[a] (DOY)	Maturity[b] (DOY)	S-Ti (days)	Ti-M (days)	S-M (days)
Spring							
Sialkot	37 ± 6.2	42 ± 6.0	94 ± 8.4	154 ± 7.2	57 ± 5.4	60 ± 5.0	117 ± 3.8
Lahore	34 ± 5.8	39 ± 5.5	88 ± 7.2	147 ± 6.1	54 ± 3.8	59 ± 3.8	113 ± 4.2
Okara	30 ± 4.3	35 ± 4.2	81 ± 3.8	140 ± 4.2	51 ± 7.4	59 ± 6.1	110 ± 7.9
Sheikhupura	21 ± 2.5	28 ± 2.3	76 ± 4.6	138 ± 5.3	55 ± 5.0	62 ± 3.9	117 ± 9.1
Kasur	28 ± 7.1	33 ± 6.8	77 ± 5.1	133 ± 7.7	49 ± 2.6	56 ± 4.5	105 ± 4.6
Narowal	26 ± 4.1	31 ± 3.9	73 ± 5.0	126 ± 3.6	47 ± 3.7	53 ± 4.1	100 ± 3.9
Gujranwala	24 ± 4.9	29 ± 4.7	76 ± 7.2	130 ± 5.9	52 ± 6.8	54 ± 3.9	106 ± 5.5
T.T. Singh	29 ± 3.7	34 ± 3.5	80 ± 8.0	133 ± 2.6	51 ± 4.2	53 ± 7.2	104 ± 3.7
Jhang	22 ± 2.8	27 ± 2.6	74 ± 4.6	129 ± 3.1	52 ± 4.8	55 ± 2.7	107 ± 4.8
Pakpattan	18 ± 3.0	23 ± 2.8	70 ± 7.1	126 ± 3.7	52 ± 8.3	56 ± 4.3	108 ± 4.0
Sahiwal	15 ± 5.6	19 ± 5.5	68 ± 8.7	124 ± 4.1	53 ± 3.7	56 ± 4.9	109 ± 2.9
Khanewal	19 ± 4.9	23 ± 4.7	70 ± 5.3	127 ± 6.2	51 ± 2.8	57 ± 7.1	108 ± 6.3
Fall							
Sialkot	282 ± 6.4	287 ± 6.2	343 ± 5.3	40 ± 7.5	61 ± 3.6	62 ± 5.5	123 ± 4.6
Lahore	293 ± 8.2	298 ± 8.0	349 ± 3.6	45 ± 8.2	56 ± 6.4	61 ± 4.3	117 ± 6.3
Okara	280 ± 5.3	285 ± 5.1	337 ± 5.8	33 ± 6.2	57 ± 5.3	61 ± 4.7	118 ± 7.4
Sheikhupura	279 ± 3.9	284 ± 3.8	337 ± 8.2	36 ± 5.6	58 ± 8.1	64 ± 7.5	122 ± 3.6
Kasur	286 ± 4.8	291 ± 4.4	338 ± 3.5	31 ± 4.2	52 ± 3.6	58 ± 5.3	110 ± 4.1
Narowal	284 ± 7.5	289 ± 7.3	334 ± 4.1	24 ± 3.6	50 ± 2.5	55 ± 4.8	105 ± 3.6
Gujranwala	279 ± 5.8	284 ± 5.6	334 ± 2.7	25 ± 5.3	55 ± 4.1	56 ± 5.2	111 ± 4.5
T.T. Singh	287 ± 3.9	292 ± 3.7	341 ± 7.3	31 ± 6.1	54 ± 4.7	55 ± 4.4	109 ± 2.7

(continued)

Table 10.3 (continued)

Locations	Sowing (DOY)	Emergence (DOY)	Tuber initiation[a] (DOY)	Maturity[b] (DOY)	S-Ti (days)	Ti-M (days)	S-M (days)
Jhang	285 ± 6.3	290 ± 6.0	340 ± 2.1	32 ± 7.7	55 ± 3.2	57 ± 7.1	112 ± 2.5
Pakpattan	282 ± 7.8	287 ± 7.7	337 ± 6.3	30 ± 6.1	55 ± 6.2	58 ± 4.2	113 ± 5.1
Sahiwal	280 ± 3.7	285 ± 3.5	336 ± 5.0	29 ± 3.8	56 ± 3.7	58 ± 7.5	114 ± 5.6
Khanewal	277 ± 2.9	282 ± 2.8	331 ± 5.3	25 ± 4.2	54 ± 4.8	59 ± 8.1	113 ± 7.2

Source: Naz et al. (2022)
± Standard deviation
[a]50% Tuber initiation
[b]50% Physiological maturity

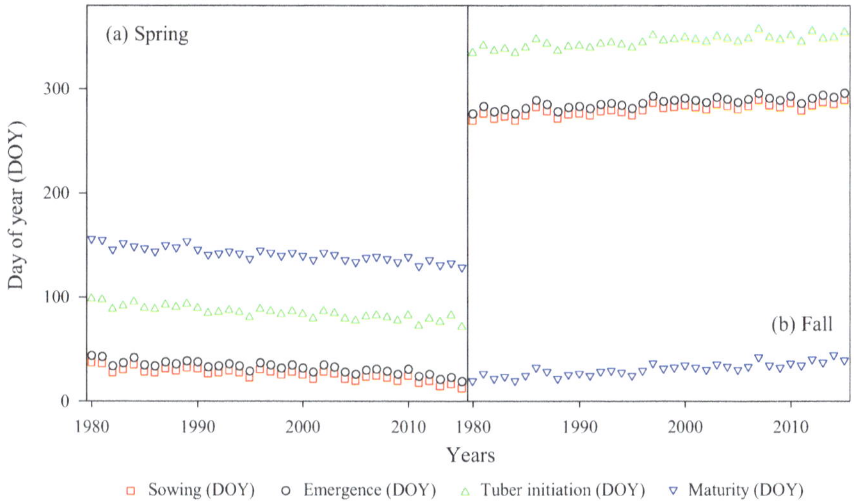

Fig. 10.5 Time series plots of farmer field observed dates of sowing, emergence, tuber initiation, and maturity of spring (**a**) and fall (**b**) season potato in Punjab, Pakistan (Source: Adopted from Naz et al. 2022)

and those that are forecasted, as well as the farmer's experience and custom. As a result of climate change, the planting of potatoes has shifted forward in time in Finland. Between 1965 and 1979, the planting of potatoes moved forward 3.4 days/decade, and the rate of advancement has accelerated to 4.5 days/decade since the 1980s (Kaukoranta and Hakala 2008). Potato planting, leafing, flowering, and harvesting dates were all advanced by 2, 3, 3.80, and 3.42 days, respectively, for every 1 °C increase in temperature, according to the research from Tryjanowski et al. (2018). This was after the researchers controlled for differences in cultivars, a noteworthy improvement over the course of time in relation to four phenology variables of potato. He found that the mean planting date of the most recent cultivar, 'Wineta', was 15 days earlier than that of the original cultivar, 'Lenino'. He also found that the leafing date, flowering date, and harvest date were all earlier than the original cultivar by a total of 11 days (Fig. 10.10).

10.7 Impact of Climate Change on Current and Future Scenarios of Potato Production Systems

The morphology and physiology of potato plants will be altered because of climate change, ultimately leading to reduced potato production. Changes in the concentrations of ozone and carbon dioxide in the atmosphere, as well as changes in the amount of land available for farming because of rising or falling sea levels, can all have an impact on crop yield. The frequency of rainfall deficits during the vegetative

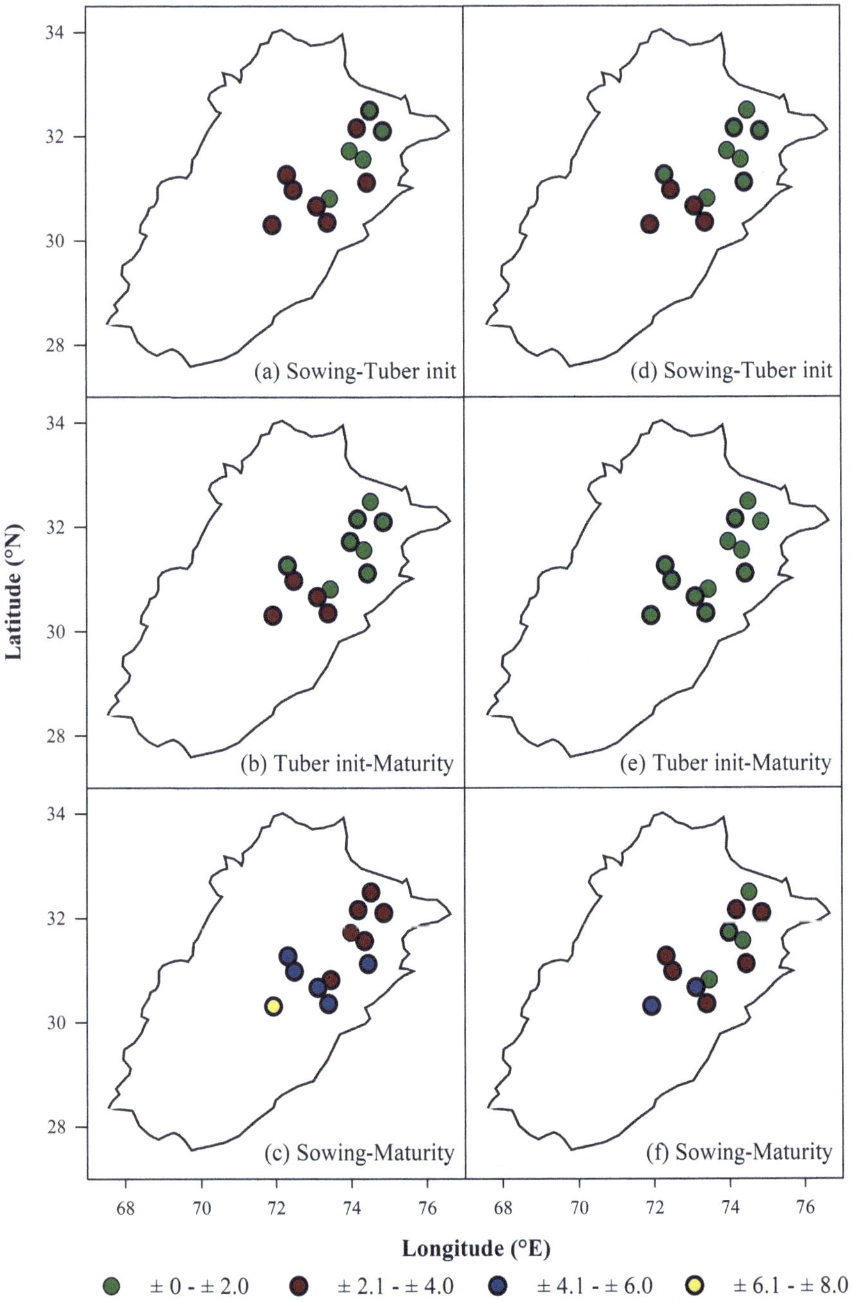

Fig. 10.6 Observed trends in phenology {(**a, d**) sowing-tuber initiation; (**b, e**) tuber initiation-maturity; and (**c, f**) sowing to maturity; circles with black borderline indicate statistically significant trends at $p = 0.05$ levels} of spring (**a–c**; left) and fall (**d–f**; right) season's potato from 1980 to 2016 in Punjab, Pakistan (Source: Adopted from Naz et al. 2022)

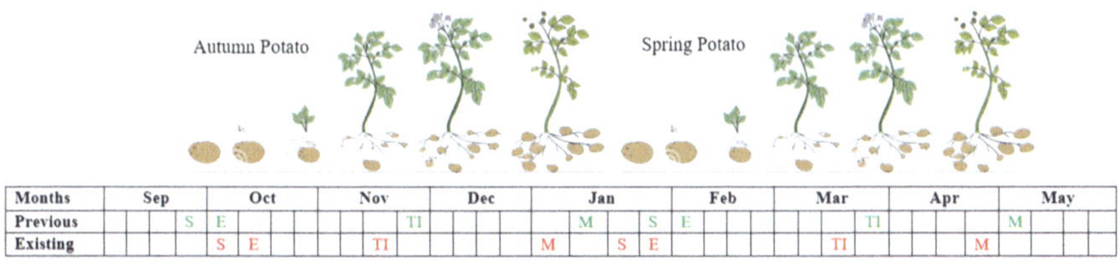

Fig. 10.7 Spatial and temporal variability of spring (left) and fall (right), respectively, potato phenological phases as effected by an increase in temperature in Punjab, Pakistan. *S* sowing, *E* emergence, *TI* tuber initiation, *M* maturity. Green and red represent previous and prevailing trends, respectively (Source: Adopted from Naz et al. 2022)

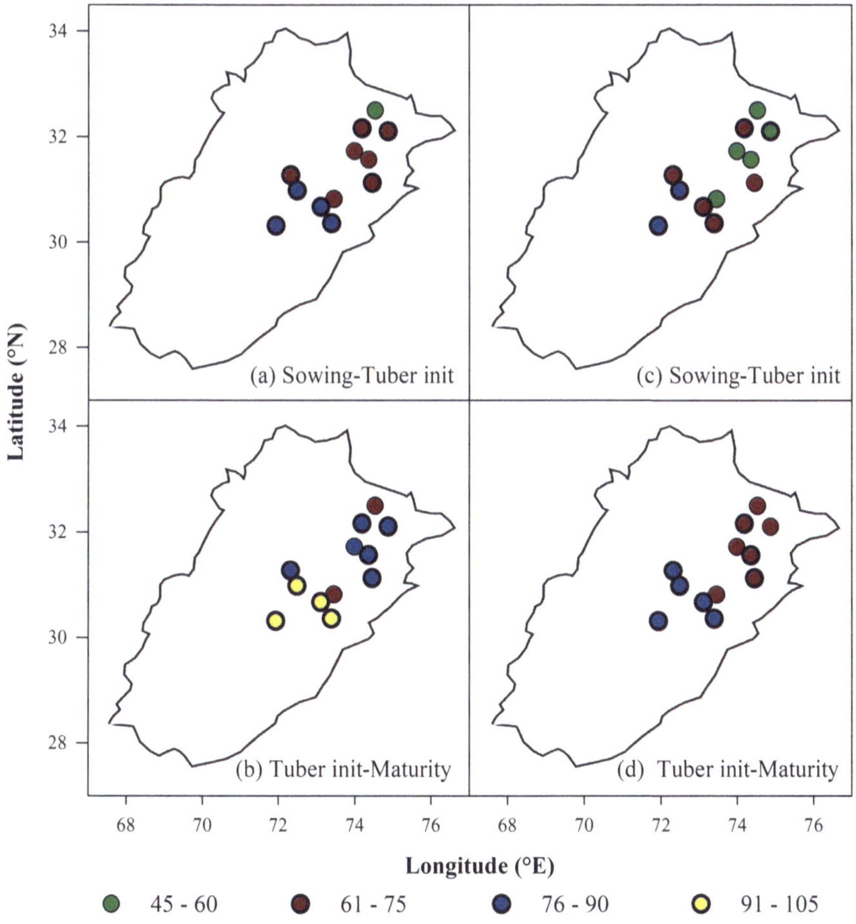

Fig. 10.8 Observed trends in thermal time required for spring (**a, b**; left) and fall (**c, d**; right) season's potato from sowing-tuber initiation (**a**) and tuber initiation to maturity (**b**); circles with black border indicate statistically significant trends at $p = 0.05$ levels in Punjab, Pakistan (Source: Adopted from Naz et al. 2022)

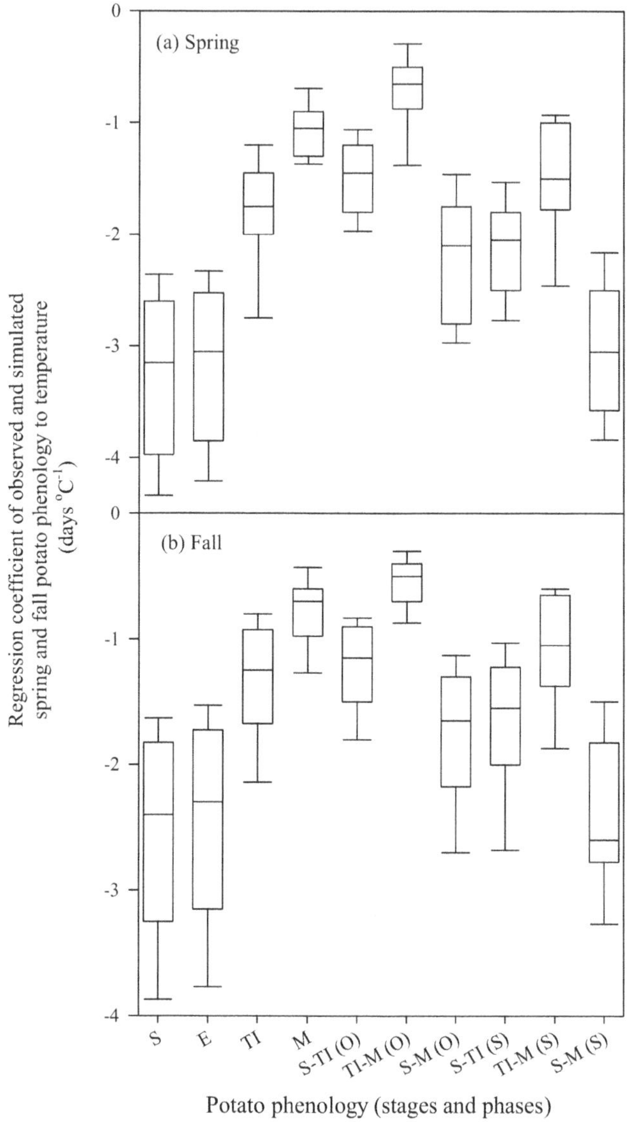

Fig. 10.9 Observed phenology (stages and phases) and simulated phenology (phases) versus temperature trends of spring (**a**) and fall (**b**) potato cultivar at each location in Punjab, Pakistan, from 1980 to 2016. (*S* sowing, *E* emergence, *TI* tuber initiation, *M* maturity, *S-TI* sowing-tuber initiation, *TI-M* tuber initiation-maturity, *S-M* sowing-maturity, *O* observed, *S* simulated). The central line represents median, box limits are 25th and 75th percentiles, and whiskers represent the minimum and maximum values (Source: Adopted from Naz et al. 2022)

Table 10.4 Summary of observed phenology response to temperature for spring and fall potato in Punjab, Pakistan, for 1980–2016

Phenology	No. of Neg.[a]	No. of Pos.[b]	No. of Sig. Neg.[c]	No. of Sig. Pos.[d]	Reg. Mean[e] (days °C^{-1})
Spring stages (observed)					
Sowing	12	0	10	0	−3.3
Emergence	12	0	10	0	−3.2
Tuber initiation	12	0	9	0	−1.8
Maturity	12	0	8	0	−1.1
Spring phases (observed)					
Sowing-tuber initiation	12	0	9	0	−1.5
Tuber initiation-maturity	12	0	9	0	−0.7
Sowing-maturity	12	0	11	0	−2.2
Spring phases (simulated)					
Sowing-tuber initiation	12	0	11	0	−2.1
Tuber initiation-maturity	12	0	10	0	−1.5
Sowing-maturity	12	0	11	0	−3.1
Fall stages (observed)					
Sowing	12	0	9	0	−2.6
Emergence	12	0	9	0	−2.5
Tuber initiation	12	0	8	0	−1.3
Maturity	12	0	8	0	−0.8
Fall phases (observed)					
Sowing-tuber initiation	12	0	8	0	−1.2
Tuber initiation-maturity	12	0	7	0	−0.5
Sowing-maturity	12	0	9	0	−1.7
Fall phases (simulated)					
Sowing-tuber initiation	12	0	10	0	−1.5
Tuber initiation-maturity	12	0	9	0	−1.1
Sowing-maturity	12	0	9	0	−2.5

Source: Naz et al. (2022)

[a]Number of locations with negative regression coefficients

[b]Number of locations with positive regression coefficients

[c]Number of locations with significant negative regression coefficients

[d]Number of locations with significant positive regression coefficients

[e]Mean of regression coefficients

Table 10.5 Comparison of the responses of spring and fall potato phenology with average temperature using the observed and simulated data in Punjab, Pakistan, during 1980–2016

Phenology	Regression coefficient (days °C^{-1})[a]				Difference between obs. and sim. data (days °C^{-1})		t-Test (p-value)	
	Obs. data		S.m. data					
Phases	Spring	Fall	Spring	Fall	Spring	Fall	Spring	Fall
Sowing-tuber initiation	−1.5	−1.2	−2.1	−1.5	0.6	0.3	0.00068**	0.00073**
Tuber initiation-maturity	−0.7	−0.5	−1.5	−1.1	0.8	0.6	0.00043**	0.00031**
Sowing-maturity	−2.2	−1.7	−3.1	−2.5	0.9	0.8	0.00082**	0.00012**

Source: Naz et al. (2022)
** Significant at the 0.01 probability level
[a]Mean of regression coefficients
[b]Duration of sowing to maturity

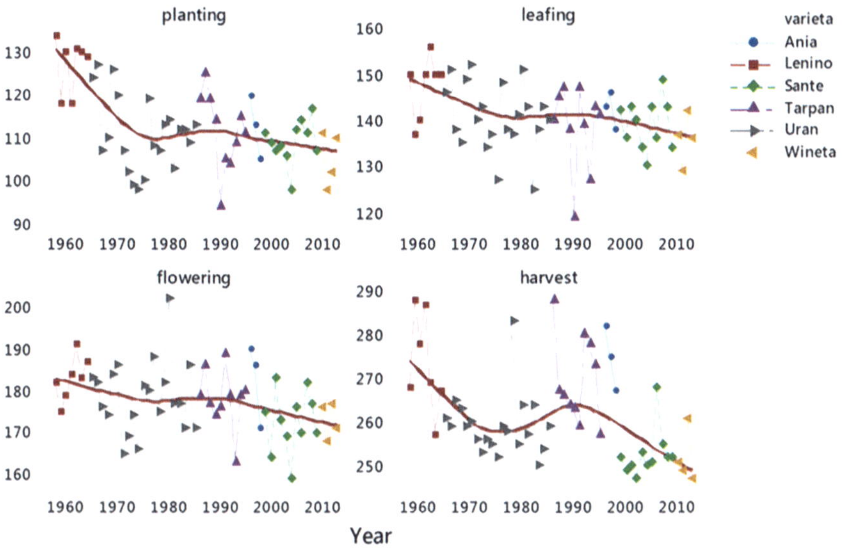

Fig. 10.10 Changes in potato phenological stages expressed as days after 1st January (Y-axis) during the years 1958–2013 (X-axis). Single smoothed (lowess) lines have been superimposed.

growth period will increase due to global warming, which will also negatively impact the growth of potato tubers and the amount of potatoes produced. It is among the most significant risks that could threaten the yield of potatoes (Miyashita et al. 2005). Potatoes are becoming an increasingly important part of the global food supply. Between 1960 and 2005, the consumption of potatoes in developing countries increased by more than twofold. For this reason, it is absolutely necessary to gain knowledge of the effects of climate change on potato yields worldwide to guarantee the continued supply of food in the future. In comparison to studies on the impacts of climate change on other key crops, studies on the effects of climate change on potatoes are relatively infrequent, especially on a worldwide basis.

The effects of climate change on agricultural productivity will vary from one region of the world to another (Parry et al. 1999). Agriculture is particularly vulnerable to climate uncertainty because its operations are wholly dependent on the prevailing weather patterns. The impact of climate on agronomic yields can differ widely depending on the country and the crop in question (Adams et al. 1998). Some regions might reap the benefits of the actions of certain aspects of climate change (for example, cold countries with increasing temperatures or "carbon fertilization effects") (Stokes and Howden 2010), while other nations might bear the brunt of the climate change's significant unfavorable effects (e.g., drought, increase in the variability of yield, reduced crop yields). According to Adger (2006) and Füssel (2007), several different types of these impacts can make a country's preexisting vulnerabilities even worse. For example, there is the possibility that an increase in temperature might have both beneficial and detrimental effects on crop yield. A country's characteristics will determine the impact's nature, which will be either positive or negative.

In spite of this, a severe rise in temperature indicates a lack of available water, which has the potential to wreak havoc on agricultural systems across the globe, particularly in semiarid regions, such as the nations of Southern Europe. In this scenario, it has been hypothesized that there is an inverse relationship between temperature rises and crop yields. Further, an increase in rainfall could be beneficial in semiarid regions because it may increase soil moisture but worsen problems in areas with abundant water. However, there is a possibility that a decrease in rainfall could have the opposite effect. Because of the availability of irrigation water in irrigated areas, the potentially detrimental effects of altered rainfall and rise in temperatures are mitigated, resulting in yields more resilient to climate vulnerability's effects.

The effects of climate uncertainty on open agroecosystems are significant, particularly on crop growth and yield in ecologically fragile areas like arid and semiarid lands (Jung et al. 2020). Understanding how crop yields will be impacted by climate change is necessary to guarantee the security of the global food supply (Godfray et al. 2011). As a direct consequence, this environmental alteration will significantly affect the expansion and development of various processes. For instance, an increase in temperature may affect several physiological processes like photosynthesis rate, respiration, and photoassimilate distribution (Chartzoulakis and Psarras 2005). Therefore, warmer conditions might benefit crop yields in areas with cold spring and summer seasons and where growing season is restricted. However, regions with warmer climates will experience decreased yields if the temperature continues to rise (Meza et al. 2008). An increase in temperature can reduce the time needed for a crop to complete its phenological stages (Boote 2011). In addition, the growing season's length can significantly impact how various cultivars of crops react to future climate change. There is a wealth of evidence in the published research that points to a negative relationship between climate change and potato production (Van der Waals et al. 2013; Sparks et al. 2014).

When the temperature is high, that is, above 25 °C, and particularly when it is higher than 30 °C, the growing period of potatoes is shortened, which results in a lower tuber yield. In some potato varieties, the degeneration of potato tubers may occur if they are subjected to elevated temperatures ranging from 30 to 40 °C during the tuber development stage (Jovovic and Kratovalieva 2015). Numerous studies have been conducted over the past few decades to investigate climate change's effects on important food crops in South Africa, such as rice, wheat, and maize. According to the findings, the fluctuations in temperature and rainfall patterns or rainfall variability in the region appear to significantly impact the yields of all three of these agricultural products. When the crop production of the 1990s was compared with that of the 2020s and 2050s, Parry et al. (2004) showed that crop yield will decline by between 2.5% and 10% in numerous parts of Asia in the 2020s, and by between 5% and 30% in the 2050s. This was established by contrasting the level of production in the 1990s, when carbon fertilization had not yet taken effect, with the level of yield projected for the 2020s and the 2050s.

Models of crop simulation are typically utilized in order to make predictions regarding the effects of a change in climate on a crop (White et al. 2011). The results of the simulation suggested that climate change would impact the growth and development of potatoes under all of the analyzed scenarios and that this would result in a loss in yield compared to the baseline years. Researchers have utilized a number of potato simulation models to examine the possible implications of climate uncertainty at local, regional, and global levels. This is in spite of the fact that the models have not been subjected to a great deal of testing (Haris et al. 2015). These models are able to perform an analysis of the impact that the climate and other factors have on the productivity of a given crop under a specific regimen of agricultural practices. The growth and production of crops can be estimated using crop models by taking into consideration variations in the soil, weather, and crop management practices (Jones et al. 2003). The impact of climate uncertainty on grain crops is the primary focus of crop models; however, tubers and root crops have been largely ignored in simulation studies, with a few notable exceptions. On the other hand, some model testing and potato simulation models have been developed and used to study possible impacts of climate change and global warming in local, regional, and global scenarios. These models have been used to simulate the growth of potatoes (Haris et al. 2015). Potato SUBSTOR-Potato is the crop model that has been found to be the most useful in simulation studies (Šťastná et al. 2010). Another popular potato model in simulation studies is SUBSTOR-Potato (Daccache et al. 2011). SUBSTOR-Potato simulates potato crop growth and evolution. This model accounts for both the growth-stimulating effects of elevated atmospheric CO_2 and the effects of temperature on crop growth and development. However, when compared to a variety of field data, the model underestimated the effects of elevated CO_2 and was unable to simulate the effects of high temperatures (Raymundo et al. 2018).

The LINTUL-Potato model was used for projection by Hijmans (2003); the results showed that potato production around the world may decrease by anywhere from 18% to 32% by the year 2050. According to the findings of Hijmans (2003), the potato yield in India could drop to 23.1 without adaptation and 22.1 with adaptation in the years 2040–2059 if climate change is allowed to continue unchecked. Potato production in India could drop by 3.16% in 2020 and 13.72% in 2050 if the country does not make any adaptations to the effects of global warming and climate change (Pradel et al. 2019). Another study conducted worldwide simulations using a modified version of the LINTUL-Potato model, and the results suggested that yield losses by the year 2050 might range anywhere from 18% to 32%. In this study, a single cultivar was simulated, and it was anticipated that there would be no limitations imposed by either water or nitrogen and no effects from atmospheric CO_2 (Hijmans 2003). Based on the current relative contribution of various states to national production, a scientist projected a decrease in potato yield due to climate change of −2.61 kg/ha in 2020 and −15.32 kg/ha in 2050. Scientists conducted a computer simulation study using the WOFOST crop growth model. The findings demonstrated that an increase in the concentration of carbon dioxide will increase potato productivity from +3.9% to +4.5%. The INFOCROP-POTATO model was used for simulation studies by Venkateswarlu and Rao (2013). Due to climate uncertainty, the results showed that total potato yield in India may decrease by 2.61% in 2020 and 15.32% in 2050 if adaptations are not made. It is anticipated that increases in temperature and concentrations of carbon dioxide may result in an increase in the production of certain crops in particular nations.

The results of studies that used a variety of potato models demonstrate that the effects of climate change vary from one location to the next. Under certain conditions, existing potato production areas would relocate toward cooler regions as a result of an increase in temperature (Supit et al. 2012). In the long run, there would be a significant drop in the amount of potatoes produced. In a model, if a high CO_2 concentration in atmosphere is considered, the decrease will be less severe because carbon dioxide will have a fertilization effect due to increased photosynthesis (Raymundo et al. 2018).

As many climate models have predicted, the frequency of droughts will increase, and as a result, the amount of potatoes that can be harvested will decrease. The crop's physiology will change as a result of the drought, and the rate of photosynthesis will decrease along with the chlorophyll content and transpiration rates. It will result in the tuber developing malformations and brown spots on the inside of it. A reduction in plant height, plant growth, plant biomass, and leaf area index are just some of the morphological changes that can be brought on by drought. The tuberization process will be slowed by drought, which hastens the onset of senescence. The plant might become wilted if the conditions are particularly harsh. If water scarcity occurs just before or during the initiation of potato tubers, the yield of potato tubers may be affected negatively by severe water constraints (Monneveux et al. 2013). The growth of the potato plant and the amount it produces will suffer if the temperature is allowed to continue to rise. If the temperature is allowed to rise above what the plant considers to be its ideal temperature, photosynthesis will slow down. While this is happening, higher temperatures will lead to physiological processes such as a decrease in the activity of photosystem II and the movement of sucrose into the tubers. It will cause malformations, necrosis, and field sprouting to develop in the potato tuber if it is allowed to continue. When morphology is taken into consideration, higher temperatures will cause a reduction in the amount of tubers produced and plant growth. Additionally, the amount of dry matter in the leaf will increase, while the amount in the tuber will decrease. Temperature increases could cause a delay in the tube formation process in potato plants.

10.8 Climate Change Impact on Potato Production

Climate change directly affected potato production in Peru. The use of seven climate scenarios from five climatic models simulated the yield of potatoes under current (1961–1990) and future climate periods (2010–2039 and 2040–2069). The average increase of predicted temperature will be from 2.1 to 3.2 °C under different climate scenarios. Global warming will probably affect the timing of planting, the adoption of later maturing cultivars, and the placement of potato production at high latitudes. Changes in potato output in many of these areas are probably minor, and occasionally even favorable. Results showed that without adaptation there was 18–32% decline in potato yield, and with adaptation, it declined by 9–18% in these periods (Hijmans 2003). Adavi et al. (2018) reported that climate change showed a negative impact on potato production in Iran with the help of three (A2, B1, and A1B) scenarios and two (IPCM4 and HadCM3) GCMs. To simulate the future and baseline potato growth and development, the SUBSTOR-Potato model was employed. Results showed that SUBSTOR-Potato and LARS-WG have an adequate level of precision to inspect potato growth and meteorological factors. Results showed reduction of 11.21% to 27.53% and from 12.60% to 30.58% yield under HadCM3 and IPCM4 model, respectively. Daccache et al. (2011) reported that irrigation requirements of potato were affected by climate change in England. Results of different global circulation models and selected emission scenarios with a SUBSTOR-Potato model showed that in the future, potato yield enhanced by 13–16% without nitrogen and water restrictions, but under the scenario of CO_2 prediction, yield increased by 14–30% in the future. Raymundo et al. (2017) reported that the SUBSTOR-Potato model was assessed under a variety of growing situations, and adjustments were made to better represent the effects of atmospheric CO_2 and high temperatures. RCP 8.5 was predicted to have the greatest adverse effects on the world's tuber crops by the end of the century. Under different climate scenarios, tuber yield decreased from −2% to −6% in 2025 and −2% to −26% in 2085 (Table 10.6).

Table 10.6 Impact of climate change on potato production across various countries in the world

Sr. No.	Crop	Country	Model	Climate scenario	Yield	References
1	Potato	Peru	LINTUL simulation model	1961–1990, 2010–2039, 2040–2069 CGCM14 CSIRO-MM ECHAM4 GFDLR15 HadCM2	Yield decreased by 18–32%	Hijmans (2003)
2	Potato	Egypt	SUBSTOR-Potato model	A1 scenario CSIRO and HadCM3	Decreased by 11–13%	Abdrabbo et al. (2010)
3	Potato	England	SUBSTOR-Potato model	SRES A1FI and B1 (1961–1990, 2050)	Increased by 14–30%	Daccache et al. (2011)
4	Potato	Iran	SUBSTOR-Potato model	(A1B, A2, and B1), HadCM3 and IPCM4 (1982–2012, 2015–2045, 2046–2075, 2076–2105)	Decreased by 11.21–30.58%	Adavi et al. (2018)
5	Potato	Maine	DSSAT	RCPs 6.0 and 8.5, 1989–2018, 2050–2079	Declined by 18% and 27%	Tooley et al. (2021)
6	Potato	USA	SUBSTOR-Potato model	RCPs 4.5 and 8.5 (1979–2009, 2040–2070, 2071–2100)	Decreased by 2–26%	Raymundo et al. (2017)
7	Potato	Ireland	SUBSTOR-Potato model	HadCM3 (1961–1990), 2041–2060) (2061–2090)	Decreasing trend observed	Holden et al. (2003)

10.9 Adaptation Strategies

There is a wealth of evidence in the published research that points to a negative relationship between climate change and potato production (Sparks et al. 2014; Kumar et al. 2015). Even though we cannot control or alter the weather conditions, certain aspects of the environment, like the soil, water, cultivar type, and management practices, could be modified to mitigate the negative effects of climate uncertainty (Moradi et al. 2013). It is well known that adaptation can help mitigate the detrimental effects that climate change has on agricultural production (Moradi et al. 2014). Adaptation aims to lessen the likelihood of adverse effects brought on by climate change while expanding available options for coping with its effects (Moradi et al. 2013). Some of the adaption techniques suggested in reports published by the Intergovernmental Panel on Climate Change (IPCC) include shifting the timing or location of cropping operations; establishing new crop regions; and improving the effectiveness of practices used to manage pests, diseases, and weeds. Other strategies include changing the sowing date, altering crop rotations, changing cultivars, and using water more efficiently (Abbas et al. 2014; Bowen et al. 2002; Jégo et al. 2008; Jones et al. 1998; Lutaladio and Castaldi 2009; Magrin et al. 1997; Mahdian and Gallichand 1995; Nadeem et al. 2022; Pasare et al. 2013; Raymundo et al. 2018; Tang et al. 2019). Agricultural production and economies are likely to experience difficulties due to climate change if adaptation and mitigation strategies are not implemented (Smit and Skinner 2002). For the purpose of estimating the growth and development of crops in an environment influenced by climate change, crop models such as SUBSTOR-Potato can be utilized. Assessments of the impact of climate change and potatoes have been done in the past using this model (Raymundo et al. 2017).

It is essential to adapt to the changing climate in order to boost yields or, at the very least, keep yields at the same levels as they are now. Some of the many available management techniques include mulching, increasing fertilizer application to offset the increased loss at higher temperatures (Kumar et al. 2015), and implementing water and soil management techniques in addition to managing pests and diseases. Other strategies include shifting cultivation to cooler seasons (Thomas-Sharma et al. 2016). One more essential strategy for adaptation is the introduction of new and improved crop varieties. Early-maturing varieties enable greater crop planting and harvesting flexibility and, if grown between two rice cycles, can generate additional revenue (Bardhan Roy 2007). Other important varietal characteristics include resistance and tolerance to biotic and abiotic stresses, such as the ability to withstand high temperatures and prolonged periods of drought (Islam et al. 2016). Consider the case of heat stress, which is a significant obstacle to potato cultivation because it causes a lag in the beginning of tuber formation and malformation and necrosis in the tubers (Levy and Veilleux 2007). Further, drought can have a negative impact on potato production because it reduces a plant's ability to absorb water and raises the concentration of salt in the soil, both of which influence the rate at which water is lost through reverse osmosis from plant cells (Basu et al. 2016). As diseases evolve (Chowdappa et al. 2015) and pests and pathogens spread, late blight and virus resistance will become more essential (Bebber et al. 2014).

10.10 Conclusion

Potato production system, which is dependent on rainfed culture, is likely to suffer losses as a result of the high temperatures, the unpredictability of the weather, and the decreased availability of irrigation. When formulating water policy, it is imperative to consider how the agricultural shifts prompted by climate change will impact water demand. In addition to increasing potato production, widespread implementation of climate change mitigation and adaptation strategies may provide a solution to the problem of climate change. Adaptation strategies include the use of cropping systems, additional inputs such as fertilizer, tillage techniques, grain drying, and other field activities. Altering planting dates and growth locations and developing new resistant potato varieties are also examples of adaptation strategies.

References

Abbas G, Ahmad S, Ahmad A, Nasim W, Fatima Z, Hussain S, Rehman MH, Khan MA, Hasanuzzaman M, Fahad S, Boote KJ, Hoogenboom G (2017) Quantification the impacts of climate change and crop management on phenology of maize-based cropping system in Punjab, Pakistan. Agric For Meteorol 247:42–55

Abbas G, Ahmad S, Hussain M, Fatima Z, Hussain S, Iqbal P, Ahmed M, Farooq M (2020a) Sowing date and hybrid choice matters production of maize-maize system. Int J Plant Prod 14:583–595

Abbas G, Fatima Z, Hussain M, Hussain S, Atique-ur-Rehman, Sarwar N, Ahmed M, Ahmad S (2020b) Nitrogen rate and hybrid selection matters productivity of maize-maize cropping system under irrigated arid environment of southern Punjab, Pakistan. Int J Plant Prod 14:309–320

Abbas G, Fatima Z, Riaz F, Iqbal P, Hussain S, Zakir I, Iqbal M, Haque E, Ahmad S (2019) Study of sowing dates and hybrids effect in maize-based cropping system under arid conditions of Southern Punjab, Pakistan. Pak J Life Soc Sci 17(1):24–31

Abbas MF, Hameed S, Rauf CA (2014) Presence of new strain of Potato virus Y in Pakistan. Int J Agric Biol 16(5):941–945

Abbas Q, Ahmad S (2018) Effect of different sowing times and cultivars on cotton fiber quality under stable cotton-wheat cropping system in southern Punjab, Pakistan. Pak J Life Soc Sci 16(2):77–84

Abbas Q, Atique-ur-Rehman, Sarwar N, Hussain S, Abbas G, Khan MN, Fatima Z, Naz S, Younis H, Ahmed M, Ullah H, Iqbal P, Iqbal M, Kan M, Ahmad S (2020c) Effect of different sowing times and cultivars on wheat grain quality under cotton-wheat cropping system in Southern Punjab, Pakistan. Pak J Life Soc Sci 18(1):41–50

Abdrabbo MAA, Khalil AA, Hassanien MKK, Abou-Hadid AF (2010) Sensitivity of potato yield to climate change. J Appl Sci Res 6(6):751–755

Adams RM, Hurd BH, Lenhart S, Leary N (1998) Effects of global climate change on agriculture: an interpretative review. Clim Res 11(1):19–30

Adavi Z, Moradi R, Saeidnejad AH, Tadayon MR, Mansouri H (2018) Assessment of potato response to climate change and adaptation strategies. Sci Hortic 228:91–102

Adger WN (2006) Vulnerability. Glob Environ Chang 16(3):268–281

Ahmad A, Abbas G, Ahmed M, Fatima Z, Anjum MA, Rasul G, Khan MA, Hoogenboom G (2019) Climate warming and management impact on the change of phenology of the rice-wheat cropping system in Punjab, Pakistan. Field Crop Res 230:46–61

Ahmad A, Ashfaq M, Rasul G, Wajid SA, Ahmad I, Khaliq T, Nasir J, Rasul F, Riaz F, Ahmad B, Ahmad S, Baig IA, Valdivia RO, Hoogenboom G (2021) Development of climate change adaptation strategies for cotton–wheat cropping system of Punjab, Pakistan. In: Rosenzweig C, Mutter CZ, Contreras EM (eds) Handbook of climate change and agroecosystems: climate change and farming system planning in Africa and South Asia: AgMIP stakeholder-driven research. World Scientific Publishing Europe Ltd, London, pp 277–327

Ahmad A, Ashfaq M, Rasul G, Wajid SA, Khaliq T, Rasul F, Saeed U, Rahman MH, Hussain J, Baig IA, Naqvi SAA, Bokhari SAA, Ahmad S, Naseem W, Hoogenboom G, Valdivia RO (2015a) Impact of climate change on the rice-wheat cropping system of Pakistan. In: Hillel D, Rosenzweig C (eds) Handbook of climate change and agro-ecosystems, vol 3. Imperial College Press and the American Society of Agronomy, London, pp 219–258

Ahmad S, Abbas G, Fatima Z, Khan RJ, Anjum MA, Ahmed M, Khan MA, Porter CH, Hoogenboom G (2017a) Quantification of the impacts of climate warming and crop management on canola phenology in Punjab, Pakistan. J Agron Crop Sci 203(5):442–452

Ahmad S, Abbas Q, Abbas G, Fatima Z, Atique-ur-Rehman, Naz S, Younis H, Khan RJ, Nasim W, Rehman MH, Ahmad A, Rasul G, Khan MA, Hasanuzzaman M (2017b) Quantification of climate warming and crop management impacts on cotton phenology. Plan Theory 6:1–16

Ahmad S, Ahmad A, Ali H, Hussain A, Garcia y Garcia A, Khan MA, Zia-Ul-Haq M, Hasanuzzaman M, Hoogenboom G (2013) Application of the CSM-CERES-Rice model for evaluation of plant density and irrigation management of transplanted rice for an irrigated semiarid environment. Irrig Sci 31(3):491–506

Ahmad S, Ahmad A, Soler CMT, Ali H, Zia-Ul-Haq M, Anothai J, Hussain A, Hoogenboom G, Hasanuzzaman G (2012a) Application of the CSM-CERES-Rice model for evaluation of plant density and nitrogen management of fine transplanted rice for an irrigated semiarid environment. Precis Agric 13(2):200–218

Ahmad S, Ahmad I, Ahmad B, Ahmad A, Wajid A, Khaliq T, Abbas G, Wilkerson CJ, Hoogenboom G (2023a) Regional integrated assessment of climate change impact on cotton production in a semi-arid environment. Clim Res 89:113–132

Ahmad S, Ali H, Atique-ur-Rehman, Khan RJ, Ahmad W, Fatima Z, Abbas G, Irfan M, Ali H, Khan MA, Hasanuzzaman M (2015b) Measuring leaf area of winter cereals by different techniques: a comparison. Pak J Life Soc Sci 13(2):117–125

Ahmad S, Ali H, Farooq U, Khan SU, Atique-ur-Rehman, Sarwar N, Shahzad AN, Doğan H, Hussain S, Sultan MT, Waheed A, Zia-ul-Haq M, Hussain K, Khan MA (2016a) Improving nitrogen and radiation-use-efficiencies of C_4 summer cereals by split nitrogen applications under irrigated arid environment. Turk J Agric For 40(2):280–289

Ahmad S, Ali H, Ismail M, Shahzad MI, Nadeem M, Anjum MA, Zia-Ul-Haq M, Firdous N, Khan MA (2012b) Radiation and nitrogen use efficiencies of C_3 winter cereals to nitrogen split application. Pak J Bot 44(1):139–149

Ahmad S, Atique-ur-Rehman, Ejaz M, Fatima Z, Kan M, Ahmad M (2018) Agricultural land-use change of major field crops in Pakistan (1961–2014). Sci Technol Dev 37(3):113–121

Ahmad S, Hasanuzzaman M (2012) Integrated effect of plant density, N rates and irrigation regimes on the biomass production, N content, PAR use efficiencies and water productivity of rice under irrigated semiarid environment. Notulae Botanicae Horti Agrobotanici Cluj-Napoca 40(1):201–211

Ahmad S, Hussain S, Fatima Z, Abbas G, Atique-ur-Rehman, Khan MR, Younis H, Naz S, Sohail M, Ajmal M, Abbas N, Akhtar M, Rauf A, Khan M, Ali Z, Hassan M, Rizwan M, Bajwa RS, Ajmal A, Naz S, Ali H, Khan AA, Ali M, Sarwar G, Khan MA, Hasanuzzaman M (2016b) Application of DSSAT model for sowing date management of C_4 summer cereals for fodder and grain crops under irrigated arid environment. Pak J Life Soc Sci 14(2):104–114

Ahmad S, Nadeem M, Abbas G, Fatima Z, Khan RJ, Ahmed M, Ahmad A, Rasul G, Khan MA (2016c) Quantification of the effects of climate warming and crop management on sugarcane phenology. Clim Res 71(1):47–61

Ahmad S, Raza I, Muhammad D, Ali H, Hussain S, Dogan H, Zia-Ul-Haq M (2015c) Radiation, water and nitrogen use efficiencies of *Gossypium hirsutum* L. Turk J Agric For 39(5):825–837

Ahmad S, Raza MA, Hussain S, Abbas G, Fatima Z, Ahmed M, Goheer MA, Garcia y Garcia A, Wilkerson CJ, Hoogenboom G (2023b) Identification of weak links in the production technology of farmer for bridging canola yield-gap. J Agric Sci 161(2):241–253. https://doi.org/10.1017/S0021859623000187

Ahmed I, Rahman MH, Ahmad A, Ahmad S, Hussain J, Ullah A, Judge J (2018a) Assessing the impact of climate variability on maize using simulation modeling under semi-arid environment of Punjab, Pakistan. Environ Sci Pollut Res 25(23):28413–28430

Ahmed I, Ullah A, Ur Rahman MH, Ahmad B, Wajid SA, Ahmad A, Hassain S (2019a) Climate change impacts and adaptation strategies for agronomic crops. In: Hussain S (ed) Climate change and agriculture. IntechOpen, London, pp 1–14

Ahmed M, Ahmad S, Tariq M, Fatima Z, Aslam Z, Raza MA, Iqbal N, Akmal M, Fayyaz-ul-Hassan, Abbasi NA, Hayat R (2020) Wastes to be the source of nutrients and energy to mitigate climate change and ensure future sustainability: options and strategies. J Plant Nutr 43(6):896–920

Ahmed M, Akram MN, Asim M, Aslam M, F-u Hassan Higgins S, Stockle CO, Hoogenboom G (2016) Calibration and validation of APSIM-wheat and CERES-wheat for spring wheat under rainfed conditions: models evaluation and application. Comput Electron Agric 123:384–401

Ahmed M, Hayat R, Ahmad M, Mahmood-ul-Hassan, Kheir AMS, Fayyaz-ul-Hassan, Rehman MH, Shaheen FA, Raza MA, Ahmad S (2022) Impact of climate change on dryland agricultural systems: a review of current status, potentials, and further work need. Int J Plant Prod 16:341–363

Ahmed M, Ijaz W, Ahmad S (2018b) Adapting and evaluating APSIM-SoilP-Wheat model for response to phosphorus under rainfed conditions of Pakistan. J Plant Nutr 41(16):2069–2084

Ahmed M, Stockle CO, Nelson R, Higgins S, Ahmad S, Raza M (2019b) Novel multimodal ensemble approach to evaluate the sole effect of elevated CO_2 on winter wheat productivity. Sci Rep 9:7813

Alexopoulos AA, Akoumianakis KA, Passam HC (2006) Effect of plant growth regulators on the tuberisation and physiological age of potato (Solanum tuberosum L.) tubers grown from true potato seed. Can J Plant Sci 86(4):1217–1225

Ali H, Iqbal N, Ahmad S, Shahzad AN, Sarwar N (2013) Performance of late sown wheat crop under different planting geometries and irrigation regimes in arid climate. Soil Tillage Res 130:109–119

Alva AK, Marcos J, Stockle C, Reddy VR, Timlin D (2010) A crop simulation model for predicting yield and fate of nitrogen in irrigated potato rotation cropping system. J Crop Improv 24(2):42–152

Amin A, Nasim W, Fahad S, Ali S, Ahmad S, Rasool A, Saleem N, Hammad HM, Sultana SR, Mubeen M, Bakhat HF, Ahmad N, Shah GM, Adnan M, Noor M, Basir A, Saud S, Rahman MH, Paz JO (2018a) Evaluation and analysis of temperature for historical (1996–2015) and projected (2030–2060) climates in Pakistan using SimCLIM climate model: ensemble application. Atmos Res 213:422–436

Amin A, Nasim W, Mubeen M, Ahmad A, Nadeem M, Urich P, Fahad S, Ahmad S, Wajid A, Tabassum F, Hammad HM, Sultana SR, Anwar S, Baloch SK, Wahid A, Wilkerson CJ, Hoogenboom G (2018b) Simulated CSM-CROPGRO-cotton yield under projected future climate by SimCLIM for southern Punjab, Pakistan. Agric Syst 167:213–222

Amin A, Nasim W, Mubeen M, Sarwar S, Urich P, Ahmad A, Wajid A, Khaliq T, Rasul F, Hammad HM, Rehmani MIA, Mubarak H, Mirza N, Wahid A, Ahmad S, Fahad S, Ullah A, Khan MN, Ameen A, Amanullah, Shahzad B, Saud S, Alharby H, Ata-Ul-Karim ST, Adnan M, Islam F, Ali QS (2018c) Regional climate assessment for temperature and precipitation in Southern Punjab (Pakistan) using SimCLIM climate model for different temporal scales. Theor Appl Climatol 131:121–131

Bakhtiyor NZ (2015) The valuation and a qualitative gradation of a human's economic perfection. Int Open Acces J Weber Econ Finance 1(1):41–49

Bardhan Roy SK (2007) Improving the livelihood of farmers by intensifying the rice-potato-rice system through double-transplanting of rice in West Bengal, India. International Potato Center

Basu PS, Singh UMMED, Kumar ANIL, Praharaj CS, Shivran RK (2016) Climate change and its mitigation strategies in pulses production. Indian J Agron 61:S71–S82

Bebber DP, Holmes T, Gurr SJ (2014) The global spread of crop pests and pathogens. Glob Ecol Biogeogr 23(12):1398–1407

Bisognin DA, Freitas STD, Brackmann A, Andriolo JL, Pereira EIP, Muller DR, Bandinelli MG (2008a) Envelhecimento fisiológico de tubérculos de batata produzidos durante o outono e a primavera e armazenados em diferentes temperaturas. Bragantia 67:59–65

Bisognin DA, Müller DR, Streck NA, Andriolo JL, Sausen D (2008b) Desenvolvimento e rendimento de clones de batata na primavera e no outono. Pesq Agrop Brasileira 43:699–705

Boote KJ (2011) Improving soybean cultivars for adaptation to climate change and climate variability. In: Yadav SS, Redden RJ, Hatfield JL, Lotze-Campen H, Hall EA (eds) Crop adaptation to climate change. Wiley, West Sussex, pp 370–395

Bowen WT, Lal SS, Govindakrishnan PM (2002) Simulation of potato-based cropping systems. In: Potato, global research & development. Proceedings of the Global Conference on Potato, New Delhi, India, 6–11 Dec 1999. Indian Potato Association, vol 2, pp 776–787

Bowen WT, Thornton PK, Hoogenboom G (1998) The simulation of cropping sequences using DSSAT. In: Tsuji GY, Hoogenboom G, Thornton PK (eds) Understanding options for agricultural production. Springer, Dordrecht, pp 313–327

Bruinsma M, Kowalchuk GA, Van Veen JA (2003) Effects of genetically modified plants on microbial communities and processes in soil. Biol Fertil Soils 37:329–337

Chakraborty S, Chakraborty N, Agrawal L, Ghosh S, Narula K, Shekhar S, Datta A (2010) Next-generation protein-rich potato expressing the seed protein gene AmA1 is a result of proteome rebalancing in transgenic tuber. Proc Natl Acad Sci 107(41):17533–17538

Chartzoulakis K, Psarras G (2005) Global change effects on crop photosynthesis and production in Mediterranean: the case of Crete, Greece. Agric Ecosyst Environ 106(2–3):147–157

Chowdappa P, Nirmal Kumar BJ, Madhura S, Mohan Kumar SP, Myers KL, Fry WE, Cooke DEL (2015) Severe outbreaks of late blight on potato and tomato in South India caused by recent changes in the Phytophthora infestans population. Plant Pathol 64(1):191–199

Daccache A, Weatherhead EK, Stalham MA, Knox JW (2011) Impacts of climate change on irrigated potato production in a humid climate. Agric For Meteorol 151(12):1641–1653

DeFauw SL, He Z, Larkin RP, Mansour SA (2012) Sustainable potato production and global food security. In: He Z, Larkin R, Honeycutt W (eds) Sustainable potato production: global case studies. Springer, Dordrecht, pp 3–19

Easterling WE, Chen X, Hays C, Brandle JR, Zhang H (1996) Improving the validation of model-simulated crop yield response to climate change: an application to the EPIC model. Clim Res 6(3):263–273

Ejaz M, Abbas G, Fatima Z, Iqbal P, Raza MA, Kheir AMS, Ahmed M, Kakar KM, Ahmad S (2022) Modelling climate uncertainty and adaptations for soybean-based cropping system. Int J Plant Prod 16:235–250

Fatima Z, Abbas G, Iqbal P, Zakir I, Khan MA, Kamal GM, Ahmad S (2021) Quantification of climate warming and crop management impacts on phenology of pulses-based cropping systems. Int J Plant Prod 15:107–123

Fatima Z, Abbas Q, Khan A, Hussain S, Ali MA, Abbas G, Younis H, Naz S, Ismail M, Shahzad MI, Nadeem M, Farooq U, Khan SU, Javed K, Khan AA, Ahmed M, Khan MA, Ahmad S (2018) Resource-use-efficiencies of C_3 and C_4 cereals under split nitrogen regimes. Agronomy 8(5):69

Fatima Z, Ahmad M, Hussain M, Abbas G, Ul-Allah S, Ahmad S, Ahmed N, Ali MA, Sarwar G, Haque E, Iqbal P, Hussain S (2020) The fingerprints of climate warming on cereal crops phenology and adaptation options. Sci Rep 10:18013

Freitas STD, Bisognin DA, Gómez ACS, Sautter CK, Costa LCD, Rampelotto MV (2006) Processing quality of potato clones during spring and autumn grown conditions of Rio Grande do Sul. Ciência Rural 36:80–85

Füssel HM (2007) Vulnerability: a generally applicable conceptual framework for climate change research. Glob Environ Chang 17(2):155–167

Gayler S, Wang E, Priesack E, Schaaf T, Maidl FX (2002) Modeling biomass growth, N-uptake and phenological development of potato crop. Geoderma 105(3–4):367–383

Godfray HCJ, Pretty J, Thomas SM, Warham EJ, Beddington JR (2011) Linking policy on climate and food. Science 331(6020):1013–1014

Gustavsen GW (2021) Sustainability and potato consumption. Potato Res 64(4):571–586

Haas BJ, Kamoun S, Zody MC, Jiang RH, Handsaker RE, Cano LM, Nusbaum C (2009) Genome sequence and analysis of the Irish potato famine pathogen Phytophthora infestans. Nature 461(7262):393–398

Haris AA, Chhabra V, Bhatt BP, Sikka AK (2015) Yield and duration of potato crop in Bihar under projected climate scenarios. J Agrometeorol 17(1):67–73

Haverkort AJ, Verhagen A (2008) Climate change and its repercussions for the potato supply chain. Potato Res 51:223–237

Hijmans RJ (2003) The effect of climate change on global potato production. Am J Potato Res 80:271–279

Hijmans RJ, Spooner DM (2001) Geographic distribution of wild potato species. Am J Bot 88(11):2101–2112

Holden NM, Brereton AJ, Fealy R, Sweeney J (2003) Possible change in Irish climate and its impact on barley and potato yields. Agric For Meteorol 116:181–196

Hoogenboom G, Porter CH, Boote KJ, Shelia V, Wilkens PW, Singh U, White JW, Asseng S, Lizaso JI, Moreno LP, Pavan W, Ogoshi R, Hunt LA, Tsuji GY, Jones JW (2019) The DSSAT crop modeling ecosystem. In: Boote KJ (ed) Advances in crop modeling for a sustainable agriculture. Burleigh Dodds Science Publishing, Cambridge, pp 173–216. https://doi.org/10.19103/AS.2019.0061.10

Hoogenboom G, Porter CH, Shelia V, Boote KJ, Singh U, White JW, Pavan W, Oliveira FAA, Moreno-Cadena LP, Lizaso JI, Asseng S, Pequeno DNL, Kimball BA, Alderman PD, Thorp KR, Jones MR, Cuadra SV, Vianna MS, Villalobos FJ, Ferreira TB, Batchelor WD, Koo J, Hunt LA, Jones JW (2021) Decision Support System for Agrotechnology Transfer (DSSAT) Version 4.8 (DSSAT.net). DSSAT Foundation, Gainesville

Hu Q, Weiss A, Feng S, Baenziger PS (2005) Earlier winter wheat heading dates and warmer spring in the US great plains. Agric For Meteorol 135(1–4):284–290

Hussain M, Ahmad S, Hussain S, Lal R, Ul-Allah S, Nawaz A (2018) Rice in saline soils: physiology, biochemistry, genetics and management. Adv Agron 148:231–287

Islam S, Cenacchi N, Sulser TB, Gbegbelegbe S, Hareau G, Kleinwechter U, Wiebe K (2016) Structural approaches to modeling the impact of climate change and adaptation technologies on crop yields and food security. Glob Food Sec 10:63–70

Jansky SH, Jin LP, Xie KY, Xie CH, Spooner DM (2009) Potato production and breeding in China. Potato Res 52:57–65

Jégo G, Martinez M, Antigüedad I, Launay M, Sanchez-Perez JM, Justes E (2008) Evaluation of the impact of various agricultural practices on nitrate leaching under the root zone of potato and sugar beet using the STICS soil–crop model. Sci Total Environ 394(2–3):207–221

Jones JW, Hoogenboom G, Porter CH, Boote KJ, Batchelor WD, Hunt LA, Wilkens PW, Singh U, Gijsman AJ, Ritchie JT (2003) The DSSAT cropping system model. Eur J Agron 18(3–4):235–265

Jones JW, Tsuji GY, Hoogenboom G, Hunt LA, Thornton PK, Wilkens PW, Singh U (1998) Decision support system for agrotechnology transfer: DSSAT v3. In: Tsuji GY, Hoogenboom G, Thornton PK (eds) Understanding options for agricultural production. Springer, Dordrecht, pp 157–177

Jovovic Z, Kratovalieva S (2015) Global strategies for sustainable use of agricultural genetic and indigenous traditional knowledge. In: Salgotra RK, Gupta BB (eds) Plant genetic resources and traditional knowledge for food security. Springer, Singapore, pp 39–72

Jung C, Schindler D (2020) Introducing a new approach for wind energy potential assessment under climate change at the wind turbine scale. Energy Convers Manag 225:113425

Kaukoranta T, Hakala K (2008) Impact of spring warming on sowing times of cereal, potato and sugar beet in Finland. Agric Food Sci 17(2):165–176

Kawchuk LM, Lynch DR, Yada RY, Bizimungu B, Lynn J (2008) Marker assisted selection of potato clones that process with light chip color. Am J Potato Res 85:227–231

Keating BA, Carberry PS, Hammer GL, Probert ME, Robertson MJ, Holzworth D, Huth NI, Hargreaves JNG, Meinke H, Hochman Z, McLean G, Verburg K, Snow V, Dimes JP, Silburn M, Wang E, Brown S, Bristow KL, Asseng S, Chapman S, McCown RL, Freebairn DM, Smith CJ (2003) An overview of APSIM: a model designed for farming systems simulation. Eur J Agron 18:267–288

Khan A, Fahad S, Khan A, Saud S, Adnan M, Wahid F, Noor M, Nasim W, Hammad HM, Bakhat HF, Ahmad S, Rehman MH, Wang D, Sönmez O (2019) Managing tillage operations and manure to restore soil carbon stocks in wheat-maize cropping system. Agron J 111:1–10

Khan MN, Hussain M, Abbas G, Fatima Z, Iqbal P, Khan A, Zakir I, Ahmed M, Mateen A, Khan MA, Ahmad S (2020) Improving resource use efficiencies of sugarcane at farmer field under arid environment. Int J Agric Biol 24(5):1279–1285

Kleinwechter U, Gastelo M, Ritchie J, Nelson G, Asseng S (2016) Simulating cultivar variations in potato yields for contrasting environments. Agric Syst 145:51–63

Kumar SN, Govindakrishnan PM, Swarooparani DN, Nitin C, Surabhi J, Aggarwal PK (2015) Assessment of impact of climate change on potato and potential adaptation gains in the Indo-Gangetic Plains of India. Int J Plant Prod 9(1):151–170
Larkin RP (2003) Characterization of soil microbial communities under different potato cropping systems by microbial population dynamics, substrate utilization, and fatty acid profiles. Soil Biol Biochem 35(11):1451–1466
Levy D, Veilleux RE (2007) Adaptation of potato to high temperatures and salinity—a review. Am J Potato Res 84:487–506
Li Z, Yang P, Tang H, Wu W, Yin H, Liu Z, Zhang L (2014) Response of maize phenology to climate warming in Northeast China between 1990 and 2012. Reg Environ Chang 14:39–48
Lutaladio N, Castaldi L (2009) Potato: the hidden treasure. J Food Compos Anal 22(6):491–493
Lutz W, Samir KC (2010) Dimensions of global population projections: what do we know about future population trends and structures? Philos Trans R Soc Lond B Biol Sci 365(1554):2779–2791
Magrin GO, Travasso MI, Díaz RA, Rodríguez RO (1997) Vulnerability of the agricultural systems of Argentina to climate change. Clim Res 9(1–2):31–36
Mahdian MH, Gallichand J (1995) Validation of the SUBSTOR model for simulating soil water content. Trans ASAE 38(2):513–520
Mahmood A, Ullah H, Shahzad AN, Ali H, Ahmad S, Zia-Ul-Haq M, Honermeier B, Hasanuzzaman M (2013) Dry matter yield and chemical composition of sorghum cultivars with varying planting density and sowing date. Sains Malaysiana 42(10):1529–1538
Manorama K, Govindakrishnan PM, Ravichandran G, Singh BP (2015) Development of a decision support system for potato crop scheduling in Nilgiri hills of Western Ghats. J Agrometeorol 17(2):219–222
Menzel A, Sparks TH, Estrella N, Koch E, Aasa A, Ahas R, Zust ANA (2006) European phenological response to climate change matches the warming pattern. Glob Chang Biol 12(10):1969–1976
Meza FJ, Silva D, Vigil H (2008) Climate change impacts on irrigated maize in Mediterranean climates: evaluation of double cropping as an emerging adaptation alternative. Agric Syst 98(1):21–30
Mishra P, Polder G, Gowen A, Rutledge DN, Roger JM (2020) Utilising variable sorting for normalisation to correct illumination effects in close-range spectral images of potato plants. Biosyst Eng 197:318–323
Miyashita K, Tanakamaru S, Maitani T, Kimura K (2005) Recovery responses of photosynthesis, transpiration, and stomatal conductance in kidney bean following drought stress. Environ Exp Bot 53(2):205–214
Monneveux P, Ramírez DA, Pino MT (2013) Drought tolerance in potato (S. tuberosum L.): can we learn from drought tolerance research in cereals? Plant Sci 205:76–86
Moradi R, Koocheki A, Nassiri Mahallati M (2014) Adaptation of maize to climate change impacts in Iran. Mitig Adapt Strateg Glob Chang 19:1223–1238
Moradi R, Koocheki A, Nassiri Mahallati M, Mansoori H (2013) Adaptation strategies for maize cultivation under climate change in Iran: irrigation and planting date management. Mitig Adapt Strateg Glob Chang 18:265–284
Mubarak MK, Hussain K, Abbas G, Altaf MT, Baloch FS, Ahmad S (2022) Productivity of sorghum (*Sorghum bicolor* L.) at diverse irrigation regimes and sowing dates in semi-arid and arid environment. Turk J Agric For 46(1):1–18
Mubeen M, Ahmad A, Wajid A, Khaliq T, Hammad HM, Sultana SR, Ahmad S, Fahad S, Nasim W (2016) Application of CSM-CERES-Maize model in optimizing irrigated conditions. Outlook Agric 45(3):173–184
Nadeem M, Khan MN, Abbas G, Fatima Z, Iqbal P, Ahmed M, Raza MA, Rehman A, Haq EU, Hayat A, Ali M, Ahmad S (2022) Application of CSM-CANEGRO model for climate change impact assessment and adaptation for sugarcane in semi-arid environment of Southern Punjab, Pakistan. Int J Plant Prod 16:443–466

Nasim W, Amin A, Fahad S, Awais M, Khan N, Mubeen M, Wahid A, Turan V, Rehman MH, Ihsan MZ, Ahmad S, Hussain S, Mian IA, Khan B, Jamal Y (2018) Future risk assessment by estimating historical heat wave trends with projected heat accumulation using SimCLIM climate model in Pakistan. Atmos Res 205:118–133

Naz S, Ahmad S, Abbas G, Fatima Z, Hussain S, Ahmed M, Hoogenboom G (2022) Modeling the impact of climate warming on potato phenology. Eur J Agron 132:126404

Parmesan C, Yohe G (2003) A globally coherent fingerprint of climate change impacts across natural systems. Nature 421(6918):37–42

Parry M, Rosenzweig C, Iglesias A, Fischer G, Livermore M (1999) Climate change and world food security: a new assessment. Glob Environ Chang 9:S51–S67

Parry ML, Rosenzweig C, Iglesias A, Livermore M, Fischer G (2004) Effects of climate change on global food production under SRES emissions and socio-economic scenarios. Glob Environ Chang 14(1):53–67

Pasare SA, Ducreux LJ, Morris WL, Campbell R, Sharma SK, Roumeliotis E, Taylor MA (2013) The role of the potato (Solanum tuberosum) CCD8 gene in stolon and tuber development. New Phytol 198(4):1108–1120

Pradel W, Gatto M, Hareau G, Pandey SK, Bhardway V (2019) Adoption of potato varieties and their role for climate change adaptation in India. Clim Risk Manag 23:114–123

Qasim M, Khalid S, Naz A, Khan MZ, Khan SA (2013) Effects of different planting systems on yield of potato crop in Kaghan Valley: a mountainous region of Pakistan. Agric Sci 4:30754

Rahman MH, Ahmad A, Wang X, Wajid A, Nasim W, Hussain M, Ahmad B, Ahmad I, Ali Z, Ishaque W, Awais M, Muddasir M, Shelia V, Ahmad S, Fahad S, Alam M, Ullah H, Hoogenboom G (2018) Multi-model projections of future climate and climate change impacts uncertainty assessment for cotton production in Pakistan. Agric For Meteorol 253–254:94–113

Rahman MH, Ahmad I, Wang D, Fahad S, Afzal M, Ghaffar A, Saddique Q, Khan MA, Saud S, Hassan S, Fahad M, Din M, Ahmad M, Yue Z, Shah RA, Sönmez O, Nasim W (2021) Influence of semi-arid environment on radiation use efficiency and other growth attributes of lentil crop. Environ Sci Pollut Res 28:13697–13711

Rao KN (2013) Evaluation of weather index insurance as an adaptation strategy to mitigate climate risks in agriculture. Doctoral dissertation, Gokhale Institute of Politics and Economics

Raymundo R, Asseng S, Prassad R, Kleinwechter U, Concha J, Condori B, Porter C (2017) Performance of the SUBSTOR-potato model across contrasting growing conditions. Field Crop Res 202:57–76

Raymundo R, Asseng S, Robertson R, Petsakos A, Hoogenboom G, Quiroz R, Wolf J (2018) Climate change impact on global potato production. Eur J Agron 100:87–98

Rehman A, Ashfaq M, Naqvi SAA, Adil SA, Bashir K, Ahmad A, Ahmad S, Ali A, Imran A (2015) Is climate change worsening the poverty of maize growers? Evidence from Punjab province of Pakistan. Ciencia e Tecnica Vitivinicola 30(1):105–116

Rezaei EE, Webber H, Gaiser T, Naab J, Ewert F (2015) Heat stress in cereals: mechanisms and modelling. Eur J Agron 64:98–113

Richardson AD, Keenan TF, Migliavacca M, Ryu Y, Sonnentag O, Toomey M (2013) Climate change, phenology, and phenological control of vegetation feedbacks to the climate system. Agric For Meteorol 169:156–173

Ritchie JT (1985) Description and performance of CERES wheat: a user-oriented wheat yield model. ARS wheat yield project, pp 159–175

Rivington M, Koo J (2010) Report on the meta-analysis of crop modelling for climate change and food security survey. https://doi.org/10.13140/RG.2.1.4599.1522

Sarwar N, Ali H, Maqsood M, Ullah E, Shahzad AN, Shahzad M, Mubeen K, Shahid MA, Ahmad S (2013) Phenological response of rice plants to different micronutrients application under water saving paddy fields on calcareous soil. Turk J Field Crops 18(1):52–57

Sarwar N, Atique-ur-Rehman, Farooq O, Wasaya A, Hussain M, El-Shehawi AM, Ahmad S, Brestic M, Mahmoud SF, Zivcak M, Farooq S (2021) Integrated nitrogen management improves productivity and economic returns of wheat-maize cropping system. J King Saud Univ Sci 33:101475

Schwartz SH (1999) A theory of cultural values and some implications for work. Appl Psychol 48(1):23–47

Shabbir RN, Waraich EA, Ali H, Nawaz F, Ashraf MY, Ahmad R, Awan MI, Ahmad S, Irfan M, Hussain S, Ahmad Z (2016) Supplemental exogenous NPK application alters biochemical processes to improve yield and drought tolerance in wheat (*Triticum aestivum* L.). Environ Sci Pollut Res 23(3):2651–2662

Singh B, Chandra A, Al-Haddad K (1999) Computer-aided modeling and simulation of active power filters. Electr Mach Power Syst 27(11):1227–1241

Singh JP, Govindakrishnan PM, Lal SS, Aggarwal PK (2005) Increasing the efficiency of agronomy experiments in potato using INFOCROP-POTATO model. Potato Res 48:131–152

Singh JP, Lal SS, Govindakrishnan PM, Dua VK, Pandey SK (2008) Climate change and potato production in India. In: Poster presented in Global Potato Conference–2008, New Delhi, pp 9–12

Smit B, Skinner MW (2002) Adaptation options in agriculture to climate change: a typology. Mitig Adapt Strateg Glob Chang 7(1):85–114

Sparks AH, Forbes GA, Hijmans RJ, Garrett KA (2014) Climate change may have limited effect on global risk of potato late blight. Glob Chang Biol 20(12):3621–3631

Šťastná M, Toman F, Dufkova J (2010) Usage of SUBSTOR model in potato yield prediction. Agric Water Manag 97(2):286–290

Stokes C, Howden M (2010) Adapting agriculture to climate change: preparing Australian agriculture, forestry and fisheries for the future. CSIRO Publishing, Collingwood

Supit I, Van Diepen CA, De Wit AJW, Wolf J, Kabat P, Baruth B, Ludwig F (2012) Assessing climate change effects on European crop yields using the Crop Growth Monitoring System and a weather generator. Agric For Meteorol 164:96–111

Tang J, Wang J, Fang Q, Dayananda B, Yu Q, Zhao P, Pan X (2019) Identifying agronomic options for better potato production and conserving water resources in the agro-pastoral ecotone in North China. Agric For Meteorol 272:91–101

Tariq M, Afzal MN, Muhammad D, Ahmad S, Shahzad AN, Kiran A, Wakeel A (2018a) Relationship of tissue potassium content with yield and fiber quality components of Bt cotton as influenced by potassium application methods. Field Crop Res 229:37–43

Tariq M, Ahmad S, Fahad S, Abbas G, Hussain S, Fatima Z, Hoogenboom G (2018b) The impact of climate warming and crop management on phenology of sunflower-based cropping systems in Punjab, Pakistan. Agric For Meteorol 256:270–282

Tariq M, Ahmad S, Fahad S, Abbas G, Hussain S, Fatima Z, Wajid N, Mubeen M, Rehman M, Khan MA, Adnan M, Wilkerson CJ, Hoogenboom G (2018c) The impact of climate warming and crop management on phenology of sunflower-based cropping systems in Punjab, Pakistan. Agric For Meteorol 256–257:270–282

Tariq M, Yasmeen A, Ahmad S, Hussain N, Afzal MN, Hasanuzzaman M (2017) Shedding of fruiting structures in cotton: factors, compensation and prevention. Trop Subtrop Agroecosyst 20(2):251–262

Tarriq M, Fatima Z, Iqbal P, Nahar K, Ahmad S, Hasanuzzaman M (2021) Sowing dates and cultivars mediated changes in phenology and yield traits of cotton-sunflower cropping system in arid environment. Int J Plant Prod 15:291–302

Thiele G, Theisen K, Bonierbale M, Walker T (2010) Targeting the poor and hungry with potato science. Potato J 37(3/4):75–86

Thomas-Sharma S, Abdurahman A, Ali S, Andrade-Piedra JL, Bao S, Charkowski AO, Forbes GA (2016) Seed degeneration in potato: the need for an integrated seed health strategy to mitigate the problem in developing countries. Plant Pathol 65(1):3–16

Tooley BE, Mallory EB, Porter GA, Hoogenboom G (2021) Predicting the response of a potato-grain production system to climate change for a humid continental climate using DSSAT. Agric For Meteorol 307:108452

Tryjanowski P, Sparks TH, Blecharczyk A, Małecka-Jankowiak I, Switek S, Sawinska Z (2018) Changing phenology of potato and of the treatment for its major pest (Colorado potato beetle)—a long-term analysis. Am J Potato Res 95:26–32

Van der Waals JE, Krüger K, Franke AC, Haverkort AJ, Steyn JM (2013) Climate change and potato production in contrasting South African agro-ecosystems 3. Effects on relative development rates of selected pathogens and pests. Potato Res 56:67–84

van Haren RJF, Haverkort AJ (1998) Agro-ecological description of LINTUL-POTATO: description and application of the LINTUL-POTATO crop growth model. In: Information technology as a tool to assess land use options in space and time, Proceedings of an international workshop, Lima Peru (CIP, ICASA, PE, AB-DLO), pp 41–49

Van Keulen H, Stol W (1995) Agro-ecological zonation for potato production. In: Haverkort AJ, MacKerron DKL (eds) Potato ecology and modelling of crops under conditions limiting growth, vol 3. Springer, Dordrecht, pp 357–371

White JW, Hoogenboom G, Kimball BA, Wall GW (2011) Methodologies for simulating impacts of climate change on crop production. Field Crop Res 124(3):357–368

Wijesinha-Bettoni R, Mouillé B (2019) The contribution of potatoes to global food security, nutrition and healthy diets. Am J Potato Res 96:139–149

Wolday K, Hruy G (2015) A review: performance evaluation of crop simulation model (APSIM) in prediction crop growth, development and yield in semi-arid tropics. J Nat Sci Res 5(21):34–39

Xiao D, Tao F (2016) Contributions of cultivar shift, management practice and climate change to maize yield in North China Plain in 1981–2009. Int J Biometeorol 60:1111–1122

Zahoor A, Ahmed M, Fayyaz-ul-Hassan SG, Ahmad S (2022) Ontogeny growth and radiation use efficiency of canola (*Brassica napus* L.) under various nitrogen management strategies and contrasting environments. Int J Plant Prod 16(1):195–208

Sweet Corn-Bell Pepper System 11

11.1 Introduction

Because they provide a means of subsistence and generate income in foreign currency, vegetables must be a part of the agricultural sector. Additionally, it is advantageous for maintaining nutritional levels and enhancing illness resistance (Ibrahim and Omotesho 2013). Vegetable production has increased by about 60% over the past 20 years (Naik et al. 2017). *Zea mays* L., a cultivar bred for human use, is used as a raw or processed ingredient in the global food business. It is well-liked by consumers for its distinctive and pleasant flavor and sweetness (Abbas et al. 2017, 2019, 2020a, b, c; Abbas and Ahmad 2018; Ahmad et al. 2012a, b, 2013, 2015a, b, c, 2016a, b, c, 2017a, b, 2018, 2019, 2021, 2023a, b). Specifically designed to maximize the sugar content, sweet corn is a hybridized species of maize that produces green ears 75–90 days after planting. About $259 million worth of sweet corn was grown for processing on farms in 1996, and 192,500 acres were used for production. Sweet corn has 5–6% sugar, 10–11% starch, 3% water-soluble polysaccharides, and 70% water (Oktem and Oktem 2005). The modern superfood for health-conscious people is sweet corn. Sweet corn has nutritional value on par with other expensive vegetables, including cauliflower, cabbage, and French beans. It also has high fiber content and is low in cholesterol. Sweet corn kernels' nutritional value is influenced by their water content (72.7%) and overall solid content (27.3%). The predominant hydrocarbon is starch. Compared to other vegetables, sugar corn kernels have somewhat high-calorie content. Sweet maize is used as human food during the milk stage of endosperm development when the kernel is tender, succulent, and sweet (Pajic et al. 2004; Ahmed et al. 2016, 2018a, b, 2019, 2020, 2022; Ahmad and Hasanuzzaman 2012; Ali et al. 2013; Amin et al. 2018a, b, c). New cultivars that are high in sugar content and have a prolonged shelf life are gaining popularity for human consumption (Lertrat and Pulam 2007). The sweet corn production in the world is presented in Fig. 11.1.

M. Ahmed et al., *Cropping Systems Modeling Under Changing Climate*,
https://doi.org/10.1007/978-981-97-0331-9_11

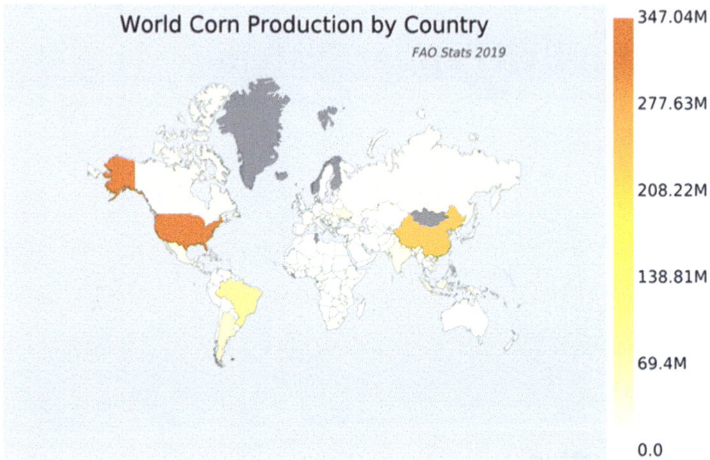

Fig. 11.1 World sweet corn production by country. (Source: FAO Stat 2017)

Capsicum (*Capsicum annum* L.), which belongs to the Solanaceae family and genus Capsicum, is also referred to as a bell pepper or hot pepper. In Pakistan, locals refer to it as "shimla mirch" (Sreedhara et al. 2013). There is an increasing tendency in their availability in many types across its global production. Vitamins A, C, E, B6, thiamine, folic acid, and beta-carotene are all found in it. People eat it in a variety of ways, including pickles, green, spices, condiments, and sauces. It gives eating dishes a distinctive flavor, taste, and beauty (Komla 2013). Pepper is grown all over the world, with the exception of Antarctica, and its history is linked to Christopher Columbus' expedition. Columbus brought hot pepper to Europe, Africa, and Asia. It quickly expanded throughout Europe before reaching Japan, India, and China (Jalu 2014). China now leads the world in the production of capsicum, contributing 50% of the total, followed by Turkey (7%) and Mexico (7%) (Table 11.1). Fruit is tiny, nonpungent, and green. The sweet pepper crop in Japan starts to grow in the final stages of spring and is harvested from May to November. The temperature rises significantly during this time, and in August, high temperatures of more than 40 °C are occasionally reported (Pagamas and Nawata 2007). Because green peppers are sometimes harvested before fully ripening, the stage of ripeness can contribute to the phytonutrient concentration (Saha et al. 2010).

Any cropping system needs a long (3-year cycle or longer) and varied crop rotation to be productive and sustainable. Cropping systems combine management methods and plant genotypes (species and variations) to create crops for specific uses and environmental benefits (Ejaz et al. 2022; Fatima et al. 2018, 2020, 2021; Nadeem et al. 2022; Naz et al. 2022; Jones et al. 2003; Keating et al. 2003; Hoogenboom et al. 2019, 2021). Around the world, a variety of vegetable cropping systems are grown, including tomato-tomato, maize-potato-okra, brinjal-cauliflower-chili, and brinjal-vegetable pea-chili. In Pakistan, there are two harvesting seasons: Rabi and Kharif. Crops for Rabi and Kharif are typically grown from November to April and May to October, respectively (Tariq et al. 2017, 2018a, b; Tarriq et al. 2021; Hussain et al. 2018; Khan et al. 2019, 2020; Zahoor et al. 2022). The Pakistani agricultural

Table 11.1 Production of sweet corn-sweet pepper systems across various regions in the world

Countries	Sweet corn production (MMT)	Sweet pepper production (MMT)	Sweet corn-sweet pepper system production (MMT)	Percent of world production (%)
China	277.2	16.7	293.9	30.7
Turkey	6.0	2.9	8.9	0.9
Indonesia	12.9	2.8	15.7	1.6
Mexico	27.6	2.7	30.3	3.2
Spain	4.6	1.4	6.0	0.6
Egypt	7.4	0.8	8.2	0.9
Nigeria	12.1	0.7	12.8	1.3
USA	348.7	0.5	349.2	36.5
Algeria	–	0.5	0.5	0.1
Netherlands	–	0.4	0.4	0.0
Brazil	125	–	125.0	13.1
Argentina	47	–	47.0	4.9
India	32	0.5	32.5	3.4
South Africa	16.7	–	16.7	1.7
Pakistan	9.2	–	9.2	1.0

economy is based on these two seasons. A common cropping system in Pakistan is sweet corn and bell pepper, in addition to these other systems.

11.2 Cropping Systems and Geographical Distribution

A cropping system is a combination of crops throughout time and area. When crops are interplanted, combinations in space and time arise when they occupy distinct growth seasons. A cropping system typically refers to the combination of crops grown during a particular year when annual crops are being discussed. Vegetable cropping methods typically take place close to metropolitan or peri-urban regions and have better cash flow. Higher cropping intensities are expected to yield higher returns, but crops with lower crop intensities and higher market values typically sell for more money than crops with higher crop intensities and lower market values.

A rise in vegetable consumption is recommended by health standards, which has resulted in a rapid growth in global vegetable output (FAO 2017). Vegetable crop production requires much higher inputs than field crop production (Congreves and Van Eerd 2015). This is due to vegetables' nutritional nature and their inherent growth requirements. India produces 133.74 million tonnes of vegetables annually from a land area of 7.98 million hectares, ranking second in the world. To fulfill the nutritional needs of a projected 1200 million people by 2020, our annual vegetable consumption has climbed to nearly 220 million tonnes (Singh 2004). Recent statistics show a rise in vegetable yield, but sufficient vegetables are still needed to feed the rapidly growing population. For ensuring a healthy and wealthy nation, we have to increase our production by altering our cropping pattern and dietary habits (Bordallo et al. 2005; Brennan and Acosta-Martinez 2017; Cautin and Augusti 2005; CIDH 2007; Deb et al. 2016; FAO 2007; FUMIAF 2005; Hatfield and Prueger 2015; Hussain and Abid 2011; Ishikawa et al. 2004; Lal 2019; Tesfaw 2013; USDA/NASS 2011; US Food and Drug Administration 2015; USDA 2003; USDA Pakistan 2017; WHO 2003).

Sweet corn can be categorized as a kind of maize that has been harvested and consumed before the physiologic maturation stage, yet fresh, while the grains are soft, and before all of the sugar has been turned into starch. Historically, one of the top five most valued vegetable crops in Florida has been sweet corn. Florida is the nation's top producer of sweet corn. The grain production of field corn has dramatically increased over the past 50 years, with national average yields in the USA rising from 5 Mt. ha^{-1} in 1966 to 11 Mt. ha^{-1} in 2016. The availability of genetically improved hybrids and better agronomic management techniques can be credited with improvements in field corn production. In contrast to field corn, which had a major yield increase, sweet corn did not. For instance, over the past 20 years (1998–2017) in the USA, grain production in field corn has increased by 31%, whereas gains in processing sweet corn have increased yields by only 16% (Dhaliwal and Williams 2019).

Bell pepper (*Capsicum annuum* L.) cultivation is one of the most important activities in the horticulture sector. Bell pepper, often referred to as sweet pepper, is commercially significant and nearly 42,000 acres of bell peppers were grown in the USA in 2018, with 4/5 of that production going to the fresh market. Georgia, one of the top three pepper-producing states, produced 5548 acres in 2017 (Wolfe and

Stubbs 2018). Although Georgia now produces few organic peppers, the value of Georgia's entire organic production is more than $47 million (USDA 2017).

India is the world's biggest producer of chilies, accounting for 25% of the total output; however, due to high domestic demand, only 4% of that output is exported. Fresh chilies, stalkless chilies, green chilies, chili powder (generally 0.7% capsaicin is the minimum quantity for the extraction market), and oleoresin are the most difficult types of chili to export.

In Pakistan, chili is regarded as a vegetable and a spice crop with high economic importance. It is Pakistan's profitable cash crop. The two species *C. annum* and *C. frutescens* are those that are most frequently grown in Pakistan. In Pakistan, on a 38.4-thousand-hectare farm, 53.7 thousand tonnes of chilies are produced, yielding an average of 1.7 tonnes/hectare and contributing 1.5% of the GDP. Punjab and Balochistan are the next largest producers of chilies, followed by Sindh.

11.3 Decision Support System for Agrotechnology Transfer

The Decision Support System for Agrotechnology Transfer (DSSAT) was made by an international group of scientists working together on the International Benchmark Sites Network for Agrotechnology Transfer project (Jones et al. 1998) to make it easier to use crop models in a systems-based approach to agronomic research. It was made early on because people needed to combine what they knew about soil, climate, crops, and management in order to make better choices about moving production technology from one place to another where soils and climates are different. The systems method gave a framework for studying how the system and its parts work. The CERES-Maize model in DSSAT V4.0 (Hoogenboom et al. 2003) models crop growth and development with a daily time step from planting to maturity. It does this by using physiological process relationships that show how maize reacts to soil and environmental conditions.

Simulation models can make it easier to compare different production methods and settings and speed up the decision-making process because trying management strategies in the field is expensive and takes a long time (Rinaldi et al. 2007). Computer-based agronomic models are described as quantitative schemes for predicting the growth, development, and yield of a crop, given a set of genotype coefficients and relevant environmental variables. The CERES-Maize model, which is part of the Decision Support System of Agrotechnology Transfer (DSSAT) V4.0 (Jones et al. 2003) for field corn output, is a well-known crop model. This model uses genotype coefficients to predict how different types of corn grow and how much corn they produce (Ritchie 1998). Literature that is currently available has not shown that this model can be used to simulate sweet corn in Florida. Also, crop parameters for sweet corn were not available in the DSSAT database, even though the model has been used for a variety of field corn varieties or hybrids around the world (Asadi and Clemente 2003).

The CERES-Maize model (V4.0; Hoogenboom et al. 2003) models crop growth and development with a daily time step from sowing to maturity based on different biological processes that show how maize plant behaves to different soil and environmental variables. Potential growth depends on photosynthesis-active radiation and how well it is absorbed, but biomass production is limited by bad weather and lack of water and nitrogen in the soil. In general, the CERES-Maize model needs information about the weather, crop, soil, and management practices. The weather inputs are the daily total global radiation (MJ m^{-2}), the daily average and highest air temperatures (°C), and the daily total rainfall (mm) (Ritchie 1998). The field experiment also gave the crop management information that was needed, such as planting density, row spacing, sowing depth, time and amount of irrigation, and amount of N fertilizer used. Crop physiological factors are shown by genotype coefficients, which show how physiological processes like development and growth work for each type of crop. Soil factors describe the way the soil looks, how it behaves chemically, and how it looks physically (Mahmood et al. 2013; Mubeen et al. 2016; Nasim et al. 2018; Mubarak et al. 2022; Rehman et al. 2015; Rahman et al. 2018, 2021; Sarwar et al. 2013, 2021; Shabbir et al. 2016).

CROPGRO is a PBM tool that applies models to many crops, such as tomato, cotton, pasture, cowpea, chickpea, dry bean, faba bean, *Mucuna*, chickpea, and cowpea (Shi et al. 2015). The soybean's CROPGRO-Bell Pepper model shows the bell pepper's genotype and phenotype. The growth survey data was used to calculate the model's parameters, which were then validated using the GLUE coefficient estimator. In order to prove that the RNN algorithm works, the weekly crop growth rate was calculated using the environmental and crop growth parameters collected during the growth time. The estimated crop growth rate was used to determine the fresh crop's weight. The PBM was checked by simulating bell pepper growth rate with Decision Support Systems for Agrotechnology Transfer v4.7 (DSSAT) (Jones et al. 2003). Because the crop growth rate from the PBM was based on dry matter, the fresh weight of each organ (leaf, stem, root, and fruit) was calculated using the

Table 11.2 Genetic coefficients, description, ranges, and guidelines for model calibration of sweetcorn in DSSAT

Genetic coefficient	Description
VAR#	Identification code or number for a specific cultivar
VAR-NAME	Name of cultivar
EXPNO	Number of experiments used to estimate cultivar parameters
ECO#	Ecotype code of this cultivar, points to the Ecotype in the ECO file
P1	Thermal time from seedling emergence to the end of the juvenile phase (expressed in degree days above a base temperature of 8 °C) during which the plant is not responsive to changes in photoperiod
P2	Extent to which development (expressed as days) is delayed for each hour increase in photoperiod above the longest photoperiod at which development proceeds at a maximum rate (which is considered to be 12.5 h)
P5	Thermal time from silking to physiological maturity (expressed in degree days above a base temperature of 8 °C)
G2	Maximum possible number of kernels per plant
G3	Kernel filling rate during the linear grain filling stage and under optimum conditions (mg/day)
PHINT	Phyllochron interval; the interval in thermal time (degree days) between successive leaf tip appearances

Source: DSSAT v4.8; Hoogenboom et al. (2021)

Table 11.3 Genetic coefficients, description, ranges, and guidelines for model calibration of bell pepper in DSSAT

Genetic coefficient	Description
EXPNO	Number of experiments used to estimate cultivar parameters
ECO#	Ecotype code of this cultivar, points to the Ecotype in the ECO file
CSDL	Critical short-day length below which reproductive development progresses with day length effect (for long-day plants) (h)
PPSEN	Slope of the relative response of development to photoperiod with time (negative for long-day plants) (1/h)
EM-FL	Time between plant emergence and flower appearance (R1) (photothermal days)
FL-SH	Time between first flower and first pod (R3) (photothermal days)
FL-SD	Time between first flower and first seed (R5) (photothermal days)
SD-PM	Time between first seed (R5) and physiological maturity (R7) (photothermal days)
FL-LF	Time between first flower (R1) and end of leaf expansion (photothermal days)
LFMAX	Maximum leaf photosynthesis rate at 30 C, 350 vpm CO_2, and high light (mg CO_2/m^2-s)
SLAVR	Specific leaf area of cultivar under standard growth conditions (cm^2/g)
SIZLF	Maximum size of full leaf (three leaflets) (cm^2)
XFRT	Maximum fraction of daily growth that is partitioned to seed + shell
WTPSD	Maximum weight per seed (g)
SFDUR	Seed filling duration for pod cohort at standard growth conditions (photothermal days)
SDPDV	Average seed per pod under standard growing conditions (#/pod)
PODUR	Time required for cultivar to reach the final pod load under optimal conditions (photothermal days)
THRSH	The maximum ratio of (seed/(seed + shell)) at maturity. Causes seed to stop growing as their dry weights increase until shells are filled in a cohort
SDPRO	Fraction of protein in seeds (g(protein)/g(seed))
SDLIP	Fraction of oil in seeds (g(oil)/g(seed))

Source: DSSAT v4.8; Hoogenboom et al. (2021)

rate based on fresh weight and growth stage. The accuracy was checked by comparing the estimated fresh crop weight from the RNN algorithm and the PBM with the real fresh crop weight. Genetic coefficients, description, ranges, and guidelines for model calibration of sweet corn and bell pepper in DSSAT v4.8 are presented in Tables 11.2 and 11.3, respectively.

11.4 Climate Change

Weather and climate changes have been noted since the beginning of the industrial revolution in the eighteenth and nineteenth centuries, which has prompted a rise in climate study and the development of climate modeling tools (IPCC 2000). Today's scientists concur that anthropogenic activity is the primary cause of these changes. Climate change, environmental deterioration, and food security are the three most pressing issues facing the world today. Enhancing nutrient (nitrogen, phosphorus) use efficiency in agricultural crop production and stabilizing yield by engaging in sustainable agriculture can increase the former and reduce the latter two (Searchinger et al. 2018). All areas of natural and human ecosystems are affected by climate change, but effects on people's basic requirements for food and water are among the most dangerous. The efficiency with which we can produce food, feed, and fiber through agricultural systems determines how well we will be able to feed the world's population in the future.

Long-term changes in the distribution patterns of climatic components lead to climate change, which is a natural process. Anomalies brought on by climate change harm the agricultural industry. Nitrogen makes up 78% of the atmosphere on earth, oxygen 21%, and argon 0.93%. These gases only interact sporadically with both outgoing infrared light and solar radiation. Water vapor ranges from 0% to 2%, while carbon dioxide only makes up 0.03–0.04%. Temperature extremes have a significant impact on the phenology of agricultural plants. More high-temperature-tolerant cultivars would be in demand as the world's temperature rises. By changing the growing season, the climate shift will impair the appropriateness and adaptability of present cultivars (Dhillon and Gill 2015).

According to Nurdin (2011), changes in cropping pattern, cropping season, agricultural production, and agricultural products can all be attributed to climate change. Rainfall is one climatic factor that directly impacts the agriculture industry. A direct effect of climate change is fluctuating rainfall, which will also cause declining yields and, in turn, a loss in farmers' revenue (Hidayati and Suryanto 2015).

Vegetable crops are sensitive to rapid temperature rises and unpredictable precipitation at any stage of growth (Afroza et al. 2010). Due to their perishability, vegetables are crucial for food and nutritional security, but the poor cannot buy them. Due to climate change, crop failures, low yields, quality decline, and pest and disease problems, olericulture becomes unprofitable (Lal et al. 2014). Additionally, compared to summer-season veggies, winter-season vegetables are more susceptible to inclement weather. Vegetable production is negatively impacted by abiotic factors such as severe temperatures (low/high), salt of the soil, droughts, and floods. So, the two main factors contributing to low vegetable yields are excessive temperatures and insufficient soil moisture. The whims of the weather have a big impact on the many stages of development such as vegetative growth, blooming, and fruiting.

Molecular level physiological activities, including photosynthesis and respiration, are crucial for healthy plant metabolism, organ growth, synthesis of essential nutrients, and stress responses. Increases in temperature, carbon dioxide (CO_2), and ozone (O_3) concentrations have a significant impact on plant physiology, but vegetables can adapt to these new circumstances. The disruption of normal plant growth and development due to high temperatures and unpredictable and irregular precipitation might ultimately lower crop output. Environmental factors have a big effect on how organic matter breaks down in the soil, how nutrients are recycled, and how easy it is for plants to get water and nutrients. But the effect on crop growth cycles, biomass buildup, and, in the end, economic return relies on how strong and long the extremes are. Plants go through a number of morphological and anatomical changes when the temperature stays high for a long time. These changes affect fruit set, size, weight, and quality, as well as seed germination, plant growth, flower dropping, pollen viability, and gametic fertilization. High temperatures also stop flower buds from growing because they cause flowers to die. Additionally, photosynthesis is significantly inhibited at temperatures above their optimal range, which results in a large loss of potential productivity (Rashid et al. 2020).

11.5 Impact of Climate Change on Crop Phenology

Global, regional, and local food security could all be impacted by climate change (Brown et al. 2015). Extreme weather events, such as heat waves, droughts, and floods, have been more often and intense in recent years. As greenhouse gas concentrations grow and cause global temperatures to rise further, these conditions will get worse (Masson-Delmotte et al. 2021). Phenology is the study of biological phenomena that are connected to particular periodic rhythms or phases and how they interact with their environment. Numerous responses are seen in various crops and kinds as a result of the ontogenetic cycle of plants, which is intimately tied to genotype, the environment in which they develop, and the interplay between these elements (Mundarain et al. 2005). For efficient management of a given crop, understanding its phenology is crucial (Soto-Ortiz and Silvertooth 2008). Agronomically, the result of a particular microclimate permits the expected response from the plant; economically, the phenological stages permit the best execution of various agricultural practices, such as hand pollination, prediction of likely pest incidence, need for a specific fertilizer, or application of hormonal substrates. From a climatological perspective, these phenomena provide the basis for interpreting changes due to bioclimatic factors.

The functions or rates of various processes, such as the identification of discrete phases and stages of development and their predicted duration under particular temperature regimes, must be ascertained to characterize crops' growth and development (Soto-Ortiz and Silvertooth 2008). Only four phenological phases are indicated in the case of *Capsicum* species: emergence, seventh leaf, flowering, and ripening. The USDA, for its part, decided in 2003 that the length of these phenological stages is based on the interval between particular phases, which depends on the type of plant (direct sowing or transplanting). There are three main stages listed as well: (1) 50% from sowing to thinning, (2) 75% from thinning or transplanting to fruit set, and (3) 100% from fruit set to harvest or the end of the process. The action that proves seed germination is the radicle's emergence. Temperature, water, oxygen, and presence of light are just a few of the variables that affect whether or not a seed will germinate. Bell pepper's seedling stage ends between 35 and 40 days after sowing, at which point the plant must be transplanted. However, this timeframe is dependent on the surrounding temperature as well as the seedling's current conformation, which includes its embryonic reserves, photosynthetic capacity, and genetic makeup (Mundarain et al. 2005). Between 9 and 13 days sees the highest concentration during the emergence period, which varies. The flowering period varies between 70 and 93 days, with larger incidence at the beginning of this period. Early types of fruit begin to ripen about 85 days, whereas later varieties do so at 107 days (Fernandes et al. 2004).

The phenological behavior of the genetic materials studied showed significant differences. Emergence started at 6 DAS for the materials that produce yellow and orange fruit, and it took until 21 DAS for those that produce red fruit. It is probable that the greenhouse's incident temperature (about 15 °C) affected this physiological process because the building was not heated to the recommended temperature of 18 °C for germination. The lowest temperature that pepper (*Capsicum* spp.) seeds can tolerate is 12–13 °C, and at this temperature, germination is slow (20–25 days), whereas at 20–25 °C, germination starts in 7 or 8 days. Temperature affects the length of phenological phases, with both suboptimal and supraoptimal temperatures lengthening the duration (Challinor et al. 2005). High daytime and nighttime temperatures have a negative impact on the sweet pepper plant's vegetative growth, blooming, fruit set, and yield as a result (Erickson and Markhart 2001).

In comparison to hot pepper, sweet pepper is more susceptible to high temperatures and sets less fruits at high nighttime temperatures. Due to inadequate pollen growth, pollen fertilization is significantly decreased below 15 °C and above 32 °C. From seed germination to seed maturation, temperature is thought to be the primary environmental factor influencing the entire phenomenon of plant growth and development.

The key to higher yields in the production of sweet peppers is to plant at the best time possible. This maximizes crop potential, ensures resource efficiency, and reduces crop competition with weeds and insect pests. A timely planting ensures improved crop growth and financial yield (Waha et al. 2012). On the other hand, improper planting dates reduce yield and produce fruits of lower quality. Proper planning time yields good yield with better quality fruits. Compared to planting late, planting early produced a good yield (Hamma et al. 2012). Due to changes in climatic conditions (temperature variations, frequency, and length of wet and dry periods), which strongly correlate with crop phenological phases, the timing of planting has a significant impact on crop yield (Wolf et al. 2015). When analyzing how different transplanting dates affected the performance of sweet pepper cultivars, a significant variation in dry biomass was discovered.

According to some authors, the amount of bioactive compounds in red pepper fruit can vary depending on its genotype and stage of ripeness (Valero et al. 2014). The timing of harvest is yet another crucial element determining the peppers' ability to retain their freshness and their antioxidant capability. At various stages of pepper development, a variety of biochemical and physiological changes take place as a result of variations in the synthesis, transportation, and degradation of different metabolites, such as variations in the concentration of organic acids and sugars, variations in the synthesis and degradation of pigments, and accumulations of volatile compounds (Manikharda et al. 2018).

Poor emergence and seedling vigor linked to changed sweet corn starch synthesis alleles are among the key problems limiting the spread of sweet corn as a crop. When adverse environmental conditions are present, these features deteriorate even more. Cold and damp soil conditions have a negative impact on the vigor and emergence of sweet corn seedlings. Drought is the most significant abiotic challenge on agriculture in general and sweet corn in particular. Drought-induced osmotic stress prevents fresh and dry biomass production, seed germination, vigor index, and water uptake (Li et al. 2017). Overall, the ability of sweet corn seeds to germinate was reduced as osmotic stress rose.

Plant drought tolerance is significantly influenced by water and nitrogen efficiency. During kernel filling, increasing the efficiency of photosynthesis and nitrogen usage can help the whole plant use nitrogen more effectively. Increased photosynthesis and nitrogen efficiency in the canopy and better use of nutrients by the whole plant when there is no enough water help sweet corn adapt to drought. Crop production is reliant on the climate when adequate irrigation is not available. Rainfall becomes more erratic as a result of climate change, which also affects groundwater supplies and lowers crop yields. In the flowering stage, sweet corn is sensitive to water stress.

The ratio of sugar and starch in the endosperm influences sweetness, which is the primary flavor element in sweet corn. Compared to regular maize, sweet corn accumulates more sugar. With lower levels of glucose, fructose, and maltose, sucrose is the main sugar. Early harvest times have sucrose levels that are roughly twice as high as normal harvest times, with quality that is close to average. Water stress can momentarily raise sugar concentration, according to the effects of irrigation on sugar concentration in sweet corn.

One of the most crucial elements in agricultural systems is the planting date. The best planting date will increase the output of maize. Grain biomass is affected by environmental circumstances throughout all growing phases. When planting maize, the temperature and the length of the day are crucial variables. The planting date has an impact on the germination rate and phonology of sweet corn. The ideal soil temperature varied from 21 to 27 °C, while the lowest soil temperature for maize germination is 13 °C. Planting often starts when the soil temperature rises to over 12 °C in the spring. In warmer areas with longer growing seasons, two crops can be planted each year. In temperate areas, however, only one crop can be planted because the growing season is shorter (April to September) and the crop has to deal with cold stress at the beginning and end of its growth. Low- or high-temperature stress impacts sweet corn, changing a variety of elements of plant growth, such as the carotenoid and vitamin E content in seedlings (Xiang et al. 2019). High temperature inhibited the accumulation of vitamin E, whereas low temperature encouraged the formation of carotenes and high temperature decreased it.

The genes that regulate the conversion of sugar to starch inside the corn kernel's endosperm have a naturally occurring recessive mutation, which is the cause of sweet corn. Because the kernels of sweet corn have a high sugar concentration in the milk during the early dough stage, they differ from other corns (field maize, popcorn, and ornamental). It is consumed when the crop is still young. Sweet corn tastes substantially sweeter than regular corn, especially at 25–30%. The type of gene responsible for sweetness determines the level and quality of sweet corn. Sweet corn matures quickly, and 75–80 days after sowing, green ears can be picked. The majority of the time, corn is manually or mechanically harvested for the fresh market when the tops of the cobs are 75% full and the moisture content is between 70% and 80%. Picking may take place over several days as the cobs seldom ripen evenly. For the best yield, plant density is a crucial consideration. Plant density affects yield components and modifies them. Up until a certain point, increasing plant density is linked to an increase in crop output; after that point, crop yield does not change as a result of increased plant density, and greater plant densities ultimately result in lower crop yield.

11.6 Effects of Climate Change on Crop Production Scenarios for the Present and the Future

The estimate of future climate change scenarios is difficult due to ambiguity in the projections of climate change. Inaccurate models and limited knowledge of the climate system, as well as the possibility of internal climatic variability, could all contribute to this uncertainty in estimates. Models can be used to understand how the climate will change in the future and to predict what will happen and how likely it is to happen. Numerous international studies have shown that agricultural production is predicted to suffer as a result of continuing climate change. The general growth, development, and production of crops are significantly impacted by high temperatures and erratic rainfall patterns. To ensure the security of the world's food supply, sweet corn is a crucial crop. However, it has been demonstrated that the effects of climate change have a negative impact on its yield. Crop development has been boosted by warmer weather, which has also shortened the growing season (Waqas et al. 2019).

Pakistan is the fifth most vulnerable nation to climate change, according to the climate risk index (Eckstein et al. 2018). According to future predictions, Pakistan's temperature will rise by 2.8 °C by 2069. As a likely result of rising temperatures, Pakistani cereals and other crops are extremely susceptible to heat stress. The availability of water and nutrients is altered by rising temperatures and erratic rainfall patterns, which can ultimately result in lower crop yields. Temperature rises hasten phenological development, shortening the growth period and lowering crop output. Similar results were seen for maize, where a 7% yield reduction was observed for every 1 °C increase in mean temperature. Compared to C3 crops, maize's photosynthesis and transpiration are less affected by increased CO_2 because it is a C4 crop. A crucial element in limiting the possible harmful effects of climate change is adaptation. Changes in sowing timing, modifications to irrigation or fertilizer management, and creation of hybrids that are climate resilient are some examples of these adaptation tactics.

The global research community has developed new socioeconomic pathways and scenario ideas to analyze the impacts and vulnerabilities of climate change that are consistent from the local to the global scale. Representative Agricultural Pathways (RAPs) scenarios help identify the degree of uncertainty surrounding climate change's effects and adaptations (Antle et al. 2015). Vegetable systems are under tremendous stress as a result of the rapidly growing human population and the pervasive consequences of climate change, which include rising temperatures and changing precipitation patterns (Marshall and Tracy 2003).

Ben-Asher et al. (2008) looked at how weather affects the growth of sweet corn in the USA. Four temperature treatments were studied in the Georgia Envirotron's climate-controlled growth rooms: 25/20, 30/25, 35/30, and 40/35 °C cover a 14/10-h light/dark cycle. During the early growth stage (V3–5), at the lowest temperature, sweet corn's net photosynthetic rate (PN) was found to be 28–34 mol m^{-2} s^{-1}, while it was 50–60% lower at the highest temperature. Every time the temperature went up by 1 °C, they saw a drop of about 1 PN unit. The highest evaporation rate (E) ranged from 0.36 to 0.54 mm h^{-1} (5.0–6.5 mm day^{-1}) during the high-temperature treatment, while the lowest E ranged from 0.25 to 0.36 mm h^{-1} during the low-temperature treatment (3.5–5.0 mm day^{-1}). The total amount of CO_2 fixed by the process at 40/35 °C was 33.7 g m^{-2} day^{-1}. As the temperature went down, this number went up by about 50%. At the lowest and highest temperatures, the water usage efficiency (WUE) went down from 14 g (CO_2) kg1 (H_2O) to 5 g (CO_2) kg1 (H_2O). The high temperature which decreased PN, the vapor pressure deficit (VPD) which was directly linked to E but did not affect PN, and the quasi-stem conductance (QC) which was directly linked to PN but did not affect E were the three main parameters that affected WUE, PN, and E of corn. So, the WUE of the 25/20 °C temperature treatment was about three times greater than that of the 40/35 °C temperature treatment.

11.7 Case Study

Dhaliwal and Williams (2022) gave evidence of sweet corn yield losses from rising temperatures. Vegetables are particularly vulnerable to climatic changes, although agricultural productivity is subject to unusual weather circumstances. Using information on sweet corn gathered over a 27-year period from 16,040 farms, our findings indicate that crop yields were negatively impacted by growing season temperatures above 30 °C. In irrigated and rainfed areas, respectively, crop yields were lowered by 0.5% and 2% for each degree day over 30 °C during anthesis. Both magnitude and direction of changes in crop yield are shown (Fig. 11.2). Generally, agricultural yield losses are caused by a unit rise in the mean growing season temperature; however, the rainfed production zone exhibits larger sensitivity in comparison to irrigated production regions. Moreover, the Northwest's irrigated output indicated a favorable impact from a rise in total precipitation, but the Midwest's irrigated production had a negative impact (Fig. 11.2). If a crop is exposed to temperatures in each unit Celsius temperature bin for a day, the model predicts changes

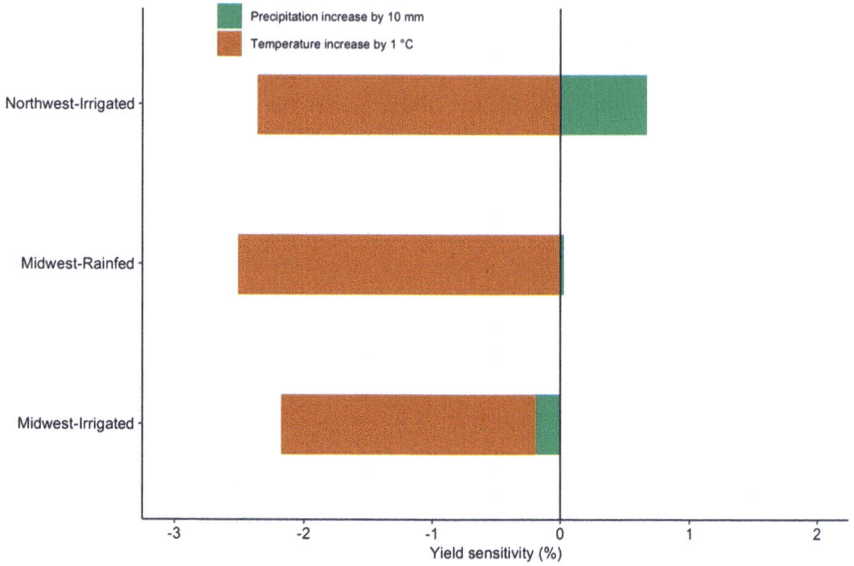

Fig. 11.2 Sweet corn yield sensitivity to changes in growing-season average temperature and total precipitation across the Midwest (Irrigated and Rainfed) and the Northwest (Irrigated) using commercial crop production data from 1992 to 2018. Yield sensitivities were estimated as percent yield losses relative to the mean yield for each production region. (Source: Modified and adopted from Dhaliwal and Williams 2022)

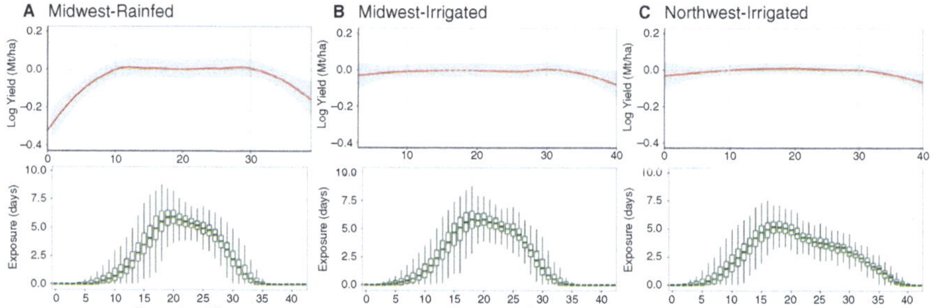

Fig. 11.3 Nonlinear relation between temperature and yield for the (**a**) Midwest-Rainfed, (**b**) Midwest-Irrigated, and (**c**) Northwest-Irrigated. Graphs at the top of each frame display changes in log yield if the crop is exposed for 1 day (24 h) to a particular 1 °C temperature interval. Exposure (days) was calculated as the sum of the fraction of a day during which temperatures fall within each interval. The 95% confidence intervals are shown by gray-shaded area. Histograms at the bottom of each frame display the average temperature exposure for all yield observations in each production region. (Source: Modified and adopted from Dhaliwal and Williams 2022)

in crop production (Fig. 11.3). In the Midwest-Rainfed production zone, a score of 0.2 indicates that an additional day at this temperature can lower crop output by 20%. Our models offer convincing evidence that there are no yield losses in the 8 °C temperature range. In comparison to irrigated production zones in the Midwest and

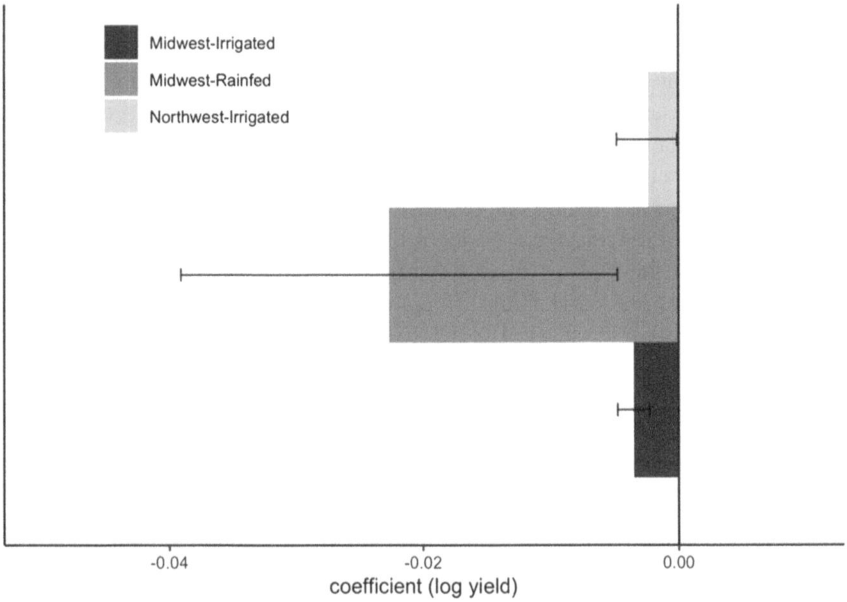

Fig. 11.4 Regression coefficient estimates for the effects of extreme degree days (EDDs) by 1 degree day during anthesis on sweet corn yield (log scale) across the Midwest (Irrigated and Rainfed) and the Northwest (Irrigated) using commercial crop production data from 1992 to 2018. Extreme degree days were calculated using a base temperature of 30 °C and no upper threshold for temperature. Error bars indicate bootstrapped 95% confidence interval using 1000 iterations. (Source: Modified and adopted from Dhaliwal and Williams 2022)

Northwest, the statistical model demonstrates that rainfed production regions experience a quick yield loss at temperatures above 30 °C (Fig. 11.3). This implies that irrigation reduces the negative impact of temperatures over 30 °C on crop yields, but the negative impact of temperatures above 35 °C is also seen in locations with irrigation. The extreme degree day measure, which was developed using hourly temperatures collected during a defined anthesis period in a field, was used to quantify additional consequences of temperatures above 30 °C. We discovered a strong impact of EDDs on sweet corn production, with rainfed farms suffering obviously greater yield losses than irrigated areas (Fig. 11.4). A coefficient of 0.02 (or 0.005) means that under rainfed circumstances, crop output is reduced by 2% (or 0.5% under irrigated conditions) for every degree day over 30 °C. This offers more proof that irrigation can help reduce the harmful effects of heat stress brought on by high temperatures (over 30 °C) during the anthesis of sweet corn. This work highlights the need of accelerating crop adaptation techniques to preserve production of this incredibly popular commodity by demonstrating evidence of sweet corn yield losses across wide geographical domains as a result of climate change.

11.8 Adaptation Techniques

Adapting to climate change is usually thought of as changing or messing with natural or human processes to deal with or prepare for climate change or its effects that have already happened or are expected to happen. Adaptation in agriculture is one of the policies that could be used to lessen the bad effects of climate change on crop production. Agriculture changes on two main levels: the family level (called micro) and the national level (macro). The focus of microlevel analysis of adaptation in agriculture is on the choices farmers make in reaction to seasonal changes in weather, economy, and other things. In crop agriculture, these microlevel tactical decisions are made at the level of individual homes. In crop agriculture, the most common microlevel adaptations are crop diversification, irrigation, mixed crop-livestock farming systems, using new crop varieties that do better in drier conditions, changing planting and harvesting dates, and combining less productive, drought-resistant varieties with high-yield crops that need a lot of water. On the other hand, national level or macro-level analysis focuses on agricultural output at the national and regional levels and how it relates to domestic and foreign policy. Crops can be changed by doing things on the supply side (like giving them more water) or on the demand side (like recovering water) or by doing both. Conservation tillage is an agricultural management approach that tries to cut down on the number or intensity of tillage operations in order to get the most economic and environmental benefits. For successful sustainable agriculture, the best crop rotation should be chosen, such as cotton-gram and sugarcane-wheat. Legumes are crucial in any rotation and should comprise 30–50% of the agricultural acreage. A mixed cropland, pasture, and livestock system is desirable or even required for the development of sustainable agriculture. Rotation of different crops can be used for the breaking of pest cycles and addition of extra nutrients into the soil. Crop rotations can increase soil fertility; protect the environment; control weeds, illnesses, and insects; as well as increase the variety of crops grown and sold (Baldwin 2006). Crop output is negatively impacted by low soil quality, weed infestation, insect pests, and disease assault under climate situations. The most significant cover crops among winter legumes are hairy vetch, subterranean clover, crimson clover, and field peas. Like velvet bean, cowpeas, soybeans, and berseem clover are summer annual legumes. The biennial and perennial legumes alfalfa, crown vetch, red clover, sweet clover, and white clover are used as cover crops. These cover crops should be introduced into current cropping systems in order to preserve crop output and to guard against soil deterioration, serious infestations of weeds, and severe attacks from insect pests and diseases brought on by climate change. Integrating cover crops into cropping systems is the best way to grow food in a way that is sustainable. In rainfed places, soil degradation is a big problem. This not only makes the soil less fertile, but it also makes the lowlands unproductive. Due to climate change, uneven rainfall can lead to soil erosion, which is very bad for soil health. Soil and water conservation has become an important part of gardening in a way that will last. When people use artificial fertilizers, the health of the land gets worse, which hurts people, animals, and plants. Nitrogenous fertilizers pollute the groundwater, which makes people and animals sick. Organic fertilizers can do things that man-made fertilizers made by people cannot, like improving the soil's chemical properties (Chrispaul et al. 2010). Because synthetic fertilizers are so expensive, many farmers now use a lot of natural sources of nutrients to grow crops, such as green manures, animal dung, composted materials, and trash from homes (Fening et al. 2010). The gradual increase in soil organic matter and nutrient supply are the major contributors to organic resources in soil (Bhattacharyya et al. 2009). The most crucial strategy is the creation of short-lived varieties, which can guard against numerous harms like attacks from insect pests and stress caused by drought. There are numerous short-duration varieties of rice, chickpea, and cassava that are resistant to illness, the cold, and other similar issues. For farmers, the most cost-effective way to address the problems of a changing environment is through improved, adapted vegetable germplasm. But most modern cultivars only show a small part of the genetic variety that is available now, including the ability to handle stress. When creating new varieties, especially for intensive, high-input production systems in developed countries, it is possible that traits that would help plants adapt to or tolerate low-input and less-than-ideal environments were chosen against. Because of the genetic and physiological complexity of this trait, normal breeding methods are not very good at making crops more tolerant of salt (Flowers 2004). Also, a plant's ability to handle salt is controlled by its development and changes at different stages. A plant's ability to handle salt at one stage of development does not always predict its ability to handle salt at other stages (Foolad 2004).

11.9 Conclusion

Sweet corn and sweet peppers are the most significant crops grown worldwide in vegetable cropping systems. Vegetables are essential for ensuring food and nutritional security, but because they are highly perishable and their prices are rising quickly due to climate change, the poor cannot afford them. Additionally, compared to summer-season veggies, winter-season vegetables are more susceptible to inclement weather. Vegetable production is negatively impacted by abiotic factors such as severe temperatures (low/high), salt of the soil, droughts, and floods. Climate change is an ongoing process that has started to be noticed in the agricultural sector in recent years, having a long-lasting impact on crop productivity. Variable climatic factors contribute to the effects of climate change on vegetable crops, and extreme weather conditions during key growth periods have a significant impact on the productivity and yields of these crops. Knowing how temperature changes affect vegetable crops is the first step in creating effective adaptation plans to deal with the negative effects of climate change. However, more focus should be placed on breeding technologies that allow for crop architecture and physiology to be genetically altered to adapt to warmer environmental conditions as well as developing other technologies that both mitigate and fully utilize the effects of climate change.

References

Abbas Q, Ahmad S (2018) Effect of different sowing times and cultivars on cotton fiber quality under stable cotton-wheat cropping system in southern Punjab, Pakistan. Pak J Life Soc Sci 16(2):77–84

Abbas G, Ahmad S, Ahmad A, Nasim W, Fatima Z, Hussain S, Rehman MH, Khan MA, Hasanuzzaman M, Fahad S, Boote KJ, Hoogenboom G (2017) Quantification the impacts of climate change and crop management on phenology of maize-based cropping system in Punjab, Pakistan. Agric For Meteorol 247:42–55

Abbas G, Fatima Z, Riaz F, Iqbal P, Hussain S, Zakir I, Iqbal M, Haque E, Ahmad S (2019) Study of sowing dates and hybrids effect in maize-based cropping system under arid conditions of Southern Punjab, Pakistan. Pak J Life Soc Sci 17(1):24–31

Abbas G, Ahmad S, Hussain M, Fatima Z, Hussain S, Iqbal P, Ahmed M, Farooq M (2020a) Sowing date and hybrid choice matters production of maize-maize system. Int J Plant Prod 14:583–595

Abbas G, Fatima Z, Hussain M, Hussain S, Atique-ur-Rehman, Sarwar N, Ahmed M, Ahmad S (2020b) Nitrogen rate and hybrid selection matters productivity of maize-maize cropping system under irrigated arid environment of southern Punjab, Pakistan. Int J Plant Prod 14:309–320

Abbas Q, Atique-ur-Rehman, Sarwar N, Hussain S, Abbas G, Khan MN, Fatima Z, Naz S, Younis H, Ahmed M, Ullah H, Iqbal P, Iqbal M, Kan M, Ahmad S (2020c) Effect of different sowing times and cultivars on wheat grain quality under cotton-wheat cropping system in Southern Punjab, Pakistan. Pak J Life Soc Sci 18(1):41–50

Afroza B, Wani KP, Khan SH, Jabeen N, Hussain K, Mufti S, Amit A (2010) Various technological interventions to meet vegetable production challenges in view of climate change. Asian J Hortic 5(2):523–529

Ahmad S, Hasanuzzaman M (2012) Integrated effect of plant density, N rates and irrigation regimes on the biomass production, N content, PAR use efficiencies and water productivity of rice under irrigated semiarid environment. Notulae Botanicae Horti Agrobotanici Cluj-Napoca 40(1):201–211

Ahmad S, Ahmad A, Soler CMT, Ali H, Zia-Ul-Haq M, Anothai J, Hussain A, Hoogenboom G, Hasanuzzaman G (2012a) Application of the CSM-CERES-Rice model for evaluation of plant density and nitrogen management of fine transplanted rice for an irrigated semiarid environment. Precis Agric 13(2):200–218

Ahmad S, Ali H, Ismail M, Shahzad MI, Nadeem M, Anjum MA, Zia-Ul-Haq M, Firdous N, Khan MA (2012b) Radiation and nitrogen use efficiencies of C_3 winter cereals to nitrogen split application. Pak J Bot 44(1):139–149

Ahmad S, Ahmad A, Ali H, Hussain A, Garcia y Garcia A, Khan MA, Zia-Ul-Haq M, Hasanuzzaman M, Hoogenboom G (2013) Application of the CSM-CERES-Rice model for evaluation of plant density and irrigation management of transplanted rice for an irrigated semiarid environment. Irrig Sci 31(3):491–506

Ahmad A, Ashfaq M, Rasul G, Wajid SA, Khaliq T, Rasul F, Saeed U, Rahman MH, Hussain J, Baig IA, Naqvi SAA, Bokhari SAA, Ahmad S, Naseem W, Hoogenboom G, Valdivia RO (2015a) Impact of climate change on the rice-wheat cropping system of Pakistan. In: Hillel D, Rosenzweig C (eds) Handbook of climate change and agro-ecosystems, vol 3. Imperial College Press and the American Society of Agronomy, London, pp 219–258

Ahmad S, Ali H, Atique-ur-Rehman KRJ, Ahmad W, Fatima Z, Abbas G, Irfan M, Ali H, Khan MA, Hasanuzzaman M (2015b) Measuring leaf area of winter cereals by different techniques: a comparison. Pak J Life Soc Sci 13(2):117–125

Ahmad S, Raza I, Muhammad D, Ali H, Hussain S, Dogan H, Zia-Ul-Haq M (2015c) Radiation, water and nitrogen use efficiencies of *Gossypium hirsutum* L. Turk J Agric For 39(5):825–837

Ahmad S, Ali H, Farooq U, Khan SU, Atique-ur-Rehman, Sarwar N, Shahzad AN, Doğan H, Hussain S, Sultan MT, Waheed A, Zia-ul-Haq M, Hussain K, Khan MA (2016a) Improving

nitrogen and radiation-use-efficiencies of C_4 summer cereals by split nitrogen applications under irrigated arid environment. Turk J Agric For 40(2):280–289
Ahmad S, Hussain S, Fatima Z, Abbas G, Atique-ur-Rehman, Khan MR, Younis H, Naz S, Sohail M, Ajmal M, Abbas N, Akhtar M, Rauf A, Khan M, Ali Z, Hassan M, Rizwan M, Bajwa RS, Ajmal A, Naz S, Ali H, Khan AA, Ali M, Sarwar G, Khan MA, Hasanuzzaman M (2016b) Application of DSSAT model for sowing date management of C_4 summer cereals for fodder and grain crops under irrigated arid environment. Pak J Life Soc Sci 14(2):104–114
Ahmad S, Nadeem M, Abbas G, Fatima Z, Khan RJ, Ahmed M, Ahmad A, Rasul G, Khan MA (2016c) Quantification of the effects of climate warming and crop management on sugarcane phenology. Clim Res 71(1):47–61
Ahmad S, Abbas G, Fatima Z, Khan RJ, Anjum MA, Ahmed M, Khan MA, Porter CH, Hoogenboom G (2017a) Quantification of the impacts of climate warming and crop management on canola phenology in Punjab, Pakistan. J Agron Crop Sci 203(5):442–452
Ahmad S, Abbas Q, Abbas G, Fatima Z, Atique-ur-Rehman, Naz S, Younis H, Khan RJ, Nasim W, Rehman MH, Ahmad A, Rasul G, Khan MA, Hasanuzzaman M (2017b) Quantification of climate warming and crop management impacts on cotton phenology. Plan Theory 6:1–16
Ahmad S, Atique-ur-Rehman, Ejaz M, Fatima Z, Kan M, Ahmad M (2018) Agricultural land-use change of major field crops in Pakistan (1961–2014). Sci Technol Dev 37(3):113–121
Ahmad A, Abbas G, Ahmed M, Fatima Z, Anjum MA, Rasul G, Khan MA, Hoogenboom G (2019) Climate warming and management impact on the change of phenology of the rice-wheat cropping system in Punjab, Pakistan. Field Crop Res 230:46–61
Ahmad A, Ashfaq M, Rasul G, Wajid SA, Ahmad I, Khaliq T, Nasir J, Rasul F, Riaz F, Ahmad B, Ahmad S, Baig IA, Valdivia RO, Hoogenboom G (2021) Development of climate change adaptation strategies for cotton–wheat cropping system of Punjab Pakistan. In: Rosenzweig C, Mutter CZ, Contreras EM (eds) Handbook of climate change and agroecosystems: climate change and farming system planning in Africa and South Asia: AgMIP stakeholder-driven research. World Scientific Publishing Europe Ltd, London, pp 277–327
Ahmad S, Ahmad I, Ahmad B, Ahmad A, Wajid A, Khaliq T, Abbas G, Wilkerson CJ, Hoogenboom G (2023a) Regional integrated assessment of climate change impact on cotton production in a semi-arid environment. Clim Res 89:113–132
Ahmad S, Raza MA, Hussain S, Abbas G, Fatima Z, Ahmed M, Goheer MA, Garcia y Garcia A, Wilkerson CJ, Hoogenboom G (2023b) Identification of weak links in the production technology of farmer for bridging canola yield-gap. J Agric Sci 161:241–253. https://doi.org/10.1017/S0021859623000187
Ahmed M, Akram MN, Asim M, Aslam M, F-u Hassan Higgins S, Stockle CO, Hoogenboom G (2016) Calibration and validation of APSIM-wheat and CERES-wheat for spring wheat under rainfed conditions: models evaluation and application. Comput Electron Agric 123:384–401
Ahmed I, Rahman MH, Ahmad A, Ahmad S, Hussain J, Ullah A, Judge J (2018a) Assessing the impact of climate variability on maize using simulation modeling under semi-arid environment of Punjab, Pakistan. Environ Sci Pollut Res 25(23):28413–28430
Ahmed M, Ijaz W, Ahmad S (2018b) Adapting and evaluating APSIM-SoilP-Wheat model for response to phosphorus under rainfed conditions of Pakistan. J Plant Nutr 41(16):2069–2084
Ahmed M, Stockle CO, Nelson R, Higgins S, Ahmad S, Raza M (2019) Novel multimodal ensemble approach to evaluate the sole effect of elevated CO_2 on winter wheat productivity. Sci Rep 9:7813
Ahmed M, Ahmad S, Tariq M, Fatima Z, Aslam Z, Raza MA, Iqbal N, Akmal M, Fayyaz-ul-Hassan, Abbasi NA, Hayat R (2020) Wastes to be the source of nutrients and energy to mitigate climate change and ensure future sustainability: options and strategies. J Plant Nutr 43(6):896–920
Ahmed M, Hayat R, Ahmad M, Mahmood-ul-Hassan, Kheir AMS, Fayyaz-ul-Hassan, Rehman MH, Shaheen FA, Raza MA, Ahmad S (2022) Impact of climate change on dryland agricultural systems: a review of current status, potentials, and further work need. Int J Plant Prod 16:341–363

Ali H, Iqbal N, Ahmad S, Shahzad AN, Sarwar N (2013) Performance of late sown wheat crop under different planting geometries and irrigation regimes in arid climate. Soil Tillage Res 130:109–119

Amin A, Nasim W, Fahad S, Ali S, Ahmad S, Rasool A, Saleem N, Hammad HM, Sultana SR, Mubeen M, Bakhat HF, Ahmad N, Shah GM, Adnan M, Noor M, Basir A, Saud S, Rahman MH, Paz JO (2018a) Evaluation and analysis of temperature for historical (1996–2015) and projected (2030–2060) climates in Pakistan using SimCLIM climate model: ensemble application. Atmos Res 213:422–436

Amin A, Nasim W, Mubeen M, Ahmad A, Nadeem M, Urich P, Fahad S, Ahmad S, Wajid A, Tabassum F, Hammad HM, Sultana SR, Anwar S, Baloch SK, Wahid A, Wilkerson CJ, Hoogenboom G (2018b) Simulated CSM-CROPGRO-cotton yield under projected future climate by SimCLIM for southern Punjab, Pakistan. Agric Syst 167:213–222

Amin A, Nasim W, Mubeen M, Sarwar S, Urich P, Ahmad A, Wajid A, Khaliq T, Rasul F, Hammad HM, Rehmani MIA, Mubarak H, Mirza N, Wahid A, Ahmad S, Fahad S, Ullah A, Khan MN, Ameen A, Amanullah, Shahzad B, Saud S, Alharby H, Ata-Ul-Karim ST, Adnan M, Islam F, Ali QS (2018c) Regional climate assessment for temperature and precipitation in Southern Punjab (Pakistan) using SimCLIM climate model for different temporal scales. Theor Appl Climatol 131:121–131

Antle JM, Valdivia RO, Boote KJ, Janssen S, Jones JW, Porter CH, Rosenzweig C, Ruane AC, Thorburn PJ (2015) AgMIP's transdisciplinary agricultural systems approach to regional integrated assessment of climate impacts, vulnerability, and adaptation. In: Handbook of climate change and agroecosystems: the agricultural model intercomparison and improvement project integrated crop and economic assessments, part 1. World Scientific, London, pp 27–44

Asadi ME, Clemente RS (2003) Evaluation of CERES-Maize of DSSAT model to simulate nitrate leaching, yield, and soil moisture content under tropical conditions. Food Agric Environ 1(3–4):270–276

Baldwin RE (2006) Globalisation: the great unbundling (s). Economic Council of Finland

Ben-Asher J, Garcia y Garcia A, Hoogenboom G (2008) Effect of high temperature on photosynthesis and transpiration of sweet corn (Zea mays L. var. rugosa). Photosynthetica 46:595–603

Bhattacharyya R, Prakash V, Kundu S, Srivastva AK, Gupta HS, Mitra S (2009) Long term effects of fertilization on carbon and nitrogen sequestration and aggregate associated carbon and nitrogen in the Indian sub-Himalayas. Nutr Cycl Agroecosyst 86:1–16

Bordallo PN, Pereira MG, Amaral JRAT, Gabriel APC (2005) Análise dialélica de genótipos de milho doce e comum para caracteres agronômicos e proteína total. Hortic Bras 23(1):123–127. https://doi.org/10.1590/S0102-05362005000100026

Brennan EB, Acosta-Martinez V (2017) Cover cropping frequency is the main driver of soil microbial changes during six years of organic vegetable production. Soil Biol Biochem 109:188–204. https://doi.org/10.1016/j.soilbio.2017.01.014

Brown ME, Antle JM, Backlund P, Carr ER, Easterling WE, Walsh MK, Ammann C, Attavanich W, Barrett CB, Bellemare MF, Dancheck V, Funk C, Grace K, Ingram JSI, Jiang H, Maletta H, Mata T, Murray A, Ngugi M, Ojima D, O'Neill B, Tebaldi C (2015) Climate change, global food security, and the U.S. food system. https://doi.org/10.7930/J0862DC7

Cautin R, Augusti M (2005) Phenological growth stages of the cherimoya tree (Annona cherimola Mill). Sci Hortic 105:491–597

Challinor AJ, Wheeler TR, Craufurd PQ, Slingo JM (2005) Simulation of the impact of high temperature stress on annual crops yields. Agric For Meteorol 135:180–189

Chrispaul M, David MM, Joseph AO, Samuel VO (2010) Effective microorganism and their influence on growth and yield of pigweed (*Amaranthus dubius*). J Agric Biol Sci 5:17–22

CIDH (2007) Consultado el 01 de octubre de 2008. Página electrónica: www.cidh.org.mx

Congreves KA, Van Eerd LL (2015) Nitrogen cycling and management in intensive horticultural systems. Nutr Cycl Agroecosyst 102:299–318. https://doi.org/10.1007/s10705-015-9704-7

Deb P, Tran DA, Udmale PD (2016) Assessment of the impacts of climate change and brackish irrigation water on rice productivity and evaluation of adaptation measures in Ca Mau province. Vietnam. Theor Appl Climatol 125:641–656

Dhaliwal DS, Williams MM (2019) Optimum plant density for crowding stress tolerant processing sweet corn. PLoS One 14:e0223107

Dhaliwal DS, Williams MM (2022) Evidence of sweet corn yield losses from rising temperatures. Sci Rep 12(1):18218

Dhillon WS, Gill PPS (2015) Climate change and fruit production. Clim Dyn Hortic Sci 1:23–31

Eckstein D, Vera K, Laura S (2018) Global climate risk index 2018: who suffers most from extreme weather events? Weather-related loss events in 2016 and 1997 to 2016. Germanwatch Nord-Süd Initiative eV

Ejaz M, Abbas G, Fatima Z, Iqbal P, Raza MA, Kheir AMS, Ahmed M, Kakar KM, Ahmad S (2022) Modelling climate uncertainty and adaptations for soybean-based cropping system. Int J Plant Prod 16:235–250

Erickson AN, Markhart AH (2001) Flower production, fruit set and physiology of bell pepper during elevated temperature and vapor pressure deficit. J Am Soc Hortic Sci 126:697–702

FAO (2007) Informe Estadístico. Consultado el 01 de octubre de 2008. Página electrónica: www.faostat.fao.org

FAO (2017) Food and Agriculture Organization of the United Nations, Statistics Division. http://www.fao.org/statistics/en/

Farmers' Choice of Adaptation Methods and Perceptions of Climate Change in the Nile Basin of Ethiopia

Fatima Z, Abbas Q, Khan A, Hussain S, Ali MA, Abbas G, Younis H, Naz S, Ismail M, Shahzad MI, Nadeem M, Farooq U, Khan SU, Javed K, Khan AA, Ahmed M, Khan MA, Ahmad S (2018) Resource-use-efficiencies of C_3 and C_4 cereals under split nitrogen regimes. Agronomy 8(5):69

Fatima Z, Ahmad M, Hussain M, Abbas G, Ul-Allah S, Ahmad S, Ahmed N, Ali MA, Sarwar G, Haque E, Iqbal P, Hussain S (2020) The fingerprints of climate warming on cereal crops phenology and adaptation options. Sci Rep 10:18013

Fatima Z, Atique-ur-Rehman, Abbas G, Iqbal P, Zakir I, Khan MA, Kamal GM, Ahmed M, Ahmad S (2021) Quantification of climate warming and crop management impacts on phenology of pulses-based cropping systems. Int J Plant Prod 15:107–123

Fening JO, Ewusi-Mensah N, Safo EY (2010) Improving the fertilizer value of cattle manure for sustaining small holder crop production in Ghana. J Agron 9:92–101

Fernandes D S D F; Carmo O. M; Pinheiro M. L. H.; Noda H.; Manoares M. F. (2004) Diversidade Fenotípica em Pimenteiras Cultivadas na Amazônia. Associacao Brasileira de Hortiucltura. Anais CBO 2004. Página electrónica: http://www.abhorticultura.com.br/CBO/

Flowers TJ (2004) Improving crop salt tolerance. J Exp Bot 55:307–319

Foolad MR (2004) Recent advances in genetics of salt tolerance in tomato. Plant Cell Tissue Organ Culture 76:101–119

FUMIAF (2005) Cultivo de Pimiento en Invernaderos de Alta Tecnología en México. Consultado 01 de octubre de 2008. Página electrónica: www.fumiaf.sagarpa.com.mx

Hamma IL, Ibrahim U, Haruna M (2012) Effect of planting date and spacing on the growth and yield of sweet pepper (Capsicum annuum L.) in Samaru area of Zaria in Nigeria. Nigerian J Agric Food Environ 8:63–66

Hatfield JL, Prueger JH (2015) Temperature extremes: effect on plant growth and development. Weather Clim Extrem 10:4–10

Hidayati IN, Suryanto (2015) Pengaruh perubahan iklim terhadap produksi pertanian dan strategi adaptasi pada lahan rawan kekeringan. J Ekonomi dan Studi Pembangunan 16:42–52

Hoogenboom G, Jones JW, Porter CH, Wilkens PW, Boote KJ, Batchelor WD, Hunt LA, Tsuji GY (2003) Decision support system for agrotechnology transfer version 4.0: overview, vol 1. University of Hawaii, Honolulu

Hoogenboom G, Porter CH, Boote KJ, Shelia V, Wilkens PW, Singh U, White JW, Asseng S, Lizaso JI, Moreno LP, Pavan W, Ogoshi R, Hunt LA, Tsuji GY, Jones JW (2019) The DSSAT crop modeling ecosystem. In: Boote KJ (ed) Advances in crop modeling for a sustainable agriculture. Burleigh Dodds Science Publishing, Cambridge, pp 173–216. https://doi.org/10.19103/AS.2019.0061.10

Hoogenboom G, Porter CH, Shelia V, Boote KJ, Singh U, White JW, Pavan W, Oliveira FAA, Moreno-Cadena LP, Lizaso JI, Asseng S, Pequeno DNL, Kimball BA, Alderman PD, Thorp KR, Jones MR, Cuadra SV, Vianna MS, Villalobos FJ, Ferreira TB, Batchelor WD, Koo J, Hunt LA, Jones JW (2021) Decision Support System for Agrotechnology Transfer (DSSAT) Version 4.8 (DSSAT.net). DSSAT Foundation, Gainesville

Hussain F, Abid M (2011) Pest and diseases of Chilli crop in Pakistan: a review. Int J Biol Biotechnol 8:325–332

Hussain M, Ahmad S, Hussain S, Lal R, Ul-Allah S, Nawaz A (2018) Rice in saline soils: physiology, biochemistry, genetics and management. Adv Agron 148:231–287

Ibrahim HY, Omotesho OA (2013) Determinant of technical efficiency in vegetable production under Fadama in northern Guinea savannah, Nigeria. J Agric Technol 9(6):1367–1379

IPCC (2000) Summary for policymakers. In Emissions scenarios: a special report of IPCC Working Group III. [Nebojša N, Ogunlade Davidson, Gerald Davis, Arnulf Grübler, Tom Kram, Emilio Lebre La Rovere, Bert Metz, eds]. Cambridge University Press, Cambridge

Ishikawa K, Sasaki S, Matsufuji H, Nunomura O (2004) High β-Carotene and capsaicinoid contents in seedless fruits of 'Shishitoh' Pepper. Hort Sci 39(1):153–155

Jalu NN (2014) Response of hot pepper (Capsicum Annuum L.) to mulching and plant spacing at Bako, West Shoa zone, Ethoipia. MSc thesis, Haramaya University

Jones JW, Tsuji GY, Hoogenboom G, Hunt LA, Thornton PK, Wilkens PW, Imamura DT, Bowen WT, Singh U (1998) Decision support system for agrotechnology transfer; DSSAT v3. In: Tsuji GY, Hoogenboom G, Thornton PK (eds) Understanding options for agricultural production. Kluwer Academic Publishers, Dordrecht, pp 157–177

Jones JW, Hoogenboom G, Porter CH, Boote KJ, Batchelor WD, Hunt LA, Wilkens PW, Singh U, Gijsman AJ, Ritchie JT (2003) The DSSAT cropping system model. Eur J Agron 18(3–4):235–265

Keating BA, Carberry PS, Hammer GL, Probert ME, Robertson MJ, Holzworth D, Huth NI, Hargreaves JNG, Meinke H, Hochman Z, McLean G, Verburg K, Snow V, Dimes JP, Silburn M, Wang E, Brown S, Bristow KL, Asseng S, Chapman S, McCown RL, Freebairn DM, Smith CJ (2003) An overview of APSIM: a model designed for farming systems simulation. Eur J Agron 18:267–288

Khan A, Fahad S, Khan A, Saud S, Adnan M, Wahid F, Noor M, Nasim W, Hammad HM, Bakhat HF, Ahmad S, Rehman MH, Wang D, Sönmez O (2019) Managing tillage operations and manure to restore soil carbon stocks in wheat-maize cropping system. Agron J 111:1–10

Khan MN, Hussain M, Abbas G, Fatima Z, Iqbal P, Khan A, Zakir I, Ahmed M, Mateen A, Khan MA, Ahmad S (2020) Improving resource use efficiencies of sugarcane at farmer field under arid environment. Int J Agric Biol 24(5):1279–1285

Komla AN (2013) Effect of organic mulch on growth and yield of sweet pepper (Capsicum Annuum L): MPhil thesis, University of Ghana, Legon

Lal R (2019) Adaptation and mitigation of climate change by improving agriculture in India. In: Climate change and agriculture in India: impact and adaptation. Springer, Cham, pp 217–227

Lal M, Nozawa S, Emori H, Harasawa K, Takahashi M, Kimoto A, Abe-ouchi T, Nakajima T, Numaguti A (2014) Future climate change implications for Indian summer monsoon and its variability. Curr Sci 81:1196–1207

Lertrat K, Pulam T (2007) Breeding for increased sweetness in sweet corn. Int J Plant Breed 1(1):27–30

Li W, Zhang XL, Ashraf U, Mo ZW, Suo HC, Li GK (2017) Dynamics of seed germination, seedling growth and physiological responses of sweet corn under peg-induced water stress. Pak J Bot 49:639–646

Mahmood A, Ullah H, Shahzad AN, Ali H, Ahmad S, Zia-Ul-Haq M, Honermeier B, Hasanuzzaman M (2013) Dry matter yield and chemical composition of sorghum cultivars with varying planting density and sowing date. Sains Malaysiana 42(10):1529–1538

Manikharda M, Takahashi M, Arakaki M, Yonamine K, Hashimoto F, Takara K, Wada K (2018) Influence of fruit ripening on color, organic acid contents, capsaicinoids, aroma compounds and antioxidant capacity of Shimatogarashi (*Capsicum frutescens*). J Oleo Sci 67:113–123

Marshall SW, Tracy WF (2003) Sweet corn. In: Ramstad PE, White P (eds) Corn chemistry and technology, 2nd edn. American Association of Cereal Chemists, Minneapolis, pp 537–569
Masson-Delmotte V et al. (2021) AR6 climate change 2021: the physical science basis. In Contribution of Working Group I to the Sixth Assessment Report of the Intergovernmental Panel on Climate Change. IPCC, Geneva
Mubarak MK, Hussain K, Abbas G, Altaf MT, Baloch FS, Ahmad S (2022) Productivity of sorghum (*Sorghum bicolor* L.) at diverse irrigation regimes and sowing dates in semi-arid and arid environment. Turk J Agric For 46(1):1–18
Mubeen M, Ahmad A, Wajid A, Khaliq T, Hammad HM, Sultana SR, Ahmad S, Fahad S, Nasim W (2016) Application of CSM-CERES-Maize model in optimizing irrigated conditions. Outlook Agric 45(3):173–184
Mundarain S, Coa M, Canizares A (2005) Fenología del crecimiento y desarrollo de plántulas de aji dulce (Capsicum frutescens L.). Revista UDO Agrícola 5(1):62–67
Nadeem M, Khan MN, Abbas G, Fatima Z, Iqbal P, Ahmed M, Raza MA, Rehman A, Haq EU, Hayat A, Ali M, Ahmad S (2022) Application of CSM-CANEGRO model for climate change impact assessment and adaptation for sugarcane in semi-arid environment of Southern Punjab, Pakistan. Int J Plant Prod 16:443–466
Naik PS, Singh M, Ranjan JK (2017) Impact of climate change on vegetable production and adaptation measures. In: Minhas P, Rane J, Pasala R (eds) Abiotic stress management for resilient agriculture. Springer, Singapore USDA, Agricultural Statistics, U.S. Department of Agriculture, U.S. Government Printing Office, Washington, DC
Nasim W, Amin A, Fahad S, Awais M, Khan N, Mubeen M, Wahid A, Turan V, Rehman MH, Ihsan MZ, Ahmad S, Hussain S, Mian IA, Khan B, Jamal Y (2018) Future risk assessment by estimating historical heat wave trends with projected heat accumulation using SimCLIM climate model in Pakistan. Atmos Res 205:118–133
Naz S, Ahmad S, Abbas G, Fatima Z, Hussain S, Khan MA, Ahmed M, Khan A, Fahad S, Nasim W, Ercisli S, Wilkerson CJ, Hoogenboom G (2022) Modeling the impact of climate warming on potato phenology. Eur J Agron 132:126404
Nurdin (2011) Antisipasi perubahan iklim untuk keberlanjutan ketahanan pangan. Universitas Negeri Gorontalo, Gorontalo
Oktem A, Oktem A (2005) Effect of nitrogen and intra spaces on sweet corn (Zea mays Sachharata Sturt) ear characteristics. Indian J Plant Sci 4(4):361–363
Pagamas P, Nawata E (2007) Effect of high temperature during seed development on quality and chemical composition of chili pepper seeds. Jp J Trop Agr 51(1):22–29
Pajic Z, Rodosavlevic M, Eric U (2004) The utilizable value of sweet corn and popcorn hybrids. Agroznanje Agroknowl J 5(4):53–60
Rahman MH, Ahmad A, Wang X, Wajid A, Nasim W, Hussain M, Ahmad B, Ahmad I, Ali Z, Ishaque W, Awais M, Muddasir M, Shelia V, Ahmad S, Fahad S, Alam M, Ullah H, Hoogenboom G (2018) Multi-model projections of future climate and climate change impacts uncertainty assessment for cotton production in Pakistan. Agric For Meteorol 253–254:94–113
Rahman MH, Ahmad I, Wang D, Fahad S, Afzal M, Ghaffar A, Saddique Q, Khan MA, Saud S, Hassan S, Fahad M, Din M, Ahmad M, Yue Z, Shah RA, Sönmez O, Nasim W (2021) Influence of semi-arid environment on radiation use efficiency and other growth attributes of lentil crop. Environ Sci Pollut Res 28:13697–13711
Rashid M, Hussain K, Malik AA, Narayan S, Nazir G, Mazahir S (2020) Impact of climate change on vegetable crops and its mitigation. Int J Curr Microbiol App Sci 11:2429–2445
Rehman A, Ashfaq M, Naqvi SAA, Adil SA, Bashir K, Ahmad A, Ahmad S, Ali A, Imran A (2015) Is climate change worsening the poverty of maize growers? Evidence from Punjab province of Pakistan. Ciencia e Tecnica Vitivinicola 30(1):105–116
Rinaldi M, Ventrella D, Gagliano C (2007) Comparison of nitrogen and irrigation strategies in tomato using CROPGRO model: a case study from southern Italy. Agric Water Manage 87(1):91–105
Ritchie JT (1998) Soil water balance and plant water stress. In: Tsuji GY, Hoogenboom G, Thornton PK (eds) Understanding options for agricultural production. Kluwer Academic, Dordrecht, pp 41–54

Saha SR, Hossain MM, Rahman MM, Kuo CG, Abdullah S (2010) Effect of high temperature stress on the performance of twelve sweet pepper genotypes. Bangladesh J Agric Res 35(3):525–534

Sarwar N, Ali H, Maqsood M, Ullah E, Shahzad AN, Shahzad M, Mubeen K, Shahid MA, Ahmad S (2013) Phenological response of rice plants to different micronutrients application under water saving paddy fields on calcareous soil. Turk J Field Crops 18(1):52–57

Sarwar N, Atique-ur-Rehman, Farooq O, Wasaya A, Hussain M, El-Shehawi AM, Ahmad S, Brestic M, Mahmoud SF, Zivcak M, Farooq S (2021) Integrated nitrogen management improves productivity and economic returns of wheat-maize cropping system. J King Saud Univ Sci 33:101475

Searchinger T, Waite R, Hanson C, Ranganathan J, Dumas P (2018) World resources report: creating a sustainable food future. World Resources Institute, Washington, DC

Shabbir RN, Waraich EA, Ali H, Nawaz F, Ashraf MY, Ahmad R, Awan MI, Ahmad S, Irfan M, Hussain S, Ahmad Z (2016) Supplemental exogenous NPK application alters biochemical processes to improve yield and drought tolerance in wheat (*Triticum aestivum* L.). Environ Sci Pollut Res 23(3):2651–2662

Shi Y, Li YN, Zhang C, Bai MJ, Wang YK (2015) Development and application of decision support system for agro-technology transfer DSSAT under water resources management. Adv Mater Res 1073:1596–1603

Singh K (2004) Transformation of vegetable science in India looking back and ahead, Financing Agriculture. (Oct–Dec), pp 15–28

Soto-Ortiz R, Silvertooth JC (2008) A crop phenology model for irrigated New Mexico Chile (*Capsicum annuum* L.) type varieties. College of Agriculture and Life Sciences. The University of Arizona. The 2007 Vegetable Report, pp 104–112. Página electrónica: http://www.azrangelands.org/pubs/crops/az1438/

Sreedhara DS, Kerutagi MG, Basavaraja H, Kunnaland LB, Dodamani MT (2013) Economics of capsicum production under protected conditions in Northern Karnataka. Karnataka J Agric Sci 26:217–219

Tariq M, Yasmeen A, Ahmad S, Hussain N, Afzal MN, Hasanuzzaman M (2017) Shedding of fruiting structures in cotton: factors, compensation and prevention. Trop Subtrop Agroecosyst 20(2):251–262

Tariq M, Afzal MN, Muhammad D, Ahmad S, Shahzad AN, Kiran A, Wakeel A (2018a) Relationship of tissue potassium content with yield and fiber quality components of Bt cotton as influenced by potassium application methods. Field Crop Res 229:37–43

Tariq M, Ahmad S, Fahad S, Abbas G, Hussain S, Fatima Z, Wajid N, Mubeen M, Rehman M, Khan MA, Adnan M, Wilkerson CJ, Hoogenboom G (2018b) The impact of climate warming and crop management on phenology of sunflower-based cropping systems in Punjab, Pakistan. Agric For Meteorol 256–257:270–282

Tarriq M, Fatima Z, Iqbal P, Nahar K, Ahmad S, Hasanuzzaman M (2021) Sowing dates and cultivars mediated changes in phenology and yield traits of cotton-sunflower cropping system in arid environment. Int J Plant Prod 15:291–302

Tesfaw A (2013) Benefit-cost analysis of growing pepper: a trial at west Gojjam, near the source of blue Nile. Int J Agric Crop Sci 6:1203–1214

US Food and Drug Administration (2015) Bell and Chile peppers. 29 May 2019. https://www.wifss.ucdavis.edu/wp-content/uploads/2016/10/Peppers_PDF.pdf

USDA (2003) Processing Chile pepper pilot loss adjustment standards handbook 2004 and succeeding crop years. USDA, Washington, DC, pp 9–13

USDA Pakistan: grain and feed annual. GAIN Report Number: PK1707 (2017). https://www.fas.usda.gov/data/pakistan-grain-and-feed-annual-3. Accessed 24 Jan 2021

USDA/NASS (2011) Vegetables, 2010 summary. Agricultural Statistics Board, National Agricultural Statistics Service, US Department of Agriculture

Valero D, Zapata PJ, Martínez-Romero D, Guillén F, Castillo S, Serrano M (2014) Pre-harvest treatments of pepper plants with nitrophenolates increase crop yield and enhance nutritive and bioactive compounds in fruits at harvest and during storage. Food Sci Technol Int 20:265–274

Waha K, Van Bussel LGL, Müller C, Bondeau A (2012) Climate-driven simulation of global crop sowing dates. Glob Ecol Biogeogr 21:247–259

Waqas MM, Awan UK, Cheema MJM, Ahmad I, Ahmad M, Ali S, Shah SHH, Bakhsh A, Iqbal M (2019) Estimation of canal water deficit using satellite remote sensing and GIS: a case study in lower Chenab canal system. J Indian Soc Remote Sens 47:1153–1162

WHO (2003) Fruit and vegetable promotion initiative: a meeting report. WHO, Geneva

Wolf J, Ouattara K, Supit I (2015) Sowing rules for estimating rainfed yield potential of sorghum and maize in Burkina Faso. Agric For Meteorol 214:208–218

Wolfe K, Stubbs K (2018) 2017 Georgia farm gate value report. The University of Georgia, Center for Agribusiness and Economic Development Publication AR-18-01

Xiang N, Li CY, Li GK, Yu YT, Hu JG, Guo XB (2019) Comparative evaluation on vitamin E and carotenoid accumulation in sweet corn (*Zea mays* L.) seedlings under temperature stress. J Agric Food Chem 67:9772–9781

Zahoor A, Ahmed M, Fayyaz-ul-Hassan, Shabbir G, Ahmad S (2022) Ontogeny growth and radiation use efficiency of canola (*Brassica napus* L.) under various nitrogen management strategies and contrasting environments. Int J Plant Prod 16(1):195–208

C_4 Cereal-Based Fodder Systems 12

12.1 Introduction

Maize, millet, and sorghum cropping system is a vast fodder system, comprising various climates and rainfed environments including Asia, America, Africa, and Australia (FAO STAT 2021). Forage is an essential sector of agriculture, because it ensures the accessibility of good feed to the animals. Changes in the climate are a main source for several environmental pressures on forage crops due to which the forage production is adversely affected (Ziervogel et al. 2006). The climate change has a great impact on the reliability and quality, requirement of water for the cultivation of fodder crop, and vegetation pattern in rangeland (Giridhar and Samireddypalle 2015). Each fodder crop, maize, millet, and sorghum, requires different environments for its production. So, a minute change in the climate causes a huge effect on the production of these fodder crops (Abbas et al. 2017; Abbas et al. 2020a, b, c; Abbas et al. 2019; Abbas and Ahmad 2018; Ahmad et al. 2019, 2021; Ahmad et al. 2015a, b, c; Ahmad et al. 2017a, b; Ahmad et al. 2013, 2018; Ahmad et al. 2012a, b; Ahmad et al. 2023a, b; Ahmad & Hasanuzzaman (2012).

Maize (*Zea mays* L.) is a C_4 crop which belongs to Poaceae family. It is an important cereal as well as fodder crop which is cultivated all over the world. It was domesticated in Mesoamerica/Southern Mexico about 9000 years ago. Globally cultivated area for the maize (dry grain) is about 197 Mha including main areas of Asia, Latin America, and sub-Saharan Africa, making it the second most sown crop in the world following wheat (FAO STAT 2021). Its grain as well as biomass are also widely used as animal feed. Maize green fodder is an energy-rich feed for ruminant animals, especially when it incorporates the stalks, leaves, and ears. While maize fodder is often ensiled in cooler climates, year-round maize cultivation in the tropics may allow for continual collection of green feed, eliminating the need for ensiling (Brewbaker 2003). In times of scarcity and harsh environment (hot and dry summer and winter), grazing entire maize plants gives green feed to cattle (Potter 2012; Newport 2006). Maize green forage is a good source of fodder for smallholder-owned animals in locations where circumstances are severe and forage is limited

M. Ahmed et al., *Cropping Systems Modeling Under Changing Climate*,
https://doi.org/10.1007/978-981-97-0331-9_12

(Methu et al. 2006). Maize is a high-energy meal that outperforms most other tropical fodder crops, whose dry matter is frequently less than 40% digestible. Besides maize, mature crop baby corn fodder and husk with silk are also a source of animal feed (Ahmed et al. 2018a, b; Ahmed et al. 2016, 2019, 2020, 2022; Ali et al. 2013; Amin et al. 2018a, b, c; Andrews and Kumar 1992; Ejaz et al. 2022; Fatima et al. 2018, 2020, 2021).

Millet (*Panicum miliaceum* L.) is also another important C_4 crop of maize, millet, and sorghum system. It is cultivated in irrigated as well as in rainfed areas. It requires hot and dry environment to grow well, although it can be cultivated in different environments like arid and semiarid. Both medium- and high-fertile soils are favorable for its production with minimum input requirements. It is used as both cereal and fodder crop. It is considered as a major crop in some parts of Asia, China, Africa, the USA, and the Soviet Union; especially in Brazil, central Asia, and the Middle East, it has got attention as a multi-cut fodder crop as both fresh and silage production (Rai et al. 2012; Jukanti et al. 2016). In systems with sorghum, maize, or frequent combination of legume products like cowpea, it is usually cultivated in rainfed (dry) regions in Africa (Mason et al. 2015). In the semiarid Southern Great Plains of the USA, pearl millet is also acquiring popularity as a fodder crop that could be substituted for summer fallow in sorghum-winter wheat animal feed cropping systems (Ismail 2012). With respect to production, India occupies the first position with 39% production of the world's millet. Because of their high nutritional value and ability to fit into crop rotations with brief growing seasons and scarce water supplies, pearl and other millets, such as foxtail, proso, and Japanese, are frequently grown for livestock feed in semiarid regions (Baumhardt and Salinas-Garcia 2006; Ismail 2012). Millet fodder comprises 0.33% crude fiber and 1.5% protein (Amanullah and Khan 2004). Due to its nutritious and palatable quality, its fresh green plant is used as animal fodder in summer and hay in winter, and the grain can be used for poultry feed (Amanullah and Khan 2004).

Sorghum (*Sorghum bicolor* (L.) Moench) also belongs to C_4 family. It is the fifth most significant crop in the world. Globally, its production is 64.58 M tonnes (FAO 2008). It is used as a raw substance to make food and animal feed, along with numerous commercial goods (Folliard et al. 2004; Fosu-Mensah 2013; Hoogenboom et al. 2019; Hussain et al. 2018; Indriani et al. 2015; Keating et al. 2003; Khan et al. 2019, 2020; Khanthavong et al. 2021; Klopfensteina et al. 2013; Ma et al. 2006; Mahmood et al. 2013; Matthews and Pilbeam 2005). Recently, it has gained special importance, with different uses, i.e., formation of bio-industrial commodities (e.g., bioplastics), beverages, and gluten-free goods for celiacs (Taylor et al. 2006). Sorghum growth is favored in warmer, tropical, semiarid areas of the globe with higher temperatures along with water stress because of its ability to withstand drought (Peacock and Heinrich 1984). The ability to tolerate water stress is more in forage sorghum as compared to the grain sorghum due to the difference in the genotypes of grain and forage varieties. Forage sorghum shows better water-use efficiency due to its no requirement of water at its critical stages of growth like the grain sorghum at its grain filling stage (Tack et al. 2017; Unger 1988). Forage sorghum also has the ability to produce a better level of heat-shock proteins, which helps it to

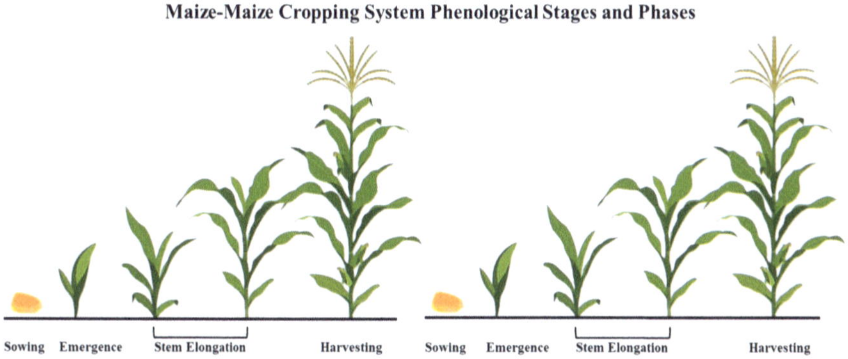

Fig. 12.1 Pictorial overview of maize-maize fodder system phenology

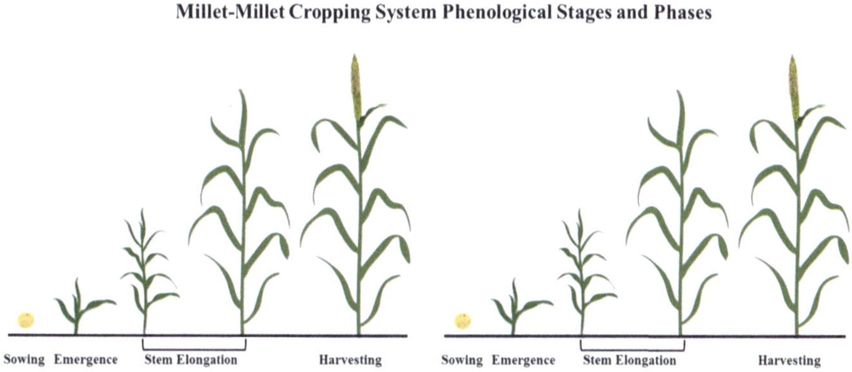

Fig. 12.2 Pictorial overview of millet-millet fodder system phenology

survive in severe heat stress (Pavli et al. 2011). Sorghum is widely produced in developing countries of Africa and Asia, where it is primarily cultivated by farmers with few resources and outdated technology. These nations produce greater than 70% of the world's sorghum. The pictorial views of maize-maize, millet-millet, sorghum-sorghum, maize-millet, maize-sorghum, and millet-sorghum are presented in Figs. 12.1, 12.2, 12.3, 12.4, 12.5, and 12.6, respectively (Mubarak et al. 2022; Mubeen et al. 2016; Nadeem et al. 2022; Nasim et al. 2018; Naz et al. 2022; Parent et al. 2008; Rahman et al. 2018, 2021; Sarwar et al. 2013, 2021; Saseendran et al. 2005; Shabbir et al. 2016; Sharad and Kumar 2012; Srichana et al. 2014; Tariq et al. 2018a, b; Tariq et al. 2017, 2021; Zahoor et al. 2022).

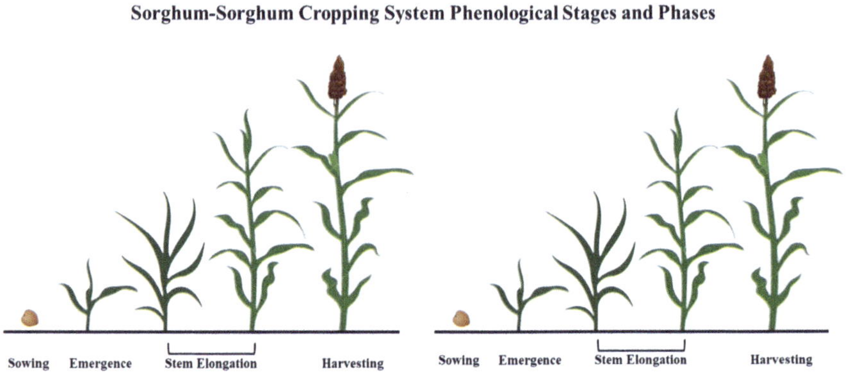

Fig. 12.3 Pictorial overview of sorghum-sorghum fodder system phenology

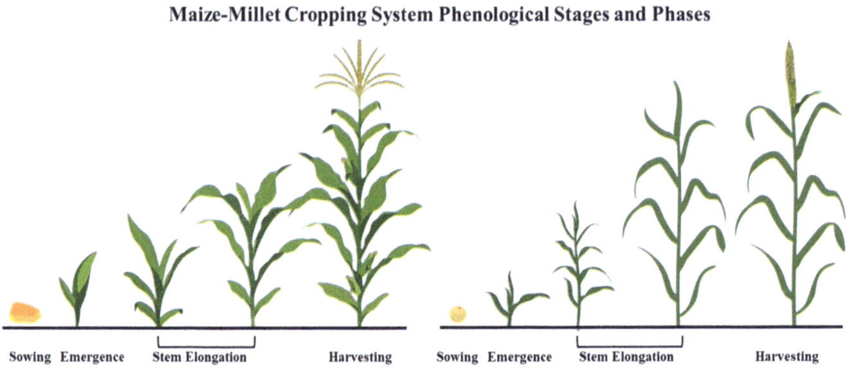

Fig. 12.4 Pictorial overview of maize-millet fodder system phenology

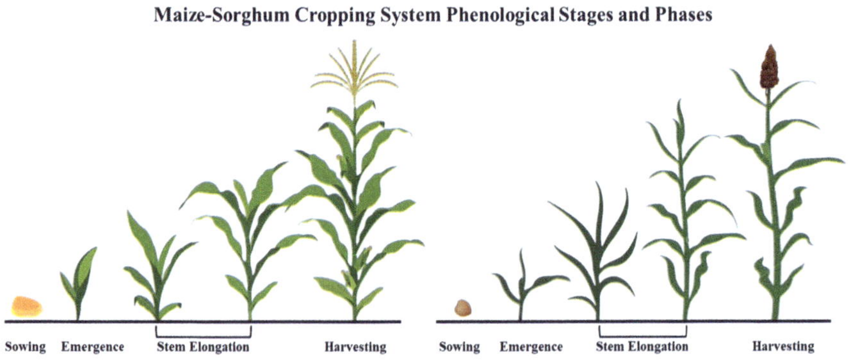

Fig. 12.5 Pictorial overview of maize-sorghum fodder system phenology

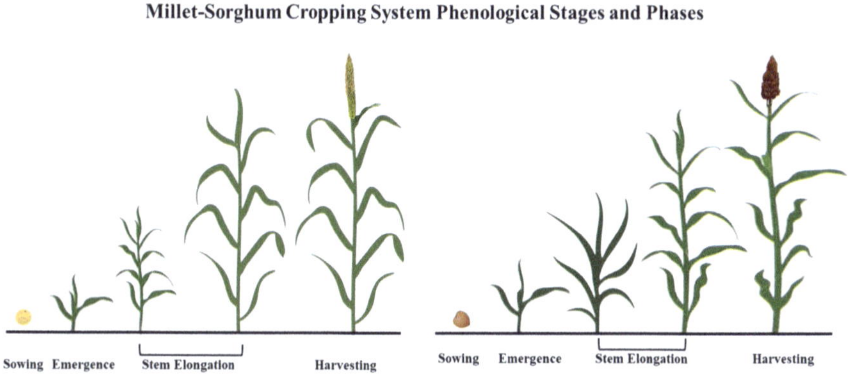

Fig. 12.6 Pictorial overview of millet-sorghum fodder system phenology

12.2 Issues of C_4 Summer Cereal-Based Fodder Cropping Systems

Globally, the climate change and weather disasters are the major issues of yield reduction in maize, millet, and sorghum system (Giorgi et al. 2019; Berg et al. 2013). Climate change-related environmental challenges such as drought, floods, and temperature fluctuation directly influence the morphology and physiology of these crops under rainfed circumstances in tropical and subtropical regions, which can result in crop failure (Ahmed and Ahmad, 2023; Ahmed 2023; Abbas et al. 2023b; Liu et al. 2023; Ahmed et al. 2022a; Ahmed et al. 2022b; Nadeem et al. 2022; Ahmed et al. 2020; Khan et al. 2020; Asseng et al. 2019; Ahmed et al. 2017; Jabeen et al. 2017; Aslam et al. 2013; Ahmed et al. 2012). It is generally known how environmental stress, such as water stress and temperature, affects crop output (Lamaoui et al. 2018; Maulana and Tesso 2013; Matsuura et al. 2016; Wang et al. 2018). Waterlogging damages more than one-third of the irrigated land on the planet. The lack of oxygen in the soil caused by ongoing flooding prevents developing roots and other living things from respiring and changes the chemical properties of the soil. Maize, millet, and sorghum are sensitive to waterlogging conditions, failing to transfer oxygen from leaves to root cells and blocking gas exchange and root respiration. Waterlogged soil is a reason to reduce stomatal conductance and respiration rate in maize (Tian et al. 2019), millet (Matsuura et al. 2016), and sorghum (Zhang et al. 2019; Promkhambu et al. 2010). So, the waterlogged soil reduces the growth and yield in maize and millet along with sorghum (Matsuura et al. 2016; Tian et al. 2019).

Temperature is another factor which controls the yield of biomass along with growth period (Abbas et al. 2023a, b; Zakir et al. 2023). Accumulation of biomass is influenced by the temperatures which disturb the photosynthesis. The enzyme Rubisco, which is in charge of plants' ability to assimilate CO_2, experiences a gross loss in carbon assimilation at temperature and dew the optimal level because of reductions in its activation state, specificity for CO_2, along with solubility of CO_2 in relation to O_2 (Jordan and Ogren 1984; Brooks and Farquhar 1985). In maize, millet, and sorghum cropping system, all crops are C_4 whose optimum temperature for photosynthesis is higher than that of C_3 crops. Range for temperature for germination of seed is 21–35 °C, for the reproductive growth is 25–28 °C, and for the vegetative development and growth is 26–34 °C (Maiti 1996). Rise in temperature affects the biomass yield in maize, millet, and sorghum system as it affects the growth duration or time period. High temperature is a reason for faster completion of the growing degree days, which causes early growth and development, resulting in decrease of phenophase period and ultimately reduction in biomass yield (Attri and Rathore 2003).

12.3 Decision Support System for Agrotechnology Transfer

Crop models are excessively employed to assess the impact of climate change upon different yields including maize (Yano et al. 2007), millet, and sorghum. The most prominent software used for crop modeling is DSSAT (Decision Support System for Agrotechnology Transmission), and it is used in many different nations. It currently includes over 42 different types of products, such as cereals, grains, grain beans, as well as root crops (Hoogenboom 2003). The first software using weather modeling makers is DSSAT. It is methodical and created to function regardless of location, season, agricultural variety, or management system. It can simulate how soil and agricultural nitrogen dynamics, genotype, soil water, and meteorological conditions affect crop development and harvests (Jones et al. 2003a, b). In just a few minutes, DSSAT combines soil, crop trait, weather, and managing choice effects. One of the models in the DSSAT, the sorghum, maize, and millet models, includes significant elements for reproductive and vegetative growth, C balance, and W balance, along with N balance (Singh and Virmani 1996). Utilizing day-to-day steps from crop planting to harvest, the model is able to replicate agricultural growth, development, and yield. Genetic coefficients that are inputs for the model and are cultivar-specific factors affect how differently crop cultivars mature, develop, and produce. The model can replicate physiological processes that explain how a crop reacts to environmental conditions like temperature, precipitation, and sun radiation as well as how the properties of the soil affect the amount of water that a crop can access to develop. CERES model for maize, millet, and sorghum combined with DSSAT (Hoogenboom et al. 2021) was used to assess the impact of climate change on the maize, millet, and sorghum system. It is a physio-biological model that is centered on crop management techniques and uses the concepts of C, N, water, and E balance for simulating the growth of millet, sorghum, and maize products for fodder purposes (Akponikpè et al. 2010; Waha et al. 2012). Crop management practices, which include crop plant genetic traits; cultivation times; phosphorus, N, and K rates; methods; ecology aspects, which include physical and chemical properties of soil; and weather conditions, which include daily minimum and maximum temperatures, rainfall, and radiation, were among the inputs required for model executions. The ultimate cereal fodder outputs are determined on harvest day, and models calculate development and growth stages as well as periods of agricultural crops on everyday stage (Murty et al. 2007; Mubeen et al. 2013). The genetic

Table 12.1 Maize genetic coefficients along with description in DSSAT model

Genetic coefficient	Description
VAR#	Identification code or number for a specific cultivar
VAR-NAME	Name of cultivar
EXPNO	Number of experiments used to estimate cultivar parameters
P1	Thermal time from seedling emergence to the end of the juvenile phase (expressed in degree days above a base temperature of 8 °C) during which the plant is not responsive to changes in photoperiod
P2	Extent to which development (expressed as days) is delayed for each hour increase in photoperiod above the longest photoperiod at which development proceeds at a maximum rate (which is considered to be 12.5 h)
P5	Thermal time from silking to physiological maturity (expressed in degree days above a base temperature of 8 °C)
G2	Maximum possible number of kernels per plant
G3	Kernel filling rate during the linear grain filling stage and under optimum conditions (mg/day)
PHINT	Phyllochron interval; the interval in thermal time (degree days) between successive leaf tip appearances

Source: DSSAT v4.8; Hoogenboom et al. (2021)

Table 12.2 Maize genetic coefficients along with description in APSIM model

Genetic coefficients	Description
tt_emerg_to_endjuv	Duration from emergence to the end of juvenile °C
est_days_endjuv_to_init	Duration—end of juvenile to flowering initiation °C day
tt_flag_to_flower	Duration—flag leaf to flowering stage °C day
Photoperiod crit_1	Hour—photoperiod critical 1 (h)
Photoperiod crit_2	Hour—photoperiod critical 2 (h)
tt_flower_to_start_grain	Duration—flowering to the start of grain filling °C day
tt_flower_to_maturity	Duration, flowering to maturity °C day
tt_maturity_to_ripe	Duration—maturity to seed ripening °C day
head_grain_no_max	Grain maximum number per head
grain_gth_rate	Grain growth rate mg/day
t_base	Base temperature °C day

Sources: Beah et al. (2021)

coefficients for DSSAT and APSIM models for maize (Tables 12.1 and 12.2), millet (Tables 12.3 and 12.4), and sorghum (Tables 12.5 and 12.6) crops are presented in respective tables.

To study the influence of climate upon maize, millet, and sorghum cropping system, except CERES-Maize, Millet, and Sorghum, various models are used including the APSIM, AquaCrop, LINTUL5, InfoCrop, SWAP, and Epic model. APSIM is a crop model which allows modules to be connected in order to simulate system productivity (McCown et al. 1996). The components in APSIM are divided into biophysical, soil, management, and economic categories (Keating et al. 2003).

Table 12.3 Millet genetic coefficients along with description in DSSAT model

Genetic coefficient	Description
VAR#	Identification code or number for a specific cultivar
VAR-NAME	Name of cultivar
EXPNO	Number of experiments used to estimate cultivar parameters
ECO#	Ecotype code for this cultivar, points to the Ecotype in the ecotype file
P1	Thermal time from seedling emergence to the end of the juvenile phase (expressed in degree days above a base temperature of 10 °C) during which the plant is not responsive to changes in photoperiod
P2O	Critical photoperiod or the longest day length (in hours) at which development occurs at a maximum rate. At values greater than P2O, the rate of development is reduced
P2R	Extent to which phasic development leading to panicle initiation (expressed in degree days) is delayed for each hour increase in photoperiod above P2O
P5	Thermal time (degree days above a base temperature of 10 °C) from beginning of grain filling (3–4 days after flowering) to physiological maturity
G1	Scaler for relative leaf size
G4	Scaler for partitioning of assimilates to the panicle (head)
PHINT	Phyllochron interval; the interval in thermal time (degree days) between successive leaf tip appearances
GT	Tillering coefficient, equivalent to G1, but on tillers
G5	Potential grain size, mg

Source: DSSAT v4.8; Hoogenboom et al. (2021)

Table 12.4 Millet genetic coefficients along with description in APSIM model

Genetic coefficients	Description
hi_incr	Rate of HI increase (optional) (1/days)
hi_max_pot	Maximum harvest index potential
grain_gth_rate	PKC95; potential grain growth rate (mg/grain/day)
tt_emerg_to_endjuv	EJvO data—TT from emergence to the end of juvenile phase (°Cd)
est_days_emerg_to_init	PSC data—estimated days from emergence to floral initiation (day)
pp_endjuv_to_init	EJvO data—photoperiod sensitivity (°Cd/h)
tt_flower_to_maturity	EJvO PKC96—TT from flowering to maturity °Cd
tt_flag_to_flower	EJvO—TT from flag leaf to flowering °Cd
tt_flower_to_start_grain	Bieler—TT from flowering to the start of grain fill °Cd
tt_maturity_to_ripe	TT from maturity to harvest ripe °Cd
y0_const	EJVO 4 plants/m^2 $n = 26$, adj $r^2 = 0.64$ largest leaf area intercept
y0_slope	EJVO 4 plants/m^2 $n = 26$, adj $r^2 = 0.64$ largest leaf area slope
x_stem_wt	Look-up table for canopy height g/stem
y_height	Plant canopy height mm
leaf_no_dead_slope2	EJVO data; dead leaf no slope 2 (after flag leaf)

Table 12.5 Sorghum genetic coefficients along with description in DSSAT model

Genetic coefficient	Description
VAR#	Identification code or number for a specific cultivar
VAR-NAME	Name of cultivar
EXPNO	Number of experiments used to estimate cultivar parameters
ECO#	Ecotype code for this cultivar, points to the ecotype in the ecotype file
P1	Thermal time from seedling emergence to the end of the juvenile phase (expressed in degree days above TBASE during which the plant is not responsive to changes in photoperiod)
P2	Thermal time from the end of the juvenile stage to tassel initiation under short days (degree days above TBASE)
P2O	Critical photoperiod or the longest day length (in hours) at which development occurs at a maximum rate. At values higher than P2O, the rate of development is reduced
P2R	Extent to which phasic development leading to panicle initiation (expressed in degree days) is delayed for each hour increase in photoperiod above P2O
PANTH	Thermal time from the end of tassel initiation to anthesis (degree days above TBASE)
P3	Thermal time from the end of flag leaf expansion to anthesis (degree days above TBASE)
P4	Thermal time from anthesis to the beginning of grain filling (degree days above TBASE)
P5	Thermal time from the beginning of grain filling to physiological maturity (degree days above TBASE)
PHINT	Phyllochron interval; the interval in thermal time between successive leaf tip appearances (degree days)
G1	Scaler for relative leaf size
G2	Scaler for partitioning of assimilates to the panicle (head)
PSAT	Critical photoperiod below which development is not delayed (optional)
PBASE	Ceiling photoperiod above which development is delayed indefinitely (optional)

Source: DSSAT v4.8; Hoogenboom et al. (2021)

Table 12.6 Sorghum genetic coefficients along with description in APSIM model

Genetic coefficients	Description
tt_emerg_to_endjuv	Thermal time accumulation from seedling emergence to the end of juvenile phase (°C days)
tt_flower_to_maturity	Thermal time accumulation from flowering to maturity (°C days)
tt_flag_to_flower	Thermal time accumulation from flag stage to flowering (°C days)
tt_flower_to_start_grain	Thermal time accumulation from flowering to the start of grain filling (°C days)
photo_crit 1	Critical photoperiod 1
photo_crit 2	Critical photoperiod 2
photo_slope	The extent to which growth is affected by photoperiod increases beyond photo_crit 1 and 2

Source: MacCarthy et al. (2009)

When considering cropping systems, it simulates mechanistic crop growth and soil processes, along with a variety of managing options. The APSIM model was created for semiarid areas which have been widely applied to smallholder agricultural systems there. It has been established that APSIM is a useful instrument for analyzing potential climate change influences on production (Keating et al. 2003; Chen et al. 2010; Wang et al. 2010). Prior to running one simulation with APSIM, the model must first be set up by deciding which module will be used in prediction and which sets will be required by the module. The initialization datasets are typically divided into two categories: exact simulation variable data, such as cultivar, husbandry, and location features, or basic data (McCown et al. 1996). The calibration and evaluation for maize, millet, and sorghum crops for DSSAT model are presented in Tables 12.7 and 12.8. The observed and simulated LAI and biomass for maize, millet, and sorghum are presented in Figs. 12.7, 12.8, and 12.9, respectively.Site parameters that are typical include soil parameters for soil modules, such as bulk density, soil characteristics, day-to-day basis weather variables (temperatures, solar radiation, rainfall), and data used in management like application time for agronomic practices, along with the kind and quantity of fertilizers for managing the module (Keating et al. 2003; McCown et al. 1996).

12.4 Climate Change and Temperature Trends

Due to the climate change, the temperature of the earth is changing day by day (Ahmad et al. 2016c). The twenty-first century is predicted to see a 1.8–4.0 C rise in surface air temperature, very likely numerous warm spells, heat waves, and heavy rains, as well as a probable rise in drought frequency, according to the IPCC (2007). The world's most susceptible areas due to the danger of climate change on agricultural output are tropics, especially semiarid areas where rising temperatures along with increased precipitation variation might have severe adverse effects (Parry et al. 2004). However, the most affected countries during 2019 were Zimbabwe, Mozambique, and Bahamas. In Bangladesh, from 1895 to 1980, temperature rise at a rate of 0.31 °C was observed annually. By using the previous data of 1961–1990, they predicted that the temperature will surge by 0.4 and 0.73 °C annually for 2050 and 2100, respectively. Due to climate change, the same positive trend was observed in Iran (1961–2010), Italy (1961–2004), and India (Toreti and Desiato 2008).

A new trend has also been discovered known as daily temperature variation. This term describes sharp variation in average temperatures amid two successive days. This change has been observed in many countries including the USA (Ebi et al. 2004), Japan, China (Cheng et al. 2014), Korea (Kim et al. 2016), Australia (Guo et al. 2011), and the Czech Republic (Plavcová and Kyselý 2010) and caused a huge damage to living things including crops. In December 2016, a huge rapid decrease of 20 °C in the day-to-day temperature was observed in the USA, whereas after a few days, a huge increase of greater than 20 °C in the day-to-day temperature was observed in Russia (Zhou et al. 2020).

Table 12.7 Calibration of DSSAT model for C_4 cereals at variable sowing dates for fodder and grain purpose crops under irrigated arid environment of Multan, Pakistan

Fodder crops						
	Maize		Millet		Sorghum	
	(DAS)		(DAS)		(DAS)	
Crops and characteristics	Obs.	Sim.	Obs.	Sim.	Obs.	Sim.
Phenology						
Anthesis	66	67	54	55	60	61
Growth	RMSE	d-value	RMSE	d-value	RMSE	d-value
LAI (m^2 m^{-2})	0.32	0.99	0.11	0.99	0.26	0.97
Biomass (kg ha^{-1})	420.25	0.99	298.04	0.99	268.36	0.99
Grain crops						
Characteristics	*(DAS)*		*(DAS)*		*(DAS)*	
	Obs.	*Sim.*	*Obs.*	*Sim.*	*Obs.*	*Sim.*
Phenology						
Anthesis	55	56	51	51	56	57
Maturity	99	100	95	96	111	112
Growth	RMSE	d-value	RMSE	d-value	RMSE	d-value
LAI (m^2 m^{-2})	0.25	0.98	0.25	0.99	0.21	0.96
Biomass (kg ha^{-1})	1169.91	0.99	801.91	0.99	396.29	0.99

Source: Ahmad et al. (2016a, b)
DAS days after sowing, *Obs.* observed, *Sim* simulated, *RMSE* root mean square error, *LAI* leaf area index

Table 12.8 Evaluation of DSSAT model for C_4 cereals at variable sowing dates for fodder and grain purpose crops under irrigated arid environment of Multan, Pakistan

Fodder crops						
	Maize		Millet		Sorghum	
	(DAS)		(DAS)		(DAS)	
Crops and characteristics	Obs.	Sim.	Obs.	Sim.	Obs.	Sim.
Phenology						
Anthesis	67	68	56	58	58	60
Growth	RMSE	d-value	RMSE	d-value	RMSE	d-value
LAI (m^2 m^{-2})	0.20	0.98	0.14	0.98	0.11	0.98
Biomass (kg ha^{-1})	617.96	0.95	216.76	0.97	239.85	0.95
Grain crops						
Characteristics	*(DAS)*		*(DAS)*		*(DAS)*	
	Obs.	*Sim.*	*Obs.*	*Sim.*	*Obs.*	*Sim.*
Phenology						
Anthesis	54	55	52	54	56	57
Maturity	104	106	92	93	103	104
Growth	RMSE	d-value	RMSE	d-value	RMSE	d-value
LAI (m^2 m^{-2})	0.12	0.98	0.17	0.99	0.19	0.98
Biomass (kg ha^{-1})	1180.21	0.94	753.43	0.97	298.36	0.95

Source: Ahmad et al. (2016a, b)*DAS* days after sowing, *Obs.* observed, *Sim* simulated, *RMSE* root mean square error, *LAI* leaf area index

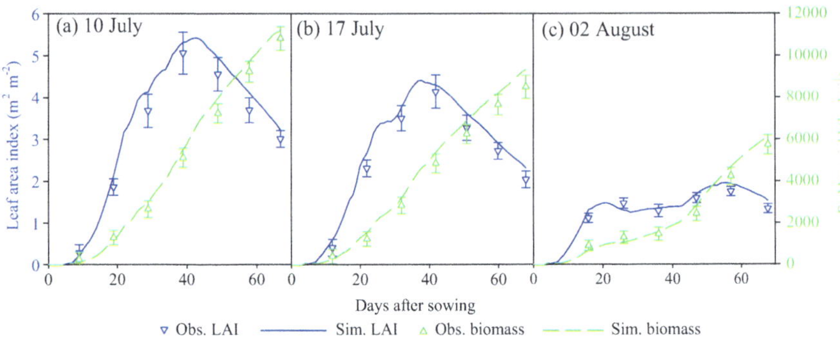

Fig. 12.7 Observed and simulated leaf area index (LAI) and biomass of fodder maize crop as effected by sowing dates under irrigated semiarid environment. (Source: Modified and adopted from Ahmad et al. 2016c)

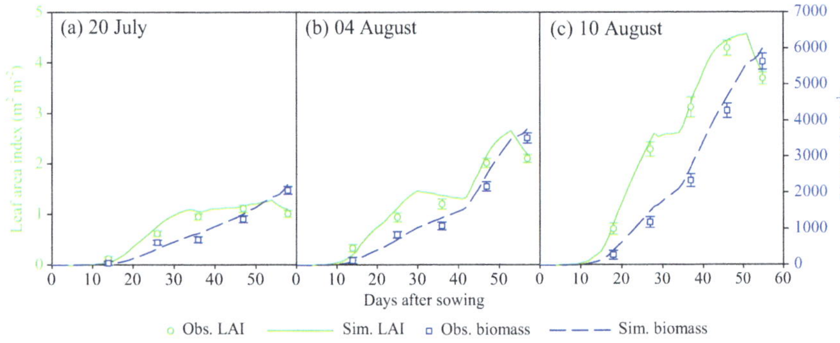

Fig. 12.8 Observed and simulated leaf area index (LAI) and biomass of millet fodder crop as effected by sowing dates under irrigated semiarid environment. (Source: Modified and adopted from Ahmad et al. 2016a, b)

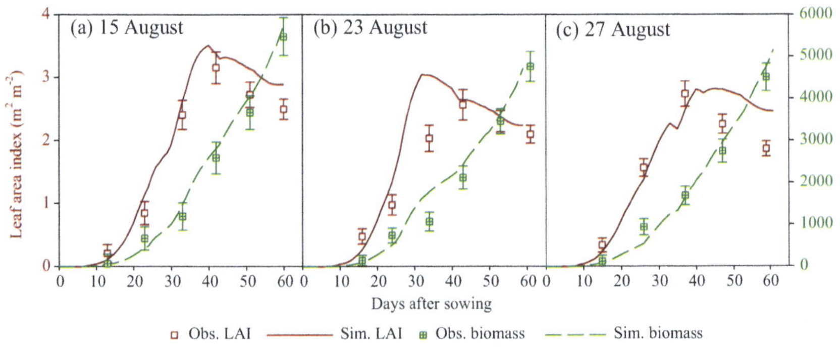

Fig. 12.9 Observed and simulated leaf area index (LAI) and biomass of sorghum fodder crop as effected by sowing dates under irrigated semiarid environment. (Source: Modified and adopted from Ahmad et al. 2016a, b)

12.5 Climate Change Impact on Maize, Millet, and Sorghum Biomass Production on Current and Future Scenarios

Climate change significantly affects the biomass yield of crops in the maize, millet, and sorghum cropping system. Climate change-related environmental stressors like drought, flooding, and temperature variability directly impact the morphology and physiology of these crops under rainfed environments in tropical and subtropical areas, which can result in crop failure. It is well known how environmental stress, such as water stress and weather, affects crop yields (Wang et al. 2018; Matsuura et al. 2016; Maulana and Tesso 2013; Lamaoui et al. 2018). The increase in atmospheric CO_2 content, upsurge in temperatures, and variations in precipitations are the main ways that climate change affects vegetation. Temperature and rainfall pattern have been changed due to change in climate. Heat waves, drought, heavy rainfall, and high concentration in ozone cause heavy damage to crops (Gömann et al. 2017).

The primary way that temperature affects biomass buildup is through photosynthesis. The enzyme Rubisco, which is responsible for plants' ability to assimilate CO_2, experiences a net loss in carbon assimilation at temperature above the optimal level because of reduction in its activation state and CO_2 specificity along with CO_2 solubility in relation to oxygen (Holiday et al. 1992). Some regions became favorable for high yield of maize, and some regions faced heavy reduction of forage yield (Maitah et al. 2021). Maize is very sensitive to the damage by frost, and the damage caused due to frost has been observed in areas where the temperature is increasing due to the climate change (Sánchez et al. 2014; Crowley 1998). For the photosynthesis, C_4 crops have more optimum temperature compared to the C_3 crops. Temperature range for maize growth is between 6.2 and 42 °C (Yin et al. 1995). Extreme reduction in temperature to below −2.2 °C just for a few minutes as well as below 0 °C just for a few hours causes huge damage to the stem, ear, and leaves of maize plant (Carter and Hesterman 1990). Similarly, high temperature beyond limit causes irreversible damage in maize plant. According to Birch et al. (1998) and Sinsawat et al. (2004), 44 °C and 45 °C cause lethal damage in the tissues of maize plant, which results in the loss of forage yield. The root growth also retards with variation in temperature. Hund et al. (2008) indicated that when the temperature is reduced by 2 °C from 17 to 15 °C, the root growth is significantly reduced. Precipitation is another factor affecting the growth of fodder maize. Climate change not only altered the temperature but also changed the precipitation pattern with respect to time and space (Jacob et al. 2017). In Europe, the summer season has more temperature and low precipitation rate, which adversely affects the maize crop (Maitah et al. 2021). Samaniego et al. (2018) stated that in Germany, 3 °C increase in temperature resulted in approximately 50% increase in the time of drought. Below 30%, low-humidity conditions cause irreversible dryness in the plant. A negative trend was also observed by Knox et al. (2016) in Germany in forage yield by increasing 3 °C temperature and decreasing the precipitation rate. To check the climate change influence on forage maize, Zizinga et al. (2022) applied AquaCrop model in Uganda. They applied climate scenario RCPs 4.5 and 8.5 using baseline

(1980–2009), current scenario (2019–2020), and future scenario (2019–2039). They explored that in Uganda, due to climate change during the future scenario, maize fodder yield will be reduced by 11% in RCP 4.5 and 5.6% in RCP 8.5. In Ireland, Holden and Brereton used DSSAT model. They predicted that in future, in western, north, and southeast areas, yield will increase by 70%, 40%, and 25%, respectively, in 2055. In Ghana environment, Srivastava et al. used RCPs 4.5 and 8.5, from General Circulation Models (GCMs) GFDL-ESM 2M, GISS-E2-H, and HadGEM2-ES. An increase in the aboveground biomass by 59% was observed under HadGEM2-ES for the near future of 2030. As compared to baseline variability, the estimated biomass yield of maize was from a lowest upsurge of 169.9 kg ha^{-1} for GFDL-ESM 2 M in 2080 to the highest upsurge of 2386.1 kg ha^{-1} for HadGEM2-ES in 2030. The Root Zone Water Quality Model (RZWQM2) was employed to model climate change impacts on maize using current managing techniques after being calibrated using 4 years' worth data from the USA, for different irrigation situations (Ma et al. 2017). Results revealed that, compared to the productivity for the current climate, predicted climate change reduced biomass up to 7% for end-century (2070–1991) under full irrigation (1992–2013). The equivalent reductions under deficit irrigation were 3%. The yield did not decline when cultivar variables were calibrated using RZWQM2, but it did model a much lesser productivity for the current climate (Table 12.9).

Many studies have evaluated the climate change effect on the production of sorghum for South Asia, Africa, North America, and Oceania (Huntington et al. 2020; McCarthy and Vlek 2012; Tack et al. 2017). For sorghum, the typical temperature range that promotes germination, growth, and reproductive phase is 21–35 °C, 26–34 °C, and 25–28 °C, respectively (Maiti 1996). An increase in temperature has an impact on the duration of plant growth. As a consequence of the faster growth and development of the produce brought on by higher temperatures and faster growth of growing degree days, the length of the phenophase is shortened, which in turn reduces output (Attri and Rathore 2003). According to Gates (1968), high-temperature stress can have an impact on osmotic adjustment in plants by raising the rate of evapotranspiration and by hindering the production and uptake of solutes, mainly glucose, that are essential for the process (Acevedo et al. 1999). In order to change the production of solutes for osmotic correction to drought, photosynthesis may decrease, dark respiration and photorespiration may rise, and leaf senescence may speed up (Paulsen 1994). High temperature has a great impact on the height of the plant (Nguyen et al. 2013; Prasad et al. 2008), phenology, and growth of leaves (Hammer et al. 2010; Nguyen et al. 2013); duration from anthesis onward the higher temperatures has major effects on the reproductive phase and germination of pollens along with the setting of seed (Nguyen et al. 2013; Singh et al. 2015, 2016). According to Machado and Paulsen (2001), sorghum shows most rapid growth at medium temperature from 25 to 35 °C. At 15–25 °C, below the medium temperature, the plant remains workable. But at a temperature of 35–40 °C, the growth in the sorghum plant ceased after 10–12 days. At 35/30 °C, withholding water caused the most harm, where leaves rolled, collected cuticular wax, and senesced after 10 days. Early development was also fast. Climate change effects on forage

Table 12.9 Future climate scenario of maize, millet, and sorghum cropping system

Crop	Country	Model	Climate scenario	Biomass yield	References
Maize	Uganda	AquaCrop model	RCP 4.5, RCP 8.5 (1980–2009, 2019–2020, 2019–2039)	Decreased by 11% in RCP 4.5 and 5.6% in RCP 8.5	Zizinga et al. (2022)
Maize	Ireland	DSSAT	1.6 °C increase in temperature (1961–1990, 2041–2070 [2055], 2061–2090 [2075])	In western, north, and southeast areas, increased by 70%, 40%, and 25% in 2055	Holden and Brereton
Maize	Ghana	LINTUL5	RCP 4.5, RCP 8.5 (1971–2000, 2016–2045 [2030], 2064–2093 [2080])	Increased by 59% with RCP 8.5 in 2030	Srivastava et al.
Maize	USA	RZWQM2, DSSAT	RCP 8.5 (1989–2009, 2070–2099 [2080])	Decreased by 7% and 3% in full and deficit irrigation, respectively	Ma et al.
Maize	Turkey	SWAP	CGCM2, ECHAM4 and MRI, RCM (1994–2003, 2070–2079, 1990–2100)	Decreased by 17% under CGCM2 and increased by 3% in RCM	Yano et al. (2007)
Maize	North Carolina	CERES-Maize and EPIC model	GFDL, UKMO (1949–1988)	Biomass decreased by 27%	Dhakhwa et al.
Maize	Germany	Mesoscale hydrologic model and statistical crop model	A1B (1971–2000, 2021–2050, 2070–2099)	Reduced by 120–1050 and 370–3910 (kg $ha^{-1}a^{-1}$) during 2021–2050 and 2070–2099, respectively	Peichl et al. (2019)
Millet	India	CERES-Millet	PRECIS model, A2 scenarios, (1961–1990, 2071–2100)	Yield decreased by 5.2–14.9%	Yadav et al. (2014)
Sorghum	Ethiopia	CERES-Sorghum model	RCP 4.5 and RCP 8.5 (1981–2010, 2020–2049, 2040–2069)	nRMSE was 4.1–5.91%	Mohammed and Misganaw

(continued)

Table 12.9 (continued)

Crop	Country	Model	Climate scenario	Biomass yield	References
Sorghum	India	InfoCrop-Sorghum model	HadCM3 A2a (1970–1995, 2020, 2050, 2080)	7% in 2020, 11% in 2050, 32% in 2080	Srivastava et al.
Sorghum	Tanzania	APSIM, DSSAT	RCP 4.5, RCP 8.5 (1980–2010, 2010–2039, 2040–2069, 2071–2099)	Increased by 5%	Msongaleli et al.
Sorghum	Machakos	APSIM	RCP 4.5, RCP 8.5 (1976–2005, 2010–2039, 2040–2069, 2070–2099)	Decreased by 0.5–6.3%	Bosire et al.
Sorghum	Italy	EPIC	A2, B1 (2011–2040, 2041–2070, 2071–2100)	Decreased by up to 35%	Rinaldi et al.
Sorghum	Argentina	ALMANAC	1, 2, 4 °C mean annual increase in temperature (1999–2018, 2016–2035, 2081–2100)	Increased by 15%	Druille et al.
Sorghum	USA	ERT model	RCP 2.6, 4.5, 6.5, 8.5 (2018–2099)	Decreased by 2.7%	Huntington et al. (2020)

sorghum yield in India were analyzed by employing the InfoCrop model (Srivastava et al. 2010). As an alternative, the baseline (1970–1995) climate data was combined with HadCM3 for the 2020, 2050, and 2080 A2a scenarios. The InfoCrop model for simulations also included the predicted carbon dioxide levels according to the Bern CC model for the scenarios. Climate change influence on crop is estimated to decline biomass yield by 7%, 11%, and 32% in 2020, 2050, and 2080, respectively (Table 12.9).

The main obstacles to millet cultivation are high temperatures and rainfall variability that can be both low and high (Ahmad et al. 2018a). Pearl millet grows best at 33 °C during the day and 28 °C at night (Ong and Monteith 1985). By shortening the growing season, temperature increases above 33 °C/28 °C reduce crop output (Ahmad et al. 2018b). By lowering the number of basal tillers, the number of grains per inflorescence, and the weight of the grains, the elevated temperature at the vegetative, stem elongation, and reproductive phases reduced grain production (Garcia-Huidobro et al. 1982). According to future climate change forecasts, the increase in temperature for arid along with semiarid areas will decrease crop output (Howden et al. 2007). The millet production declined by 41% as a result of a temperature increase of 6 °C. Due to climate change, the increase in the CO_2 concentration for fodder millet production is beneficial as it helps to mitigate the influence of higher temperature on crop production (Mitchell et al. 1993). As compared to maize and sorghum, the root system in the millet is better and well adopted for water-lodging conditions. The main environmental element that negatively affects the development of foliage and roots, stomatal conductance, photosynthesis, and buildup of dry matter is the growing water shortage in millet (Aminian et al. 2011). Crop metabolism and photosynthesis traits are primarily affected by drought stress (Maghsoudi and Maghsoudi 2008). It causes the chlorophyll concentration of leaves and photosynthesis to decline (Peng et al. 2006). Numerous studies have shown that the stress of drought has a significant impact on plant physiology in general, and on output and development in particular. The speed, intensity, and length of the drought occurrence affect plant physiological processes like photosynthesis and transpiration (Zhang et al. 2012). The first sign of dry stress typically appears at the stomatal level. To avoid desiccation, stomata expand to a lesser extent (Yan et al. 2011). Lu et al. (2015) concluded that the drought stress declined the basal diameter, plant height, LA, total biomass, spike mass, spike length, spike number, and spike diameter. Yadav et al. (2014) used PRECIS output from A2 and baseline (1960–1990) and studied the effects of projected climate change (2071–2100) of A2 scenario and its probable influence on millet yield of Saurashtra region of Gujarat, India. By using the DSSAT (v4.5) model, baseline and A2 scenario yields were evaluated. The complete calibration and validation of CERES-Millet model simulation findings gave suitable and acceptable range (mean error < ±15%). The model predicted a decline in biomass yield of millet from 5.9% to 14.9% (Table 12.9).

12.6 Adaptation Strategies

To mitigate the various climate change influences on productivity and growth of fodder crops like maize and millet, along with sorghum cropping system, different strategies can be adopted. To reduce climate change influences, adaptation strategies such as 15 days' earlier transplanting than usual, variety changes, improved water management with additional fertilizer, and early transplanting with variety changes were attempted. In case of millet, these strategies would give a positive result in increasing the yield of the forage millet (Yadav et al. 2014). To enhance and improvement in extension services by providing the knowledge of new technologies of agriculture. Planting densities and planting dates with improving fertilizer management should be better adapted strategies for each crop.

By using different management techniques and creating new hybrids that are adaptable to climate changes in the future, crop output in the future can be made to adapt to climate change. Early crop sowing in semiarid regions may support plants in completing their life cycles before the summer's rising temperatures arrive. An excellent adaptation technique will be to conserve water more effectively (Ko et al. 2012). Vernalization and photoperiod requirements are sufficiently flexible, and suitable cultivar selection is possible to accommodate warming temperatures (Masle et al. 1989). Enhancing the heat tolerance of highland germplasm and bringing lowland subtropical and tropical genotypes are more accessible to farmers (Cairns et al. 2013a, b). More adaptations to response to the future changes in climate include modifications in genotypes to waterlogging conditions, heat and drought stress, and plant diseases and pests (Cairns et al. 2013a, b).

12.7 Conclusion

C_4 crops have supremacy over C_3 crops being a more efficient utilizer of inputs like water, nutrients, and sunlight. However, the production and productivity of these crops are being affected due to climate change and climate variability. There is a dire need to adopt climate-smart practices for enhancing the productivity of maize, millet, and sorghum crops as fodder for feeding the livestock.

References

Abbas Q, Ahmad S (2018) Effect of different sowing times and cultivars on cotton fiber quality under stable cotton-wheat cropping system in southern Punjab, Pakistan. Pak J Life Soc Sci 16(2):77–84

Abbas G, Ahmad S, Ahmad A, Nasim W, Fatima Z, Hussain S, Rehman MH, Khan MA, Hasanuzzaman M, Fahad S, Boote KJ, Hoogenboom G (2017) Quantification the impacts of climate change and crop management on phenology of maize-based cropping system in Punjab, Pakistan. Agric For Meteorol 247:42–55

Abbas G, Fatima Z, Riaz F, Iqbal P, Hussain S, Zakir I, Iqbal M, Haque E, Ahmad S (2019) Study of sowing dates and hybrids effect in maize-based cropping system under arid conditions of Southern Punjab, Pakistan. Pak J Life Soc Sci 17(1):24–31

Abbas G, Ahmad S, Hussain M, Fatima Z, Hussain S, Iqbal P, Ahmed M, Farooq M (2020a) Sowing date and hybrid choice matters production of maize-maize system. Int J Plant Prod 14:583–595

Abbas G, Fatima Z, Hussain M, Hussain S, Atique-ur-Rehman, Sarwar N, Ahmed M, Ahmad S (2020b) Nitrogen rate and hybrid selection matters productivity of maize-maize cropping system under irrigated arid environment of southern Punjab, Pakistan. Int J Plant Prod 14:309–320

Abbas Q, Atique-ur-Rehman, Sarwar N, Hussain S, Abbas G, Khan MN, Fatima Z, Naz S, Younis H, Ahmed M, Ullah H, Iqbal P, Iqbal M, Kan M, Ahmad S (2020c) Effect of different sowing times and cultivars on wheat grain quality under cotton-wheat cropping system in Southern Punjab, Pakistan. Pak J Life Soc Sci 18(1):41–50

Abbas G, Atique-ur-Rehman, Sarwar N, Fatima Z, Hussain S, Ahmed M, Raza MA, Kan M, DoĞan H, Khan MA, Ahmad S (2023a) Deciding sowing-window for maize-based cropping system in arid and semi-arid environments in Punjab, Pakistan. Turk J Agric For 47(6):1078–1098

Abbas G, Ahmed M, Fatima Z, Hussain S, Kheir AMS, Ercişli S, Ahmad S (2023b) Modeling the potential impact of climate change on maize-maize cropping system in semi-arid environment and designing of adaptation options. Agric For Meteorol 341:109674

Acevedo EH, Silva PC, Silva HR, Solar BR (1999) Wheat production in mediterranean environments. In: Satorre EH, Slafer GA (eds) Wheat: ecology and physiology of yield determination. Food Products Press, Binghamton, pp 295–331

Ahmad S, Hasanuzzaman M (2012) Integrated effect of plant density, N rates and irrigation regimes on the biomass production, N content, PAR use efficiencies and water productivity of rice under irrigated semiarid environment. Notulae Botanicae Horti Agrobotanici Cluj-Napoca 40(1):201–211

Ahmad S, Ahmad A, Soler CMT, Ali H, Zia-Ul-Haq M, Anothai J, Hussain A, Hoogenboom G, Hasanuzzaman G (2012a) Application of the CSM-CERES-Rice model for evaluation of plant density and nitrogen management of fine transplanted rice for an irrigated semiarid environment. Precis Agric 13(2):200–218

Ahmad S, Ali H, Ismail M, Shahzad MI, Nadeem M, Anjum MA, Zia-Ul-Haq M, Firdous N, Khan MA (2012b) Radiation and nitrogen use efficiencies of C_3 winter cereals to nitrogen split application. Pak J Bot 44(1):139–149

Ahmad S, Ahmad A, Ali H, Hussain A, Garcia y Garcia A, Khan MA, Zia-Ul-Haq M, Hasanuzzaman M, Hoogenboom G (2013) Application of the CSM-CERES-Rice model for evaluation of plant density and irrigation management of transplanted rice for an irrigated semiarid environment. Irrig Sci 31(3):491–506

Ahmad A, Ashfaq M, Rasul G, Wajid SA, Khaliq T, Rasul F, Saeed U, Rahman MH, Hussain J, Baig IA, Naqvi SAA, Bokhari SAA, Ahmad S, Naseem W, Hoogenboom G, Valdivia RO (2015a) Impact of climate change on the rice-wheat cropping system of Pakistan. In: Hillel D, Rosenzweig C (eds) Handbook of climate change and agro-ecosystems, vol 3. Imperial College Press and the American Society of Agronomy, London, pp 219–258

Ahmad S, Ali H, Atique-ur-Rehman, Khan RJ, Ahmad W, Fatima Z, Abbas G, Irfan M, Ali H, Khan MA, Hasanuzzaman M (2015b) Measuring leaf area of winter cereals by different techniques: a comparison. Pak J Life Soc Sci 13(2):117–125

Ahmad S, Raza I, Muhammad D, Ali H, Hussain S, Dogan H, Zia-Ul-Haq M (2015c) Radiation, water and nitrogen use efficiencies of *Gossypium hirsutum* L. Turk J Agric For 39(5):825–837

Ahmad S, Ali H, Farooq U, Khan SU, Atique-ur-Rehman, Sarwar N, Shahzad AN, Doğan H, Hussain S, Sultan MT, Waheed A, Zia-ul-Haq M, Hussain K, Khan MA (2016a) Improving nitrogen and radiation-use-efficiencies of C_4 summer cereals by split nitrogen applications under irrigated arid environment. Turk J Agric For 40(2):280–289

Ahmad S, Hussain S, Fatima Z, Abbas G, Atique-ur-Rehman, Khan MR, Younis H, Naz S, Sohail M, Ajmal M, Abbas N, Akhtar M, Rauf A, Khan M, Ali Z, Hassan M, Rizwan M, Bajwa RS, Ajmal A, Naz S, Ali H, Khan AA, Ali M, Sarwar G, Khan MA, Hasanuzzaman M (2016b) Application of DSSAT model for sowing date management of C_4 summer cereals for fodder and grain crops under irrigated arid environment. Pak J Life Soc Sci 14(2):104–114

Ahmad S, Nadeem M, Abbas G, Fatima Z, Khan RJ, Ahmed M, Ahmad A, Rasul G, Khan MA (2016c) Quantification of the effects of climate warming and crop management on sugarcane phenology. Clim Res 71(1):47–61

Ahmad S, Abbas G, Fatima Z, Khan RJ, Anjum MA, Ahmed M, Khan MA, Porter CH, Hoogenboom G (2017a) Quantification of the impacts of climate warming and crop management on canola phenology in Punjab, Pakistan. J Agron Crop Sci 203(5):442–452

Ahmad S, Abbas Q, Abbas G, Fatima Z, Atique-ur-Rehman, Naz S, Younis H, Khan RJ, Nasim W, Rehman MH, Ahmad A, Rasul G, Khan MA, Hasanuzzaman M (2017b) Quantification of climate warming and crop management impacts on cotton phenology. Plan Theory 6:1–16

Ahmad S, Atique-ur-Rehman, Ejaz M, Fatima Z, Kan M, Ahmad M (2018) Agricultural land-use change of major field crops in Pakistan (1961–2014). Sci Technol Dev 37(3):113–121

Ahmad I, Saeed U, Fahad M et al (2018a) Yield forecasting of spring maize using remote sensing and crop modeling in Faisalabad Punjab, Pakistan. J Indian Soc Remote Sens:1–11

Ahmad I, Wajid SA, Ahmad A et al (2018b) Optimizing irrigation and nitrogen requirements for maize through empirical modeling in semi-arid environment. Environ Sci Pollut Res 26:1227–1237

Ahmad A, Abbas G, Ahmed M, Fatima Z, Anjum MA, Rasul G, Khan MA, Hoogenboom G (2019) Climate warming and management impact on the change of phenology of the rice-wheat cropping system in Punjab, Pakistan. Field Crop Res 230:46–61

Ahmad A, Ashfaq M, Rasul G, Wajid SA, Ahmad I, Khaliq T, Nasir J, Rasul F, Riaz F, Ahmad B, Ahmad S, Baig IA, Valdivia RO, Hoogenboom G (2021) Development of climate change adaptation strategies for cotton–wheat cropping system of Punjab, Pakistan. In: Rosenzweig C, Mutter CZ, Contreras EM (eds) Handbook of climate change and agroecosystems: climate change and farming system planning in Africa and South Asia: AgMIP stakeholder-driven research. World Scientific Publishing Europe Ltd, London, pp 277–327

Ahmad S, Ahmad I, Ahmad B, Ahmad A, Wajid A, Khaliq T, Abbas G, Wilkerson CJ, Hoogenboom G (2023a) Regional integrated assessment of climate change impact on cotton production in a semi-arid environment. Clim Res 89:113–132

Ahmad S, Raza MA, Hussain S, Abbas G, Fatima Z, Ahmed M, Goheer MA, Garcia y Garcia A, Wilkerson CJ, Hoogenboom G (2023b) Identification of weak links in the production technology of farmer for bridging canola yield-gap. J Agric Sci 161:241–253. https://doi.org/10.1017/S0021859623000187

Ahmed M (2023) The science of climate change. In: Brinkmann R (ed) The Palgrave handbook of global sustainability. Springer International Publishing, Cham, pp 195–222. https://doi.org/10.1007/978-3-031-01949-4_22

Ahmed M, Ahmad S (eds) (2023) Disaster risk reduction in agriculture. Springer Nature. https://doi.org/10.1007/978-981-99-1763-1. ISBN 978-981-99-1762-4, Published: 03 November 2023

Ahmed M, Hassan FU, Aslam M, Aslam MA (2012) Physiological attributes based resilience of wheat to climate change. Int J Agric Biol 14(3):407–412

Ahmed M, Akram MN, Asim M, Aslam M, F-u Hassan Higgins S, Stockle CO, Hoogenboom G (2016) Calibration and validation of APSIM-wheat and CERES-wheat for spring wheat under rainfed conditions: models evaluation and application. Comput Electron Agric 123:384–401

Ahmed M, Stöckle CO, Nelson R, Higgins S (2017) Assessment of climate change and atmospheric CO_2 impact on winter wheat in the Pacific Northwest using a multimodel ensemble. Front Ecol Evol 5. https://doi.org/10.3389/fevo.2017.00051

Ahmed I, Rahman MH, Ahmad A, Ahmad S, Hussain J, Ullah A, Judge J (2018a) Assessing the impact of climate variability on maize using simulation modeling under semi-arid environment of Punjab-Pakistan. Environ Sci Pollut Res 25(23):28413–28430

Ahmed M, Ijaz W, Ahmad S (2018b) Adapting and evaluating APSIM-SoilP-Wheat model for response to phosphorus under rainfed conditions of Pakistan. J Plant Nutr 41(16):2069–2084

Ahmed M, Stockle CO, Nelson R, Higgins S, Ahmad S, Raza M (2019) Novel multimodal ensemble approach to evaluate the sole effect of elevated CO_2 on winter wheat productivity. Sci Rep 9:7813

Ahmed M, Ahmad S, Tariq M, Fatima Z, Aslam Z, Raza MA, Iqbal N, Akmal M, Fayyaz-ul-Hassan, Abbasi NA, Hayat R (2020) Wastes to be the source of nutrients and energy to mitigate climate change and ensure future sustainability: options and strategies. J Plant Nutr 43(6):896–920

Ahmed M, Hayat R, Ahmad M, Mahmood-ul-Hassan, Kheir AMS, Fayyaz-ul-Hassan, Rehman MH, Shaheen FA, Raza MA, Ahmad S (2022) Impact of climate change on dryland agricultural systems: a review of current status, potentials, and further work need. Int J Plant Prod 16:341–363

Ahmed M, Hayat R, Ahmad M, ul-Hassan M, Kheir AMS, ul-Hassan F, ur-Rehman MH, Shaheen FA, Raza MA, Ahmad S (2022a) Impact of climate change on dryland agricultural systems: a review of current status, potentials, and further work need. Int J Plant Prod. https://doi.org/10.1007/s42106-022-00197-1

Ahmed M, Ahmad S, Kheir AMS (2022b) Climate change: an overview. In: Ahmed M (ed) Global agricultural production - resilience to climate change. Springer Nature Switzerland AG. https://doi.org/10.1007/978-3-031-14973-3_1

Akponikpè PBI, Gérard B, Michels K, Bielders C (2010) Use of the APSIM model in long term simulation to support decision making regarding nitrogen management for pearl millet in the Sahel. Eur J Agron 32:144–154

Ali H, Iqbal N, Ahmad S, Shahzad AN, Sarwar N (2013) Performance of late sown wheat crop under different planting geometries and irrigation regimes in arid climate. Soil Tillage Res 130:109–119

Amanullah, Khan A (2004) Growth characters and fodder production potential of millet varieties under irrigated conditions. Pak J Soil Sci 23(3–4):64–67

Amin A, Nasim W, Fahad S, Ali S, Ahmad S, Rasool A, Saleem N, Hammad HM, Sultana SR, Mubeen M, Bakhat HF, Ahmad N, Shah GM, Adnan M, Noor M, Basir A, Saud S, Rahman MH, Paz JO (2018a) Evaluation and analysis of temperature for historical (1996–2015) and projected (2030–2060) climates in Pakistan using SimCLIM climate model: ensemble application. Atmos Res 213:422–436

Amin A, Nasim W, Mubeen M, Ahmad A, Nadeem M, Urich P, Fahad S, Ahmad S, Wajid A, Tabassum F, Hammad HM, Sultana SR, Anwar S, Baloch SK, Wahid A, Wilkerson CJ, Hoogenboom G (2018b) Simulated CSM-CROPGRO-cotton yield under projected future climate by SimCLIM for southern Punjab, Pakistan. Agric Syst 167:213–222

Amin A, Nasim W, Mubeen M, Sarwar S, Urich P, Ahmad A, Wajid A, Khaliq T, Rasul F, Hammad HM, Rehmani MIA, Mubarak H, Mirza N, Wahid A, Ahmad S, Fahad S, Ullah A, Khan MN, Ameen A, Amanullah, Shahzad B, Saud S, Alharby H, Ata-Ul-Karim ST, Adnan M, Islam F, Ali QS (2018c) Regional climate assessment for temperature and precipitation in Southern Punjab (Pakistan) using SimCLIM climate model for different temporal scales. Theor Appl Climatol 131:121–131

Aminian R, Mohammadi S, Hoshmand S et al (2011) Chromosomal analysis of photosynthesis rate and stomatal conductance and their relationships with grain yield in wheat (*Triticum aestivum* L.) under water-stressed and well-watered conditions. Acta Physiol Plant 33:755–764

Andrews DL, Kumar KA (1992) Pearl millet for food, feed, and forage. Adv Agron 48:89–139

Aslam Z, Ahmed M, Sajad M, Asif M, Akmal M, Awan FK, Ijaz W, Ashraf R, Khattak JZK (2013) A comparison of statistical and dynamic modeling of wheat (*Triticum aestivum* L.) fungal diseases under the climate change. J Food Agric Environ 11(2):1327–1332

Asseng S, Martre P, Maiorano A, Rötter RP, O'Leary GJ, Fitzgerald GJ, Girousse C, Motzo R, Giunta F, Babar MA, Reynolds MP, Kheir AMS, Thorburn PJ, Waha K, Ruane AC, Aggarwal PK, Ahmed M, Balkovič J, Basso B, Biernath C, Bindi M, Cammarano D, Challinor AJ, De Sanctis G, Dumont B, Eyshi Rezaei E, Fereres E, Ferrise R, Garcia-Vila M, Gayler S, Gao Y, Horan H, Hoogenboom G, Izaurralde RC, Jabloun M, Jones CD, Kassie BT, Kersebaum K-C, Klein C, Koehler A-K, Liu B, Minoli S, Montesino San Martin M, Müller C, Naresh Kumar S, Nendel C, Olesen JE, Palosuo T, Porter JR, Priesack E, Ripoche D, Semenov MA, Stöckle C, Stratonovitch P, Streck T, Supit I, Tao F, Van der Velde M, Wallach D, Wang E, Webber H, Wolf J, Xiao L, Zhang Z, Zhao Z, Zhu Y, Ewert F (2019) Climate change impact and adaptation for wheat protein. Glob Chang Biol 25(1):155–173. https://doi.org/10.1111/gcb.14481

Attri SD, Rathore LS (2003) Simulation of impact of projected climate change on wheat in India. Int J Climatol 23:693–705

Baumhardt RL, Salinas-Garcia J (2006) Dryland agriculture in Mexico and the U.S. Southern Great Plains. In: Dryland agriculture, Agronomy monograph no. 23, 2nd edn. American Society of Agronomy, Crop Science Society of America, Soil Science Society of America, Madison

Beah A, Kamara AY, Jibrin JM, Akinseye FM, Tofa AI, Ademulegun TD (2021) Simulation of the optimum planting windows for early and intermediate-maturing maize varieties in the Nigerian Savannas using the APSIM model. Front Sustain Food Syst 5:624886

Berg A, De Noblet-Ducoudre N, Sultan B, Lengaigne M, Guimberteau M (2013) Projections of climate change impacts on potential C4 crop productivity over tropical regions. Agric For Meteorol 170:89–102

Birch CJ, Hammer GL, Rickert KG (1998) Modelling leaf production and crop development in maize (*Zea mays* L.) after tassel initiation under diverse conditions of temperature and photoperiod. Field Crop Res 58:81–95

Brewbaker JL (2003) Corn production in the tropics: the Hawaii experience. College of Tropical Agriculture and Human Resources, University of Hawaii, Honolulu

Brooks A, Farquhar GD (1985) Effect of temperature on the CO_2/O_2 specificity of ribulose-1,5-bisphosphate carboxylase/oxygenase and the rate of respiration in the light. Planta 165:397–406

Cairns JE, Hellin J, Sonder K, Araus JL, MacRobert JF, Thierfelder C, Prasanna BM (2013a) Adapting maize production to climate change in sub-Saharan Africa. Food Secur 5:345–360

Cairns JE, Sonder K, Zaidi PH, Verhulst N, Mahuku G, Babu R, Nair SK, Das B, Govaerts B, Vinayan MT, Rashid Z, Noor JJ, Devi P, Vicente FS, Prasanna BM (2013b) Maize production in a changing climate: impacts, adaptation, and mitigation strategies. Adv Agron 114:1–58

Carter PR, Hesterman OB (1990) Handling Corn Damage by Autumn Frost. Purdue University Cooperative Extension Service Climate and Weather Shaykewich CF (1994) An appraisal of cereal crop phenology modelling. Can J Plant Sci 75:329–341

Chen C, Wang E, Qiang Y (2010) Modelling the effects of climate variability and water management on crop water productivity and water balance in the North China Plain. Agric Water Manag 97:1175–1184

Cheng J, Zhu R, Xu Z, Xu X, Wang X, Li K, Su H (2014) Temperature variation between neighboring days and mortality: a distributed lag non-linear analysis. Int J Public Health 59:923–931

Crowley JG (1998) Improving the yield and quality of forage maize. End of project report. Teagasc, Oak Park, 10 pp

Ebi KL, Exuzides KA, Lau E, Kelsh M, Barnston A (2004) Weather changes associated with hospitalizations for cardiovascular diseases and stroke in California, 1983–1998. Int J Biometeorol 49:48–58

Ejaz M, Abbas G, Fatima Z, Iqbal P, Raza MA, Kheir AMS, Ahmed M, Kakar KM, Ahmad S (2022) Modelling climate uncertainty and adaptations for soybean-based cropping system. Int J Plant Prod 16:235–250

FAO (2008). http://faostat.fao.org/site/567/default.aspx#ancor. Accessed 19 Nov 2022

FAO STAT (2021) Statistic database. Rome, Food and Agriculture Organization of the United Nations

Fatima Z, Abbas Q, Khan A, Hussain S, Ali MA, Abbas G, Younis H, Naz S, Ismail M, Shahzad MI, Nadeem M, Farooq U, Khan SU, Javed K, Khan AA, Ahmed M, Khan MA, Ahmad S (2018) Resource-use-efficiencies of C_3 and C_4 cereals under split nitrogen regimes. Agronomy 8(5):69

Fatima Z, Ahmad M, Hussain M, Abbas G, Ul-Allah S, Ahmad S, Ahmed N, Ali MA, Sarwar G, Haque E, Iqbal P, Hussain S (2020) The fingerprints of climate warming on cereal crops phenology and adaptation options. Sci Rep 10:18013

Fatima Z, Atique-ur-Rehman, Abbas G, Iqbal P, Zakir I, Khan MA, Kamal GM, Ahmed M, Ahmad S (2021) Quantification of climate warming and crop management impacts on phenology of pulses-based cropping systems. Int J Plant Prod 15:107–123

Folliard A, Traoré PCS, Vaksmann M, Kouressy M (2004) Modeling of sorghum response to photoperiod: a threshold–hyperbolic approach. Field Crop Res 89:59–70

Fosu-Mensah BY (2013) Modeling the impact of climate change on maize (*Zea mays* L.) yield under rainfed conditions in sub-humid Ghana. UNU-INRA Working Paper 1, United Nations University-Institute for Natural Resources in Africa, Accra, p 36

Garcia-Huidobro J, Monteith JL, Squire GR (1982) Time, temperature and germination of pearl millet (*Pennisetum typhoides* S. & H.) I. Constant temperature. J Exp Bot 33:288–296

Gates DM (1968) Transpiration and leaf temperature. Annu Rev Plant Physiol 19:211–238

Giorgi F, Raffaele F, Coppola E (2019) The response of precipitation characteristics to global warming from climate projections. Earth Syst Dynam 10:73–89

Giridhar K, Samireddypalle A (2015) Impact of climate change on forage availability for livestock. In: Sejian V, Gaughan J, Baumgard L, Prasad C (eds) Climate change impact on livestock: adaptation and mitigation. Springer, New Delhi, pp 51–60

Gömann H, Frühauf C, Lüttger A, Weigel H-J (2017) Landwirtschaft. In: Brasseur GP, Jacob D, Schuck-Zöller S (eds) Klimawandel in Deutschland. Springer, Berlin, pp 183–191. https://doi.org/10.1007/978-3-662-50397-3_18

Guo Y, Barnett AG, Yu W, Pan X, Ye X, Huang C, Tong S (2011) A large change in temperature between neighbouring days increases the risk of mortality. PLoS One 6:e16511

Hammer GL, van Oosterom EJ, McLean G, Chapman SC, Broad I, Harland P, Muchow RC (2010) Adapting APSIM to model the physiology and genetics of complex adaptive traits in field crops. J Exp Bot 61:2185–2202

Holiday AS, Martindale W, Alred R, Brooks AL, Leegood RC (1992) Changes in activities of enzymes of carbon metabolism in leaves during exposure of plants to low temperature. Plant Physiol 98:1105–1114

Hoogenboom G (2003) Crop growth and development. In: Bendi DK, Nieder R (eds) Handbook of processes and modeling in the soil-plant system. Haworth Press, New York, pp 655–691

Hoogenboom G, Porter CH, Boote KJ, Shelia V, Wilkens PW, Singh U, White JW, Asseng S, Lizaso JI, Moreno LP, Pavan W, Ogoshi R, Hunt LA, Tsuji GY, Jones JW (2019) The DSSAT crop modeling ecosystem. In: Boote KJ (ed) Advances in crop modeling for a sustainable agriculture. Burleigh Dodds Science Publishing, Cambridge, pp 173–216. https://doi.org/10.19103/AS.2019.0061.10

Hoogenboom G, Porter CH, Shelia V, Boote KJ, Singh U, White JW, Pavan W, Oliveira FAA, Moreno-Cadena LP, Lizaso JI, Asseng S, Pequeno DNL, Kimball BA, Alderman PD, Thorp KR, Jones MR, Cuadra SV, Vianna MS, Villalobos FJ, Ferreira TB, Batchelor WD, Koo J, Hunt LA, Jones JW (2021) Decision Support System for Agrotechnology Transfer (DSSAT) Version 4.8 (DSSAT.net). DSSAT Foundation, Gainesville

Howden SM, Soussana J-F, Tubiello FN et al (2007) Adapting agriculture to climate change. Proc Natl Acad Sci 104:1969

Hund A, Fracheboud Y, Soldati A, Stamp P (2008) Cold tolerance of maize seedlings as determined by root morphology and photosynthetic traits. Eur J Agron 28:178–185

Huntington T, Cui X, Mishra U, Scown CD (2020) Machine learning to predict biomass sorghum yields under future climate scenarios. Biofuels Bioprod Biorefin 14:566–577

Hussain M, Ahmad S, Hussain S, Lal R, Ul-Allah S, Nawaz A (2018) Rice in saline soils: physiology, biochemistry, genetics and management. Adv Agron 148:231–287

Indriani NP, Yuwariah Y, Rochana A, Djuned H (2015) Effect of intercropping between corn (*Zea mays*) and peanut (*Arachis hypogaea*) with arbuscular mycorrhizal (amf) on the yield and forage mineral content. Pak J Nutr 14(6):362–365

IPCC (2007) Summary for policymakers. In: Parry ML, Canziani OF, Palutikof JP, van der Linden PJ, Hanson CE (eds) Climate change 2007: impacts, adaptation and vulnerability, Contribution of Working Group II to the Fourth Assessment Report of the Intergovernmental Panel on Climate Change. Cambridge University Press, Cambridge, pp 7–22

Ismail SM (2012) Optimizing productivity and irrigation water use efficiency of pearl millet as a forage crop in arid regions under different irrigation methods and stress. Afr J Agric Res 7:2509–2518

Jabeen M, Gabriel HF, Ahmed M, Mahboob MA, Iqbal J (2017) Studying impact of climate change on wheat yield by using DSSAT and GIS: a case study of Pothwar Region. In: Ahmed M, Stockle OC (eds) Quantification of climate variability, adaptation and mitigation for agricultural sustainability. Springer International Publishing, Cham, pp 387–411

Jacob D, Kottmeier C, Petersen J, Rechid D, Teichmann C (2017) Regionale Klima modellierung. In: Brasseur GP, Jacob D, Schuck-Zöller S (eds) Klimawandel in Deutschland. Springer, Berlin, pp 27–35. https://doi.org/10.1007/978-3-662-50397-3_4

Jones JW, Hoogenboom G, Porter CH, Boote KJ, Batchelor WD, Hunt LA, Wilkens PW, Singh U, Gijsman AJ, Ritchie JT (2003a) The DSSAT cropping system model. Eur J Agron 18(3–4):235–265

Jones JW, Hoogenboom G, Porter CH, Boote KJ, Batchelor WD, Hunt LA, Wilkens PW, Singh U, Gijsman AJ, Ritchie JT (2003b) DSSAT cropping system model. Eur J Agron 18:235–265

Jordan DB, Ogren WL (1984) The CO2/O2 specificity of ribulose-1,5-bisphosphate carboxylase/oxygenase. Planta 161:308–313

Jukanti AK, Gowda CL, Rai KN, Manga VK, Bhatt RK (2016) Crops that feed the world 11. Pearl Millet (*Pennisetum glaucum* L.): an important source of food security, nutrition and health in the arid and semi-arid tropics. Food Secur 8:307–329

Keating BA, Carberry PS, Hammer GL, Probert ME, Robertson MJ, Holzworth D, Huth NI, Hargreaves JNG, Meinke H, Hochman Z, McLean G, Verburg K, Snow V, Dimes JP, Silburn M, Wang E, Brown S, Bristow KL, Asseng S, Chapman S, McCown RL, Freebairn DM, Smith CJ (2003) An overview of APSIM: a model designed for farming systems simulation. Eur J Agron 18:267–288

Khan A, Fahad S, Khan A, Saud S, Adnan M, Wahid F, Noor M, Nasim W, Hammad HM, Bakhat HF, Ahmad S, Rehman MH, Wang D, Sönmez O (2019) Managing tillage operations and manure to restore soil carbon stocks in wheat-maize cropping system. Agron J 111:1–10

Khan MA, Tahir A, Khurshid N, Husnain MI, Ahmed M, Boughanmi H (2020) Economic effects of climate change-induced loss of agricultural production by 2050: a case study of Pakistan. Sustain For 12(3):1216. https://doi.org/10.3390/su12031216

Khanthavong P, Yabuta S, Asai H, Hossain MA, Akagi I, Sakagami J-I (2021) Root response to soil water status via interaction of crop genotype and environment. Agronomy 11:708

Kim J, Shin J, Lim YH, Honda Y, Hashizume M, Guo YL, Kan H, Yi S, Kim H (2016) Comprehensive approach to understand the association between diurnal temperature range and mortality in East Asia. Sci Total Environ 539:313–321

Klopfensteina TJ, Ericksona GE, Berger LL (2013) Maize is a critically important source of food, feed, energy and forage in the USA. Field Crop Res 153:5–11

Knox J, Daccache A, Hess T, Haro D (2016) Meta-analysis of climate impacts and uncertainty on crop yields in Europe. Environ Res Lett 11:113004

Ko J, Ahuja LR, Saseendran SA, Green TR, Ma L, Nielsen DC, Walthall CL (2012) Climate change impacts on dryland cropping systems in the Central Great Plains, USA. Clim Chang 111:445–472

Lamaoui M, Jemo M, Datla R, Bekkaoui F (2018) Heat and drought stresses in crops and approaches for their mitigation. Front Chem 6:26

Liu B, Martre P, Ewert F, Webber H, Waha K, Thorburn PJ, Ruane AC, Aggarwal PK, Ahmed M, Balkovič J, Basso B, Biernath C, Bindi M, Cammarano D, Cao W, Challinor AJ, De Sanctis G, Dumont B, Espadafor M, Eyshi Rezaei E, Fereres E, Ferrise R, Garcia-Vila M, Gayler S, Gao Y, Horan H, Hoogenboom G, Izaurralde RC, Jabloun M, Jones CD, Kassie BT, Kersebaum KC, Klein C, Koehler A-K, Maiorano A, Minoli S, Montesino San Martin M, Müller C, Naresh Kumar S, Nendel C, O'Leary GJ, Eivind Olesen J, Palosuo T, Porter JR, Priesack E, Ripoche D, Rötter RP, Semenov MA, Stöckle C, Stratonovitch P, Streck T, Supit I, Tao F, Van der Velde M, Wang E, Wolf J, Xiao L, Zhang Z, Zhao Z, Zhu Y, Asseng S (2023) AgMIP-Wheat multi-model simulations on climate change impact and adaptation for global wheat. Open Data J Agric Res 9:10–25. https://doi.org/10.18174/odjar.v9i0.18092

Lu HB, Qiao YM, Gong XC, Li HQ, Zhang Q, Zhao ZH, Meng LL (2015) Influence of drought stress on the photosynthetic characteristics and dry matter accumulation of hybrid millet. Photosynthetica 53:306–311

Ma L, Hoogenboom G, Ahuja LR, Ascough JC, Saseendran SA (2006) Evaluation of the RZWQM-CERES-Maize hybrid model for maize production. Agric Syst 87:274–295

MacCarthy DS, Sommer R, Vlek PLG (2009) Modeling the impacts of contrasting nutrient and residue management practices on grain yield of sorghum (*Sorghum bicolor* (L.) Moench) in a semi-arid region of Ghana using APSIM. Field Crop Res 113(2):105–115

Machado S, Paulsen GM (2001) Combined effects of drought and high temperature on water relations of wheat and sorghum. Plant Soil 233:179–187

Maghsoudi K, Maghsoudi A (2008) Analysis of the effects of stomatal frequency and size on transpiration and yield of wheat (*Triticum aestivum* L.). Am Eur J Agric Environ Sci 3:865–872

Mahmood A, Ullah H, Shahzad AN, Ali H, Ahmad S, Zia-Ul-Haq M, Honermeier B, Hasanuzzaman M (2013) Dry matter yield and chemical composition of sorghum cultivars with varying planting density and sowing date. Sains Malaysiana 42(10):1529–1538

Maitah M, Malec K, Maitah K (2021) Influence of precipitation and temperature on maize production in The Czech Republic from 2002 to 2019. Sci Rep 11:10467

Maiti RK (1996) Sorghum science. Science Publishers, Lebanon

Masle J, Doussinault G, Sun B (1989) Responses of wheat genotypes to temperature and photoperiod in natural conditions. Crop Sci 29:712–721

Mason SC, Maman N, Pale S (2015) Pearl millet production practices in semi-arid West Africa: a review. Exp Agric 15:501–541

Matsuura A, An P, Murata K, Inanaga S (2016) Effect of pre- and post-heading waterlogging on growth and grain yield of four millets. Plant Prod Sci 19:348–359

Matthews RB, Pilbeam C (2005) Modelling the long-term productivity and soil fertility of maize/millet cropping systems in the mid-hills of Nepal. Agric Ecosyst Environ 111:119–139

Maulana F, Tesso TT (2013) Cold temperature episode at seedling and flowering stages reduces growth and yield components in sorghum. Crop Sci 53:564–574

McCarthy DS, Vlek PLG (2012) Impact of climate change on sorghum production under different nutrient and crop residue management in semi-arid region of Ghana: a modeling perspective. Afr Crop Sci J 20:243–259

McCown RL, Hammer GL, Hargreaves JNG, Holzworth DP, Freebairn DM (1996) APSIM: a novel software system for model development, model testing and simulation in agricultural systems research. Agric Syst 50:255–271

Methu JN, Kiruiro EM, Abate AN (2006) Your feed shortage problem: use maize forage. KARI Resource Center, Nairobi

Mitchell RAC, Mitchell VJ, Driscoll SP et al (1993) Effects of increased CO_2 concentration and temperature on growth and yield of winter wheat at two levels of nitrogen application. Plant Cell Environ 16:521–529

Mubarak MK, Hussain K, Abbas G, Altaf MT, Baloch FS, Ahmad S (2022) Productivity of sorghum (*Sorghum bicolor* L.) at diverse irrigation regimes and sowing dates in semi-arid and arid environment. Turk J Agric For 46(1):1–18

Mubeen M, Ahmad A, Wajid A, Khaliq T, Bakhsh A (2013) Evaluating CSM-CERES-Maize model for irrigation scheduling in semiarid conditions of Punjab, Pakistan. Int J Agric Biol 15:1–10

Mubeen M, Ahmad A, Wajid A, Khaliq T, Hammad HM, Sultana SR, Ahmad S, Fahad S, Nasim W (2016) Application of CSM-CERES-Maize model in optimizing irrigated conditions. Outlook Agric 45(3):173–184

Murty MVR, Singh P, Wani SP, Khairwal IS, Srinivas K (2007) Yield gap analysis of sorghum and pearl millet in India using simulation modeling. Global Theme on Agroecosystems Report No. 37, International Crops Research Institute for the Semi-Arid Tropics, Patencheru, AP, India, p 82

Nadeem M, Nazer Khan M, Abbas G, Fatima Z, Iqbal P, Ahmed M, Ali Raza M, Rehman A, Ul Haq E, Hayat A, Ali M, Ahmad S (2022) Application of CSM-CANEGRO model for climate change impact assessment and adaptation for sugarcane in semi-arid environment of Southern Punjab, Pakistan. Int J Plant Prod. https://doi.org/10.1007/s42106-022-00192-6

Nasim W, Amin A, Fahad S, Awais M, Khan N, Mubeen M, Wahid A, Turan V, Rehman MH, Ihsan MZ, Ahmad S, Hussain S, Mian IA, Khan B, Jamal Y (2018) Future risk assessment by estimating historical heat wave trends with projected heat accumulation using SimCLIM climate model in Pakistan. Atmos Res 205:118–133

Naz S, Ahmad S, Abbas G, Fatima Z, Hussain S, Khan MA, Ahmed M, Khan A, Fahad S, Nasim W, Ercisli S, Wilkerson CJ, Hoogenboom G (2022) Modeling the impact of climate warming on potato phenology. Eur J Agron 132:126404

Newport A (2006) Grain on the stump. Beef, Beef Magazine, Oct 2006

Nguyen TC, Singh V, van Oosterom EJ, Chapman SC, Jordan DR, Hammer GL (2013) Genetic variability in high temperature effects on seed-set in sorghum. Funct Plant Biol 40:439–448

Ong CK, Monteith JL (1985) Response of pearl millet to light and temperature. Field Crop Res 11:141–160

Parent C, Capelli N, Berger A, Crevecoeur M, Dat JF (2008) An overview of plant responses to soil waterlogging. Plant Stress 2:20–27

Parry ML, Rosenzweig C, Iglesias A, Livermore M, Fischer G (2004) Effects of climate change on global food production under SRES emissions and socioeconomic scenarios. Glob Environ Chang 14:53–67

Paulsen GM (1994) High temperature responses of crop plants. In: Boote KT, Bennett JM, Sinclair TR, Paulsen GM (eds) Physiology and determination of crop yield. ASA, SSA and SSSA, Madison, pp 365–389

Pavli OI, Ghikas DV, Katsiotis A, Skaracis GN (2011) Differential expression of heat shock protein genes in sorghum (*Sorghum bicolor* L.) genotypes under heat stress. Aust J Crop Sci 5:511–515

Peacock JM, Heinrich GM (1984) Light and temperature responses in Sorghum. ICRISAT (International Crop Research Institute for the Semi-arid Tropics). Agrometeorology of

Sorghum and Millet in the Semi-arid Tropics. Proceedings of the International Symposium, ICRISAT Center, India, 15–20 November 1982. Patancheru, AP, India

Peichl M, Thober S, Samaniego L, Hansjürgens B, Marx A (2019) Climate impacts on long-term silage maize yield in Germany. Sci Rep 9:7674

Peng Y, Peng ZS, Song HX et al (2006) Chromosomal location of the genes associated with photosynthesis of *Lophopyrum elongatum* in Chinese Spring background. Agric Sci China 5:579–586

Plavcová E, Kyselý J (2010) Relationships between sudden weather changes in summer and mortality in The Czech Republic, 1986–2005. Int J Biometeorol 54:539–551

Potter B (2012) Grazing corn in the Northwest. OMAFRA Virtual Beef, West. Beef Dev. Center, Factsheet 2012.03

Prasad PVV, Pisipati SR, Mutava RN, Tuinstra MR (2008) Sensitivity of grain sorghum to high temperature stress during reproductive development. Crop Sci 48:1911–1917

Promkhambu A, Younger A, Polthanee A, Akkasaeng C (2010) Morphological and physiological responses of sorghum (*Sorghum bicolor* L. Moench) to waterlogging. Asian J Plant Sci 9:183–193

Rahman MH, Ahmad A, Wang X, Wajid A, Nasim W, Hussain M, Ahmad B, Ahmad I, Ali Z, Ishaque W, Awais M, Muddasir M, Shelia V, Ahmad S, Fahad S, Alam M, Ullah H, Hoogenboom G (2018) Multi-model projections of future climate and climate change impacts uncertainty assessment for cotton production in Pakistan. Agric For Meteorol 253–254:94–113

Rahman MH, Ahmad I, Wang D, Fahad S, Afzal M, Ghaffar A, Saddique Q, Khan MA, Saud S, Hassan S, Fahad M, Din M, Ahmad M, Yue Z, Shah RA, Sönmez O, Nasim W (2021) Influence of semi-arid environment on radiation use efficiency and other growth attributes of lentil crop. Environ Sci Pollut Res 28:13697–13711

Rai KN, Blümmel M, Singh AK, Rao AS (2012) Variability and relationships among forage yield and quality traits in pearl millet. Eur J Plant Sci Biotechnol 6:118–124

Rehman A, Ashfaq M, Naqvi SAA, Adil SA, Bashir K, Ahmad A, Ahmad S, Ali A, Imran A (2015) Is climate change worsening the poverty of maize growers? Evidence from Punjab province of Pakistan. Ciencia e Tecnica Vitivinicola 30(1):105–116

Samaniego L, Thober S, Kumar R, Wanders N, Rakovec O, Pan M, Zink M, Sheffield J, Wood EF, Marx A (2018) Anthropogenic warming exacerbates European soil moisture droughts. Nat Clim Chang 8:421–426

Sánchez B, Rasmussen A, Porter JR (2014) Temperatures and the growth and development of maize and rice: a review. Glob Chang Biol 20:408–417

Sarwar N, Ali H, Maqsood M, Ullah E, Shahzad AN, Shahzad M, Mubeen K, Shahid MA, Ahmad S (2013) Phenological response of rice plants to different micronutrients application under water saving paddy fields on calcareous soil. Turk J Field Crops 18(1):52–57

Sarwar N, Atique-ur-Rehman, Farooq O, Wasaya A, Hussain M, El-Shehawi AM, Ahmad S, Brestic M, Mahmoud SF, Zivcak M, Farooq S (2021) Integrated nitrogen management improves productivity and economic returns of wheat-maize cropping system. J King Saud Univ Sci 33:101475

Saseendran SA, Ma L, Nielsen DC, Vigil MF, Ahuja LR (2005) Simulating planting date effects on corn production using RZWQM and CERES-Maize models. Agron J 97:58–71

Shabbir RN, Waraich EA, Ali H, Nawaz F, Ashraf MY, Ahmad R, Awan MI, Ahmad S, Irfan M, Hussain S, Ahmad Z (2016) Supplemental exogenous NPK application alters biochemical processes to improve yield and drought tolerance in wheat (*Triticum aestivum* L.). Environ Sci Pollut Res 23(3):2651–2662

Sharad KJ, Kumar V (2012) Trend analysis of rainfall and temperature data for India. Curr Sci 102(1):37–49

Singh P, Virmani SM (1996) Modeling growth and yield of chickpea (*Cicer arietinum* L.). Field Crop Res 46:41–59

Singh V, Nguyen CT, van Oosterom EJ, Chapman SC, Jordan DR, Hammer GL (2015) Sorghum genotypes differ in high temperature responses for seed set. Field Crop Res 171:32–40

Singh V, Nguyen CT, Yang Z, Chapman SC, van Oosterom EJ, Hammer GL (2016) Genotypic differences in effects of short episodes of high-temperature stress during reproductive development in sorghum. Crop Sci 56:1–12

Sinsawat V, Leipner J, Stamp P, Fracheboud Y (2004) Effect of heat stress on the photosynthetic apparatus in maize (Zea mays L.) grown at control or high temperature. Environ Exp Bot 52:123–129

Srichana D, Suttitham W, Thongsunthiah P, Panja P, Jariyapamornkoon N (2014) Nutrients and ruminal digestibility of baby corn by-product silages under different harvesting methods. Thammasat Int J Sci Technol 19(2):30–37

Tack J, Lingenfelser J, Jagadish SVK (2017) Disaggregating sorghum yield reductions under warming scenarios exposes narrow genetic diversity in US breeding programs. Proc Natl Acad Sci U S A 114:9296–9301

Tariq M, Yasmeen A, Ahmad S, Hussain N, Afzal MN, Hasanuzzaman M (2017) Shedding of fruiting structures in cotton: factors, compensation and prevention. Trop Subtrop Agroecosyst 20(2):251–262

Tariq M, Afzal MN, Muhammad D, Ahmad S, Shahzad AN, Kiran A, Wakeel A (2018a) Relationship of tissue potassium content with yield and fiber quality components of Bt cotton as influenced by potassium application methods. Field Crop Res 229:37–43

Tariq M, Ahmad S, Fahad S, Abbas G, Hussain S, Fatima Z, Wajid N, Mubeen M, Rehman M, Khan MA, Adnan M, Wilkerson CJ, Hoogenboom G (2018b) The impact of climate warming and crop management on phenology of sunflower-based cropping systems in Punjab, Pakistan. Agric For Meteorol 256–257:270–282

Tarriq M, Fatima Z, Iqbal P, Nahar K, Ahmad S, Hasanuzzaman M (2021) Sowing dates and cultivars mediated changes in phenology and yield traits of cotton-sunflower cropping system in arid environment. Int J Plant Prod 15:291–302

Taylor JRN, Schober TJ, Bean SR (2006) Novel food and non-food uses for sorghum and millets. J Cereal Sci 44:252–271

Tian L, Li J, Bi W, Zuo S, Li L, Li W, Sun L (2019) Effects of waterlogging stress at different growth stages on the photosynthetic characteristics and grain yield of spring maize (Zea mays L.) under field conditions. Agric Water Manag 218:250–258

Toreti A, Desiato F (2008) Temperature trend over Italy from 1961 to 2004. Theor Appl Climatol 91:51–58

Unger PW (1988) Grain and forage sorghum production with no-tillage on dryland. Agron J 80:193–197

Waha K, Van Bussel LGJ, Müller C, Bondeau A (2012) Climate-driven simulation of global crop sowing dates. Glob Ecol Biogeogr 21:247–259

Wang E, Chen C, Yu Q (2010) Modeling wheat and maize productivity as affected by climate variation and irrigation supply in North China plain. Agron J 102(3):1037–1049

Wang J, Vanga SK, Saxena R, Orsat V, Raghavan V (2018) Effect of climate change on the yield of cereal crops: a review. Climate 6:41

Yadav S, Patel H, Lunagaria M, Parmar P, Chaudhari N, Karande B, Pandey V. (2014) Impact assessment of projected climate change on pearl millet in Gujarat

Yan HX, Fang LB, Huang DZ (2011) Effects of drought stress on the biomass distribution and photosynthetic characteristics of cluster mulberry. China J Appl Ecol 22:3365–3370

Yano T, Aydin M, Haraguchi T (2007) Impact of climate change on irrigation demand and crop growth in a Mediterranean environment of Turkey. Sensors 7:2297–2315

Yin X, Kropff L, Martin J, McLaren G, Visperas RM (1995) A nonlinear model for crop development as a function of temperature. Agric For Meteorol 77:1–16

Zahoor A, Ahmed M, Fayyaz-ul-Hassan, Shabbir G, Ahmad S (2022) Ontogeny growth and radiation use efficiency of canola (*Brassica napus* L.) under various nitrogen management strategies and contrasting environments. Int J Plant Prod 16(1):195–208

Zakir I, Wahab AA, Abbas T, Haider ST, Irum S, Sabir S, Hussain S, Ahmad S (2023) Cereal-vegetable intercropping: hindrances and strategies to increase intercropping. Turk J Agric For 47(6):1115–1129

Zhang CS, Liu GB, Xue S et al (2012) Photosynthetic characteristics of Bothriochloa ischaemum under drought stress and elevated CO_2 concentration. China J Appl Ecol 23:3009–3015
Zhang F, Zhu K, Wang Y, Zhang Z, Lu F, Yu H, Zou J (2019) Changes in photosynthetic and chlorophyll fluorescence characteristics of sorghum under drought and waterlogging stress. Photosynthetica 57:1156–1164
Zhou X, Wang Q, Yang T (2020) Decreases in days with sudden day-to-day temperature change in the warming world. Glob Planet Chang 192:103239
Ziervogel G, Nyong A, Osman B, Conde C, Cortes S, Dowing T (2006) Climate variability and change: implications for household food security. Assessments of Impacts and Adaptations to Climate Change (AIACC) Working Paper No. 20, January 2006. The AIACC Project Office, International START Secretariat, Washington, DC
Zizinga A, Mwanjalolo J-GM, Tietjen B, Bedadi B, Pathak H, Gabiri G (2022) Climate change and maize productivity in Uganda: simulating the impacts and alleviation with climate smart agriculture practices. Agric Syst 199:103407

13 Alfalfa System

13.1 Introduction

Alfalfa (*Medicago sativa* L. or *Medicago falcata* L.) (Figs. 13.1 and 13.2) is the most important forage legume in the world and is sometimes referred to as the "forage's queen" because of its extensive adaptability, high production, and better quality (Putnam and Orloff 2014). Alfalfa is an annual forage legume that is widely used in cultivated fields and grasslands around the world. Because of its high yield and nutritional characteristics, it is utilized for ruminant food (Ghaleb et al. 2021), highest concentration of protein feed (Huyghe 2003), nutritive value, and high yield (Martin 2014; Biazzi et al. 2017; Bélanger et al. 2014). It originated from Southwest Asia (Afghanistan and Iran) and the Mediterranean Basin and was domestically used as the first forage crop (Cook et al. 2005). It is a fodder crop that is suited for cropping systems and agriculture production. It has a deep root system like legumes that helps to improve the structure of soil and supply of nitrogen to future crops due to its effective biological nitrogen-fixation capabilities (Huyghe et al. 2014). Alfalfa is cultivated on 35 million ha and grows in more than 80 countries (Radovic et al. 2009). In 2009, alfalfa was cultivated on 11.9 million ha in North America, 7.12 million ha in Europe, 7 million ha in South America, and 2.23 million ha in Asia and other areas of Africa and Oceania (Yuegao and Cash 2009). It is a crucial leguminous feed that is raised as a perennial crop in arid, cold climates. Once sown, it remains to exhibit good regeneration for 6–8 years with less management practices. One of the miracle plants of the frigid deserts, it provides nourishment for the ground, animals, and people. The crop offers safety and stability to farmers through guaranteed returns while being well adapted to the demands of cold, arid conditions. It is a crop that is more evenly distributed throughout the growing season than berseem and is drought and heat resistant, productive, and persistent (Iqbal et al. 2018). In temperate areas, where the temperature is a key element for determining crop growth, alfalfa grows widely in the world. Because of its nutritional value and high production, it is the most common perennial legume and forage species used to

M. Ahmed et al., *Cropping Systems Modeling Under Changing Climate*,
https://doi.org/10.1007/978-981-97-0331-9_13

Fig. 13.1 Pictorial view of *Medicago sativa* L. (Source: http://www.mdidea.com/products/new/Alfalfa_photo05.jpg)

Fig. 13.2 Pictorial view of *Medicago falcata* L. (Source: https://en.wikipedia.org/wiki/Medicago_falcata)

provide feed in Canada (Jing et al. 2020). The largest global alfalfa producers are the USA and Argentina; however, a large amount of land is dedicated to alfalfa in Australia, South and North Africa, Southern Europe, Chile, Mexico, Canada, and the Middle East (Putnam and Orloff 2014). It is mostly grown in China (Nasar et al. 2020; Xu et al. 2022a, b), New Zealand (Moot et al. 2015), Jammu and Kashmir (Iqbal et al. 2018), Pakistan (Ahmad et al. 2016), some parts of Europe like Finland

(Seppanen et al. 2018), France (Strullu et al. 2020), the Czech Republic (Hakl et al. 2019), and several parts in Ethiopia (Alemayehu et al. 2020). The maximum export of alfalfa was reported in America than Australia, Spain, Canada, and Italy (Tables 13.1 and 13.2), while import of alfalfa was higher by Japan than China, the UAE, Republic of Korea, and Saudi Arabia (Tables 13.3 and 13.4). Alfalfa is subjected to difficult growth conditions, such as dryness in northern China and Argentina, cold temperatures in Canada and Russia, and soil problems in Australia and California (America). Various cultivars (*Medicago sativa* L. or *Medicago falcata* L.) are types of wild, landrace, and variety that grow under different environmental conditions globally (Table 13.5). Climate change has a negative impact on alfalfa's productivity and quality in certain places (Collino et al. 2005; Cornacchione and Suarez 2015; Zhang et al. 2018; Dear et al. 2008).

Table 13.1 Alfalfa exported quantity in tonnes

Country	2017	2018	2019	2020	2021	2017–2021
USA	5,032,045	4,600,845	4,873,957	4,888,031	5,140,364	4,907,048.4
Australia	1,095,931	1,286,727	1,084,999	1,056,902	1,691,614	1,243,234.6
Spain	1,255,453	1,195,129	1,089,745	1,193,297	1,627,913	1,272,307.4
Canada	473,279	500,342	492,846	521,258	673,036	532,152.2
Italy	359,228	449,260	611,249	691,521	582,648	538,781.2
Others	1,447,996	1,814,457	2,052,130	1,931,906	1,686,721	1,786,642

Source: FAOSTAT

Table 13.2 Alfalfa export percent value

Countries	2017	2018	2019	2020	2021	2017–2021
USA	54.7	49.8	50.4	50.5	48.5	50.78
Australia	13.6	12.6	12.5	12.8	14.7	13.24
Spain	8.3	10.4	8.5	8.1	12.0	9.46
Canada	5.0	5.5	5.1	5.4	6.1	5.42
Italy	3.0	3.8	4.7	4.7	3.8	4.00
Others	15.4	17.9	18.8	18.5	14.9	17.10

Source: FAOSTAT

Table 13.3 Alfalfa imported quantity in tonnes

Countries	2017	2018	2019	2020	2021	2017–2021
Japan	2,195,586	2,220,142	2,269,543	2,233,116	2,234,346	2,230,546.6
China	1,855,946	1,707,104	1,627,174	1,721,993	2,044,440	1,791,331.4
UAE	1,673,994	1,717,506	1,929,689	1,738,708	1,610,795	1,734,138.4
Republic of Korea	1,354,120	1,080,904	1,062,316	1,002,549	1,016,580	1,103,293.8
Saudi Arabia	480,816	868,963	659,552	870,206	482,969	672,501.2
Others	2,272,289	2,530,822	2,444,299	2,347,600	2,552,517	2,429,505.4

Source: FAOSTAT

Table 13.4 Alfalfa import value in percent

Countries	2017	2018	2019	2020	2021	2017–2021
Japan	25.9	25.0	25.7	25.1	24.6	25.26
China	18.5	17.1	17.1	19.1	22.4	18.84
UAE	15.0	13.5	14.8	13.1	11.0	13.48
Republic of Korea	14.0	11.0	10.7	10.1	10.0	11.16
Saudi Arabia	6.7	9.5	8.6	9.3	8.1	8.44
Others	19.9	23.9	23.1	23.3	23.9	22.82

Source: FAOSTAT

Table 13.5 Alfalfa cultivars grown under different environmental conditions globally

Sr. No.	Subspecies	Type	Ploidy	Name	Country	Longitude	Latitude
1	Falcata	Wild	2×	'Romanica', 'Quasifalcata'	Russia	55.74	37.61
	Falcata	Wild	4×	'Krasnokutskaya'	Russia	50.91	47.05
3	Falcata	Wild	4×	'Maron', 'Malzeville'	France	48.63	6.18
4	Glomerata	Wild	2×	'Glomerata'	France	43.60	5.75
5	Sativa	Wild	4×	'Oscuro', 'Jara', 'Villamajor', 'Pancrudo'	Spain	41.72	−1.95
6	Falcata	Variety	2×	'Anik'	Canada		
7	Sativa	Landrace	4×	'Flamande'	France	48.85	2.35
8	Sativa	Landrace	4×	'Poitou'	France	46.58	0.34
9	Sativa	Landrace	4×	'Provence'	France	43.53	5.44
10	Sativa	Landrace	4×	'Gabès'	Tunisia	33.90	10.1
11	Sativa	Landrace	4×	'Cremonese'	Italy	45.13	10.02
12	Sativa	Landrace	4×	'Crau'	France	43.55	4.85
13	Sativa	Landrace	4×	'Demnate3', 'Atlas'	Morocco	31.73	−7.87
14	Sativa	Landrace	4×	'Dra15'	Morocco	30.33	−5.83
15	Sativa	Landrace	4×	'Ziz10'	Morocco	31.93	−4.42
16	Sativa	Landrace	4×	'Baghdadi'	Iran	32.13	48.18
17	Sativa	Variety	4×	'Banat VS'	Serbia		
18	Sativa	Variety	4×	'SW Nexus'	Sweden		
19	Sativa	Variety	4×	'Luzelle', 'Lukal', 'Ludelis', 'Barmed', 'Harpe', 'Orca'	France		
20	Sativa	Variety	4×	'Holyna'	Czech Republic		
21	Sativa	Variety	4×	'Radius'	Poland		
22	Sativa	Variety	4×	'FG-CO416C4164', 'Alforex6', 'Magna790'	USA		

Table 13.5 (continued)

Sr. No.	Subspecies	Type	Ploidy	Name	Country	Longitude	Latitude
23	Sativa	Variety	4×	'Gongnong1', 'Bauding'	China		
24	Sativa	Variety	4×	'Picena GR'	Italy		
25	Sativa	Variety	4×	'Sarghoda Alfalfa 2002'	Pakistan		

Source: Modified and adapted from Ghaleb et al. (2021)
The subspecies, population type (variety, landrace, or wild population), name of accession, and country of origin are given. For the wild populations, the geographic coordinates of the collection site are indicated. Autumn dormancy is given when known

13.2 Issues of Alfalfa Cropping System

Soil N and C concentration was increased due to continuous cultivation of alfalfa, but it lowered the phosphorus quantity in soil and decreased the water content in deep soil (Jiang et al. 2006). On arable lands, the vegetation of alfalfa is a complicated process, and different environmental components (soil properties, topography, etc.) affect it (Cimalova and Lososova 2009) and also human approaches (Yuan et al. 2015). Griggs (2002) reported agricultural infrastructure losses due to the shifting of rural agricultural businesses and fields to suburban and urban ones, including producers, equipment manufacturers and maintenance technicians, and commercial outlets. Managing farmland units that are surrounded by residential and commercial buildings and reached by highways with rising traffic volumes and speeds is difficult. Water problems include repeated salinity, drought and other impurities, public criticism of water infrastructural development and their cost-effectiveness, and growing demand for scarce water supplies from nonagricultural sectors. There is less access to skilled workers, special operators, and technical support. There is also limited and less research on the development and availability of genetically modified seeds of alfalfa, fertilizers, and pesticides. Increasing noxious weed population with the continuous growing of alfalfa decreases the alfalfa hay quality. Increasing cost of electricity and fuel and other cultural practices lower the profitability.

13.3 Alfalfa Modeling

The genetic coefficients of alfalfa in DSSAT and APSIM models are presented in Tables 13.6 and 13.7, respectively. To determine the effect of climate change on alfalfa in Eastern Canada, CSM-CROPGRO-Perennial Forage model was used to simulate the alfalfa regrowth. Under the model simulation, a number of harvests

Table 13.6 Alfalfa genetic coefficients along with description in DSSAT model

Genetic coefficient	Description
ECO#	Ecotype code of this cultivar, points to the Ecotype in the ECO file
CSDL	Critical short-day length below which reproductive development progresses with no day length effect (for short-day plants) (h)
PPSEN	Slope of the relative response of development to photoperiod with time (positive for short-day plants) (1/h)
EM-FL	Time between plant emergence and flower appearance (R1) (photothermal days)
FL-SH	Time between first flower and first pod (R3) (photothermal days)
FL-SD	Time between first flower and first seed (R5) (photothermal days)
SD-PM	Time between first seed (R5) and physiological maturity (R7) (photothermal days)
FL-LF	Time between first flower (R1) and end of leaf expansion (photothermal days)
LFMAX	Maximum leaf photosynthesis rate at 30 C, 350 vpm CO_2, and high light
	(mg CO_2/m^2-s)—default was 1.98—set to 1.76 = 40 μmol m^{-2} s^{-1} yielding
	Boote et al.'s 35.5 μmol m^{-2} s^{-1} @ 2000 μmol photons m^{-2} s^{-1}
SLAVR	Specific leaf area of cultivar under standard growth conditions (cm^2/g)
SIZLF	Maximum size of full leaf (three leaflets) (cm^2)
XFRT	Maximum fraction of daily growth that is partitioned to seed + shell
WTPSD	Maximum weight per seed (g)
SFDUR	Seed filling duration for pod cohort at standard growth conditions (photothermal days)
SDPDV	Average seed per pod under standard growing conditions (#/pod)
PODUR	Time required for cultivar to reach the final pod load under optimal conditions (photothermal days)

Source: DSSAT v4.8; Hoogenboom et al. (2021)

Table 13.7 Alfalfa genetic coefficients along with description in APSIM model

Genetic coefficients	Description
hi_incr	Rate of HI increase (optional) (1/days)
hi_max_pot_stress	Maximum harvest index potential
tt_emergence units	TT from emergence to the end of juvenile phase (°C d)
est_days_emerg_to_init	Estimated days from emergence to floral initiation (d)
x_pp_end_of_juvenile	Photoperiod
y_tt_end_of_juvenile	TT from end of juvenile to floral initiation
y_tt_floral_initiation	TT from initiation to flowering (°C d)
y_tt_flowering	TT from flowering to the start of grain fill (°C d)
tt_maturity units	TT from maturity to harvest ripe (°C d)
x_stem_wt	Stem weight (g/plant)
y_height	Plant height (mm)

were not affected, rising temperature declined the simulated biomass, an increase in atmospheric CO_2 concentration increased the yield, and more rainfall showed little change. The hot temperature may enhance the simulated harvest biomass due to harvest increases (Jing et al. 2020). The physiological activities of an alfalfa plant, as described by crop initialization settings, are simulated by the APSIM-Plant module on a daily basis with data on climatic conditions, soil conditions, and management practices. APSIM model calibrated the alfalfa radiation-use efficiency for shoot improved to the radiation-use efficiency for whole plant showed improvement in root and shoot biomass in New Zealand. The root mean standard deviation of root improved from 29% to 18% and 53% to 38% of the shoot. The root mean standard deviation of shoot biomass improved from 914 kg ha^{-1} to 653 kg ha^{-1} (Moot et al. 2015). The CSM-CROPGRO-PFM for alfalfa simulates winterkill whenever the lowest air temperature falls below −25 °C (Hoogenboom et al. 2019; Jones et al. 2003; Malik et al. 2018). The CSM-CROPGRO-Perennial Forage model was used to simulate the alfalfa yield and growth in Northeast Spain. The observed total biomass value improved from 1706 to 5617 kg DM ha^{-1}, but the range of simulated biomass was from 1983 to 6463 kg DM ha^{-1}. LAI values ranged from 0.1 to 6.7 in the observed data and from 0.41 to 9.6 in simulated data. Low LAI levels indicate early regeneration, while high values indicate full vegetative stage (before cutting). The range of observed values for the herbage crude protein and nitrogen concentrations was from 16% to 27% and 2.5% to 5.5%, respectively. For herbage CP and N, the range of simulated values was from 14% to 26% and from 1.8% to 3.7%, respectively (Malik et al. 2018). The impacts of cultivation techniques on dry matter yield and nitrogen accumulation by alfalfa perennial crop, were simulated by STICS agro-environmental model. The model simulation showed the model efficiency (0.70 and 0.75) of aboveground biomass, and total biomass and rRMSE were 42% and 36%, respectively. It also showed the model efficiency value (0.90 and 0.60) and relative root mean square error value (29% and 31%) of total nitrogen content and aboveground biomass content (Strullu et al. 2020) (Table 13.8).

Table 13.8 Impact of climate change on alfalfa production across countries in the world

Sr. No.	Country	Model	Climatic scenarios and study periods	Yield (increase or decrease)	References
1	New Zealand	APSIM-Plant	2000–2004	RMSD value was improved by 53% to 38% in shoot biomass and 29% to 18% in root biomass by improving the shoot radiation-use efficiency to whole-plant radiation-use efficiency	Moot et al. (2015)
2	Western Australia	APSIM	1995–1998	Root mean square deviation for biomass was 1.3 t ha^{-1} for the mean observed biomass of 4.17 t ha^{-1}	Dolling et al. (2005)
3	Eastern Austria	CropSyst	2007–2008	Scenario analysis reported that in the drier year 2007, root depth was 15% higher than standard cultivar and optimum for aboveground biomass, and in the year 2008, 1.3 m root length for sufficient aboveground root biomass	Raza et al. (2013)
4	Eastern Canada	CROPGRO-Perennial Forage model	1992–2016	A little increase in precipitation and an increase in CO_2 and temperature decreased the aboveground biomass	Jing et al. (2020)
5	Eastern Canada (Ottawa and Quebec City)	DNDC	RCPs 4.5 and 8.5 (1994–2011)	Yield increased by 60.3% and 81.8% under RCPs 4.5 and 8.5, respectively	He et al. (2019)
6	China	DNDC	SSP 1–2.6, SSP 2–4.5, SSP 5–8.5 (2025–2035)	In rainfed and irrigated areas, dry matter yield between 4.5 and 9.0 Mg ha^{-1} and grazing intensity decreased by 8–13% and 20–33%	Xu et al. (2022a, b)
7	France	STICS v9		rRMSE values of simulated aboveground biomass and total biomass of crop were 36% and 42%	Strullu et al. (2020)

8	Northeast Spain	CROPGRO-Perennial Forage model	2012–2016	Results showed that aboveground biomass ranged from 1706 to 5617 kg DM ha^{-1} and simulated value ranged from 1983 to 6434 kg DM ha^{-1}	Malik et al. (2018)
9	Ethiopia		3 GCMs, RCP 4.5 (2041–2060)	Projected rise of temperature by 1.1–3 °C in Ethiopia and increased (3–5%) in cultivated land of alfalfa under climatic models from total land	Alemayehu et al. (2020)
10	Mexico		11 GCMs, RCPs 4.5 and 8.5 (2021–2040, 2041–2060, 2061–2080)	Yield decreased by up to 17.8–60.0% and 40.1–69.7% under RCPs 4.5 and 8.5	Medina-Garcia et al. (2020)
11	North America (Eastern Canada)	Integrated Farm System Model (IFSM)	RCPs 4.5 and 8.5 (1971–2000, 2020–2049, 2050–2079)	Yield increased by 5–35%	Thivierge et al. (2016)

13.4 Modeling of the Dynamics of Seed Lot Germination

The cumulative percent of germination rates over time was suited to a simplified non-rectangular curve for both the impact of the variation in genetics and seed deterioration (Thornley et al. 1976).

Formula:

$$y = \left(\frac{1}{2\theta}\right) \cdot \left(\alpha \cdot (t - t_c) + y_{\max} - \sqrt{\left(\alpha \cdot (t - t_c) + y_{\max}\right)^2 - \left(4 \cdot \theta \cdot \alpha \cdot (t - t_c) \cdot y_{\max}\right)}\right)$$

where y is the cumulated seed germination (%); θ is a unitless parameter that determines the sharpness of the knee of the curve; α is the maximum germination rate (% of germinating seeds per hour); t is the time (hour); t_c is the lag to start germination (hour) during which no seed germinates; and y max is the asymptotic maximum seed germination (%). After several tests, the number of parameters to be estimated was limited to 0.97.

13.5 Impact of Climate Change on Alfalfa Phenology

Small-seeded perennial alfalfa forage plants demand a lot of attention when first being established (Fig. 13.3). Phenology modules are accessible in APSIM (Robertson et al. 2002), CropSyst (Confalonieri and Bechini 2004), DSSAT CSM-CROPGRO-Perennial model (Malik et al. 2018), and STICS framework (Strullu et al. 2020). The main factors that affect the phenology of alfalfa are temperature and photoperiod in which every stage of development i.e., plant height, nodes per stem, etc. (Hanson et al. 1988; Teixeira 2006) is affected. In changing climatic conditions, alfalfa germination is affected by cool or hot temperature conditions of 5–40 °C. At a high temperature of 40 °C, germination was minimum, but at low temperature, germination was affected less. The optimum temperature for alfalfa germination is 11.7–21.1 °C (Ahmed et al. 2019). Durr et al. (2015) reported on the impact of temperature on alfalfa cultivars; cultivated species develop more quickly in a wide temperature range than wild species, and the behavior of various cultivars to temperature during germination has always become a major issue. Increasing temperature affects the *Medicago sativa*, *Medicago falcata*, and wild sativa cultivars. Alfalfa germination was optimum at 15 and 22 °C, but at 40 °C, it showed less germination. Maximum germination was at 5–34 °C temperature for sativa cultivar, while wild sativa and falcata cultivars showed less germination at 5 and 34 °C. At both temperatures, germination was enhanced by seed scarification in wild sativa and falcata cultivars. In Fig. 13.3, germination percentage of alfalfa is shown at

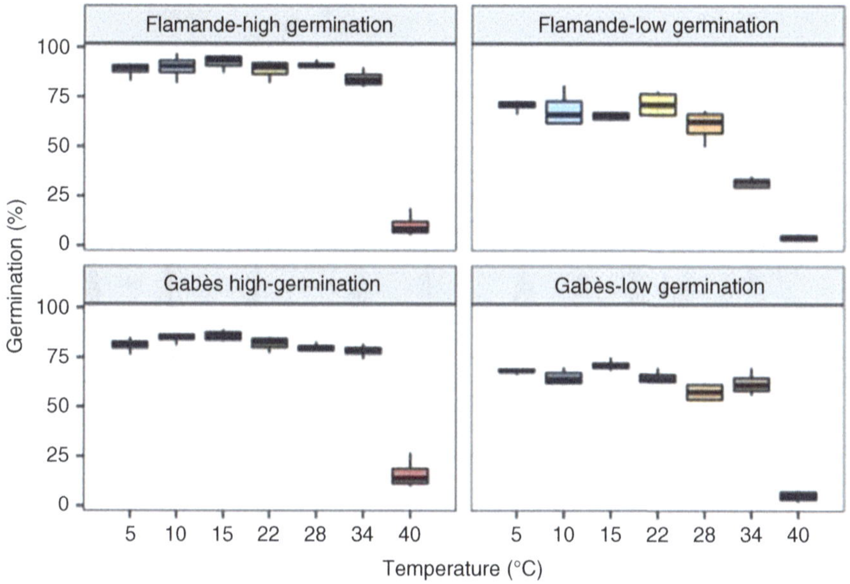

Fig. 13.3 Temperature impact on alfalfa germination. (Source: Adapted from Ghaleb et al. 2021)

different temperature ranges between 5 and 40 °C (Ghaleb et al. 2021). The optimum temperature for lucerne growth is 25 °C, and high production can occur between 10 and 30 °C; yield dramatically declines below 10 °C (McDonald et al. 2003). The scheduling of management activities and the defoliation strategy must be optimized based on the morphological and physiological development of alfalfa. Identifying the seasonal alfalfa development processes in relation to management, climatic factors, and genotype is a problem for the module on alfalfa phenology. Under environmental conditions, the phenological response is simulated by APSIM NextGen Alfalfa model. The simulation's conclusions for predicting stages of development were consistent that Nash-Sutcliffe efficiency (NSE) value for days to visible bud (0.77) ranged from 0.53 to 0.84 for main stem node and days to flowering stage (0.67) (Yang et al. 2021). Photoperiod reaction was examined in ten alfalfa cultivar seedlings, and different vegetative development stages were separated based on photoperiod response. This featured a juvenile phase, which did not exhibit any PP reaction but which must be finished even before the plant exhibits photoperiod response. The optimal maximum photoperiod is 18 h; rather, it is alfalfa flowering that is affected by the photoperiod (Major et al. 1991). The establishment and phenology of alfalfa crop are presented in Figs. 13.4 and 13.5, respectively.

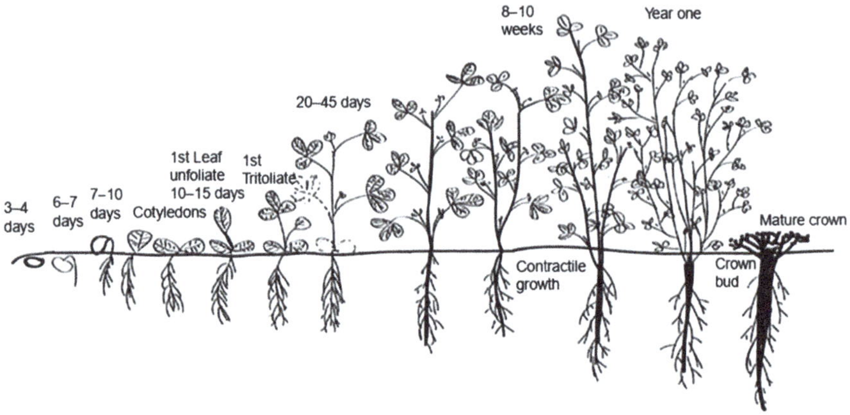

Fig. 13.4 Establishment of an alfalfa plant. (Source: Adapted from Putnam and Orloff 2014)

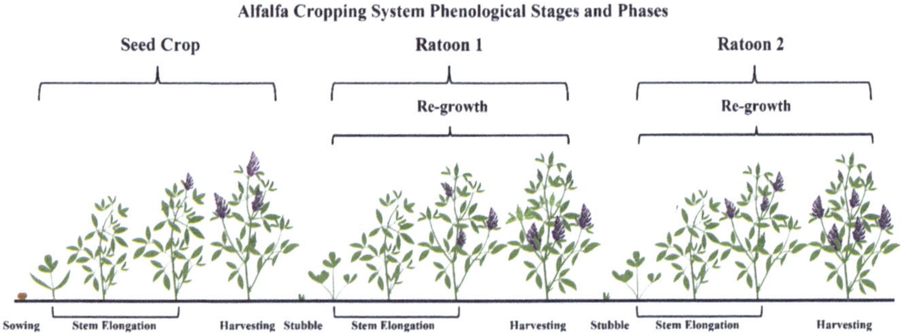

Fig. 13.5 Overview of phenological stages and phases of alfalfa crop

13.6 Climate Change Impact on Alfalfa Production

In New Zealand, the APSIM model was used to calibrate the development of alfalfa crop under the photoperiod and temperature response that sowed in a temperate cool climate. Simulated results showed that changing the RUE for the "shoot" to a "whole-plant" radiation-use efficiency of 1.6 g/MJ total solar radiation and incorporating a seasonal rate of root biomass turnover (i.e., respiration plus senescence) accounting the beginning offset between observed and simulated root biomass. These adjustments reduced the mean belowground biomass RMSD (29% to 18%) and mean aboveground biomass RMSD (53% to 38%) (Moot et al. 2015). The DeNitrification-DeComposition (DNDC) model with future scenarios (RCPs 4.5 and 8.5) was used to assess the effect of climate change on alfalfa biomass in Eastern Canada from 1994 to 2011. Under the DNDC model, alfalfa's dry matter yield improved due to the winterkill effect. At Elora, the average of DNDC models was less than 2%, but in 1996–1997 overwinter, it was 18.6%, and at Normandin, in

2009–2010 overwinter, it was 27.1%. Under future scenarios, decreased snow cover, which results in colder soil temperatures and lessened autumn hardening, may increase the occurrence of winterkill for areas around Quebec City and Ottawa in the future. Due to earlier sowing and spring regrowth with extra cuttings, improved C absorption, and decreased water stress under higher CO_2 concentrations, annual average yield increased 60.3% and 81.8% under the future scenario RCPs 4.5 and 8.5, respectively, and it was reported that for increasing production, the impact of winterkill will be mitigated by hardy cultivars of winter (He et al. 2019). Alemayehu et al. (2020) used three global circulation models (GCMs) (CCSM4, MIROC5, and HadGEM2-AO) to estimate the yield changes with two climate scenario RCPs 4.5 and 8.5 for 2041–2060. In Ethiopia, under RCP 4.5, temperature rose by 1.1–3 °C in 2050, but in some northeastern parts, temperature decreased by −0.1 °C. Two climatic models projected that in the cooling areas of eastern lowland and central part of Ethiopia, temperature rose by 0.9–3 °C, and in the southern part, temperature increased by more than 3 °C. The climatic model projected that in some parts of the country, alfalfa's cultivated area slightly increased in the 2050s as compared to a base period. Medina-Garcia et al. (2020) used 11 global circulation models under RCPs 4.5 and 8.5 to estimate alfalfa production for future scenarios (2030, 2050, 2070). Due to increasing temperatures, alfalfa production decreases in the future by up to 17.8–60.0% and 40.1–69.7% under RCPs 4.5 and 8.5, but in 2070, under the projection of RCP 4.5, yield will reduce to 24.7%. Cardinal temperatures provide a summary of how temperature affects seed germination (Durr et al. 2015). It is regarded as among the most high-quality forage species for massive feed production in grassland (Nasar et al. 2020; Xu et al. 2022a, b). In China, meat production was 77.6 million tonnes, being the second largest meat producer in the world, in 2020, but increased grazing capacity under rainfed and irrigated conditions will decline by 8–13% and 20–33% in the natural grassland in China due to climate change in future. Alfalfa dry production yield varied from 4.5 to 9.0 Mg ha^{-1} in both conditions, respectively. In the future (2025–2035), the DNDC model shows that alfalfa cultivation will reduce the natural grazing grassland, along with an increase in soil organic carbon concentration (8.5 and 9.9 g kg^{-1}) in alfalfa pasture management under both conditions, respectively (Xu et al. 2022a, b). Forage schedule of harvest is expected to be affected by climate change (Jing et al. 2013). Annual aboveground biomass yield of forage increased with an extra cut of forage in Eastern Canada (Jing et al. 2014). In the USA, changes in temperature and atmospheric CO_2 concentration and reduced precipitation were found to affect alfalfa yield (Izaurralde et al. 2011). Belanger et al. (2002) reported that elevated CO_2 and temperature might be in alfalfa's favor during summer but rising in temperature during winter damaged alfalfa crop with increasing trends. The model Hydrus-1D was utilized to simulate soil water dynamics behavior in the 0–1500 cm profile and calculate the period of recovery in the alfalfa-corn and alfalfa-wheat cropping system using the calibrated parameters of soil hydraulic, in which 53 climatic simulations were derived using random non-repeat cropping sequences under 53 years of data of observed climatic (1968–2020) and ten global climatic models under climate scenario RCPs 4.5 and 8.5 from 2021 to 2100. The findings revealed that, under the 53 climatic simulations, the recovery period required for the alfalfa-wheat and alfalfa-corn cropping system was 14–44 years and 11–31 years, respectively. The median recovery period was 37.5–42.5 years and 30–31.5 years under climate scenario RCPs 4.5 and 8.5 for alfalfa-wheat and alfalfa-corn cropping systems (Zhao et al. 2022). At the end of this century, the temperature is expected to rise by 1.3–5 °C in China. Due to this global warming, the time required for the recovery of soil water will increase after alfalfa cutting because the rate of potential evapotranspiration increases (Zhao et al. 2022). The nutritive and dry matter yield of forage was simulated with the use of an Integrated Farm System Model (IFSM) prediction with the scenarios of two greenhouse gas concentrations under climate scenario RCPs 4.5 and 8.5 from 2020 to 2080 in Eastern Canada. Results of climatic projections showed that in warmer areas, yield decreased, but in colder regions, it increased. Due to the high accumulation of growing degree days, yield of the first cut increased, but after that, yields decreased because of high temperature and water stress. The nutritive value of forage reduced in warmer areas under climatic scenario RCP 8.5 (2050–2079) due to increasing temperature and CO_2 concentration, but yield increased (Thivierge et al. 2016). According to weather conditions, the harvested cycle for lucerne is 40 days, and its annual DM production is between 10 and 12 t/ha (Limagrain 2011).

13.7 Climate Change's Impact on Alfalfa Quality

The digestibility of alfalfa dry matter is reduced with the rise in temperature. Changes in CO_2 concentration (700 mmol mol^{-1} CO_2) and increased temperature (4 °C higher than ambient temperature) will reduce the crude protein content and in vitro dry matter digestibility (IVDMR) but will increase the fiber content. Cellulose concentration in alfalfa is increased due to an increase in CO_2 and temperature (Sanz-Saez et al. 2012). Changing climate and CO_2 concentration reduce alfalfa nutritive values (Thivierge et al. 2016). The additional cut and increased dry matter yield of forage have resulted in a minute effect on forage nutritional value (Jing et al. 2014). In high-CO_2 environments, inadequate leaf N supply immobilizes carbon, which increases carbohydrate buildup, decreases photosynthetic regulation, and limits the yield (Sanz-Sáez et al. 2010). Reduction in crude protein brought on by losses in shoot nitrogen content under high CO_2 results in a loss in forage quality (Campbell et al. 2000). During elevated carbon dioxide exposure, the in vitro dry matter digestibility (IVDMD) of various forages decreases, resulting in an increase in fiber percentage in plants (Owensby et al. 1996; Morgan et al. 2004). Alfalfa shoot biomass is more affected by elevated carbon dioxide than root biomass (Erice et al. 2007). Due to global warming, neutral detergent fiber (NDF) content has increased and digestibility has decreased because of increasing temperature (Thorvaldsson et al. 2007), but according to Bertrand et al. (2008), under high temperatures, neutral detergent fiber content has remained unaltered. Under conditions of elevated CO_2, the source organ function of alfalfa roots during regrowth may change (Erice et al. 2006).

13.8 Adaptation Strategies

Alfalfa has a variable reaction to temperature for germination, which allows for the development of new cultivars that are adapted to upcoming global conditions brought by the changing climate (Ahmed et al. 2019). Fodder quality should be improved including feeding value parameters such as digestibility and fiber content (Julier et al. 2000). An essential method for adapting to climate change is the fodder harvest strategy with the potential for future cuts. Because it is mostly used with grasses in combination, the percentage of each species in these combinations may shift as a result of climate change. Alfalfa may profit more from the upcoming growing circumstances than grasses because it is a legume with the ability to fix nitrogen (Thivierge et al. 2016). Alfalfa should be rotated with annual crops, particularly cereals, to lower disease inoculum rates. Insect infestations that spread infection or stress plants should be monitored and controlled. Cultivars should be resistant to diseases (https://www.gov.mb.ca/agriculture/crops/crop-management/forages/management-of-diseases-of-alfalfa-seed.html).

Improvements should be made in nutrient budgeting, irrigation management, and technology for increased alfalfa productivity and feed composition (Griggs 2002). Modeling strategies should be taken to minimize the yield gap in the future.

13.9 Conclusion

The sustained productivity of alfalfa is very critical for livestock production. The decreasing trends of yield are being reported across various countries in the world. Climate change is also one of the reasons for lower productivity. The Decision Support System for Agrotechnology Transfer is an emerging approach to tackle the issue of climate change for determining climate-smart packages for alfalfa production under current and future climate scenarios.

References

Ahmad J, Iqbal A, Ayub M, Akhtar J (2016) Forage yield potential and quality attributes of Alfalfa (*Medicago sativa* L.) under various agro-management techniques. J Animal Plant Sci 26(2):465–474

Ahmed LQ, Durand JL, Gutierrez AJE (2019) Genetic diversity of Alfalfa (*Medicago sativa*) in response to temperature during germination. Seed Sci Technol 47(3):351–356

Alemayehu S, Ayana EK, Dile YT, Demissie T, Yimam Y, Girvetz E, Aynekulu E, Solomon D, Worqlul AW (2020) Evaluating land suitability and potential climate change impacts on Alfalfa (*Medicago sativa*) production in Ethiopia. Atmos 11:1124

Belanger G, Rochette P, Castonguay Y, Bootsma A, Mongrain D, Ryan DAJ (2002) Climate change and winter survival of perennial forage crops in eastern Canada. Agron J 94:1120–1130

Bélanger G, Castonguay Y, Lajeunesse J (2014) Benefits of mixing timothy with Alfalfa for forage yield, nutritive value, and weed suppression in northern environments. Can J Plant Sci 94:51–60

Bertrand A, Tremblay GF, Pelletier S, Castonguay Y, Belanger G (2008) Yield and nutritive value of timothy as affected by temperature, photoperiod and time of harvest. Grass Forage Sci 63:421–432

Biazzi E, Nazzicari N, Pecetti L, Brummer EC, Palmonari A, Tava A, Annicchiarico P (2017) Genome-wide association mapping and genomic selection for Alfalfa (*Medicago sativa*) forage quality traits. PLoS One 12:e0169234

Campbell BD, Stafford-Smith DM, GCTE (2000) Pastures and Rangelands network members. A synthesis of recent global change research on pasture and rangeland production: reduced uncertainties and their management implications. Agric Ecosyst Environ 82:39–55

Cimalova S, Lososova Z (2009) Arable weed vegetation of the northeastern part of the Czech Republic: effects of environmental factors on species composition. Plan Ecol 203:45–57

Collino DJ, Dardanelli JL, De Luca MJ, Racca RW (2005) Temperature and water availability effects on radiation and water use efficiencies in Alfalfa (*Medicago sativa* L.). Aust J Exp Agric 45(4):383–390

Confalonieri R, Bechini L (2004) A preliminary evaluation of the simulation model CropSyst for Alfalfa. Eur J Agron 21:223–237

Cook BG, Pengelly BC, Brown SD, Donnelly JL, Eagles DA, Franco MA, Hanson J, Mullen BF, Partridge IJ, Peters M, Schultze-Kraft R (2005) Tropical forages. CSIRO, DPI&F(Qld), CIAT and ILRI, Brisbane

Cornacchione MV, Suarez DL (2015) Emergence, forage production, and ion relations of Alfalfa in response to saline waters. Crop Sci 55(1):444–457

Dear BS, Reed KFM, Craig AD (2008) Outcomes of the search for new perennial and salt tolerant pasture plants for southern Australia. Aust J Exp Agric 48(4):578–588

Dolling PJ, Robertson MJ, Asseng S, Ward PR, Latta RA (2005) Simulating Alfalfa growth and water use on diverse soil types in a Mediterranean-type environment. Aust J Agric Res 56:503–515

Durr C, Dickie JB, Yang XY, Pritchard HW (2015) Ranges of critical temperature and water potential values for the germination of species worldwide: contribution to a seed trait database. Agric Forest Meteorol 200:222–232

Erice G, Irigoyen JJ, Perez P, Martínez-Carrasco R, Sanchez-Díaz M (2006) Effect of elevated CO_2, temperature and drought on dry matter partitioning and photosynthesis before and after cutting of nodulated Alfalfa. Plant Sci 170:1059–1067

Erice G, Irigoyen JJ, Sanchez-Diaz M, Avice JC, Ourry A (2007) Effect of drought, elevated CO_2 and temperature on accumulation of N and vegetative storage proteins (VSP) in taproot of nodulated Alfalfa before and after cutting. Plant Sci 172:903–912

Ghaleb W, Ahmed LQ, Escobar-Gutiérrez AJ, Julier B (2021) The history of domestication and selection of Alfalfa: a new perspective from the genetic diversity for seed germination in response to temperature and scarification. Front Plant Sci 11:578121

Griggs TC (2002) Extension Agronomy Specialist, Dept. of Plants, Soils, & Biometeorology, Utah State Univ., Logan, UT 84322-4820; W.W. Riggs, Extension Educator, University of Nevada Cooperative Extension, Eureka County, P.O. Box 613, Eureka, NV 89316; In: Proceedings, Western Alfalfa and Forage Conference, 11–13 Dec 2002, Sparks, NV, UC Cooperative Extension, University of California, Davis

Hakl J, Mofidian SMA, Kozova Z, Fuksa P, Jaromir S (2019) Estimation of Alfalfa yield stability for enabling effective cultivar selection under rainfed conditions. Grass Forage Sci 74:687–695

Hanson AA, Barnes D, Hill RR, Heichel GH, Leath K, Hunt O, Marten G, Tesar M, Mickelson S, Holtgraver K (1988) Alfalfa and Alfalfa improvement. American Society of Agronomy, Madison

He W, Grant B, Smith WN, VanderZaag AC, Piquette S, Qian B, Jing Q, Rennie TJ, Belanger G, Jego G, Deen B (2019) Assessing Alfalfa production under historical and future climate in eastern Canada: DNDC model development and application. Environ Model Softw 122:104540

Hoogenboom G, Porter CH, Boote KJ, Shelia V, Wilkens PW, Singh U, White JW, Asseng S, Lizaso JI, Moreno LP, Pavan W, Ogoshi R, Hunt LA, Tsuji GY, Jones JW (2019) The DSSAT crop modeling ecosystem. In: Boote KJ (ed) Advances in crop modeling for a sustainable agriculture. Burleigh Dodds Science Publishing, Cambridge, pp 173–216. https://doi.org/10.19103/AS.2019.0061.10

Hoogenboom G, Porter CH, Shelia V, Boote KJ, Singh U, White JW, Pavan W, Oliveira FAA, Moreno-Cadena LP, Lizaso JI, Asseng S, Pequeno DNL, Kimball BA, Alderman PD, Thorp KR, Jones MR, Cuadra SV, Vianna MS, Villalobos FJ, Ferreira TB, Batchelor WD, Koo J, Hunt LA, Jones JW (2021) Decision Support System for Agrotechnology Transfer (DSSAT) version 4.8 (DSSAT.net), DSSAT Foundation, Gainesville

Huyghe C (2003) Forages and protein production. In: Proceedings of the AFPF days. French Association for Forage Production, Paris, pp 3–16

Huyghe C, Delaby L, Rosellini D, Veronesi F (2014) Perspectives of Alfalfa in Europe. Legume Perspect 4:5–6

Iqbal S, Ahmad M, Zaffar G, Ahmad F, Dar EA (2018) Alfalfa (Medicago sativa L): a valuable source of forage/fodder crop under cold arid conditions of Ladakh (Jammu and Kashmir). Res J Chem Environ Sci 6(4):74–84

Izaurralde RC, Thomson AM, Morgan JA, Fay PA, Polley HW, Hatfield JL (2011) Climate impacts on agriculture: Implications for forage and rangeland production. Agron J 103:371–381

Jiang HM, Jiang JP, Jia Y, Li FM, Xu JZ (2006) Soil carbon pool and effects of soil fertility in seeded Alfalfa fields on the semi-arid losses Plateau in China. Soil Biol Biochem 38(8):2350–2358

Jing Q, Belanger G, Qian B, Baron V (2013) Timothy yield and nutritive value under climate change in Canada. Agron J 105:1683–1694

Jing Q, Belanger G, Qian B, Baron V (2014) Timothy yield and nutritive value with a three-harvest system under the projected future climate in Canada. Can J Plant Sci 94:213–222

Jing Q, Qian B, Belanger G, VanderZaag A, Jego G, Smith W, Grant B, Shang J, Liu J, He W, Boote K, Hoogenboom G (2020) Simulating Alfalfa regrowth and biomass in eastern Canada using the CSMCROPGRO-perennial forage model. Eur J Agron 113:125971

Jones JW, Hoogenboom G, Porter CH, Boote KJ, Batchelor WD, Hunt LA, Wilkens PW, Singh U, Gijsman AJ, Ritchie JT (2003) DSSAT cropping system model. Eur J Agron 18:235–265

Julier B, Huyghe C, Ecalle C (2000) Within- and among-cultivar genetic variation in Alfalfa: forage quality, morphology and yield. Crop Sci 40(2):365–369

Limagrain (2011) Growers guide to Lucerne. Rothwell, Market Rasen, Lincolnshire

Major DJ, Hanna MR, Beasley BW (1991) Photoperiod response characteristics of Alfalfa (Medicago sativa L) cultivars. Can J Plant Sci 71:87–93

Malik W, Boote KJ, Hoogenboom G, Cavero J, Dechmi F (2018) Adapting the CROPGRO model to simulate Alfalfa growth and yield. Agron J 110(5):1777–1790

Martin NP (2014) Perspectives of Alfalfa in North America. Legume Perspect 4:7–8

McDonald W, Nikandrow A, Bishop A, Lattimore M, Gardner P, Williams R, Hysons L (2003) Lucerne for pasture and fodder. Agfact.2.2.2.25, 3rd edn. NSW Agriculture, Orange

Medina-Garcia G, Echavarria-Chairez FG, Ruiz-Corral JA, Rodriguez-Moreno VM, Soria-Ruiz J, La Mora-Orozco CD (2020) Global warming effect on Alfalfa production in Mexico. Mexican J Livestock Sci 11(2):34–48

Moot DJ, Hargreaves J, Brown HE, Teixeira EI (2015) Calibration of the APSIM-Alfalfa model for 'Grasslands Kaituna' Alfalfa crops grown in New Zealand. New Zealand J Agric Res 58(2):190–202

Morgan JA, Mosier AR, Milchunas DG, LeCain DR, Nelson JA, Parton WJ (2004) CO_2 enhances productivity, alters species composition, and reduces digestibility of shortgrass steppe vegetation. Ecol Appl 14:208–219

Nasar J, Shao Z, Arshad A, Jones FG, Liu S, Li C, Khan MZ, Khan T, Banda JSK, Zhou X, Gao Q (2020) The effect of maize-Alfalfa intercropping on the physiological characteristics, nitrogen uptake and yield of maize. Plant Biol 22(6):1140–1149

Owensby CE, Cochran RM, Auen LM (1996) Effects of elevated carbon dioxide on forage quality for ruminants. In: Korner C, Bazzaz F (eds) Carbon dioxide, populations and communities, Physiological ecology series. Academic Press, San Diego, pp 363–371

Putnam DH, Orloff SB (2014) Forage crop. In: van Alfen NK (ed) Encyclopedia of agriculture and food systems. Elsevier, London, pp 381–405

Radovic J, Sokolovic D, Markovic J (2009) Alfalfa-most important perennial forage legume in animal husbandry. Biotechnol Animal Husbandry 25:465–475

Raza A, Friedel JK, Moghaddam A, Ardakani MR, Loiskandl W, Himmelbauer M, Bodner G (2013) Modeling growth of different Alfalfa cultivars and their effect on soil water dynamics. Agric Water Manag 119:100–110

Robertson MJ, Carberry PS, Huth NI, Turpin JE, Probert ME, Poultron PL, Bell MJ, Wright GC, Yeates SJ, Brinsmead RB (2002) Simulation of growth and development of diverse legumes species in APSIM. Aust J Agric Res 53:429–446

Sanz-Sáez A, Erice G, Aranjuelo I, Nogues S, Irigoyen JJ, Sánchez-Díaz M (2010) Photosynthetic down-regulation under elevated CO_2 exposure can be prevented by nitrogen supply in nodulated Alfalfa. J Plant Physiol 167:1558–1565

Sanz-Saez A, Erice G, Aguirreolea J, Munoz F, Sanchez-Díaz M, Irigoyen JJ (2012) Alfalfa forage digestibility, quality and yield under future climate change scenarios vary with *Sinorhizobium meliloti* strain. J Plant Physiol 169(8):782–788

Seppanen MM, Alitalo V, Bäckström HK, Mäkiniemi K, Jokela V, Falghera-Winseman L, Khazaei H (2018) Growth, freezing tolerance, and yield performance of Alfalfa (Medicago sativa L.) cultivars grown under controlled and field conditions in northern latitudes. Can J Plant Sci 98:1109–1118

Strullu L, Beaudoin N, Thiebeau P, Julier B, Mary B, Ruget F, Ripoche D, Rakotovololona L, Louarn G (2020) Simulation using the STICS model of C&N dynamics in Alfalfa from sowing to crop destruction. Eur J Agron 112:125948

Teixeira EI (2006) Understanding growth and development of Alfalfa crops (Medicago Sativa L.) with contrasting levels of perennial reserves. PhD thesis, Lincoln University

Thivierge MN, Jego G, Belanger G, Bertrand A, Tremblay GF, Rotz CA, Qian B (2016) Predicted yield and nutritive value of an Alfalfa–Timothy mixture under climate change and elevated atmospheric carbon dioxide. Agron J 108(2):585–603

Thornley JHM, Sutcliffe JF, Mahlberg P (1976) Mathematical models in plant physiology. Academic Press, London

Thorvaldsson G, Tremblay GF, Kunelius HT (2007) The effects of growth temperature on digestibility and fibre concentration of seven temperate grass species. Acta Agric Scand B Soil Plant Sci 57:322–328

Xu L, Li D, Wang D, Ye L, Nie Y, Fang H et al (2022a) Achieving the dual goals of biomass production and soil rehabilitation with sown pasture on marginal cropland: evidence from a multiyear field experiment in northeast inner Mongolia. Front Plant Sci 13:985864

Xu L, Ye L, Nie Y, Yang G, Xin X, Yuan B, Yang X (2022b) Sown Alfalfa pasture decreases grazing intensity while increasing soil carbon: experimental observations and DNDC model predictions. Front Plant Sci 13:1019966

Yang X, Brown HE, Teixeira EI, Moot DJ (2021) Development of a Alfalfa model in APSIM next generation: 1 phenology and morphology of genotypes with different fall dormancies. Eur J Agron 130:126372

Yuan ZQ, Epstein H, Li FM (2015) Factors affecting the recovery of abandoned semi-arid fields after legume introduction on the Loess Plateau. Ecol Eng 79:86–93

Yuegao H, Cash D (2009) Global status and development trends of Alfalfa. In: Cash D (ed) Alfalfa management guide for Ningxia. United Nations Food and Agriculture Organization, Roman, pp 1–12

Zhang W, Hou L, Yang J, Song S, Mao X, Zhang Q et al (2018) Establishment and management of Alfalfa pasture in cold regions of China. Chin Sci Bull 63(17):1651–1663

Zhao X, Huang M, Yan X, Yang Y (2022) The impacts of climate change and cropping systems on soil water recovery in the 0–1500 cm soil profile after Alfalfa. Agric Water Manag 272:107878

Groundnut-Canola System 14

14.1 Introduction

Peanut is an imperative crop grown across the world. One of India's major sources of food, oil, and fodder is groundnut. The protein content of peanuts, which are technically a legume, is higher than that of any other nut and is on par with or higher than that of a serving of beans (Chang et al. 2013). Commercially, peanuts are mostly used to produce oil, but outside oil, they also produce a variety of other useful by-products, including proteins, fiber, polyphenols, antioxidants, vitamins, and minerals, all of which may be employed as functional ingredients in a variety of processed meals (Bertioli et al. 2019). If the peanut oil is removed, the cake's protein level might increase to 50%. Due to soil and rainfall variability over space, it is cultivated in marginal dry and semiarid agroecosystems with significant yield swings. Each year, 44.01 million tonnes of peanuts are produced globally. With an annual production volume of 16.68 million tonnes, China is the world's top producer of peanuts. With an annual production of 6.85 million tonnes, India ranks in second. Together, China and India produce more than 50% of the world's production. In addition to local and international economic, social, and regulatory changes, climate change, which is anticipated to increase intra- and interannual rainfall variability, may further restrict India's groundnut sector (Abbas et al. 2017, 2019, 2020a, b, c; Abbas and Ahmad 2018; Ahmad et al. 2012a, b, 2013, 2015a, b, c, 2016a, b, c, 2017a, b, 2018, 2019, 2021, 2023a, b; Zhen et al. 2023).

For the production of edible oils, canola (*Brassica napus* L.) is crucial to guaranteeing global food security. It is cultivated in more than 50 countries and contributes around 18% of the world's total grain production of oilseed crop. Moreover, it serves as a crucial intercalary crop for controlling cereal root infections (Chmielewska et al. 2021). The primary sources of edible oil in the nation are rapeseed and mustard. Rapeseed is a great source of both protein and oil, with its meal comprising 38–40% of protein and 40–44% of high-quality edible oil. It contains a wide range of amino acids, including lysine, cysteine, and methionine. Double-zero (00) kinds

M. Ahmed et al., *Cropping Systems Modeling Under Changing Climate*, https://doi.org/10.1007/978-981-97-0331-9_14

of rapeseed and mustard are those with less than 2% erucic acid in the oil and fewer than 30 moles/g glucosinolates in oil-free meal. Rapeseed (00) oil is appropriate for human consumption, and its meal is suitable for use in animal feed and for poultry in particular (Kirkegaard et al. 2021; Mustafa et al. 2022). Across the world, rapeseed was grown on 35 million acres in 2019. 85% of the world's seeded areas, 29.8 million hectares, were planted with rapeseed by the top five leaders in terms of acreage (Tariq et al. 2017, 2018a, b; Tarriq et al. 2021; Hussain et al. 2018; Khan et al. 2019, 2020; Zahoor et al. 2022). The European Union produced 19.5 million metric tonnes of rapeseed, making it the most in the world. In 2022–2023, the total amount of rapeseed oil produced worldwide was 31.8 million metric tonnes.

14.2 Issues of Groundnut-Canola Cropping System

While high temperatures and drought stress are the key climatic elements that restrict the yield of the groundnut-canola cropping system, it is reasonable to predict that productivity pressure would only increase as a result of climate change. Low soil fertility is one of the main issues contributing to low yield in barani cropping systems (Lilley et al. 2019). Poor and erratic rainfall contributes to peanut production reductions, poor seed quality, and aflatoxin contamination (Gul et al. 2022). Smallholder farmers worldwide have low yields of groundnut-canola due to a number of biotic and abiotic variables, including pests like termites and aphids as well as diseases like groundnut rosette disease, leaf spot, and rust. Abiotic causes include drought and low soil quality (Majola et al. 2021). Low-yield cultivars and uneven nitrogen and phosphate fertilization are the main causes of the yield gap, among other agronomic issues. Because of high moisture and heat stressors during the reproductive period, late seeding might result in low yield. The construction of an effective groundnut-canola cropping system is anticipated to become more difficult as a result of a changing environment and limitations on the use of crop protection measures (Ahmed et al. 2016, 2018a, b, 2019, 2020, 2022; Ahmad and Hasanuzzaman 2012; Ali et al. 2013; Amin et al. 2018a, b, c; Raza et al. 2019).

14.3 Decision Support System for Agrotechnology Transfer

14.3.1 Groundnut

Comprehending growth and yield of groundnut in cropping system by using computer simulation models is very necessary (Mahmood et al. 2013; Mubeen et al. 2016; Nasim et al. 2018; Mubarak et al. 2022; Rehman et al. 2015; Rahman et al. 2018, 2021; Sarwar et al. 2013, 2021; Shabbir et al. 2016). Crop growth simulation models can simulate the longer period of time-based influences of climate fluctuations as well as effects of different soil characteristics and crop husbandry practices on water equilibrium in soil, nutrient dynamics, growth, phenology, and yield of groundnut. Crop modeling has been acknowledged as an effective method for estimating yield potential, identifying yield gaps and limits, assessing genotype by environment by management interactions, and evaluating agronomic choices to boost productivity (Ejaz et al. 2022; Fatima et al. 2018, 2020, 2021; Nadeem et al. 2022; Naz et al. 2022; Jones et al. 2003; Keating et al. 2003; Hoogenboom et al. 2019, 2021; Kumar et al. 2012). By adjusting management decisions to the present and anticipated weather, crop growers will be capable to lower yield perils and boost groundnut production. The use of growth simulation models must increase the effectiveness of better verdict making for the soil, irrigation, and management practices of crop. Crop models may integrate the soil-plant-atmosphere continuum and trace how various climatic parameters interact with crop development and production, making them important tools for measuring the effects of climatic changes on agriculture. Faye et al. (2018) stated that crop models for groundnut are practical tools for examining how climate change may affect crops and developing effective adaptation measures. As in the assessment process, the model had strong agreement with the observed seed yield for both kinds; the model accurately re-created the seed yield under the no heat stress and drought stress conditions after calibration. The seed yield predicted by Tc matched the data obtained quite well. Haerani et al. (2023) used groundnut model in Australia. An influence of climate change upon the possible spread of groundnut crop in Australia was assessed with the help of the model. With the help of CLIMatic indEX's Species Distribution Models (SDMs), the distributions of peanut crops both currently and in the future were projected (CLIMEX). Different GCMs were used to simulate the distributions of future peanut crops in Australia for the years 2030, 2050, 2070, and 2100 under the SRES A2 climate change scenarios from the CliMond 10 database. Across the two GCMs' anticipated years, the data showed a growth in places in Australia that are unsuited for peanut farming. In 2100, more unsuitable regions were predicted by the CSIRO-Mk3.0 model (76% of the Australian continent) than by the MIROC-H model (48% of the Australian continent). It was discovered that locations that are now appropriate for the development of peanut crops may be constrained by the expected rise in dry stress in the future. Zhen et al. (2023) used a model for peanut, and simulated results showed that the model produced accurate yield and maturity date predictions for the three assessment seasons, with RMSE values of 429 kg/ha and 5 days, respectively, and an overall R^2 of 0.76 and 0.83. Xu et al. (2017) determined the evaluations of the calibrated DSSAT model using flowering and maturity dates in the 12 observation places. The simulation's anthesis and maturity dates closely match the data, according to the results, which have R^2 values of 0.74 and 0.82, respectively. The production of peanuts under irrigation and rain conditions is then simulated over the main peanut-growing regions in China using the calibrated and verified DSSAT model. Mokhtar and Abo Elenein (2014) used a model for the peanut cultivar. Peanut phenological stages were closely simulated to field measured values for flowering stage, seed filling initiation stage, and maturity stage; slight variations occurred at the maturity to the harvest phase. The model predicted that harvesting stage was almost 7 days delayed than the actual date for the 2010 season. Even yet, subsequent growth stages responded strongly to the model, even though overestimation happened at this level. The percentage discrepancy between the measured and expected length of the growing season varied from 0.00% to 6.01%. According to the statistical study, the length of the growth season was accurately predicted with an R^2 value of 0.999. Since peanut production was simulated with a satisfactory level of accurateness for the study's purposes, validation results were generally satisfactory. This shows that the CSM-CROPGRO-Peanut crop model is very helpful for evaluation and simulation of the groundnut yield and irrigation requirements using environment circumstance data. Singh et al. (2012) used CSM-CROPGRO-Peanut model in India and West Africa. The groundnut varieties 'JL-24' and 'M-335' were previously calibrated by the model for Indian locations, and it discovered a significant correlation between predicted and measured pod yields (cv. JL-24: R^2 = 0.90; cv. M-335: R^2 = 0.82). For the cultivars, the d-values were 0.97 and 0.92 for 'JL-24' and 'M-335', respectively. According to the variety 'P-55-437' calibration for Western African locations, there is a significant correlation between predicted and actual pod production due to acceptable value of R^2 = 0.73, and the degree of Willmott index = 0.91. These results demonstrate the reliability of the CROPGRO model in simulating groundnut, phenology, leaf area index, biomass, and production in retort to the climate variability dynamics and genetic variation for several soil-climate environmental conditions in studied locations. They also support the accuracy of genetical based coefficients of three varieties of groundnut. The genetic coefficients of groundnut crop in DSSAT and APSIM models are presented in Tables 14.1 and 14.2, respectively.

Table 14.1 Genetic coefficients, description, ranges, and guidelines for model calibration of groundnut in DSSAT

Genetic coefficient	Description
EXPNO	Number of experiments used to estimate cultivar parameters
ECO#	Ecotype code of this cultivar, points to the Ecotype in the ECO file
CSDL	Critical short-day length below which reproductive development progresses with no day length effect (for short-day plants) (h)
PPSEN	Slope of the relative response of development to photoperiod with time (positive for short-day plants) (1/h)
EM-FL	Time between plant emergence and flower appearance (R1) (photothermal days)
FL-SH	Time between first flower and first pod (R3) (photothermal days)
FL-SD	Time between first flower and first seed (R5) (photothermal days)
SD-PM	Time between first seed (R5) and physiological maturity (R7) (photothermal days)
FL-LF	Time between first flower (R1) and end of leaf expansion (photothermal days)
LFMAX	Maximum leaf photosynthesis rate at 30 C, 350 vpm CO_2, and high light (mg CO_2/m^2-s)
SLAVR	Specific leaf area of cultivar under standard growth conditions (cm^2/g)
SIZLF	Maximum size of full leaf (three leaflets) (cm^2)
XFRT	Maximum fraction of daily growth that is partitioned to seed + shell
WTPSD	Maximum weight per seed (g)

(continued)

Table 14.1 (continued)

Genetic coefficient	Description
SFDUR	Seed filling duration for pod cohort at standard growth conditions (photothermal days)
SDPDV	Average seed per pod under standard growing conditions (#/pod)
PODUR	Time required for cultivar to reach the final pod load under optimal conditions (photothermal days)
THRSH	The maximum ratio of (seed/(seed + shell)) at maturity. Causes seed to stop growing as their dry weights increase until shells are filled in a cohort. (Threshing percentage)
SDPRO	Fraction of protein in seeds (g(protein)/g(seed))
SDLIP	Fraction of oil in seeds (g(oil)/g(seed))

Source: DSSAT v4.8; Hoogenboom et al. (2021)

Table 14.2 Genetic coefficients, description, ranges, and guidelines for model calibration of groundnut in APSIM

Genetic coefficients	Description
x_pp_hi_incr	(h)
y_hi_incr	Rate of HI increase (optional) 1/day
x_hi_max_pot_stress	Maximum harvest index potential
tt_emergence	TT from emergence to the end of juvenile phase (°Cd)
est_days_emerg_to_init	Estimated days from emergence to floral initiation h
x_pp_end_of_juvenile	Photoperiod (day)
y_tt_end_of_juvenile	TT from end juvenile to floral initiation
x_pp_floral_initiation	(h)
y_tt_floral_initiation	TT from initiation to flowering (°Cd)
x_pp_flowering	(h)
y_tt_flowering	TT from flowering to start grain fill (°Cd)
x_pp_start_grain_fill	(h)
y_tt_start_grain_fill	(°Cd)
tt_end_grain_fill	(°Cd)
tt_maturity	TT from maturity to harvest ripe (°Cd)
x_stem_wt	Stem weight (g/plant)
y_height	Plant height (mm)

14.3.2 Canola

Qian et al. (2018) evaluated CSM-CROPGRO-Canola model. Prior to modeling effects of climate change, CSM-CROPGRO-Canola model was initially assessed with cultivars 'InVigor 5440' and '5030LL' at experimental locations (Brandon and Normandin). For both sites, the simulated biomass throughout the growth season closely matched the measured values with an nRMSE less than 17% and an EF and d near to 1.0. Evaluation's findings suggested accurate models of canola biomass's dynamic expansion. With an RMSE of 247 kg ha^{-1}, an nRMSE of 8%, and both EF and d near to 1.0, the model accurately predicted the seed yields in the studies. All of these statistics demonstrated how well predicted yields and observed values matched each other. El-Hamdi et al. (2020) calibrated and validated the AquaCrop model, and field experiments were carried out at Sakha Experimental Research Station in the Kafr El-Sheikh Governorate of Egypt over the two subsequent winter seasons of 2018/2019 and 2019/2020 (Version 5). To investigate the effects of two irrigation regimes (50% and 70% of soil moisture depletion, SMD), two cultivars of canola ('Serw4' and 'Serw6') were seeded. The AquaCrop model was then calibrated using dataset from the first season and verified using dataset from the second season using several statistical indicators, such as coefficient of determination (R^2) and degree of agreement (d). Also, a simulation of how climate change may affect canola yields was conducted. Findings indicate that, on average, irrigation at 50% soil moisture depletion results in greater seed and biomass yields (Mg ha^{-1}) and canola water productivity values than irrigation at 70% soil moisture depletion. Both canola cultivars exhibit this pattern, although 'Serw4' records values more accurately than 'Serw6'. Additionally, the results demonstrate that the AquaCrop model (Version 5) can accurately simulate the seed and biomass yield of canola cultivars as well as the water productivity of canola in studied region, with values of R^2 and d indicating a strong agreement between predicted and observed values. Heidarybeni et al. (2018) evaluated a CSM-APSIM model in Iran. The simulation of the days after sowing to maturity (DTM) stage had the largest RMSE (5 days), while the bias error was −0.7 days. The days after seeding to emergence simulation had the most bias inaccuracy (DTE). While this indication in the other phases (p-value = 0.01) is significant, the correlation coefficient between the DTG and DTE was not statistically significant. The observed and simulated days after sowing to floral initiation (DTFI) and days after sowing to flowering (DASF) showed the highest correlations. Wang et al. (2012) validated APSIM-Canola model. With the use of information from three field experiment locations in Yangtze River Basin, the APSIM-Canola model was calibrated and put to the test. The main phenological phases, biomass, and grain yield were measured in these tests, which used various cultivars and planting dates. The model was able to predict commencement of phenological phases with varied sowing dates and to explain 75% of difference in biomass and yield brought on by late sowings after calibration of phenological parameters and maximum harvest index. Under late planting dates, the model overstated canola production. The genetic coefficients of canola crop in DSSAT and APSIM models are presented in Tables 14.3 and 14.4, respectively.

Table 14.3 Genetic coefficients, description, ranges, and guidelines for model calibration of canola in DSSAT

Genetic coefficient	Description
ECO#	Ecotype code of this cultivar, points to the Ecotype in the ECO file
CSDL	Critical short-day length below which reproductive development progresses with no day length effect (for short-day plants) (h)
PPSEN	Slope of the relative response of development to photoperiod with time (positive for short-day plants) (1/h)
EM-FL	Time between plant emergence and flower appearance (R1) (photothermal days)
FL-SH	Time between first flower and first pod (R3) (photothermal days)
FL-SD	Time between first flower and first seed (R5) (photothermal days)
SD-PM	Time between first seed (R5) and physiological maturity (R7) (photothermal days)
FL-LF	Time between first flower (R1) and end of leaf expansion (photothermal days)
LFMAX	Maximum leaf photosynthesis rate at 30 C, 350 vpm CO_2, and high light (mg CO_2/m^2-s)
SLAVR	Specific leaf area of cultivar under standard growth conditions (cm^2/g)
SIZLF	Maximum size of full leaf (three leaflets) (cm^2)
XFRT	Maximum fraction of daily growth that is partitioned to seed + shell
WTPSD	Maximum weight per seed (g)
SFDUR	Seed filling duration for pod cohort at standard growth conditions (photothermal days)
SDPDV	Average seed per pod under standard growing conditions (#/pod)
PODUR	Time required for cultivar to reach the final pod load under optimal conditions (photothermal days)

Source: DSSAT v4.8; Hoogenboom et al. (2021)

Table 14.4 Genetic coefficients, description, ranges, and guidelines for model calibration of canola in APSIM

Genetic coefficients	Description
x_pp_hi_incr	Photoperiod (h)
y_hi_incr	Rate of HI increase (1/day)
x_hi_max_pot_stress	Average stress at flowering
y_hi_max_pot	Maximum harvest index potential
tt_emergence	(°Cd)
est_days_emerg_to_init	Estimated days from emergence to floral initiation (day)
x_pp_endjuv_to_init	Photoperiod (h)
y_tt_endjuv_to_init	TT from end juvenile to floral initiation (°Cd)
x_pp_floral_initiation	Photoperiod (h)
y_tt_floral_initiation	TT from initiation to flowering (°Cd)
x_pp_flowering	(h)
y_tt_flowering	(°Cd)
x_pp_start_grain	(h)
y_tt_start_grain_fill	(°Cd)
tt_end_grain	(°Cd)
x_stem_wt	Stem weight per plant (g/plant)
y_height	Plant canopy height (mm)

14.4 Climate Change and Temperature Trend

As per Acharya et al. (2021), according to the Mann-Kendall trend study of 1990–2018, average annual temperatures increased by 0.04 °C/year, relative humidity decreased by 0.21% year^{-1}, and annual precipitation decreased by 29.28 mm/year. Anwar et al. (2015) investigated the temperature increasing trend due to climate change in Australia. The average yearly temperatures at Cunderdin, Katanning, Wagga Wagga, and Hamilton are 18, 16, 16, and 13 °C, respectively, for the historical weather data from 1961 to 2010. By 2030, 2060, and 2090, projected yearly temperatures will rise steadily. For 2030, 2060, and 2090 in Cunderdin, respectively, the expected upsurges in yearly temperature are 1.02, 2.04, and 3.51 °C relative to baseline, with comparable rises for Katanning, Wagga Wagga, and Hamilton. Upsurges are in line with predictions made by Hennessy et al. (2010) for the entirety of Australia, which showed comparatively small variations for study locations compared to warming additional inland, where temperatures topped 1 °C by 2030 and hovered about 3.4 °C by 2070.

14.5 Climate Change Impact on Groundnut-Canola Morphophysiological Responses

14.5.1 Groundnut

One of the main abiotic stresses that plants experience now as a result of climate change is heat stress. Over-35 °C temperatures impede down plant development. Excessive daytime temperatures that coincide with the reproductive stage can seriously harm the processes involved. Heat-acclimated (HA) and non-acclimated (NA) plants both had significantly lower stomatal conductance (Gs), net photosynthetic rate (Pn), and transpiration rates (Tr) after heat treatment (Kalarani et al. 2023). Photosynthesis is one of the primary physiological processes that is impacted by heat stress and drought in plants. Reduced leaf expansion, incorrect operation of the photosynthetic apparatus, and leaf aging are the key factors affecting it. High temperatures can lower a seed's ability to germinate, which leads to poor germination and stand establishment. The timing, severity, and duration of the heat stress all influence how negatively high temperatures affect grain harvests. Rice yield was decreased by temperature stress because several growth and yield attributes performed worse (Shekhar et al. 2019). The quantity, mass, and development of the roots are also decreased by heat stress, which eventually limits the amount of water and nutrients that can reach the plant's aboveground portions. During heat stress, source and washbasin activities are reduced, which has a significant impact on growth and, ultimately, economic production (Aravind et al. 2022). In very high temperatures, groundnut plants do produce some blooms, albeit not always viable blossoms. Heat stress considerably decreased the proportion of flowers that produced pegs but had no impact on flower production or the percentage of pegs that became pods. Studies indicate that the peg number is determined by heat stress during the development of the floral bud (Puli et al. 2021).

14.5.2 Canola

Heat stress does not appear to impair flower production, but it can significantly affect the fertility of those blooms, which in turn can affect the development of pods and seeds. As a result, empty pods may be misshaped or absent. Hormone balance and typical pod production may not occur for a week (Lohani et al. 2020). High temperatures are a significant stress, and in recent decades, global warming has increased the rise in air temperatures. The processes by which plants adapt to extreme temperatures are therefore quite interesting. Plants that are subjected to heat stress (HS) have significant, and occasionally fatal, negative impacts. Plants have developed complex HS response systems to deal with such circumstances. For instance, HS affects a number of fundamental physiological processes in plants, such as photosynthesis, respiration, and water metabolism (Secchi et al. 2023). Since HS decreases photosynthetic efficiency, plants live shorter lives and produce less. One of the physiological processes that is most susceptible to heat is photosynthesis. Under HS, carbon metabolism in the stroma of chloroplasts and photochemical processes in thylakoid lamellae are vulnerable to damage. Thylakoid membranes are damaged by heat stress, which prevents the activity of electron carriers and enzymes that are linked to membranes (Viana et al. 2022). This lowers the rate of photosynthesis. Because the PSII complex is the least heat tolerant, activity of photosystem II (PSII) is specifically significantly decreased or even completely stopped under HS. Other effects of HS include lowering ribulose-1,5-bisphosphate carboxylase/oxygenase (Rubisco) activity, levels of photosynthetic pigments, and ability to fix carbon. These effects also affect chloroplast shape and the thermal stability of photosynthetic system components (Kumar and Srivastava 2020). These elements greatly lower photosynthetic efficiency when HS is present. In order to research the thermostability of plants and the negative consequences of warming on agricultural productivity, it is important to have a basic understanding of how photosynthetic physiology reacts to environmental changes. HS inhibits cellular activity and lowers thermotolerance through increasing cellular membrane permeability and the loss of cellular electrolytes. Moreover, the buildup of reactive oxygen species brought on by HS damages membranes and lowers thermotolerance (Uppal et al. 2019).

14.6 Impact of Climate Change on Groundnut-Canola System Phenology

14.6.1 Groundnut

Noorhosseini et al. (2018) indicated that days until peanut harvest maturity will also shorten in all synoptic stations with a 2 °C upsurge in temperature. In Gilan Province, the average days to harvest maturity forecast for current situation (142 days) and the 2 °C increase in temperature (123 days) differ by a substantial amount, according to the results of a t-test. The Astara synoptic station has the most variation in days before harvest maturity, at 23 days, and Kiashahr and Rudsar have the lowest variation in days until harvest maturity, at 16 days. In Gilan, the average shift in peanut harvest maturity days with a 2 °C temperature rise is typically 19 days fewer than the current situation.

14.6.2 Canola

14.6.2.1 Case Study in Pakistan in Asia

Ahmad et al. (2017a) studied the impact of climate warming and crop management practices on canola phenology in Punjab, Pakistan. Results revealed that the period of canola crop sowing in the autumn season, that is controlled via the meteorological conditions and crop husbandry activities such as cultivar switching and sowing dates, affects yield. To create plans for adjusting agricultural systems to climatic changes, it is essential to comprehend how crop husbandry techniques and climate factors affect phenology. This study's objective was to ascertain how warming trends from 1980 to 2014 affected the phenology of canola in the middle and south Punjab in Pakistan. Canola crop is grown from the last week of September to the last week of October in the study regions (Table 14.5). The CSM-CROPGRO-Canola model was well calibrated and evaluated. Generally, the canola model executed fit for all sites (slope = 0.92, $R^2 = 0.81$, for flowering stage; slope = 0.81, $R^2 = 0.78$, for the maturity stage) (Fig. 14.1). During baseline weather data period 1980–2014, increasing temperature trend was revealed at entirely ten sites (Figs. 14.2 and 14.3). Mean values of 6.02, 3.14, 3.31, and 1.89 days/decade were advanced for planting, emergence, flowering, and maturity, respectively (Figs. 14.4 and 14.5). For each of the ten sites included in this study, the length of the planting to flowering, planting to maturity, and anthesis to maturity phases lessened by mean values of 2.71, 4.13, and 1.42 days/decade (Fig. 14.6). The number of growing degree days demand of canola varieties from planting to flowering period enhanced at wholly ten places, which ranged in 65–88 °C days/decade (Fig. 14.7). A rise in temperature was positively connected with the dates of planting, emergence, flowering, and maturity by mean values of 2.71, 1.41, 1.49, and 0.85 days/°C, respectively (Table 14.6). Nevertheless, for all ten places, there was a negative relationship between the phenology periods of planting to flowering, flowering to maturity, and planting to harvesting and an increase in temperature of, respectively, by 1.22, 0.64, and 1.86 days/°C (Table 14.7). The model was applied using a usually grown cultivar of each site that had been field validated for entire places, and results showed that the warming trend caused the simulated phenological phases to occur sooner than the observed phenological stages (Fig. 14.8). By developing new cultivars with longer thermal time needs, one-fourth of the unfavorable consequences of this thermal trend were offset. Therefore, it should be a top priority to develop new canola genotypes with high temperature tolerance and a higher number of growing degree days needed.

Table 14.5 Average observed phenology of canola in Punjab, Pakistan, during the period of 1980–2014

Locations	S (DOY)	E (DOY)	A[a] (DOY)	M[b] (DOY)	S-A (days)	A-M (days)	S-M (days)
Sialkot	282 ± 6.0	286 ± 5.5	338 ± 8.4	84 ± 8.1	56 ± 8.6	111 ± 6.4	161 ± 6.3
Gujranwala	280 ± 6.6	284 ± 6.2	336 ± 7.6	80 ± 6.7	54 ± 7.3	112 ± 4.7	165 ± 5.5
Hafizabad	284 ± 5.7	288 ± 5.1	338 ± 5.7	85 ± 7.0	55 ± 5.7	111 ± 4.8	163 ± 5.2
Sheikhupura	286 ± 6.2	290 ± 6.0	340 ± 8.6	87 ± 7.5	54 ± 8.1	101 ± 5.5	156 ± 7.5
Nankana Sahib	287 ± 6.9	291 ± 6.4	351 ± 7.3	84±	64 ± 7.2	98 ± 4.9	162 ± 5.8
Multan	289 ± 6.4	293 ± 6.1	342 ± 9.0	83 ± 8.8	54 ± 5.7	105 ± 2.9	159 ± 6.1
Lodhran	271 ± 5.6	275 ± 5.3	344 ± 8.2	88 ± 9.4	65 ± 9.8	86 ± 6.0	152 ± 7.3
Bahawalpur	294 ± 6.9	298 ± 6.6	343 ± 10.2	80 ± 9.1	66 ± 10.2	85 ± 4.9	151 ± 9.6
Bahawalnagar	284 ± 6.7	288 ± 5.2	339 ± 11.5	86 ± 13.2	60 ± 8.9	89 ± 6.2	150 ± 8.4
Rahim Yar khan	276 ± 6.1	280 ± 5.8	354 ± 7.6	92 ± 13.8	76 ± 13.7	84 ± 12.1	161 ± 5.4

Source: Modified and adapted from Ahmad et al. (2017a)
DOY day of year, *S* sowing, *E* emergence, *A* anthesis, *M* maturity
[a]50% anthesis
[b]Physiological maturity

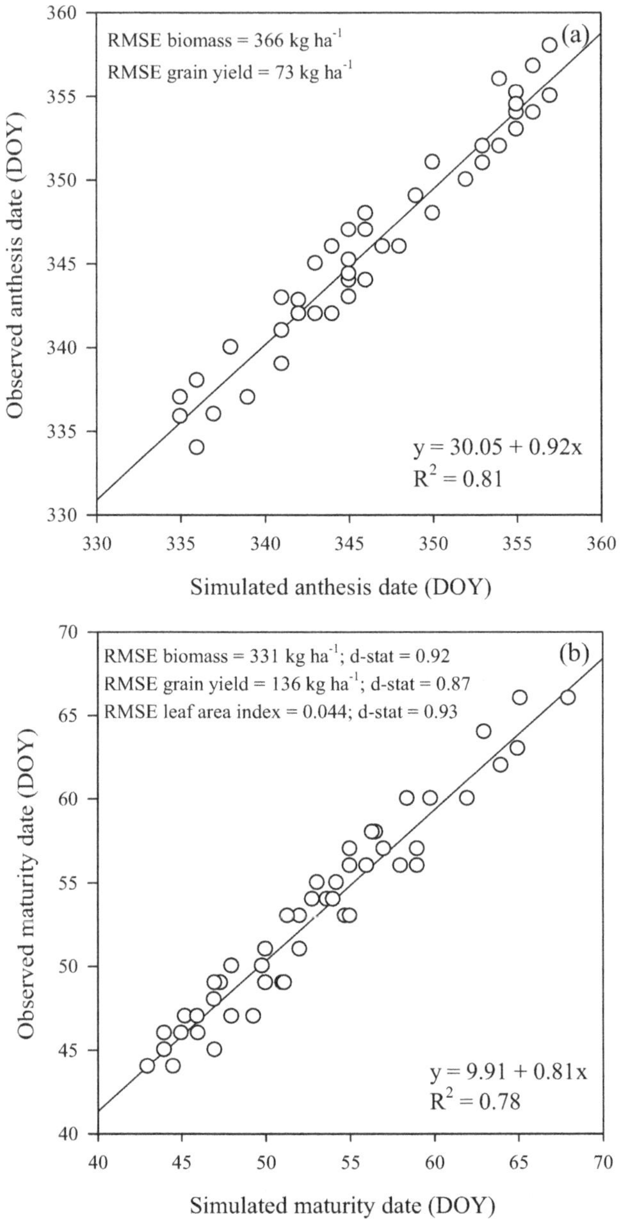

Fig. 14.1 Comparison of simulated and observed anthesis dates (**a**) and maturity dates (**b**) for canola during 1983–1985 at ten locations in Punjab, Pakistan. (Source: Modified and adapted from Ahmad et al. 2017a)

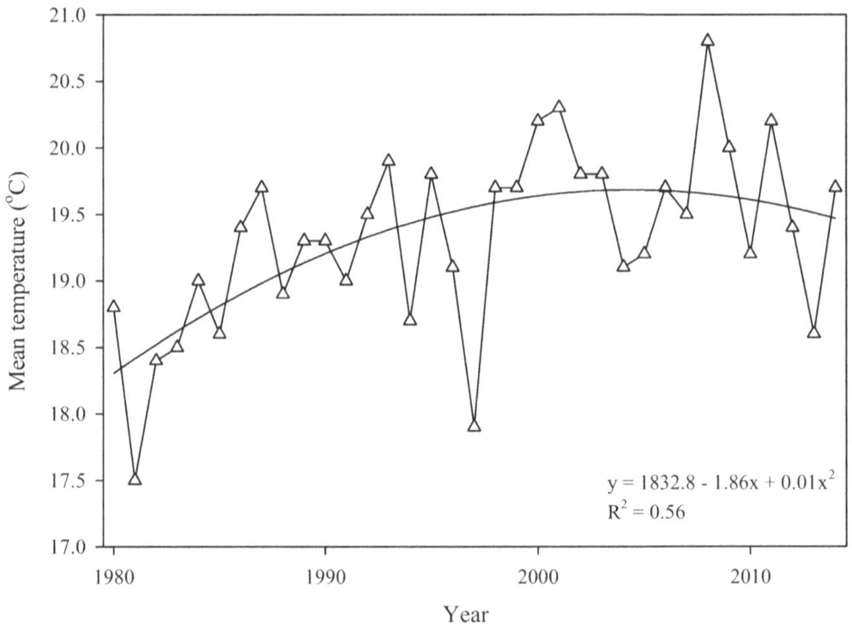

Fig. 14.2 Mean temperature trend from 1980 to 2014 for Punjab, Pakistan. (Source: Modified and adapted from Ahmad et al. 2017a)

Heidarybeni et al. (2018) studied in Iran that the days to emergence, DTFI, DTFL, DTEGF, and days to maturity stages will be shortened from 1 to 13 days based on the scenario RCP 8.5, with the blooming and grain filling phases seeing the most shortening. According to statistics from RCP 4.5, DTFL and DTEGF phases will shorten from 2 to 3 days, with the blooming season showing the highest rate of reduction. After that, the DTEJ, DTFI, and DTM phases will advance. Stages DTFI and DTM will be extended by up to 3 days. The DTEJ will be extended up to 9 days, similar to RCP 8.5. Li et al. (2022) stated that averaged sowing date for canola crop was advanced by 1.1–9.5 days/decade, the averaged harvesting date was earlier by 4.4–9.9 days/decade, and the averaged thermal time requirement was reduced through 6.0–19.6 days/decade.

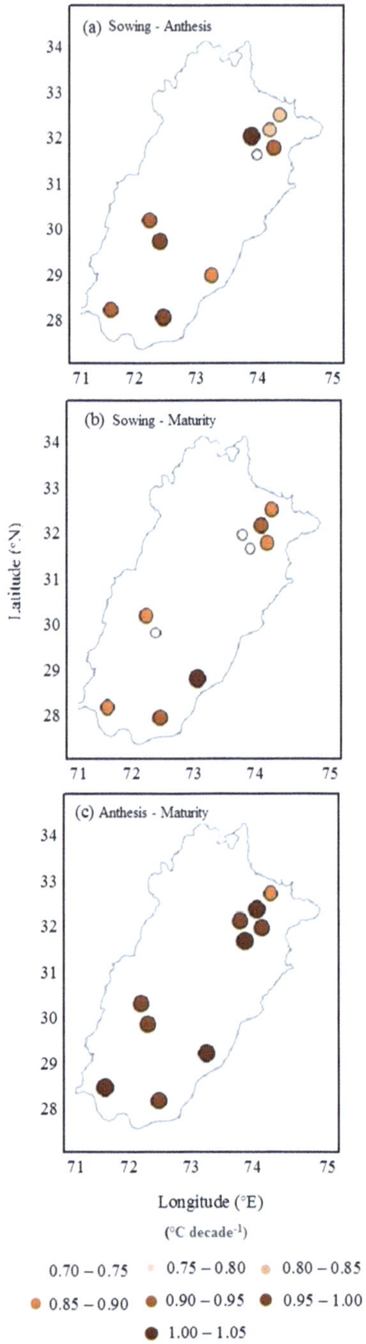

Fig. 14.3 Observed trend in mean temperature (°C/decade) during phenological phases: (**a**) sowing to anthesis, (**b**) sowing to maturity, and (**c**) anthesis to maturity of canola in Punjab, Pakistan, during 1980–2014. (Source: Modified and adapted from Ahmad et al. 2017a)

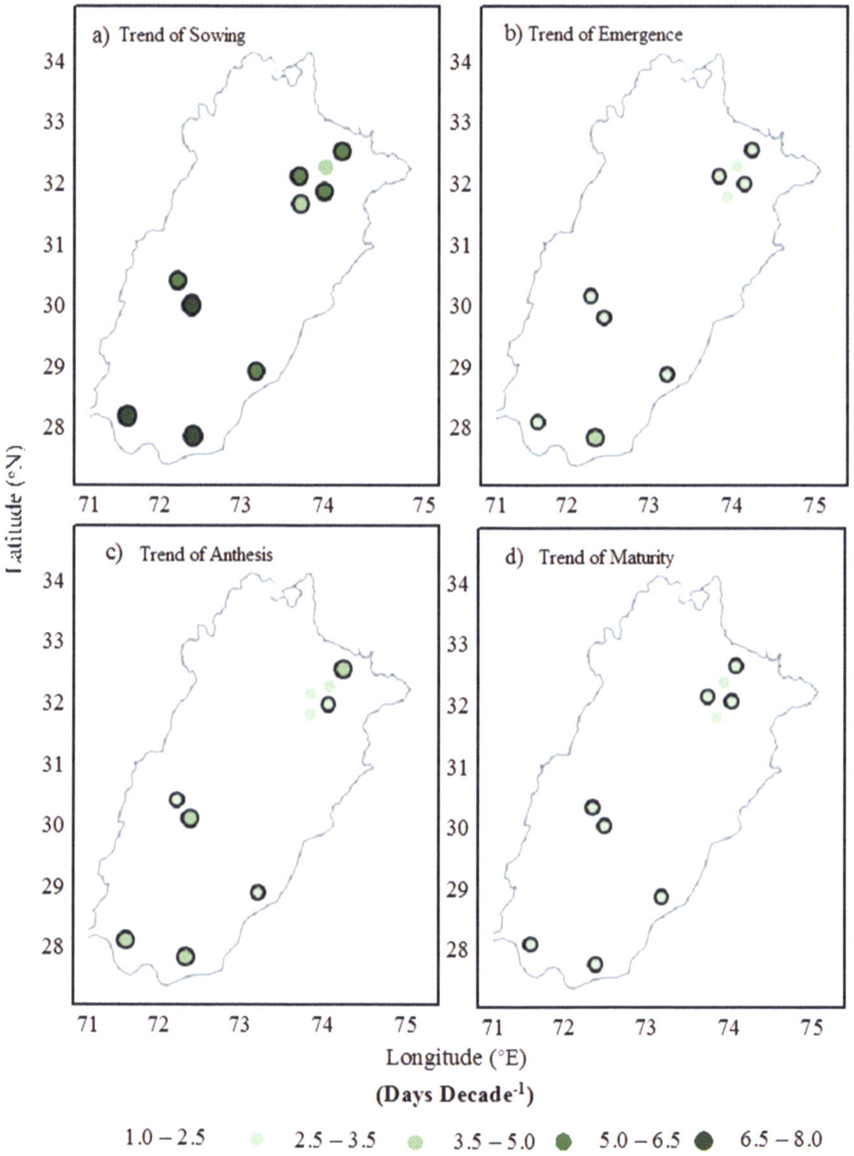

Fig. 14.4 Observed trends in phenological stages (days/decade) for canola in Punjab, Pakistan, during 1980–2014. (Source: Modified and adapted from Ahmad et al. 2017a)

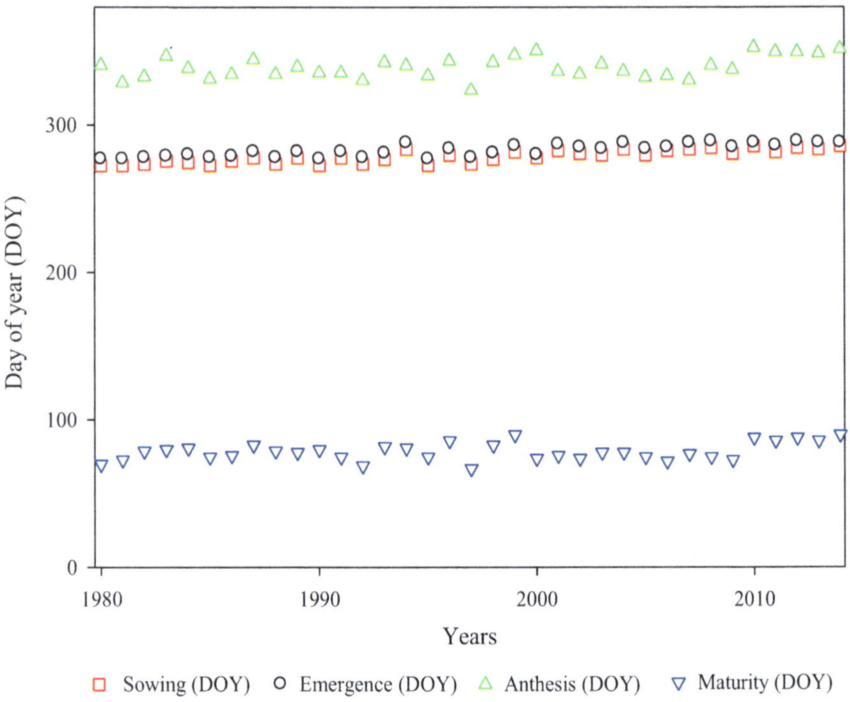

Fig. 14.5 Farmers' annual sowing, emergence, anthesis, and maturity dates for canola in Punjab, Pakistan. (Source: Modified and adapted from Ahmad et al. 2017a)

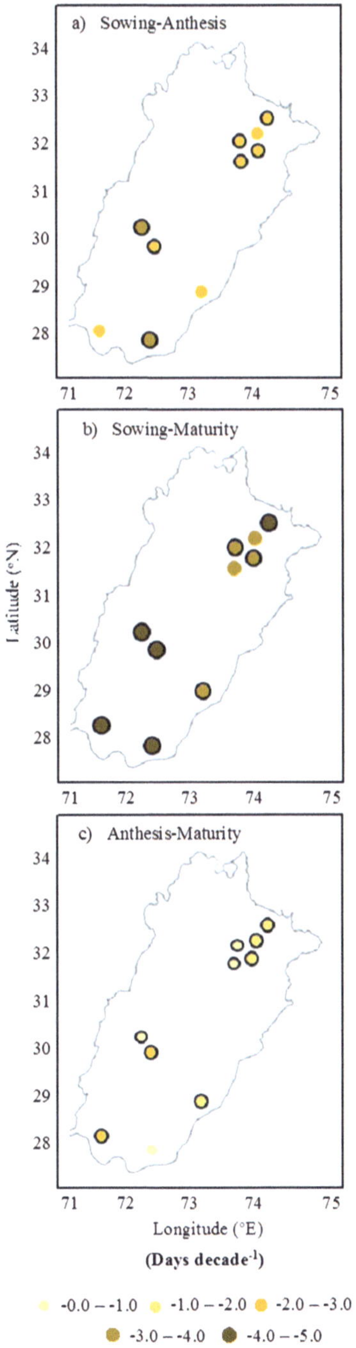

Fig. 14.6 Observed trends in phenology phases (days/decade) of canola in Punjab, Pakistan, during 1980–2014. (Source: Modified and adapted from Ahmad et al. 2017a)

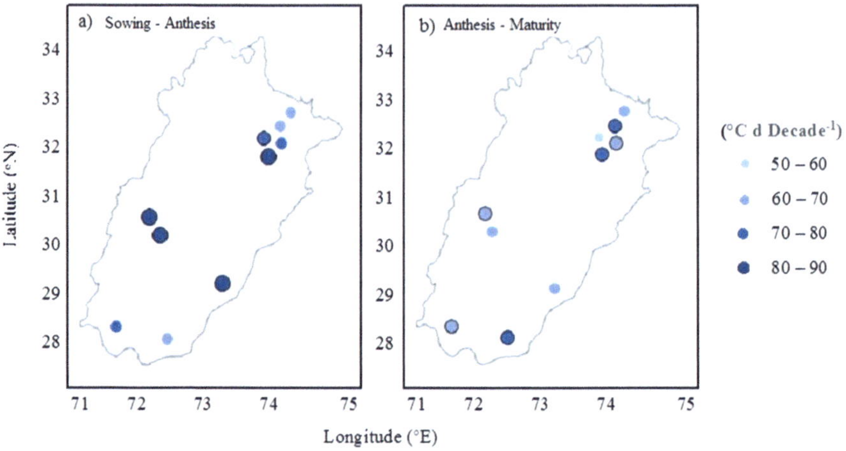

Fig. 14.7 Observed trend in thermal time (°C days/decade) requirement of canola cultivars in Punjab, Pakistan, during 1980–2014. (Source: Modified and adapted from Ahmad et al. 2017a)

Table 14.6 Summary of the observed and simulated phenological responses to the increase in temperature for canola in Punjab, Pakistan, for 1980–2014

Phenology	No. of neg.[a]	No. of pos.[b]	No. of sig. neg.[c]	No. of sig. pos.[d]	Reg. mean[e] (days °C^{-1})
Observed stages					
Sowing	0	10	0	9	2.71
Emergence	10	0	0	8	1.41
Anthesis	10	0	0	7	1.49
Maturity	10	0	0	8	0.85
Observed phases					
Sowing-anthesis	10	0	7	0	−1.22
Sowing-maturity	10	0	8	0	−1.86
Anthesis-maturity	10	0	9	0	−0.64
Simulated phases					
Sowing-anthesis	10	0	9	0	−1.82
Sowing-maturity	10	0	10	0	−2.45
Anthesis-maturity	10	0	10	0	−1.20

Source: Modified and adapted from Ahmad et al. (2017a)
[a]Number of locations with negative regression coefficients
[b]Number of locations with positive regression coefficients
[c]Number of locations with significant negative regression coefficients
[d]Number of locations with significant positive regression coefficients
[e]Mean of regression coefficients

Table 14.7 Comparison of the responses of canola phenology with average temperature using the observed and simulated data in Punjab, Pakistan, during 1980–2014

	Regression coefficient (days °C^{-1})			
Phenology	Obs. data	Sim. data	Difference between obs. and sim. data (days °C^{-1})	t-Test (p-value)
Sowing-anthesis	−1.22	−1.82	0.60	0.0014**
Sowing-maturity	−1.86	−2.45	0.59	0.0017**
Anthesis-maturity	−0.64	−1.20	0.56	0.0021**

Source: Modified and adapted from Ahmad et al. (2017a)
Obs. observed, *Sim.* simulated
**Significant at the 0.01 probability level

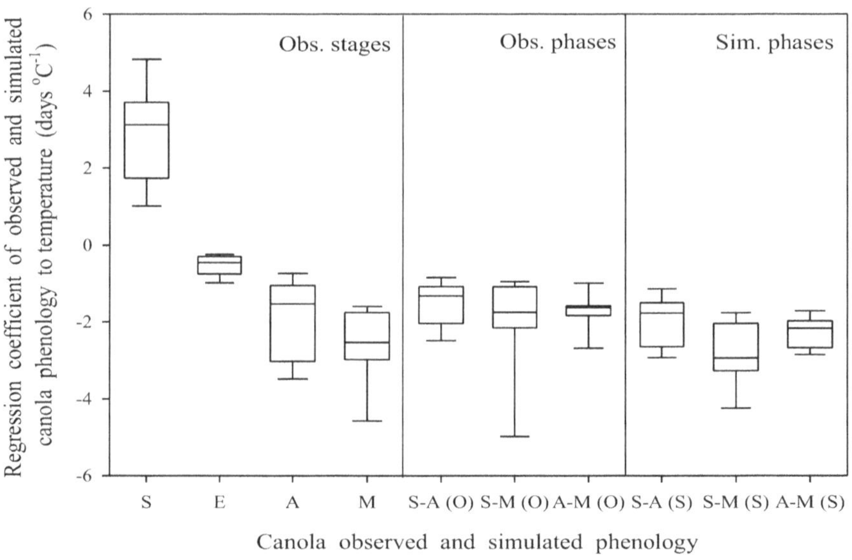

Fig. 14.8 Observed phenology (stages and phases) and simulated phenology (phases) versus temperature trends using a standard canola cultivar at each location in Punjab, Pakistan, from 1980 to 2014. (*S* sowing, *E* emergence, *A* anthesis, *M* maturity, *S-A* sowing-anthesis, *S-M* sowing-maturity, *A-M* anthesis-maturity). Horizontal line: median; box: 25th and 75th percentiles; whiskers: minimum and maximum values. (Source: Modified and adapted from Ahmad et al. 2017a)

14.7 Impact of Climate Change on Current and Future Scenarios of Crop Production

14.7.1 Groundnut

Extreme weather events are occurring more frequently, air temperatures are rising, there is a vapor pressure deficit in the atmosphere, and rainfall patterns are shifting spatially and temporally as a result of global warming. The increasing adaptability of peanuts, a significant oil crop, is impacted by climate change. Faye et al. (2018) determined the impact of climate change in Senegal. When rising [CO_2] is taken into consideration over the period of 2016–2045 in the dry season, positive increases of up to 2.4% for RCP 4.5 and 8.3% for RCP 8.5 for seed yield were identified. The seed yield rose by 19.0% for RCP 8.5 and 11.0% for RCP 4.5 during the wet season, respectively. Seed production increased by 19.0% for RCP 8.5 and 11.0% for RCP 4.5 during the wet season. When compared to the rainy season, especially when Ta is used to simulate the effects of heat stress, the effect of climate change on seed production was negative (−45.3%) in the dry season where maximum Ta is frequently greater than 38 °C. Because of the effect of increasing [CO_2], climate change may have only minor adverse effects on peanut production in Senegal. Wei et al. (2022) predicted under changing climate conditions that in the Huang-Hai zone, locations west of Haihe River Basin are best places to cultivate peanuts. According to the statistics for future scenario, utmost appropriate zones have grown for peanut crop. It may be demonstrated that model validation findings have good accurateness and may be utilized to direct and optimize groundnut yield under appropriate management practices. Sarr and Camara (2018) stated that for the two scenarios taken into consideration, analysis of the yield change in the near future reveals a decline for all RCMs except the RACMO22T model. All RCMs forecast a decline in the production of peanuts in the distant future. Moreover, this decline is more pronounced under the RCP 8.5 scenario. These findings suggested that the peanut crop would be severely impacted by climate change, necessitating the development of adaptation techniques to safeguard this industry. Xu et al. (2017) indicated that the yield of irrigated peanuts will decrease by 2.6% under the RCP 2.6 scenario, 9.9% under the RCP 4.5 scenario, and 29% under the RCP 8.5 scenario. Similar to this, there will be a decline in the yield of rainfed peanuts, with decreases of 2.5%, 11.5%, and 30% under the RCP 2.6, 4.5, and 8.5 scenarios, respectively. Noorhosseini et al. (2018) revealed that in Gilan, the average change in peanut yield with a 2-degree increase in temperature is 8.73% more than it is at the moment. Mokhtar and Abo Elenein (2014) revealed that the consumptive use of peanuts increased more under the climate HadCM3 B2 scenario compared to A2. In addition, compared to the B2 scenario, the A2 scenario indicated a higher decline in peanut grain and biomass yields. Kadiyala et al. (2021) projected by model application in India that in different areas, groundnut yields would fluctuate by 2.3–43.2% by 2050 when climate alone is taken into account, according to an optimistic scenario. Nevertheless, when economic (population and income) and market factors (elasticities, trade, etc.) were also taken into account, the change in groundnut yields

Table 14.8 Seed yield reduction/increase in groundnut cropping systems across various regions in the world

Country/ continent	Model	Climatic scenarios	Seed yield reduction/ increase (%)	References
Senegal	DSSAT	2021–2050	7	Sarr and Camara (2018)
China	DSSAT	2021–2090	−29	Xu et al. (2017)
Iran	SSM-model	1992–2015	9	Noorhosseini et al. (2018)
Egypt	CROPGRO	A2 and B2	33	Mokhtar and Abo Elenein (2014)
India	CROPGRO	2040–2069	34	Kadiyala et al. (2021)
Southern Africa	AquaCrop	2030–2060	15	Mabhaudhi et al. (2018)
Senegal	Lintul5	2016–2045	−45	Faye et al. (2018)
Niger	SAISON	1951–1998	25	Van Duivenbooden et al. (2002)

varied from 0.9% to 16.2%. Van Duivenbooden et al. (2002) explored that groundnut production is anticipated to decrease by 11–25% in 2025 (Table 14.8).

14.7.2 Canola

The production and profitability of canola are severely hampered by climate change and its associated effects, such as rising temperatures and water scarcity at critical periods of plant growth. Qian et al. (2018) predicted in Canada that under RCP 4.5, the projected seed yield drop for Brandon, West Nipissing, and Normandin was 42%, 21%, and 24% in the near future and 37%, 27%, and 23% in the distant future, respectively. RCP 8.5 predicted a greater decline, particularly in the far future at Brandon and West Nipissing. In rainfed circumstances with current N fertilizer application rates, the simulated seed production drop was correlated with an increase in heat and water stressors. Xing et al. (2019) simulated canola yield and WUE in comparison to 1961–2000 under various future climate projections. Without taking into account the effects of CO_2, yields fall by 5.2% and 6.6% for RCP 4.5 and RCP 8.5, respectively, by the 2040s, with the greater decline occurring by the 2080s. In turn, WUE falls by 1.0% and 1.1% by the 2040s for RCP 4.5 and RCP 8.5 and falls by 3.9% and 4.2% by the 2080s, respectively. Yet, when taking into account the effects of CO_2 concentrations on canola growth and water usage, canola yield and its WUE clearly rise. Heidarybeni et al. (2018) expected that under the RCP 8.5 scenario, the average yield will climb to 18% when nutrition and management are ideal, but it will increase to 18% and 13.6% when management is subpar. In terms of optimal nutrition and management, RCP 4.5 will grow by 13.4%, whereas it will rise by 14.3% and 13.6% in terms of intermediate and bad management, respectively. This indicates that if there is no water constraint, global warming will benefit the canola production in this region. Kushal et al. (2009) applied canola model, in which, in order to observe the effects of climate change, the model was run after

Table 14.9 Seed yield reduction/increase in canola cropping systems across various regions in the world

Country/ continent	Model	Climatic scenarios	Seed yield reduction/ increase (%)	References
Canada	CROPGRO	2041–2070	42	Qian et al. (2018)
Australia	APSIM	2039–2069	7	Xing et al. (2019)
Iran	APSIM	2020–2050	18	Heidarybeni et al. (2018)
India	InfoCrop	A1	23	Kushal et al. (2009)
Iran	CROPGRO	2021–2050	33	Asgari et al. (2022)
Niger	SAISON	1951–1998	25	Van Duivenbooden et al. (2002)

validation while taking into account modified climatic scenarios (1 and 2 °C temperature rise over the norm). The yield of the B9 variety of mustard decreased by 300 kg ha^{-1}, compared to the usual value of 1417 kg ha^{-1}, when the temperature increased by 1 °C (at both the maximum and lowest). The potential production might drop to roughly 978 kg ha^{-1} with a 2 °C increase in temperature. Kc et al. (2020) demonstrated that during the growing season, which is defined as the 3 months of June, July, and August, temperature, rainfall, access to high-speed Internet, use of chemical fertilizer, application of fungicides, and average age of farm operators could all account for about 38% of the variability in canola yield. In accordance with the three RCP scenarios, crop production will decline by 8–33% at the same planting date in various drainage systems. However, in various drainage systems, planting date management will boost canola yield and WUE by 20–64% and 19–58%, respectively, mostly because of a higher transpiration/precipitation ratio. Asgari et al. (2022) revealed on the basis of the findings that it is feasible to prevent waterlogging by effective drainage management and agronomic practices, which enables one to produce the best canola yield and acceptable WUE under climate change (Table 14.9).

14.8 Adaptation Strategies

Singh et al. (2014) studied in India that in Anantapur, Junagadh, and Sadore, productivity benefits from combining both drought and heat tolerance climbed to 13%, 12%, and 31%, respectively, under climate change. The increased yields from heat or drought tolerance at the Samanko location were insignificant. CROPGRO-Groundnut model may be used to assess potential combinations of features that will be required to boost and maintain groundnut production under climate change at target locations. Acharya et al. (2021) revealed that in the mid-hills of central Nepal, more farmers were found to be implementing conservation agriculture strategies such as crop mulching as a response to climate change, such as adjusting planting times, switching crop cultivars, balanced fertilizing, and conservation agricultural practices. Groundnut and canola yields in cropping systems can be less affected by climate change through the application of adaptation methods such as optimized sowing dates, fertilizer use, and climate-resilient cultivars. Carr et al. (2022) showed that optimized planting dates and cultivars with longer crop cycle duration might actually counteract adverse influences of climate change on agricultural productivity. Common adaptation measures could enhance crop yields influenced by climate change by 13%. Influences of climate change on crops have not been alleviated by increased fertilizer usage, but it might significantly boost yields in the present and future. Findings imply that a combination of increasing fertilizer use along with implementing planting techniques that benefit from favorable climate circumstances has a significant potential to safeguard and improve West Africa's crop production in the future. Reduced crop nutrient stress can affect how sensitive yields are to climate change, although increased mineral and organic fertilization mitigates this effect. A hypothetical cultivar is needed with greater tolerance to temperature and water stress during the most vulnerable phenological stages. It should also have a higher rooting density. Extended crop cycle cultivar is needed. Crop life cycles are becoming shorter as temperatures increase, but this can be offset by lengthening them by raising their thermal need (Ahmed and Ahmad 2023; Ahmed 2023; Abbas et al. 2023; Liu et al. 2023; Ahmed et al. 2022a; Ahmed et al. 2022b; Nadeem et al. 2022; Ahmed 2020a, b; Khan et al. 2020; Asseng et al. 2019; Ahmed et al. 2017; Jabeen et al. 2017; Aslam et al. 2013; Ahmed et al. 2012; Kushal et al. 2009). This gives vegetative growth and grain development more time. Crop life cycles that are short might lessen the likelihood that crops will be negatively impacted by intra-seasonal rainfall and temperature changes.

14.9 Conclusion

Low-yield cultivars and uneven nitrogen and phosphate fertilization are the main causes of the yield gap, among other agronomic issues for groundnut-canola cropping system. Because of high moisture and heat stressors during the reproductive period, late seeding might result in low yield. The construction of an effective groundnut-canola cropping system is anticipated to become more difficult as a result of a changing environment and limitations on the use of crop protection measures. The ability to extrapolate the effects of climate change on groundnut-canola cropping system in a number of field studies and controlled trials to different climatic zones, soils, management practices, crops, and climate change scenarios has been made possible by physiology-based crop simulation models. These models have been used to examine the effects of the various climate change scenarios and their combined effects on groundnut-canola cropping system. Adaptation management strategies are very useful in order to alleviate the adverse influences of the climatic changes and seize some of the new possibilities presented by these changes on groundnut-canola cropping system.

References

Abbas Q, Ahmad S (2018) Effect of different sowing times and cultivars on cotton fiber quality under stable cotton-wheat cropping system in southern Punjab, Pakistan. Pak J Life Soc Sci 16(2):77–84

Abbas G, Ahmad S, Ahmad A, Nasim W, Fatima Z, Hussain S, Rehman MH, Khan MA, Hasanuzzaman M, Fahad S, Boote KJ, Hoogenboom G (2017) Quantification the impacts of climate change and crop management on phenology of maize-based cropping system in Punjab, Pakistan. Agric For Meteorol 247:42–55

Abbas G, Fatima Z, Riaz F, Iqbal P, Hussain S, Zakir I, Iqbal M, Haque E, Ahmad S (2019) Study of sowing dates and hybrids effect in maize-based cropping system under arid conditions of Southern Punjab, Pakistan. Pak J Life Soc Sci 17(1):24–31

Abbas G, Ahmad S, Hussain M, Fatima Z, Hussain S, Iqbal P, Ahmed M, Farooq M (2020a) Sowing date and hybrid choice matters production of maize-maize system. Int J Plant Prod 14:583–595

Abbas G, Fatima Z, Hussain M, Hussain S, Atique-ur-Rehman, Sarwar N, Ahmed M, Ahmad S (2020b) Nitrogen rate and hybrid selection matters productivity of maize-maize cropping system under irrigated arid environment of Southern Punjab, Pakistan. Int J Plant Prod 14:309–320

Abbas Q, Atique-ur-Rehman, Sarwar N, Hussain S, Abbas G, Khan MN, Fatima Z, Naz S, Younis H, Ahmed M, Ullah H, Iqbal P, Iqbal M, Kan M, Ahmad S (2020c) Effect of different sowing times and cultivars on wheat grain quality under cotton-wheat cropping system in Southern Punjab, Pakistan. Pak J Life Soc Sci 18(1):41–50

Abbas G, Ahmed M, Fatima Z, Hussain S, Kheir AMS, Ercişli S, Ahmad S (2023) Modeling the potential impact of climate change on maize-maize cropping system in semi-arid environment and designing of adaptation options. Agric For Meteorol 341:109674. https://doi.org/10.1016/j.agrformet.2023.109674

Acharya M, Amgain LP, Khanal S, Mishra SR, Shrestha A (2021) Assessments of climate change adaptation measures on agro-climatic indices and productivity of late planted rapeseed in Nepalese mid-hills. Plant Biol Crop Res 4(2):1037

Ahmad S, Hasanuzzaman M (2012) Integrated effect of plant density, N rates and irrigation regimes on the biomass production, N content, PAR use efficiencies and water productivity of rice under irrigated semiarid environment. Notulae Botanicae Horti Agrobotanici Cluj-Napoca 40(1):201–211

Ahmad S, Ahmad A, Soler CMT, Ali H, Zia-Ul-Haq M, Anothai J, Hussain A, Hoogenboom G, Hasanuzzaman G (2012a) Application of the CSM-CERES-Rice model for evaluation of plant density and nitrogen management of fine transplanted rice for an irrigated semiarid environment. Precis Agric 13(2):200–218

Ahmad S, Ali H, Ismail M, Shahzad MI, Nadeem M, Anjum MA, Zia-Ul-Haq M, Firdous N, Khan MA (2012b) Radiation and nitrogen use efficiencies of C_3 winter cereals to nitrogen split application. Pak J Bot 44(1):139–149

Ahmad S, Ahmad A, Ali H, Hussain A, Garcia y Garcia A, Khan MA, Zia-Ul-Haq M, Hasanuzzaman M, Hoogenboom G (2013) Application of the CSM-CERES-Rice model for evaluation of plant density and irrigation management of transplanted rice for an irrigated semiarid environment. Irrig Sci 31(3):491–506

Ahmad A, Ashfaq M, Rasul G, Wajid SA, Khaliq T, Rasul F, Saeed U, Rahman MH, Hussain J, Baig IA, Naqvi SAA, Bokhari SAA, Ahmad S, Naseem W, Hoogenboom G, Valdivia RO (2015a) Impact of climate change on the rice-wheat cropping system of Pakistan. In: Hillel D, Rosenzweig C (eds) Handbook of climate change and agro-ecosystems, vol 3. Imperial College Press and the American Society of Agronomy, London, pp 219–258

Ahmad S, Ali H, Atique-ur-Rehman, Khan RJ, Ahmad W, Fatima Z, Abbas G, Irfan M, Ali H, Khan MA, Hasanuzzaman M (2015b) Measuring leaf area of winter cereals by different techniques: a comparison. Pak J Life Soc Sci 13(2):117–125

Ahmad S, Raza I, Muhammad D, Ali H, Hussain S, Dogan H, Zia-Ul-Haq M (2015c) Radiation, water and nitrogen use efficiencies of *Gossypium hirsutum* L. Turk J Agric For 39(5):825–837

Ahmad S, Ali H, Farooq U, Khan SU, Atique-ur-Rehman, Sarwar N, Shahzad AN, Doğan H, Hussain S, Sultan MT, Waheed A, Zia-ul-Haq M, Hussain K, Khan MA (2016a) Improving nitrogen and radiation-use-efficiencies of C_4 summer cereals by split nitrogen applications under irrigated arid environment. Turk J Agric For 40(2):280–289

Ahmad S, Hussain S, Fatima Z, Abbas G, Atique-ur-Rehman, Khan MR, Younis H, Naz S, Sohail M, Ajmal M, Abbas N, Akhtar M, Rauf A, Khan M, Ali Z, Hassan M, Rizwan M, Bajwa RS, Ajmal A, Naz S, Ali H, Khan AA, Ali M, Sarwar G, Khan MA, Hasanuzzaman M (2016b) Application of DSSAT model for sowing date management of C_4 summer cereals for fodder and grain crops under irrigated arid environment. Pak J Life Soc Sci 14(2):104–114

Ahmad S, Nadeem M, Abbas G, Fatima Z, Khan RJ, Ahmed M, Ahmad A, Rasul G, Khan MA (2016c) Quantification of the effects of climate warming and crop management on sugarcane phenology. Clim Res 71(1):47–61

Ahmad S, Abbas G, Fatima Z, Khan RJ, Anjum MA, Ahmed M, Khan MA, Porter CH, Hoogenboom G (2017a) Quantification of the impacts of climate warming and crop management on canola phenology in Punjab, Pakistan. J Agron Crop Sci 203(5):442–452

Ahmad S, Abbas Q, Abbas G, Fatima Z, Atique-ur-Rehman, Naz S, Younis H, Khan RJ, Nasim W, Rehman MH, Ahmad A, Rasul G, Khan MA, Hasanuzzaman M (2017b) Quantification of climate warming and crop management impacts on cotton phenology. Plan Theory 6:1–16

Ahmed M, Hassan FU, Aslam M, Aslam MA (2012) Physiological attributes based resilience of wheat to climate change. Int J Agric Biol 14(3):407–412

Ahmad S, Atique-ur-Rehman, Ejaz M, Fatima Z, Kan M, Ahmad M (2018) Agricultural land-use change of major field crops in Pakistan (1961–2014). Sci Technol Dev 37(3):113–121

Ahmad A, Abbas G, Ahmed M, Fatima Z, Anjum MA, Rasul G, Khan MA, Hoogenboom G (2019) Climate warming and management impact on the change of phenology of the rice-wheat cropping system in Punjab, Pakistan. Field Crop Res 230:46–61

Ahmed M, Stöckle CO, Nelson R, Higgins S (2017) Assessment of climate change and atmospheric CO_2 impact on winter wheat in the Pacific Northwest using a multimodel ensemble. Front Ecol Evol 5. https://doi.org/10.3389/fevo.2017.00051

Ahmed M (2020a) Introduction to modern climate change. Andrew E. Dessler: Cambridge University Press, 2011, 252 pp, ISBN-10: 0521173159. Sci Total Environ 734:139397. https://doi.org/10.1016/j.scitotenv.2020.139397

Ahmed M (2020b) Systems modeling. Springer Nature Singapore Pte Ltd., p 409. https://doi.org/10.1007/978-981-15-4728-7. https://link.springer.com/book/10.1007%2F978-981-15-4728-7#toc

Ahmed M (2023) The science of climate change. In: Brinkmann R (ed) The Palgrave handbook of global sustainability. Springer International Publishing, Cham, pp 195–222. https://doi.org/10.1007/978-3-031-01949-4_22

Ahmed M, Ahmad S (eds) (2023) Disaster risk reduction in agriculture. Springer Nature. https://doi.org/10.1007/978-981-99-1763-1. ISBN 978-981-99-1762-4, Published: 03 November 2023

Ahmad A, Ashfaq M, Rasul G, Wajid SA, Ahmad I, Khaliq T, Nasir J, Rasul F, Riaz F, Ahmad B, Ahmad S, Baig IA, Valdivia RO, Hoogenboom G (2021) Development of climate change adaptation strategies for cotton–wheat cropping system of Punjab, Pakistan. In: Rosenzweig C, Mutter CZ, Contreras EM (eds) Handbook of climate change and agroecosystems: climate change and farming system planning in Africa and South Asia: AgMIP stakeholder-driven research. World Scientific Publishing Europe Ltd, London, pp 277–327

Ahmed M, Hayat R, Ahmad M, ul-Hassan M, Kheir AMS, ul-Hassan F, ur-Rehman MH, Shaheen FA, Raza MA, Ahmad S (2022a) Impact of climate change on dryland agricultural systems: a review of current status, potentials, and further work need. Int J Plant Prod. https://doi.org/10.1007/s42106-022-00197-1

Ahmed M, Ahmad S, Kheir AMS (2022b) Climate change: an overview. In: Ahmed M (ed) Global agricultural production - resilience to climate change. Springer Nature Switzerland AG. https://doi.org/10.1007/978-3-031-14973-3_1

Ahmad S, Ahmad I, Ahmad B, Ahmad A, Wajid A, Khaliq T, Abbas G, Wilkerson CJ, Hoogenboom G (2023a) Regional integrated assessment of climate change impact on cotton production in a semi-arid environment. Clim Res 89:113–132

Ahmad S, Raza MA, Hussain S, Abbas G, Fatima Z, Ahmed M, Goheer MA, Garcia y Garcia A, Wilkerson CJ, Hoogenboom G (2023b) Identification of weak links in the production technology of farmer for bridging canola yield-gap. J Agric Sci 161(2):241–253. https://doi.org/10.1017/S0021859623000187

Ahmed M, Akram MN, Asim M, Aslam M, F-u Hassan Higgins S, Stockle CO, Hoogenboom G (2016) Calibration and validation of APSIM-wheat and CERES-wheat for spring wheat under rainfed conditions: models evaluation and application. Comput Electron Agric 123:384–401

Ahmed I, Rahman MH, Ahmad A, Ahmad S, Hussain J, Ullah A, Judge J (2018a) Assessing the impact of climate variability on maize using simulation modeling under semi-arid environment of Punjab-Pakistan. Environ Sci Pollut Res 25(23):28413–28430

Ahmed M, Ijaz W, Ahmad S (2018b) Adapting and evaluating APSIM-SoilP-Wheat model for response to phosphorus under rainfed conditions of Pakistan. J Plant Nutr 41(16):2069–2084

Ahmed M, Stockle CO, Nelson R, Higgins S, Ahmad S, Raza M (2019) Novel multimodal ensemble approach to evaluate the sole effect of elevated CO_2 on winter wheat productivity. Sci Rep 9:7813

Ahmed M, Ahmad S, Tariq M, Fatima Z, Aslam Z, Raza MA, Iqbal N, Akmal M, Fayyaz-ul-Hassan, Abbasi NA, Hayat R (2020) Wastes to be the source of nutrients and energy to mitigate climate change and ensure future sustainability: options and strategies. J Plant Nutr 43(6):896–920

Ahmed M, Hayat R, Ahmad M, Mahmood-ul-Hassan, Kheir AMS, Fayyaz-ul-Hassan, Rehman MH, Shaheen FA, Raza MA, Ahmad S (2022) Impact of climate change on dryland agricultural systems: a review of current status, potentials, and further work need. Int J Plant Prod 16:341–363

Ali H, Iqbal N, Ahmad S, Shahzad AN, Sarwar N (2013) Performance of late sown wheat crop under different planting geometries and irrigation regimes in arid climate. Soil Tillage Res 130:109–119

Amin A, Nasim W, Fahad S, Ali S, Ahmad S, Rasool A, Saleem N, Hammad HM, Sultana SR, Mubeen M, Bakhat HF, Ahmad N, Shah GM, Adnan M, Noor M, Basir A, Saud S, Rahman MH, Paz JO (2018a) Evaluation and analysis of temperature for historical (1996–2015) and projected (2030–2060) climates in Pakistan using SimCLIM climate model: ensemble application. Atmos Res 213:422–436

Amin A, Nasim W, Mubeen M, Ahmad A, Nadeem M, Urich P, Fahad S, Ahmad S, Wajid A, Tabassum F, Hammad HM, Sultana SR, Anwar S, Baloch SK, Wahid A, Wilkerson CJ, Hoogenboom G (2018b) Simulated CSM-CROPGRO-cotton yield under projected future climate by SimCLIM for Southern Punjab, Pakistan. Agric Syst 167:213–222

Amin A, Nasim W, Mubeen M, Sarwar S, Urich P, Ahmad A, Wajid A, Khaliq T, Rasul F, Hammad HM, Rehmani MIA, Mubarak H, Mirza N, Wahid A, Ahmad S, Fahad S, Ullah A, Khan MN, Ameen A, Amanullah SB, Saud S, Alharby H, Ata-Ul-Karim ST, Adnan M, Islam F, Ali QS (2018c) Regional climate assessment for temperature and precipitation in Southern Punjab (Pakistan) using SimCLIM climate model for different temporal scales. Theor Appl Climatol 131:121–131

Anwar MR, Li Liu D, Farquharson R, Macadam I, Abadi A, Finlayson J, Wang B, Ramilan T (2015) Climate change impacts on phenology and yields of five broadacre crops at four climatologically distinct locations in Australia. Agric Syst 132:133–144

Aravind B, Nayak SN, Choudhary RS, Gandhadmath SS, Prasad PVV, Pandey MK, Bhat RS, Puppala N, Latha P, Sudhakar P, Varshney RK (2022) Integration of genomics approaches in abiotic stress tolerance in groundnut (*Arachis hypogaea* L.): an overview. In: Kole C (ed) Genomic designing for abiotic stress resistant oilseed crops. Springer, Cham, pp 149–197. https://doi.org/10.1007/978-3-030-90044-1_4

Asgari A, Darzi-Naftchali A, Saberali SF, Nadi M (2022) Assessing DSSAT performance for predicting yield and water productivity of rainfed canola in a subsurface-drained field. Theor Appl Climatol 149(3–4):1659–1670

Aslam Z, Ahmed M, Sajad M, Asif M, Akmal M, Awan FK, Ijaz W, Ashraf R, Khattak JZK (2013) A comparison of statistical and dynamic modeling of wheat (*Triticum aestivum* L.) fungal diseases under the climate change. J Food Agric Environ 11(2):1327–1332

Asseng S, Martre P, Maiorano A, Rötter RP, O'Leary GJ, Fitzgerald GJ, Girousse C, Motzo R, Giunta F, Babar MA, Reynolds MP, Kheir AMS, Thorburn PJ, Waha K, Ruane AC, Aggarwal PK, Ahmed M, Balkovič J, Basso B, Biernath C, Bindi M, Cammarano D, Challinor AJ, De Sanctis G, Dumont B, Eyshi Rezaei E, Fereres E, Ferrise R, Garcia-Vila M, Gayler S, Gao Y, Horan H, Hoogenboom G, Izaurralde RC, Jabloun M, Jones CD, Kassie BT, Kersebaum K-C, Klein C, Koehler A-K, Liu B, Minoli S, Montesino San Martin M, Müller C, Naresh Kumar S, Nendel C, Olesen JE, Palosuo T, Porter JR, Priesack E, Ripoche D, Semenov MA, Stöckle C, Stratonovitch P, Streck T, Supit I, Tao F, Van der Velde M, Wallach D, Wang E, Webber H, Wolf J, Xiao L, Zhang Z, Zhao Z, Zhu Y, Ewert F (2019) Climate change impact and adaptation for wheat protein. Glob Chang Biol 25(1):155–173. https://doi.org/10.1111/gcb.14481

Bertioli DJ, Jenkins J, Clevenger J, Dudchenko O, Gao D, Seijo G, Leal-Bertioli SC, Ren L, Farmer AD, Pandey MK, Samoluk SS (2019) The genome sequence of segmental allotetraploid peanut Arachis hypogaea. Nat Genet 51(5):877–884

Carr TW, Mkuhlani S, Segnon AC, Ali Z, Zougmoré R, Dangour AD, Green R, Scheelbeek PF (2022) Climate change impacts and adaptation strategies for crops in West Africa: a systematic review. Environ Res Lett 17:053001

Chang AS, Sreedharan A, Schneider KR (2013) Peanut and peanut products: a food safety perspective. Food Control 32(1):296–303

Chmielewska A, Kozłowska M, Rachwał D, Wnukowski P, Amarowicz R, Nebesny E, Rosicka-Kaczmarek J (2021) Canola/rapeseed protein–nutritional value, functionality and food application: a review. Crit Rev Food Sci Nutr 61(22):3836–3856

Ejaz M, Abbas G, Fatima Z, Iqbal P, Raza MA, Kheir AMS, Ahmed M, Kakar KM, Ahmad S (2022) Modelling climate uncertainty and adaptations for soybean-based cropping system. Int J Plant Prod 16:235–250

El-Hamdi KH, Khair AS, Abu Ishwayrib HM (2020) Simulation of water stress and climate change impacts on canola yield, seed quality and water productivity at North Delta, Egypt. J Soil Sci Agric Eng 11(8):377–383

Fatima Z, Abbas Q, Khan A, Hussain S, Ali MA, Abbas G, Younis H, Naz S, Ismail M, Shahzad MI, Nadeem M, Farooq U, Khan SU, Javed K, Khan AA, Ahmed M, Khan MA, Ahmad S (2018) Resource-use-efficiencies of C_3 and C_4 cereals under split nitrogen regimes. Agronomy 8(5):69

Fatima Z, Ahmad M, Hussain M, Abbas G, Ul-Allah S, Ahmad S, Ahmed N, Ali MA, Sarwar G, Haque E, Iqbal P, Hussain S (2020) The fingerprints of climate warming on cereal crops phenology and adaptation options. Sci Rep 10:18013

Fatima Z, Atique-ur-Rehman AG, Iqbal P, Zakir I, Khan MA, Kamal GM, Ahmed M, Ahmad S (2021) Quantification of climate warming and crop management impacts on phenology of pulses-based cropping systems. Int J Plant Prod 15:107–123

Faye B, Webber H, Diop M, Mbaye ML, Owusu-Sekyere JD, Naab JB, Gaiser T (2018) Potential impact of climate change on peanut yield in Senegal, West Africa. Field Crop Res 219:148–159

Gul A, Chandio AA, Siyal SA, Rehman A, Xiumin W (2022) How climate change is impacting the major yield crops of Pakistan? an exploration from long-and short-run estimation. Environ Sci Pollut Res 29(18):26660–26674

Haerani H, Apan A, Nguyen-Huy T, Basnet B (2023) Modelling future spatial distribution of peanut crops in Australia under climate change scenarios. Geo-spatial Inf Sci. https://doi.org/10.1080/10095020.2022.2155255

Heidarybeni M, Yazdanpanh H, Mehnatkesh A (2018) Impacts of climate change on canola yields and phenology (case study: Chaharmahal Va Bakhtiari, Iran). Phys Geogr Res Q 50(2):373–389

Hoogenboom G, Porter CH, Boote KJ, Shelia V, Wilkens PW, Singh U, White JW, Asseng S, Lizaso JI, Moreno LP, Pavan W, Ogoshi R, Hunt LA, Tsuji GY, Jones JW (2019) The DSSAT crop modeling ecosystem. In: Boote KJ (ed) Advances in crop modeling for a sustainable agriculture. Burleigh Dodds Science Publishing, Cambridge, pp 173–216. https://doi.org/10.19103/AS.2019.0061.10

Hoogenboom G, Porter CH, Shelia V, Boote KJ, Singh U, White JW, Pavan W, Oliveira FAA, Moreno-Cadena LP, Lizaso JI, Asseng S, Pequeno DNL, Kimball BA, Alderman PD, Thorp KR, Jones MR, Cuadra SV, Vianna MS, Villalobos FJ, Ferreira TB, Batchelor WD, Koo J, Hunt LA, Jones JW (2021) Decision Support System for Agrotechnology Transfer (DSSAT) Version 4.8 (DSSAT.net). DSSAT Foundation, Gainesville

Hussain M, Ahmad S, Hussain S, Lal R, Ul-Allah S, Nawaz A (2018) Rice in saline soils: physiology, biochemistry, genetics and management. Adv Agron 148:231–287

Jabeen M, Gabriel HF, Ahmed M, Mahboob MA, Iqbal J (2017) Studying impact of climate change on wheat yield by using DSSAT and GIS: a case study of Pothwar Region. In: Ahmed M, Stockle OC (eds) Quantification of climate variability, adaptation and mitigation for agricultural sustainability. Springer International Publishing, Cham, pp 387–411

Jones JW, Hoogenboom G, Porter CH, Boote KJ, Batchelor WD, Hunt LA, Wilkens PW, Singh U, Gijsman AJ, Ritchie JT (2003) The DSSAT cropping system model. Eur J Agron 18(3–4):235–265

Kadiyala MDM, Nedumaran S, Padmanabhan J, Gumma MK, Gummadi S, Srigiri SR, Robertson R, Whitbread A (2021) Modeling the potential impacts of climate change and adaptation strategies on groundnut production in India. Sci Total Environ 776:145996

Kalarani MK, Senthil A, Punitha S, Sowmyapriya S, Umapathi M, Geethalakshmi V (2023) Abiotic stress responses in groundnut (*Arachis hypogaea* L.): mechanisms and adaptations. In: Legumes: physiology and molecular biology of abiotic stress tolerance. Springer Nature Singapore, Singapore, pp 279–290

Kc KB, Montocchio D, Berg A, Fraser ED, Daneshfar B, Champagne C (2020) How climatic and sociotechnical factors influence crop production: a case study of canola production. SN Appl Sci 2:1–9

Keating BA, Carberry PS, Hammer GL, Probert ME, Robertson MJ, Holzworth D, Huth NI, Hargreaves JNG, Meinke H, Hochman Z, McLean G, Verburg K, Snow V, Dimes JP, Silburn M, Wang E, Brown S, Bristow KL, Asseng S, Chapman S, McCown RL, Freebairn DM, Smith CJ (2003) An overview of APSIM: a model designed for farming systems simulation. Eur J Agron 18:267–288

Khan A, Fahad S, Khan A, Saud S, Adnan M, Wahid F, Noor M, Nasim W, Hammad HM, Bakhat HF, Ahmad S, Rehman MH, Wang D, Sönmez O (2019) Managing tillage operations and manure to restore soil carbon stocks in wheat-maize cropping system. Agron J 111:1–10

Khan MA, Tahir A, Khurshid N, Husnain MI, Ahmed M, Boughanmi H (2020) Economic effects of climate change-induced loss of agricultural production by 2050: a case study of Pakistan. Sustain For 12(3):1216. https://doi.org/10.3390/su12031216

Kirkegaard JA, Lilley JM, Berry PM, Rondanini DP (2021) Canola. In: Crop physiology case histories for major crops. Academic Press, London, pp 518–549

Kumar P, Srivastava DK (2020) Molecular and biotechnological interventions for improving Brassicaceae crops for abiotic stress tolerance. In: Hasanuzzaman M (ed) The plant family Brassicaceae: biology and physiological responses to environmental stresses. Springer, Singapore, pp 437–450

Kumar U, Singh P, Boote KJ (2012) Effect of climate change factors on processes of crop growth and development and yield of groundnut (*Arachis hypogaea* L.). Adv Agron 116:41–69

Kushal S, Saon B, Huda AKS (2009) Simulating the impact of climate change on the performance of rapeseed and mustard through crop simulation model. Int J Agric Environ Biotechnol 2(4):318–322

Li X, Chen C, Yang X, Xiong J, Ma N (2022) The optimisation of rapeseed yield and growth duration through adaptive crop management in climate change: evidence from China. Ital J Agron 17:2104

Liu B, Martre P, Ewert F, Webber H, Waha K, Thorburn PJ, Ruane AC, Aggarwal PK, Ahmed M, Balkovič J, Basso B, Biernath C, Bindi M, Cammarano D, Cao W, Challinor AJ, De Sanctis G, Dumont B, Espadafor M, Eyshi Rezaei E, Fereres E, Ferrise R, Garcia-Vila M, Gayler S, Gao Y, Horan H, Hoogenboom G, Izaurralde RC, Jabloun M, Jones CD, Kassie BT, Kersebaum KC, Klein C, Koehler A-K, Maiorano A, Minoli S, Montesino San Martin M, Müller C, Naresh

Kumar S, Nendel C, O'Leary GJ, Eivind Olesen J, Palosuo T, Porter JR, Priesack E, Ripoche D, Rötter RP, Semenov MA, Stöckle C, Stratonovitch P, Streck T, Supit I, Tao F, Van der Velde M, Wang E, Wolf J, Xiao L, Zhang Z, Zhao Z, Zhu Y, Asseng S (2023) AgMIP-Wheat multi-model simulations on climate change impact and adaptation for global wheat. Open Data J Agric Res 9:10–25. https://doi.org/10.18174/odjar.v9i0.18092

Lilley JM, Flohr BM, Whish JP, Farre I, Kirkegaard JA (2019) Defining optimal sowing and flowering periods for canola in Australia. Field Crop Res 235:118–128

Lohani N, Jain D, Singh MB, Bhalla PL (2020) Engineering multiple abiotic stress tolerance in canola, Brassica napus. Front Plant Sci 11:3

Mabhaudhi T, Chibarabada TP, Chimonyo VGP, Modi AT (2018) Modelling climate change impact: a case of Bambara groundnut (*Vigna subterranea*). Phys Chem Earth Parts A/B/C 105:25–31

Mahmood A, Ullah H, Shahzad AN, Ali H, Ahmad S, Zia-Ul-Haq M, Honermeier B, Hasanuzzaman M (2013) Dry matter yield and chemical composition of sorghum cultivars with varying planting density and sowing date. Sains Malaysiana 42(10):1529–1538

Majola NG, Gerrano AS, Shimelis H (2021) Bambara groundnut (*Vigna subterranea* [L.] Verdc.) production, utilisation and genetic improvement in Sub-Saharan Africa. Agronomy 11(7):1345

Mokhtar NAY, Abo Elenein RA (2014) Evaluate impact of climate change on yield and water consumptive use for peanut under North Egypt environmental condition. J Soil Sci Agric Eng 5(12):1809–1821

Mubarak MK, Hussain K, Abbas G, Altaf MT, Baloch FS, Ahmad S (2022) Productivity of sorghum (*Sorghum bicolor* L.) at diverse irrigation regimes and sowing dates in semi-arid and arid environment. Turk J Agric For 46(1):1–18

Mubeen M, Ahmad A, Wajid A, Khaliq T, Hammad HM, Sultana SR, Ahmad S, Fahad S, Nasim W (2016) Application of CSM-CERES-Maize model in optimizing irrigated conditions. Outlook Agric 45(3):173–184

Mustafa HSB, Ejaz UH, Tariq M, Muhammad A, Sundas S, Rizwana Q, Ahmad NG, Qamar AT (2022) Sandal canola (RBN-13028): rapeseed cultivar with better qualitative and quantitative traits under the scenario of climate change. Plant Cell Biotechnol Mol Biol 23(1&2):111–120

Nadeem M, Nazer Khan M, Abbas G, Fatima Z, Iqbal P, Ahmed M, Ali Raza M, Rehman A, Ul Haq E, Hayat A, Ali M, Ahmad S (2022) Application of CSM-CANEGRO model for climate change impact assessment and adaptation for sugarcane in semi-arid environment of Southern Punjab, Pakistan. Int J Plant Prod. https://doi.org/10.1007/s42106-022-00192-6

Nasim W, Amin A, Fahad S, Awais M, Khan N, Mubeen M, Wahid A, Turan V, Rehman MH, Ihsan MZ, Ahmad S, Hussain S, Mian IA, Khan B, Jamal Y (2018) Future risk assessment by estimating historical heat wave trends with projected heat accumulation using SimCLIM climate model in Pakistan. Atmos Res 205:118–133

Naz S, Ahmad S, Abbas G, Fatima Z, Hussain S, Khan MA, Ahmed M, Khan A, Fahad S, Nasim W, Ercisli S, Wilkerson CJ, Hoogenboom G (2022) Modeling the impact of climate warming on potato phenology. Eur J Agron 132:126404

Noorhosseini SA, Soltani A, Ajamnoroozi H (2018) Modeling the Impact of climate change on peanut production on the basis of Increasing 2°C temperature in Future Environmental conditions of Gilan Province, Iran. Int J Agric Manage Dev 8(2):257–273

Puli COR, Akila CS, Pandit V, Konduru S, Kandi SR, Chinta S (2021) Peanut (*Arachis hypogaea* L.) transgenic plants for abiotic stress tolerance. Genetically modified crops: current status. Prospects Challenges Vol 1:139–173

Qian B, Jing Q, Bélanger G, Shang J, Huffman T, Liu J, Hoogenboom G (2018) Simulated canola yield responses to climate change and adaptation in Canada. Agron J 110(1):133–146

Rahman MH, Ahmad A, Wang X, Wajid A, Nasim W, Hussain M, Ahmad B, Ahmad I, Ali Z, Ishaque W, Awais M, Muddasir M, Shelia V, Ahmad S, Fahad S, Alam M, Ullah H, Hoogenboom G (2018) Multi-model projections of future climate and climate change impacts uncertainty assessment for cotton production in Pakistan. Agric For Meteorol 253–254:94–113

Rahman MH, Ahmad I, Wang D, Fahad S, Afzal M, Ghaffar A, Saddique Q, Khan MA, Saud S, Hassan S, Fahad M, Din M, Ahmad M, Yue Z, Shah RA, Sönmez O, Nasim W (2021) Influence

of semi-arid environment on radiation use efficiency and other growth attributes of lentil crop. Environ Sci Pollut Res 28:13697–13711
Raza A, Razzaq A, Mehmood SS, Zou X, Zhang X, Lv Y, Xu J (2019) Impact of climate change on crops adaptation and strategies to tackle its outcome: a review. Plan Theory 8(2):34
Rehman A, Ashfaq M, Naqvi SAA, Adil SA, Bashir K, Ahmad A, Ahmad S, Ali A, Imran A (2015) Is climate change worsening the poverty of maize growers? Evidence from Punjab province of Pakistan. Ciencia e Tecnica Vitivinicola 30(1):105–116
Sarr AB, Camara M (2018) Simulation of the impact of climate change on peanut yield in Senegal. Int J Phys Sci 13(5):79–89
Sarwar N, Ali H, Maqsood M, Ullah E, Shahzad AN, Shahzad M, Mubeen K, Shahid MA, Ahmad S (2013) Phenological response of rice plants to different micronutrients application under water saving paddy fields on calcareous soil. Turk J Field Crop 18(1):52–57
Sarwar N, Atique-ur-Rehman, Farooq O, Wasaya A, Hussain M, El-Shehawi AM, Ahmad S, Brestic M, Mahmoud SF, Zivcak M, Farooq S (2021) Integrated nitrogen management improves productivity and economic returns of wheat-maize cropping system. J King Saud Univ Sci 33:101475
Secchi MA, Fernandez JA, Stamm MJ, Durrett T, Prasad PV, Messina CD, Ciampitti IA (2023) Effects of heat and drought on canola (*Brassica napus* L.) yield, oil, and protein: a meta-analysis. Field Crop Res 293:108848
Shabbir RN, Waraich EA, Ali H, Nawaz F, Ashraf MY, Ahmad R, Awan MI, Ahmad S, Irfan M, Hussain S, Ahmad Z (2016) Supplemental exogenous NPK application alters biochemical processes to improve yield and drought tolerance in wheat (*Triticum aestivum* L.). Environ Sci Pollut Res 23(3):2651–2662
Shekhar S, Rustagi A, Kumar D, Yusuf MA, Sarin NB, Lawrence K (2019) Groundnut AhcAPX conferred abiotic stress tolerance in transgenic banana through modulation of the ascorbate–glutathione pathway. Physiol Mol Biol Plants 25:1349–1366
Singh P, Boote KJ, Kumar U, Srinivas K, Nigam SN, Jones JW (2012) Evaluation of genetic traits for improving productivity and adaptation of groundnut to climate change in India. J Agron Crop Sci 198:399–413
Singh P, Nedumaran S, Ntare BR, Boote KJ, Singh NP, Srinivas K, Bantilan MCS (2014) Potential benefits of drought and heat tolerance in groundnut for adaptation to climate change in India and West Africa. Mitig Adapt Strateg Glob Chang 19:509–529
Tariq M, Yasmeen A, Ahmad S, Hussain N, Afzal MN, Hasanuzzaman M (2017) Shedding of fruiting structures in cotton: factors, compensation and prevention. Trop Subtrop Agroecosyst 20(2):251–262
Tariq M, Afzal MN, Muhammad D, Ahmad S, Shahzad AN, Kiran A, Wakeel A (2018a) Relationship of tissue potassium content with yield and fiber quality components of Bt cotton as influenced by potassium application methods. Field Crop Res 229:37–43
Tariq M, Ahmad S, Fahad S, Abbas G, Hussain S, Fatima Z, Wajid N, Mubeen M, Rehman M, Khan MA, Adnan M, Wilkerson CJ, Hoogenboom G (2018b) The impact of climate warming and crop management on phenology of sunflower-based cropping systems in Punjab, Pakistan. Agric For Meteorol 256–257:270–282
Tarriq M, Fatima Z, Iqbal P, Nahar K, Ahmad S, Hasanuzzaman M (2021) Sowing dates and cultivars mediated changes in phenology and yield traits of cotton-sunflower cropping system in arid environment. Int J Plant Prod 15:291–302
Uppal RK, Brill R, Bromfield J (2019) Effect of heat stress on canola grain yield and quality. Grains Research and Development Corporation, Barton
Van Duivenbooden N, Abdoussalam S, Ben Mohamed A (2002) Impact of climate change on agricultural production in the Sahel–Part 2. Case study for groundnut and cowpea in Niger. Clim Chang 54(3):349–368
Viana VE, Aranha BC, Busanello C, Maltzahn LE, Panozzo LE, de Oliveira AC, Rombaldi CV, Pegoraro C (2022) Metabolic profile of canola (*Brassica napus* L.) seedlings under hydric, osmotic and temperature stresses. Plant Stress 3:100059

Wang S, Wang E, Wang F, Tang L (2012) Phenological development and grain yield of canola as affected by sowing date and climate variation in the Yangtze River Basin of China. Crop Pasture Sci 63(5):478–488

Wei S, Li K, Yang Y, Wang C, Liu C, Zhang J (2022) Comprehensive climatic suitability evaluation of peanut in Huang-Huai-Hai region under the background of climate change. Sci Rep 12(1):11350

Xing H, Brill R, Li G, Li Liu D, Blake A, Slinger D (2019) Future climate change increases canola productivity and water use efficiency in the rainfed cropping systems of Southern Australia. In: 19th Australian Agronomy Conference. Australian Society for Agronomy, pp 1–4

Xu H, Tian Z, Zhong H, Fan D, Shi R, Niu Y, He X, Chen M (2017) Impacts of climate change on peanut yield in China simulated by CMIP5 multi-model ensemble projections. In: Remote sensing and modeling of ecosystems for sustainability XIV, vol 10405. SPIE, San Diego, pp 223–229

Zahoor A, Ahmed M, Fayyaz-ul-Hassan SG, Ahmad S (2022) Ontogeny growth and radiation use efficiency of canola (*Brassica napus* L.) under various nitrogen management strategies and contrasting environments. Int J Plant Prod 16(1):195–208

Zhen X, Huo W, Tian D, Zhang Q, Sanz-Saez A, Chen CY, Batchelor WD (2023) County level calibration strategy to evaluate peanut irrigation water use under different climate change scenarios. Eur J Agron 143:126693

Guar-Wheat System 15

15.1 Introduction

Guar is primarily cultivated as a seed crop, but it can be used as forage, vegetable, and green manure crop (Kuravadi et al. 2013). No doubt, guar is an industrial crop, but our farmers grow it on poor soils or fields. It is a deep-rooted, drought-hardy annual legume crop of summer season, and its incorporation in the cropping system could help to improve soil health. Dwarf and tall varieties are common that can bear four to ten branches with a height of 50–100 cm. Flowers have long pedicels with axillary racemes. Guar seeds contain hull, endosperm, and germ. Guar seed endosperm has polysaccharides called galactomannans, and they are utilized as gelling as well as thickening agents across diverse industries (Jukanti et al. 2015; Tahmouzi et al. 2023) (Figs. 15.1 and 15.2). Gresta et al. (2014) reported increased demand of guar by various stakeholders (Fig. 15.3). The largest consumer of guar gum is the USA, and its maximum import is from India. Guar is a biological N-fixing crop, and it can provide substantial quantity of N for the next crop (Zamir et al. 2016; Ul'yanich et al. 2022). The production of grain legume like guar is vulnerable to diverse stresses that include water and temperature (Vyas 2014). Deepika and Dhingra (2014) concluded that guar can tolerate drought as well as salinity. It can fit easily into crop rotation due to its shorter growing season (90–120 days). Based on these features, guar can be a viable crop for semiarid regions where issues of water scarcity along with salinity prevail (Ashraf et al. 2005). Kuravadi et al. (2013) studied adaptability of guar at sites where annual rainfall was <250 mm and proved guar as a potential drought-tolerant annual crop. However, different earlier researchers reported a negative impact of drought for legumes, and this effect was stronger when drought prevailed during grain filling called as terminal drought. Agronomic practices, i.e., adjusting sowing window and planting geometry along with nutrients (macro and micro), can improve legumes' drought resistance (Waraich et al. 2011; Siddique et al. 2012). Ahmad et al. (2016) stated selenium (Se) importance for increasing the root system's water uptake capacity under drought. Silicon (Si) and

M. Ahmed et al., *Cropping Systems Modeling Under Changing Climate*,
https://doi.org/10.1007/978-981-97-0331-9_15

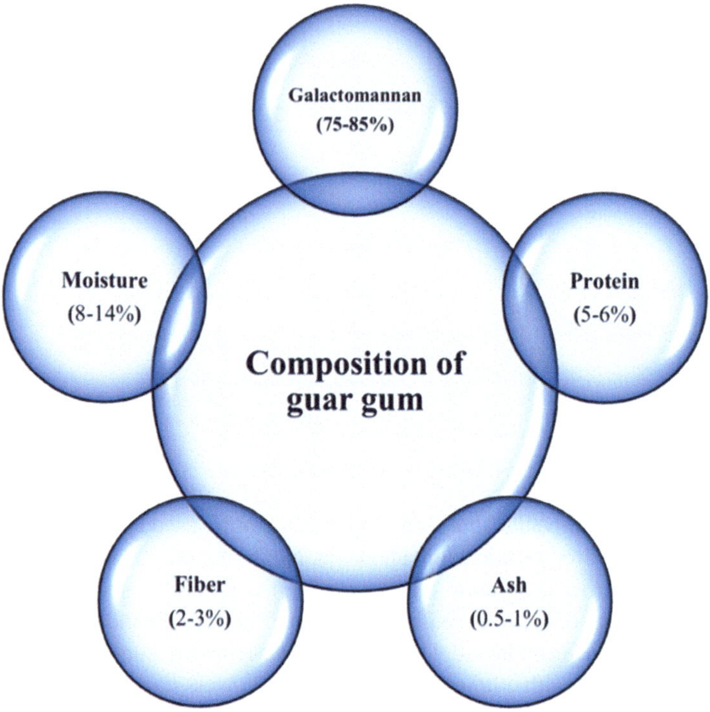

Fig. 15.1 Composition of guar (*Cyamopsis tetragonoloba* L.) gum. (Source: Tahmouzi et al. 2023)

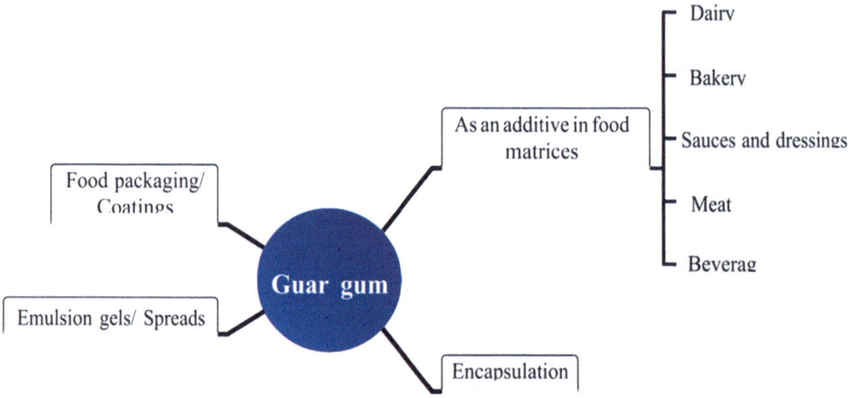

Fig. 15.2 Usage of guar gum in food science. (Source: Tahmouzi et al. 2023)

zinc (Zn) can also be helpful for improving tolerance against drought in crops (Almeselmani et al. 2012; Alonso et al. 2018; Ashraf et al. 2002; Asseng et al. 2015; Asseng et al. 2019; Boote et al. 2003; Boote et al. 1998; Boote et al. 2001; Buttar et al. 2012; Guerra et al. 2008; Ihsan et al. 2019; Fahad et al. 2016; Hattori et al. 2005; Hoogenboom et al. 2017 Jones et al. 2003; Lubbe & Verpoorte 2011; Messina

Fig. 15.3 Features of guar plant and its products. (Source: Tahmouzi et al. 2023)

et al. 2006; Shaban et al. 2012; Slafer & Savin, 1994; Suriharn et al. 2007; Suriharn et al. 2008; Tao et al. 2017; Wang & Zhang 2009; White & Hoogenboom 2003; Bao et al. 2015; Zambrano et al. 2004 ; Ali et al. 2023; Bakhat et al. 2018; Etesami and Jeong 2018; Abidi et al. 2015). India is a top guar producer (80%) followed by Pakistan where its contribution to the total world production is only 15% (Fig. 15.4).

Gradual rise in intensity as well as duration of global temperature has emerged as a possible risk to agrarian yield along with sustainability all around the globe as reported by the IPCC (IPCC 2023; Ahmed 2020, 2023; Abbas et al. 2023b). Hansen et al. (2012) reported an intensification in mean temperature change with an occurrence of 0.18 °C/decade. Intensification in temperatures has significant

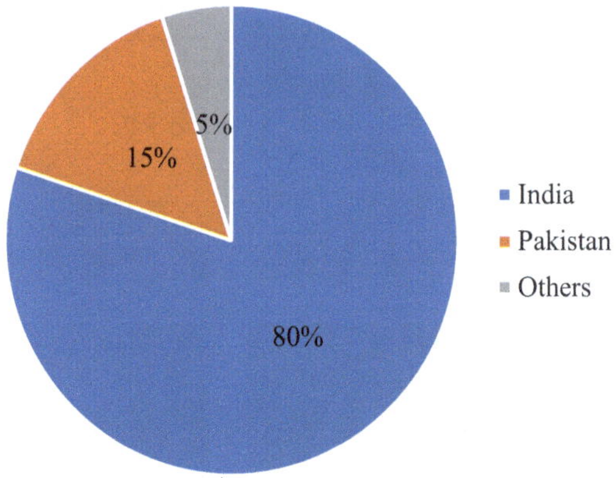

Fig. 15.4 Major guar-producing countries

impacts on yield losses in crops (Bakhat et al. 2018; Akram et al. 2019). Higher temperatures impact crops in numerous ways, which include physiological and structural modifications, decreased photosynthesis, impairment of chlorophyll biosynthesis, and disturbance in enzyme activity (Saud et al. 2017; Awan et al. 2014). Temperature has a noteworthy impact on growth and development, and optimum temperature needed for guar germination is 30 °C; however, minimum is 21 °C. Therefore, sowing time has a significant effect on the stand establishment of crop as concluded by Patil (2014) who reported significant difference in guar phenology. Gresta et al.'s (2013) study showed earlier phenological development in guar due to late June sowing as compared to mid-May. Ali et al. (2004) concluded similar findings in which they found taller, higher aboveground dry plant biomass and grain yield for early June sowing as compared to other sowings in dryland regions of Pakistan. Singla et al. (2016) in their research showed that mid-June sowing produced better plant stand, more transpiration, LAI, clusters, pods, seeds plant^{-1}, 37% more aboveground dry plant biomass, and 45% higher seed yield. Therefore, favorable weather conditions are very important to have a good yield of guar as more photosynthates moved from source to sink are only possible at favorable weather circumstances (Asseng et al. 2017; Erice et al. 2011; Shrestha and Adams 2023). Climate variability and change are anticipated to exaggerate drought as well as heat stresses on crops of rainfed regions (Ahmed and Ahmad 2023; Ahmed 2023; Abbas et al. 2023a; Liu et al. 2023; Ahmed et al. 2022a; Ahmed et al. 2022b; Nadeem et al. 2022; Ahmed 2020a, b; Khan et al. 2020; Asseng et al. 2019; Ahmed et al. 2017; Jabeen et al. 2017; Aslam et al. 2013; Ahmed et al. 2012; IPCC 2023; Abbas et al. 2023a, b; Menzel et al. 2020). This will require improvement in the agro-management practices and development of more productive guar varieties, which are drought and heat tolerant and resistant so that they can be fitted easily in cropping systems (Gresta et al. 2014; Abidi et al. 2015).

Table 15.1 Guar crop calendar

Months	Jun	Jul	Aug	Sep	Oct	Nov	Dec
Guar							
Planting							
Harvesting							

Table 15.2 Growth stage descriptions for guar

Stage	Stage title	Stage description
Vegetative (V) stages		
VE	Emergence	Plants are visible at the soil surface
V0	Cotyledonary node	Cotyledonary node is developed with cotyledons flat and open at the soil surface
V1	First node	One developed node on the main axis of the plant with associated leaves unfolded and flat
V(*N*)	*N*th node	*N* developed nodes on the main axis
Reproductive (R) stages		
R1	First flower	The first flower is visible anywhere on the plant
R2	First pod	The first pod is visible and in early development with a length of approximately 10–15 mm
R3	First seed/full pod	Immature seeds with a diameter of at least 3 mm are present within the first pod. This is also the point at which the first pod is fully expanded, with a length characteristic to variety
R4	Full seed	The first pod cavity cross section is apparently filled with seed
R5	First maturity	The first pod is naturally desiccated and is fully brown in color
R6	50% Maturity	50% of pods are desiccated and brown in color
R7	Harvest maturity	At least 50% of the plant stem has desiccated, which may be achieved by the natural maturation process, frost, or chemical harvest aids in addition to achieving at least 75% pod maturity

Source: Adams et al. (2020)

15.2 Guar Phenology

Sowing of guar starts in June and completes in July, and plants reach the flowering stage in August depending upon varieties. Generally, long-duration varieties take more time as compared to short-duration varieties. Harvesting starts in the month of October in case of short-duration varieties, while it extends to November if varieties are indeterminate (Table 15.1). Guar growth and developmental patterns have been elaborated by Adams et al. (2020) (Fig. 15.2 and Table 15.2). Meftahizadeh et al. (2019) investigated the effect of diverse sowing dates on Indian along with Iranian landraces of guar in an arid environment. Results showed that planting dates have shown a significant impact on guar crop phenology, seed yield, and physiological and metabolic traits (Fig. 15.5). Suthar et al. (2018) conducted comparative experiment regarding earlier growth stage study of 25 guar cultivars at higher salinity levels and observed genotypic variation at different salinity levels. Similarly,

Fig. 15.5 Guar phenological stages. (Source: Adams et al. 2020)

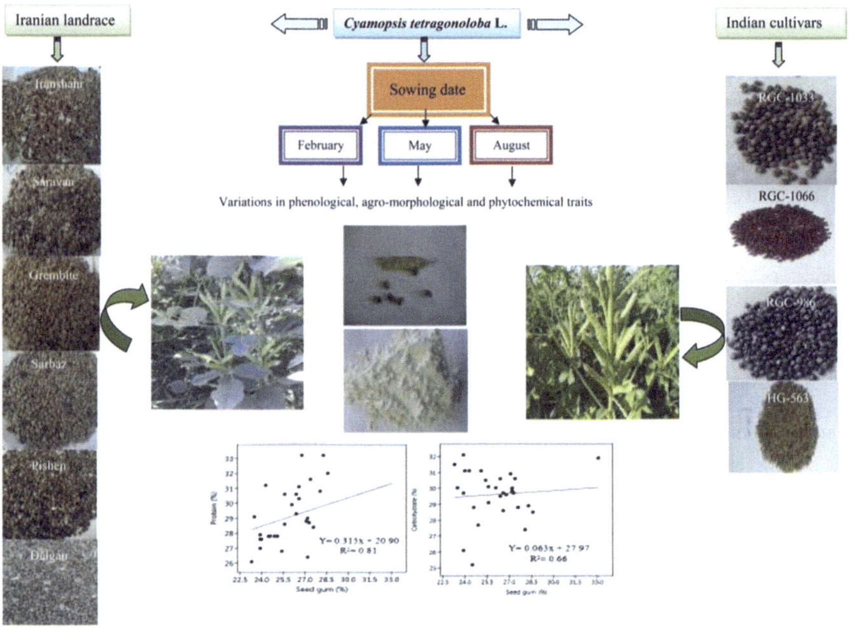

Fig. 15.6 Sowing dates' impact on guar cultivars in an arid environment. (Source: Meftahizadeh et al. 2019)

phenotypic variation and its association with root nodulation along with growth characteristics of 50 guar genotypes were evaluated by Shrestha et al. (2021) (Fig. 15.6). Results showed that nodule masses/plant varied among genotypes with moderate heritability, while low heritability was observed for nodule number. Correlation was observed between nodule masses versus total N, as well as positive correlation was noted for stem diameters versus nodule masses. High heritability was observed for branch numbers as well as stem diameters, and both have a very high agronomic importance. In case of branched genotypes, reproductive development was delayed which can be helpful for breeding guar for variable environments. The obtained results can be used to develop future desirable guar plant ideotypes.

15.3 Decision Support System for Agrotechnology Transfer

Simulation model for guar (DSSAT-CROPGRO-Bean Model) can be employed for quantifying probable benefits of different agro-management practices and incorporation of heat and drought resistance as well as productivity increasing traits into guar crop (Alagarswamy et al. 2006; Bosi et al. 2020; Jing et al. 2020; Malik and Dechmi 2019). Since breeders are now aiming to have new crops with specific plant traits, early assessment of these traits is necessary to minimize economic losses. This is only possible by using crop simulation models, which can calculate growth as well as productivity benefits owing to newer technologies across diverse environments by employing location-specific soil, weather, and agronomic data (Boote 2011; Martre et al. 2017). Many scientists in the past have projected the use of crop models in genetic improvement of crops for higher yields (Tardieu et al. 2018; Singh et al. 2014a, b, c) and DSSAT v4.8 (Hoogenboom et al. 2024) is one sucg crop model that also have CROPGRO-Guar model (Hoogenboom et al. 2024). Crop growth in this model is simulated on a daily basis from sowing to maturity, as well as it also involves all physiological processes. These processes describe responses of legume crops against soil as well as aerial weather conditions. Potential growths are modeled by PAR along with its interception, while stresses related to temperatures, soil water, as well as soil N deficiencies have been incorporated in the model to show their impacts on crop growth. Minimum data needed to run DSSAT model includes meteorological data (Tmax, Tmin, precipitation, and sunshine hours or solar radiation), soil data (physicochemical properties), cultivar coefficients (Hoogenboom et al. 2024), and agro-management information. Detailed explanation of DSSAT-CROPGRO-Guar Model is available (www.dssat.net). Basso et al. (2016) presented a comprehensive review about the global performances of the CERES family of DSSAT, which proves its importance under a wide range of agro-managements to improve crop production.

Crop growth simulation models have emerged as a tool for enhancing the understanding of crop ecology along with physiology. Heat tolerance is an important feature that needs to be considered for adaptation under rainfed conditions. Water along with high-temperature stress is a grave restriction in arid areas of the world. Houghton et al. (2012) reported 1.5–6 °C future increase in temperature. Therefore, heat-tolerant cultivars will be desired for sustaining productivity as well as production (Khan et al. 2020; Hussain et al. 2021). Rainfed farming is mainly weather dependent. This model can help to design ideotypes for target environments. CSM-CROPGRO-Bean (Boote et al. 2011), besides being a widely tested legume model (Singh et al. 2012) that was coupled to DSSAT, has a great potential to understand climate variability impacts on peanut yield. Likewise, Nigam et al. (2001) stated that poor yield of legume crop in developing republics was mostly due to the biotic as well as abiotic stresses like unreliable rainfall, drought, unavailability of high-productivity varieties, damages by insects, pests, and diseases, and poor cultural practices, along with limited input usage. Guar production gaps under extreme climatic conditions could be filled by using crop modeling approaches to design agro-management strategies and crop traits under a specific set of environmental and management conditions.

15.4 Application of CROPGRO Model for Simulating Guar Growth and Productivity

DSSAT-CROPGRO model has been applied for simulating guar growth and productivity by Boote et al. (2023). Field experiment data from Texas and New Mexico was used to run CROPGRO-Soybean model so that it can be adapted for guar. Results showed that guar was more sensitive to cold temperatures, compared to soybean, and parameters adapted for guar simulation were life cycle period, duration to the end of leaf area expansion, pod adding period, leaf photosynthesis, seed (size, protein, lipid), as well as single seed-filling period. Cardinal temperature was increased by 2–4 °C as compared to soybean species file (Table 15.3). Furthermore, genetic coefficients of CROPGRO-Guar were modified in guar cultivars by considering default value of CROPGRO-Soybean cultivar, i.e., 'William 821' (Table 15.4). Similarly, guar drought tolerance feature was highlighted in the model as it shows slow rates of leaf senescence as well as more tolerance toward N fixation at drought. The model also simulated time course growth and pod or seed yield of guar cultivars with good accuracy (Fig. 15.7).

Table 15.3 CROPGRO-Guar cardinal temperatures (°C) in species file in comparison with CROPGRO-Soybean

	Guar				Soybean			
Growth/development process	T_b	T_{opt1}	T_{opt2}	T_{ceil}	T_b	T_{opt1}	T_{opt2}	T_{ceil}
Vegetative development	9	29	35	45	7	28	35	45
Reproduction to flowering and first seed	9	26.5	28.5	45	6	26	30	45
Reproduction of first seed to mature	−5	24	26	45	−15	26	34	45
Light saturated leaf photosynthesis	10	40	442	48	8	40	442	48
T_{min} effect on the next day's leaf photosynthesis	33	233	–	–	0	193	–	–
Leaf area expansion	9	23	23	4	8.7	22	22	4
Height/width increase	9	26	26	4	1	26	26	4
Pod and seed addition	14	23	26.5	40	14	21	26.5	40
Seed growth rate	10	23.5	25	42	6	21	23.5	41

Source: Boote et al. (2023)
Where T^b is the base temperature (°C), T_{opt1} is the first optimum temperature (°C), T_{opt2} is the second optimum temperature (°C), T_{ceil} is the ceiling temperature (°C)

Table 15.4 Calibrated CROPGRO-Guar genetic coefficients in comparison with values of CROPGRO-Soybean default value

Genetic coefficients	'Kinman' (guar)	'Monument' (guar)	'Williams 821' (soybean)
Critical long day length above which reproductive development progresses as rapidly as possible with no day length effect (CLDL) (h)	13.59	13.59	13.4
Slope of the relative response of development vs. photoperiod (PP-SEN) (1 h^{-1})	0.249	0.249	0.285
Time between emergence and flower appearance (EM-FL) (ptd)	14.5	14.5	19
Time between first flower and beginning pod (FL-SH) (ptd: photothermal days)	8	8	8.3
Time between first flower and beginning seed (FL-SD) (ptd: photothermal days)	12	12	14.2
Time between beginning seed and (beginning) physiological maturity (SD-PM) (ptd: photothermal days)	63	63	32.2
Time between first flower and last leaf expansion (FL-LF) (ptd: photothermal days)	60	58	28
Maximum leaf photosynthetic rate at 30 °C, 350 μL L^{-1} CO^2, and high light (LFMAX) (mg CO_2 m^2 s^{-1})	1.33	1.11	0.99
Specific leaf area of cultivar under standard growth conditions (SLAVAR) (cm^2 g^{-1})	350	350	385
Maximum size of fifth full leaf (SIZLF) (cm^2)	140	140	180
Maximum fraction of daily growth that is partitioned to seed + shell (XFRT)	0.7	0.7	1
Maximum weight per seed (WTPSD) (g)	0.03	0.03	0.18
Seed-filling duration for seed cohort under standard conditions (SFDUR) (ptd: photothermal days)	55	55	26
Seed number per pod (SDPDV)	8	8	2.4
Duration of pod addition under standard conditions (PODUR) (ptd: photothermal days)	20.5	19.5	12
Threshing percentage [seed (seed + shell)$^{-1}$] (THRSH)	57.8	56.8	77
Fraction of protein in seeds (SDPRO) [g (protein) g (seed)$^{-1}$]	0.3	0.3	0.405
Fraction of oil in seeds (SDLIP) [g (oil) g (seed)$^{-1}$]	0.06	0.06	0.205
Rate of leaf appearance (TRIFL), per ptd (ecotype)	0.44	0.44	0.32
Time from first flower to last leaf on main stem (FLVS) (ptd: photothermal days) (ecotype)	65	65	26
Proportion of time between first seed and physiological maturity that the last seed can be added (PM09) (ecotype)	0.5	0.5	0.35

Source: Boote et al. (2023)

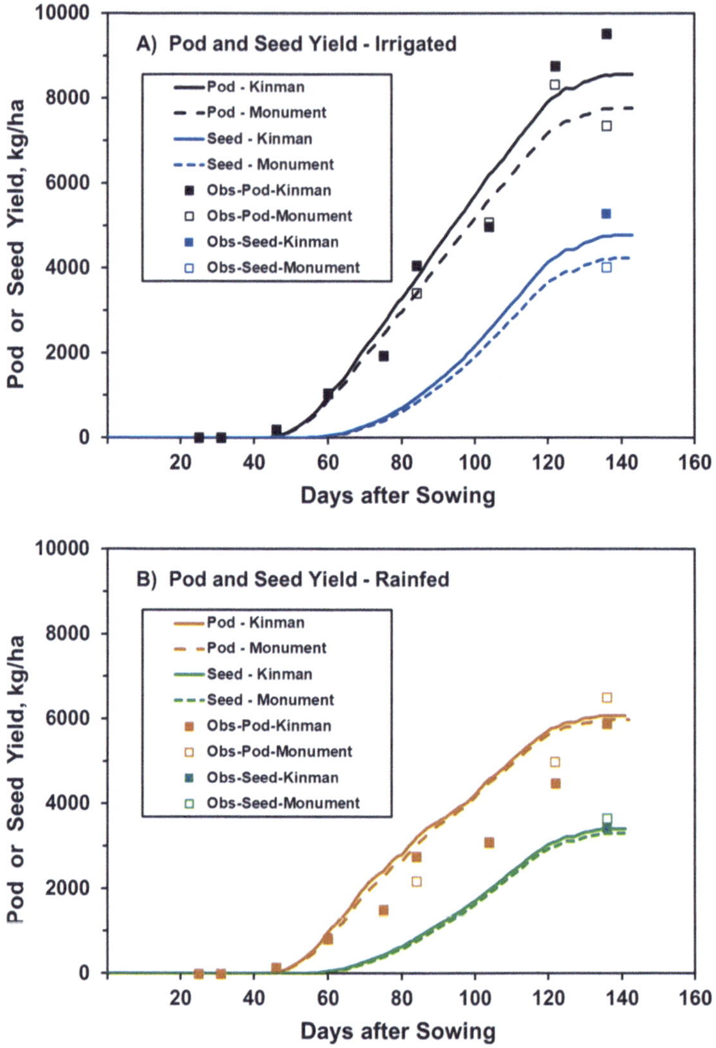

Fig. 15.7 Guar-CROPGRO observed (points) and simulated (lines) and pod and seed mass over time for (**a**) irrigated and (**b**) rainfed treatments. (Source: Boote et al. 2023)

15.5 Integration of Guar in Dryland Wheat System

Agriculture is the major GDP contributor in South Asia while showing an increasing trend during the past decades being the outcome of green revolution, but the region is yet neglected and most of the people in the region are poor. Two-thirds of the people live in villages. It is the poorest area followed by Africa where most republics are least developed. The World Bank has indicated that 40% of people in this area live under the International Poverty Line ($1.25/day). The highest GDP/

capita in the region is depicted by Sri Lanka, while the lowest is shown by Nepal. However, 82% of the South Asian economy is dominated by India followed by Pakistan. Pakistan is the fifth largest with reference to GDP in the region, and trade between countries is only 2%. Population of South Asian countries is one-quarter more than the developing world, which results in huge pressure on natural resources. The topography of the South Asian region is steeply sloping mountains which comprises 4%, while 42% is humid moist lowland. The dry subhumid areas comprise 29%, whereas the remaining 24% is arid to semiarid. Hills and mountains are the dominant features in most of the countries, and these suffer from extensive deforestation and soil erosion. Most of the subhumid zones are benefited from monsoon rains having alluvial type of soil. The area under forest in the region is very low (14% of the total area), resulting in poor grazing land and scarcity of fresh water.

Pakistan's economy is agro-based, which provides employment to 45% of people. It also provides raw material to diverse industries while at the same time ensuring food security to burgeoning population. The IPCC (2023) has reported severe threats of temperature, floods, droughts, and extreme weather events to agriculture in the developing countries. Most of the lands in Pakistan are designated as arid as well as semiarid, where precipitation is low to fulfill the crop requirements for sustainable farming. Around 68% of Pakistan receives 250 mm annual rainfall, while 24% comes in the range of 250–500 mm, and only 8% area in north Punjab has rainfall over 500 mm. The crops growing in rainfed region are severely affected due to scanty rainfall, higher temperature, and sloppy topography (Fig. 15.8). However, there are further issues, i.e., water stress, erosion, crust formation, shortage of nutrients, less nutrient-use efficiency, weed problem, and marginal soils. Wheat, being the staple food in Pakistan, is sown on larger areas as compared to other crops. The precipitation pattern in the area is South Asian monsoon (van Ogtrop et al. 2014; Ahmed et al. 2014a, b, c).

Farmers sow wheat in winter months, which starts from end October to December, while it ends from mid-April to May. However, preseason and in-season precipitation result in large variations in crop yields from year to year. Therefore, the main root cause of low farm productivity in Pothwar is poor rainwater management and unsustainable farming, which results in no crop in some years. Similarly, increased land degradation is also observed due to the use of exhaustive cropping system and traditional cultural operations. More importantly, due to the long-term extensive use of cereal-based cropping system, crop yield and soil health have deteriorated significantly. The summer fallow that is adopted by more than 80% of growers has resulted in loss of water, N, and carbon, showing a significant effect on biogeochemical cycles. Furthermore, use of conventional tillage technologies (i.e., moldboard ploughing) has worsened the scenario and resulted in loss of water, organic C, and nutrients and soil erosion. Due to the limitations mentioned above, severe environmental damages have often made rainfed farming a nonprofitable business, so farmers are shifting to other sectors. Therefore, cost-effective agrarian deeds in dry regions must be aimed upon well-organized and justifiable exploitation of inadequate water resources (Ahmed et al. 2013). Guar crop is considered as one of the potential crops to cope with the challenges of various types of stresses like drought,

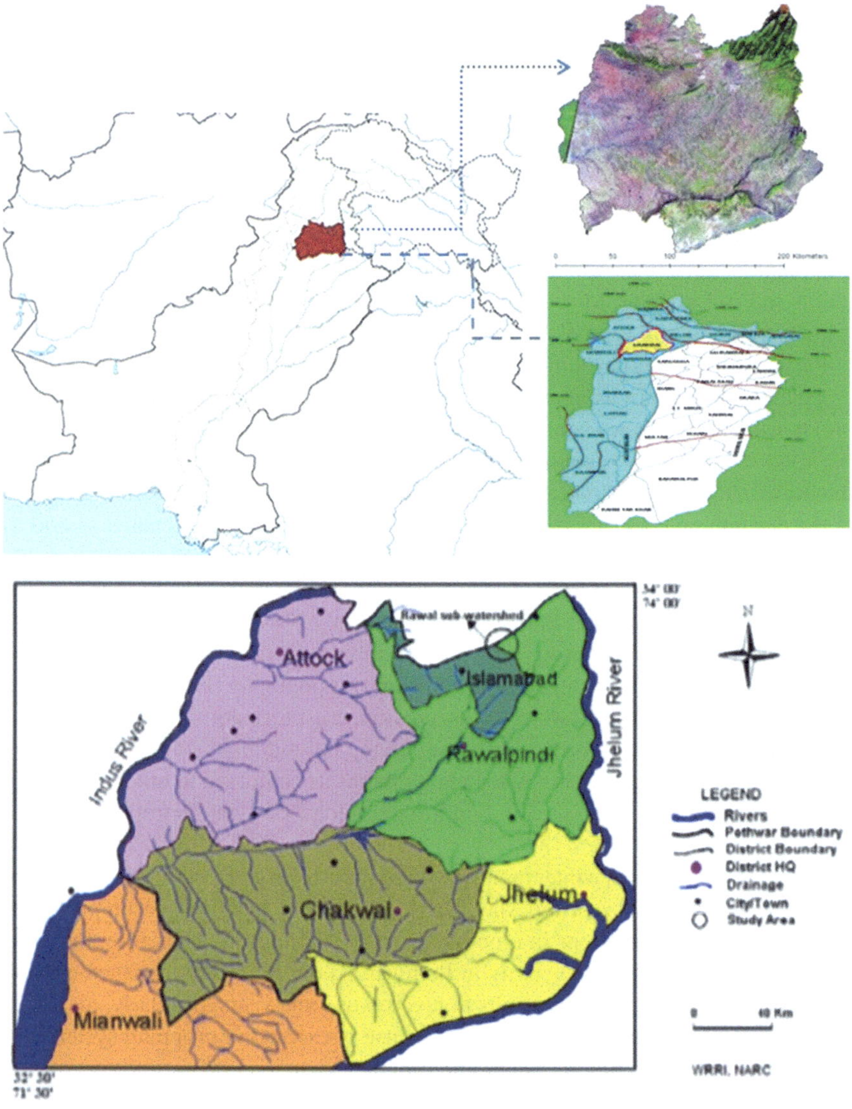

Fig. 15.8 Map of rainfed Pothwar

salt, as well as rise in temperature (heat) (Shrestha and Adams 2023; Boote et al. 2023). Therefore, in view of the above issues, it is being felt that future crops should be tolerant to these abiotic stresses. This can be easily achieved by sowing crops like guar in rainfed regions.

Drought and salinity are major issues faced by farmers, scientists, and policymakers across the globe. Even growers have to buy fodder for their animals; however, salt- and drought-resistant fodder crops could be sown. However, farmers can grow guar as a multipurpose crop to obtain fodder, BNF, and industrial by-products. The demand of this crop is increasing year by year, and it can be potential plants for rainfed Pothwar (Suthar et al. 2018; Shrestha et al. 2021). It is grown as an industrial crop as well as for fodder purpose because of high protein concentration. Guar can fix nitrogen from atmosphere because it is a legume crop, so needs less input of fertilizers. It is one of the major sources of galactomannans for the industry utilization. Galactomannans from guar have been used for various industrial purposes such as viscosifier, emulsion stabilizer, and suspending and gelling agent and utilized in various fields like oil operation, agro-chemistry, cosmetics, and food (Mudgil et al. 2014). Guar gum is a unique addition to the different food products as a fiber source and balances the food. Therefore, we recommend adoption of guar for effective utilization of natural and human resource for ensuring food security as well as incomes of rural Pothwar farming populations in Northern Pakistan. Being an environment-friendly legume crop, guar can fit very well in our cropping systems and has a vital role in improving economy, conserving water, and preventing land degradation. The existing pathetic status of guar in Pakistan is stagnant and can largely be attributed to the ignorance of farmers about its use, nonavailability of specific varieties for specific farming conditions, and lack of advanced production technology.

Shrestha et al. (2023) did systems research to see possible benefit of guar in Chillicothe, TX. Wheat-guar system with different intensities, i.e., WG1.3, WG2, G1, and W1, was used. Their findings showed that yield was highest for WG2, while it was lowest for G1. Similar trend of mean net income/year was shown by G1 < W1 < WG1.3 < WG2. Furthermore, they noted no direct indications regarding BNF from guar on the next wheat crop.

15.6 Conclusion

Guar is a drought-hardy plant due to its ability to withstand periods of water scarcity and thrive in arid as well as semiarid areas. It can grow with little as 300–600 mm (12–24 inches) of annual rainfall, making it well suited to regions with erratic or limited rainfall. It can survive extended periods of drought and resume growth when moisture becomes available. This resilience allows it to bounce back after drought conditions. Guar can grow in all soil types, i.e., sandy soils to clayey soils, which provides flexibility for cultivation in different drought-prone regions. Guar inclusion in cropping systems could help to improve soil health, increase productivity, and improve WUE of subsequent crops. Moreover, adoption of legume-based cropping pattern would improve soil fertility after some time. Since legumes enhance N-supplying power of soils, they increase the organic matter and soil recovery, stimulate soil biological activity, make the soil easier to till, reduce soil erosion by wind and water, and improve the physical characteristics, i.e., soil aeration as well as water-holding capacity, thus increasing yields of succeeding rotation. Guar-wheat rotation could result in the highest yield, good protein content, and better yield components. Diversification of rainfed cropping systems through introduction of guar in the system could be a viable option and climate-smart intervention. Hence, our reviewed work suggests introducing high-yielding guar genotypes with advanced production technologies complemented with simulation modeling to significantly increase rainfed farm income. Crop modeling for guar can be a valuable tool for optimizing agricultural practices, improving resource management, and mitigating risks associated with climate variability.

References

Abbas G, Ahmed M, Fatima Z, Hussain S, Kheir AMS, Ercişli S, Ahmad S (2023a) Modeling the potential impact of climate change on maize-maize cropping system in semi-arid environment and designing of adaptation options. Agric For Meteorol 341:109674. https://doi.org/10.1016/j.agrformet.2023.109674

Abbas G, Raza MA, Ahmed M, Saeed A, Khalid MHB, Manzoor A, Awan TH, Kheir AMS, Nasim W, Ahmad S (2023b) Global framework on climate change. In: Jatoi WN, Mubeen M, Hashmi MZ, Ali S, Fahad S, Mahmood K (eds) Climate change impacts on agriculture: concepts, issues and policies for developing countries. Springer International Publishing, Cham, pp 3–22. https://doi.org/10.1007/978-3-031-26692-8_1

Abidi N, Liyanage S, Auld D, Norman L, Grover K, Augadi S, Singla S, Trostle C (2015) Challenges and opportunities for increasing guar production in the United States to support unconventional oil and gas production. In: Uddameri V, Morse A, Tindle KJ (eds) Hydraulic fracturing impacts and technologies: a multidisciplinary perspective. CRC Press, Boca Raton, pp 207–225

Adams CB, Boote KJ, Shrestha R, MacMillan J, Hinson PO, Trostle C (2020) Growth stages and developmental patterns of guar. Agron J 112(6):4990–5001. https://doi.org/10.1002/agj2.20415

Ahmad R, Waraich EA, Nawaz F, Ashraf MY, Khalid M (2016) Selenium (Se) improves drought tolerance in crop plants—a myth or fact? J Sci Food Agric 96(2):372–380. https://doi.org/10.1002/jsfa.7231

Ahmed M (2020) Introduction to modern climate change. Andrew E. Dessler: Cambridge University Press, 2011, 252 pp, ISBN-10: 0521173159. Sci Total Environ 734:139397. https://doi.org/10.1016/j.scitotenv.2020.139397

Ahmed M (2023) The science of climate change. In: Brinkmann R (ed) The Palgrave handbook of global sustainability. Springer International Publishing, Cham, pp 195–222. https://doi.org/10.1007/978-3-031-01949-4_22

Ahmed M, Hassan FU, Aslam M, Aslam MA (2012) Physiological attributes based resilience of wheat to climate change. Int J Agric Biol 14(3):407–412

Ahmed M, Kamran A, Asif M, Qadeer U, Ahmed ZI, Goyal A (2013) Silicon priming: a potential source to impart abiotic stress tolerance in wheat: a review. Aust J Crop Sci 7(4):484

Ahmed M, Asif M, Hassan F-U (2014a) Augmenting drought tolerance in sorghum by silicon nutrition. Acta Physiol Plant 36(2):473–483. https://doi.org/10.1007/s11738-013-1427-2

Ahmed M, Fayyaz ul H, Asif M (2014b) Amelioration of drought in sorghum (Sorghum bicolor L.) by silicon. Commun Soil Sci Plant Anal 45(4):470–486. https://doi.org/10.1080/00103624.2013.863907

Ahmed M, Fayyaz Ul H, Van Ogtrop FF (2014c) Can models help to forecast rainwater dynamics for rainfed ecosystem? Weather Clim Extremes 5–6:48–55. https://doi.org/10.1016/j.wace.2014.07.001

Ahmed M, Stöckle CO, Nelson R, Higgins S (2017) Assessment of climate change and atmospheric CO_2 impact on winter wheat in the Pacific Northwest using a multimodel ensemble. Front Ecol Evol 5. https://doi.org/10.3389/fevo.2017.00051

Ahmed M (2020a) Introduction to modern climate change. Andrew E. Dessler: Cambridge University Press, 2011, 252 pp, ISBN-10: 0521173159. Sci Total Environ 734:139397. https://doi.org/10.1016/j.scitotenv.2020.139397

Ahmed M (2020b) Systems modeling. Springer Nature Singapore Pte Ltd., p 409. https://doi.org/10.1007/978-981-15-4728-7. https://link.springer.com/book/10.1007%2F978-981-15-4728-7#toc

Ahmed M (2023) The science of climate change. In: Brinkmann R (ed) The palgrave handbook of global sustainability. Springer International Publishing, Cham, pp 195–222. https://doi.org/10.1007/978-3-031-01949-4_22

Ahmed M, Ahmad S (eds) (2023) Disaster risk reduction in agriculture. Springer Nature. https://doi.org/10.1007/978-981-99-1763-1. ISBN 978-981-99-1762-4, Published: 03 November 2023

Ahmed M, Hayat R, Ahmad M, ul-Hassan M, Kheir AMS, ul-Hassan F, ur-Rehman MH, Shaheen FA, Raza MA, Ahmad S (2022a) Impact of climate change on dryland agricultural systems: a review of current status, potentials, and further work need. Int J Plant Prod. https://doi.org/10.1007/s42106-022-00197-1

Ahmed M, Ahmad S, Kheir AMS (2022b) Climate change: an overview. In: Ahmed M (ed) Global agricultural production - resilience to climate change. Springer Nature Switzerland AG. https://doi.org/10.1007/978-3-031-14973-3_1

Akram R, Fahad S, Masood N, Rasool A, Ijaz M, Ihsan MZ, Maqbool MM, Ahmad S, Hussain S, Ahmed M, Kaleem S, Sultana SR, Mubeen M, Saud S, Kamran M, Nasim W (2019) Plant growth and morphological changes in rice under abiotic stress. In: Hasanuzzaman M, Fujita M, Nahar K, Biswas JK (eds) Advances in rice research for abiotic stress tolerance. Woodhead Publishing, Duxford, pp 69–85. https://doi.org/10.1016/B978-0-12-814332-2.00004-6

Alagarswamy G, Boote KJ, Allen LH, Jones JW (2006) Evaluating the CROPGRO–Soybean model ability to simulate photosynthesis response to carbon dioxide levels. Agron J 98(1):34–42. https://doi.org/10.2134/agronj2004-0298

Ali Z, Zahid MS, Zia-ul-Hassan M, Bashir M (2004) Sowing dates effects on growth, development and yield of guar (Cyamopsis tetragonoloba L.) under rainfed conditions of Pothwar region [Pakistan]. J Agric Res 42(1):33–40

Ali H, Ahmad M, Alvi MH, Ali MF, Mahmood I, Ahmad S, Sameen A (2023) Foliar application of silicon to boost biochemical and physiological response in oat under water stress. Silicon 15:5317–5329. https://doi.org/10.1007/s12633-023-02443-1

Almeselmani M, Deshmukh P, Chinnusamy V (2012) Effects of prolonged high temperature stress on respiration, photosynthesis and gene expression in wheat (Triticum aestivum L.) varieties differing in their thermotolerance. Plant Stress 6(1):25–32

Alonso MP, Abbate PE, Mirabella NE, Aramburu Merlos F, Panelo JS, Pontaroli AC (2018) Analysis of sink/source relations in bread wheat recombinant inbred lines and commercial cultivars under a high yield potential environment. Eur J Agron 93:82–87. https://doi.org/10.1016/j.eja.2017.11.007

Ashraf MY, Akhtar K, Sarwar G, Ashraf M (2002) Evaluation of arid and semi-arid ecotypes of guar (Cyamopsis tetragonoloba L.) for salinity (NaCl) tolerance. J Arid Environ 52(4):473–482. https://doi.org/10.1006/jare.2002.1017

Ashraf MY, Akhtar K, Sarwar G, Ashraf M (2005) Role of the rooting system in salt tolerance potential of different guar accessions. Agron Sustain Dev 25(2):243–249

Aslam Z, Ahmed M, Sajad M, Asif M, Akmal M, Awan FK, Ijaz W, Ashraf R, Khattak JZK (2013) A comparison of statistical and dynamic modeling of wheat (*Triticum aestivum* L.) fungal diseases under the climate change. J Food Agric Environ 11(2):1327–1332

Asseng S, Ewert F, Martre P, Rotter RP, Lobell DB, Cammarano D, Kimball BA, Ottman MJ, Wall GW, White JW, Reynolds MP, Alderman PD, Prasad PVV, Aggarwal PK, Anothai J, Basso B, Biernath C, Challinor AJ, De Sanctis G, Doltra J, Fereres E, Garcia-Vila M, Gayler S, Hoogenboom G, Hunt LA, Izaurralde RC, Jabloun M, Jones CD, Kersebaum KC, Koehler AK, Muller C, Naresh Kumar S, Nendel C, O'Leary G, Olesen JE, Palosuo T, Priesack E, Eyshi Rezaei E, Ruane AC, Semenov MA, Shcherbak I, Stockle C, Stratonovitch P, Streck T, Supit I, Tao F, Thorburn PJ, Waha K, Wang E, Wallach D, Wolf J, Zhao Z, Zhu Y (2015)

Rising temperatures reduce global wheat production. Nat Clim Change 5(2):143–147. https://doi.org/10.1038/nclimate2470

Asseng S, Kassie BT, Labra MH, Amador C, Calderini DF (2017) Simulating the impact of source-sink manipulations in wheat. Field Crop Res 202:47–56. https://doi.org/10.1016/j.fcr.2016.04.031

Asseng S, Martre P, Maiorano A, Rötter RP, O'Leary GJ, Fitzgerald GJ, Girousse C, Motzo R, Giunta F, Babar MA, Reynolds MP, Kheir AMS, Thorburn PJ, Waha K, Ruane AC, Aggarwal PK, Ahmed M, Balkovič J, Basso B, Biernath C, Bindi M, Cammarano D, Challinor AJ, De Sanctis G, Dumont B, Eyshi Rezaei E, Fereres E, Ferrise R, Garcia-Vila M, Gayler S, Gao Y, Horan H, Hoogenboom G, Izaurralde RC, Jabloun M, Jones CD, Kassie BT, Kersebaum K-C, Klein C, Koehler A-K, Liu B, Minoli S, Montesino San Martin M, Müller C, Naresh Kumar S, Nendel C, Olesen JE, Palosuo T, Porter JR, Priesack E, Ripoche D, Semenov MA, Stöckle C, Stratonovitch P, Streck T, Supit I, Tao F, Van der Velde M, Wallach D, Wang E, Webber H, Wolf J, Xiao L, Zhang Z, Zhao Z, Zhu Y, Ewert F (2019) Climate change impact and adaptation for wheat protein. Glob Chang Biol 25(1):155–173. https://doi.org/10.1111/gcb.14481

Awan FK, Khurshid MY, Afzal O, Ahmed M, Chaudhry AN (2014) Agro-morphological evaluation of some exotic common bean (Phaseolus vulgaris L.) genotypes under rainfed conditions of Islamabad, Pakistan. Pak J Bot 46(1):259–264

Bakhat HF, Bibi N, Zia Z, Abbas S, Hammad HM, Fahad S, Ashraf MR, Shah GM, Rabbani F, Saeed S (2018) Silicon mitigates biotic stresses in crop plants: a review. Crop Prot 104:21–34. https://doi.org/10.1016/j.cropro.2017.10.008

Bao Y, Hoogenboom G, McClendon R, Urich P (2015) Soybean production in 2025 and 2050 in the southeastern USA based on the SimCLIM and the CSM-CROPGRO-Soybean models. Clim Res 63(1):73–89. https://doi.org/10.3354/cr01281

Basso B, Liu L, Ritchie JT (2016) A comprehensive review of the CERES-Wheat, -Maize and -Rice Models' performances. In: Sparks DL (ed) Advances in agronomy, vol 136. Academic Press, London, pp 27–132. https://doi.org/10.1016/bs.agron.2015.11.004

Boote KJ (2011) Improving soybean cultivars for adaptation to climate change and climate variability. In: Yadav SS, Redden RJ, Hatfield JL, Lotze-Campen H, Hall EA (eds) Crop adaptation to climate change. Wiley-Blackwell, West Sussex, pp 370–395. https://doi.org/10.1002/9780470960929.ch26

Boote KJ, Jones JW, Hoogenboom G, Pickering NB (1998) The CROPGRO model for grain legumes. In: Tsuji G, Hoogenboom G, Thornton P (eds) Understanding options for agricultural production, Systems approaches for sustainable agricultural development, vol 7. Springer, Dordrecht, pp 99–128. https://doi.org/10.1007/978-94-017-3624-4_6

Boote KJ, Kropff MJ, Bindraban PS (2001) Physiology and modelling of traits in crop plants: implications for genetic improvement. Agric Syst 70(2):395–420. https://doi.org/10.1016/S0308-521X(01)00053-1

Boote KJ, Jones JW, Batchelor WD, Nafziger ED, Myers O (2003) Genetic coefficients in the CROPGRO–Soybean model: links to field performance and genomics. Agron J 95(1):32–51. https://doi.org/10.2134/agronj2003.3200

Boote KJ, Allen LH, Prasad PVV, Jones JW (2011) Testing effects of climate change in crop models. In: Handbook of climate change and agroecosystems. Imperial College Press, London, pp 109–129. https://doi.org/10.1142/9781848166561_0007

Boote KJ, Hoogenboom G, Ale S, Adams C, Shrestha R, Mvuyekure RF, Himanshu SK, Grover K, Angadi S (2023) Adapting the CROPGRO model to simulate growth and yield of guar, Cyamopsis tetragonoloba L, an industrial legume crop. Ind Crop Prod 197:116596. https://doi.org/10.1016/j.indcrop.2023.116596

Bosi C, Sentelhas PC, Pezzopane JRM, Santos PM (2020) CROPGRO-Perennial Forage model parameterization for simulating Piatã palisade grass growth in monoculture and in a silvopastoral system. Agric Syst 177:102724. https://doi.org/10.1016/j.agsy.2019.102724

Buttar G, Grover K, Thind H, Saroa G (2012) N-sparing benefit of clusterbean (Cyamopsis tetragonoloba) to Indian mustard (Brassica juncea) under semi-arid conditions of Northwestern India. Vegetos 25(2):313–319

Deepika, Dhingra HR (2014) Effect of salinity stress on morpho-physiological, biochemical and yield characters of cluster bean [Cyamopsis tetragonoloba (L.) Taub.]. Ind J Plant Physiol 19:393–398

Erice G, Sanz-Sáez A, Aranjuelo I, Irigoyen JJ, Aguirreolea J, Avice JC, Sánchez-Díaz M (2011) Photosynthesis, N_2 fixation and taproot reserves during the cutting regrowth cycle of alfalfa under elevated CO_2 and temperature. J Plant Physiol 168(17):2007–2014. https://doi.org/10.1016/j.jplph.2011.07.007

Etesami H, Jeong BR (2018) Silicon (Si): review and future prospects on the action mechanisms in alleviating biotic and abiotic stresses in plants. Ecotoxicol Environ Saf 147:881–896. https://doi.org/10.1016/j.ecoenv.2017.09.063

Fahad S, Hussain S, Saud S, Hassan S, Tanveer M, Ihsan MZ, Shah AN, Ullah A, Khan F, Ullah S (2016) A combined application of biochar and phosphorus alleviates heat-induced adversities on physiological, agronomical and quality attributes of rice. Plant Physiol Biochem 103:191–198

Gresta F, Sortino O, Santonoceto C, Issi L, Formantici C, Galante YM (2013) Effects of sowing times on seed yield, protein and galactomannans content of four varieties of guar (Cyamopsis tetragonoloba L.) in a Mediterranean environment. Ind Crop Prod 41:46–52

Gresta F, De Luca AI, Strano A, Falcone G, Santonoceto C, Anastasi U, Gulisano G (2014) Economic and environmental sustainability analysis of guar (Cyamopsis tetragonoloba L.) farming process in a Mediterranean area: two case studies. Ital J Agron 9(1):20–24

Guerra LC, Hoogenboom G, Garcia y Garcia A, Banterng P, Beasley JP Jr (2008) Determination of cultivar coefficients for the CSM-CROPGRO-peanut model using variety trial data. Trans ASABE 51(4):1471–1481. https://doi.org/10.13031/2013.25227

Hansen J, Ruedy R, Sato M, Lo K (2012) Global temperature in 2011, trends, and prospects. GISS surface temperature analysis. National Aeronautics and Space Administration

Hattori T, Inanaga S, Araki H, An P, Morita S, Luxová M, Lux A (2005) Application of silicon enhanced drought tolerance in Sorghum bicolor. Physiol Plant 123(4):459–466

Hoogenboom G, Porter CH, Shelia V, Boote KJ, Singh U, White JW, Hunt LA, Ogoshi R, Lizaso JI, Koo J, Asseng S, Singels A, Moreno LP, Jones JW (2017) Decision Support System for Agrotechnology Transfer (DSSAT) Version 4.7 (https://DSSAT.net). DSSAT Foundation, Gainesville

Hoogenboom G, Porter CH, Shelia V, Boote KJ, Singh U, Pavan W, Oliveira FAA, Moreno-Cadena LP, Ferreira TB, White JW, Lizaso JI, Pequeno DNL, Kimball BA, Alderman PD, Thorp KR, Cuadra SV, Vianna MS, Villalobos FJ, Batchelor WD, Asseng S, Jones MR, Hopf A, Dias HB, Hunt LA, Jones JW (2024) Decision Support System for Agrotechnology Transfer (DSSAT) version 4.8.2 (www.DSSAT.net). DSSAT Foundation, Gainesville, FL

Houghton RA, House JI, Pongratz J, van der Werf GR, DeFries RS, Hansen MC, Le Quéré C, Ramankutty N (2012) Carbon emissions from land use and land-cover change. Biogeosciences 9(12):5125–5142. https://doi.org/10.5194/bg-9-5125-2012

Hussain T, Akram Z, Shabbir G, Manaf A, Ahmed M (2021) Identification of drought tolerant chickpea genotypes through multi trait stability index. Saudi J Biol Sci. https://doi.org/10.1016/j.sjbs.2021.07.056

Ihsan MZ, Daur I, Alghabari F, Alzamanan S, Rizwan S, Ahmad M, Waqas M, Shafqat W (2019) Heat stress and plant development: role of sulphur metabolites and management strategies. Acta Agric Scand B Soil Plant Sci 69:332–342. https://doi.org/10.1080/09064710.2019.1569715

IPCC (2023) Summary for policymakers. In: Lee H, Romero J (eds) Climate change 2023: synthesis report. A report of the Intergovernmental Panel on Climate Change, Contribution of Working Groups I, II and III to the Sixth Assessment Report of the Intergovernmental Panel on Climate Change. IPCC, Geneva

Jabeen M, Gabriel HF, Ahmed M, Mahboob MA, Iqbal J (2017) Studying impact of climate change on wheat yield by using DSSAT and GIS: a case study of Pothwar Region. In: Ahmed M, Stockle OC (eds) Quantification of climate variability, adaptation and mitigation for agricultural sustainability. Springer International Publishing, Cham, pp 387–411

Jing Q, Qian B, Bélanger G, VanderZaag A, Jégo G, Smith W, Grant B, Shang J, Liu J, He W, Boote K, Hoogenboom G (2020) Simulating alfalfa regrowth and biomass in eastern Canada using the CSM-CROPGRO-perennial forage model. Eur J Agron 113:125971. https://doi.org/10.1016/j.eja.2019.125971

Jones JW, Hoogenboom G, Porter CH, Boote KJ, Batchelor WD, Hunt LA, Wilkens PW, Singh U, Gijsman AJ, Ritchie JT (2003) The DSSAT cropping system model. Eur J Agron 18(3–4):235–265. https://doi.org/10.1016/S1161-0301(02)00107-7

Jukanti AK, Bhatt R, Sharma R, Kalia RK (2015) Morphological, agronomic, and yield characterization of cluster bean (Cyamopsis tetragonoloba L.) germplasm accessions. J Crop Sci Biotechnol 18:83–88

Khan A, Ahmad M, Shah MKN, Ahmed M (2020) Performance of wheat genotypes for morpho-physiological traits using multivariate analysis under terminal heat stress. Pak J Bot 52(6). https://doi.org/10.30848/PJB2020-6(30)

Kuravadi NA, Verma S, Pareek S, Gahlot P, Kumari S, Tanwar UK, Bhatele P, Choudhary M, Gill KS, Pruthi V, Tripathi SK, Dhugga KS, Randhawa GS (2013) Guar: an industrial crop from marginal farms. In: Bhullar GS, Bhullar NK (eds) Agricultural sustainability. Academic Press, San Diego, pp 47–60. https://doi.org/10.1016/B978-0-12-404560-6.00003-4

Liu B, Martre P, Ewert F, Webber H, Waha K, Thorburn PJ, Ruane AC, Aggarwal PK, Ahmed M, Balkovič J, Basso B, Biernath C, Bindi M, Cammarano D, Cao W, Challinor AJ, De Sanctis G, Dumont B, Espadafor M, Eyshi Rezaei E, Fereres E, Ferrise R, Garcia-Vila M, Gayler S, Gao Y, Horan H, Hoogenboom G, Izaurralde RC, Jabloun M, Jones CD, Kassie BT, Kersebaum KC, Klein C, Koehler A-K, Maiorano A, Minoli S, Montesino San Martin M, Müller C, Naresh Kumar S, Nendel C, O'Leary GJ, Eivind Olesen J, Palosuo T, Porter JR, Priesack E, Ripoche D, Rötter RP, Semenov MA, Stöckle C, Stratonovitch P, Streck T, Supit I, Tao F, Van der Velde M, Wang E, Wolf J, Xiao L, Zhang Z, Zhao Z, Zhu Y, Asseng S (2023) AgMIP-Wheat multi-model simulations on climate change impact and adaptation for global wheat. Open Data J Agric Res 9:10–25. https://doi.org/10.18174/odjar.v9i0.18092

Lubbe A, Verpoorte R (2011) Cultivation of medicinal and aromatic plants for specialty industrial materials. Ind Crop Prod 34(1):785–801

Malik W, Dechmi F (2019) DSSAT modelling for best irrigation management practices assessment under Mediterranean conditions. Agric Water Manag 216:27–43. https://doi.org/10.1016/j.agwat.2019.01.017

Martre P, Yin X, Ewert F (2017) Modeling crops from genotype to phenotype in a changing climate. Field Crop Res 202:1–4. https://doi.org/10.1016/j.fcr.2017.01.002

Meftahizadeh H, Ghorbanpour M, Asareh MH (2019) Comparison of morphological and phytochemical characteristics in guar (Cyamopsis tetragonoloba L.) landraces and cultivars under different sowing dates in an arid environment. Ind Crop Prod 140:111606. https://doi.org/10.1016/j.indcrop.2019.111606

Menzel A, Yuan Y, Matiu M, Sparks T, Scheifinger H, Gehrig R, Estrella N (2020) Climate change fingerprints in recent European plant phenology. Glob Chang Biol 26(4):2599–2612. https://doi.org/10.1111/gcb.15000

Messina CD, Jones JW, Boote KJ, Vallejos CE (2006) A gene-based model to simulate soybean development and yield responses to environment. Crop Sci 46(1):456–466. https://doi.org/10.2135/cropsci2005.04-0372

Mudgil D, Barak S, Khatkar BS (2014) Guar gum: processing, properties and food applications—a review. J Food Sci Technol 51:409–418

Nadeem M, Nazer Khan M, Abbas G, Fatima Z, Iqbal P, Ahmed M, Ali Raza M, Rehman A, Ul Haq E, Hayat A, Ali M, Ahmad S (2022) Application of CSM-CANEGRO model for climate change impact assessment and adaptation for sugarcane in semi-arid environment of Southern Punjab, Pakistan. Int J Plant Prod. https://doi.org/10.1007/s42106-022-00192-6

Nigam S, Upadhyaya H, Chandra S, Rao RN, Wright G, Reddy A (2001) Gene effects for specific leaf area and harvest index in three crosses of groundnut (Arachis hypogaea). Ann Appl Biol 139(3):301–306 van Ogtrop F, Ahmad M, Moeller C (2014) Principal components of sea surface temperatures as predictors of seasonal rainfall in rainfed wheat growing areas of Pakistan. Meteorol Appl 21(2):431–443. https://doi.org/10.1002/met.1429

Patil D (2014) Genetic variability and sowing dates effect of cluster bean (Cyamopsis tetragonoloba L., Taub) genotypes in semi arid region of Maharashtra, India. Plant Arch 14(1):1–6

Saud S, Fahad S, Yajun C, Ihsan MZ, Hammad HM, Nasim W, Amanullah J, Arif M, Alharby H (2017) Effects of nitrogen supply on water stress and recovery mechanisms in Kentucky bluegrass plants. Front Plant Sci 8:983–983. https://doi.org/10.3389/fpls.2017.00983

Shaban M, Lak M, Hamidvand Y, Nabaty E, Khodaei F, Yarahmadi M, Azimi S, Rahmat M, Motlagh Z (2012) Response of chickpea (Cicer arietinum L.) cultivars to integrated application of zinc nutrient with water stress. Int J Agric Crop Sci 4(15):1074–1082

Shrestha R, Adams CB (2023) Photosynthesis in guar: recovery from water stress, basic parameter estimates, and intrinsic variation among germplasm. J Crop Improv 37(5):626–646. https://doi.org/10.1080/15427528.2022.2121348

Shrestha R, Adams CB, Ravelombola W, MacMillan J, Trostle C, Ale S, Hinson P (2021) Exploring phenotypic variation and associations in root nodulation, morphological, and growth character traits among 50 guar genotypes. Ind Crop Prod 171:113831. https://doi.org/10.1016/j.indcrop.2021.113831

Shrestha R, Adams CB, Abello F, DeLaune PB, Trostle C, Rajan N, Ale S, Ravelombola W (2023) Intensifying dryland wheat systems by integrating guar increased production and profitability. Ind Crop Prod 197:116608. https://doi.org/10.1016/j.indcrop.2023.116608

Siddique KH, Johansen C, Turner NC, Jeuffroy M-H, Hashem A, Sakar D, Gan Y, Alghamdi SS (2012) Innovations in agronomy for food legumes. A review. Agron Sustain Dev 32:45–64

Singh P, Boote KJ, Kumar U, Srinivas K, Nigam SN, Jones JW (2012) Evaluation of genetic traits for improving productivity and adaptation of groundnut to climate change in India. J Agron Crop Sci 198(5):399–413. https://doi.org/10.1111/j.1439-037X.2012.00522.x

Singh P, Nedumaran S, Boote KJ, Gaur PM, Srinivas K, Bantilan MCS (2014a) Climate change impacts and potential benefits of drought and heat tolerance in chickpea in South Asia and East Africa. Eur J Agron 52(Part B):123–137. https://doi.org/10.1016/j.eja.2013.09.018

Singh P, Nedumaran S, Ntare BR, Boote KJ, Singh NP, Srinivas K, Bantilan MCS (2014b) Potential benefits of drought and heat tolerance in groundnut for adaptation to climate change in India and West Africa. Mitig Adapt Strat Glob Change 19(5):509–529. https://doi.org/10.1007/s11027-012-9446-7

Singh P, Nedumaran S, Traore PCS, Boote KJ, Rattunde HFW, Prasad PVV, Singh NP, Srinivas K, Bantilan MCS (2014c) Quantifying potential benefits of drought and heat tolerance in rainy season sorghum for adapting to climate change. Agric For Meteorol 185:37–48. https://doi.org/10.1016/j.agrformet.2013.10.012

Singla S, Grover K, Angadi SV, Schutte B, VanLeeuwen D (2016) Guar stand establishment, physiology, and yield responses to planting date in southern New Mexico. Agron J 108(6):2289–2300

Slafer GA, Savin R (1994) Source—sink relationships and grain mass at different positions within the spike in wheat. Field Crop Res 37(1):39–49. https://doi.org/10.1016/0378-4290(94)90080-9

Suriharn B, Patanothai A, Pannangpetch K, Jogloy S, Hoogenboom G (2007) Determination of cultivar coefficients of peanut lines for breeding applications of the CSM-CROPGRO-Peanut model. Crop Sci 47(2):607–619. https://doi.org/10.2135/cropsci2006.01.0050

Suriharn B, Patanothai A, Pannangpetch K, Jogloy S, Hoogenboom G (2008) Yield performance and stability evaluation of peanut breeding lines with the CSM-CROPGRO-Peanut model. Crop Sci 48(4):1365–1372. https://doi.org/10.2135/cropsci2007.09.0523

Suthar JD, Rajpar I, Ganjegunte GK, Zia ul H (2018) Comparative study of early growth stages of 25 guar (Cyamopsis tetragonoloba L.) genotypes under elevated salinity. Ind Crop Prod 123:164–172. https://doi.org/10.1016/j.indcrop.2018.06.045

Tahmouzi S, Meftahizadeh H, Eyshi S, Mahmoudzadeh A, Alizadeh B, Mollakhalili-Meybodi N, Hatami M (2023) Application of guar (Cyamopsis tetragonoloba L.) gum in food technologies: a review of properties and mechanisms of action. Food Sci Nutr 11(9):4869–4897. https://doi.org/10.1002/fsn3.3383

Tao F, Rötter RP, Palosuo T, Díaz-Ambrona CGH, Mínguez MI, Semenov MA, Kersebaum KC, Nendel C, Cammarano D, Hoffmann H, Ewert F, Dambreville A, Martre P, Rodríguez L, Ruiz-Ramos M, Gaiser T, Höhn JG, Salo T, Ferrise R, Bindi M, Schulman AH (2017) Designing future barley ideotypes using a crop model ensemble. Eur J Agron 82:144–162. https://doi.org/10.1016/j.eja.2016.10.012

Tardieu F, Simonneau T, Muller B (2018) The physiological basis of drought tolerance in crop plants: a scenario-dependent probabilistic approach. Annu Rev Plant Biol 69(1):733–759. https://doi.org/10.1146/annurev-arplant-042817-040218

Ul'yanich P, Belimov A, Kuznetsova I, Sazanova A, Yuzikhin O, Laktionov YV, Karlov D, Vishnyakova M, Safronova V (2022) Effectiveness of nitrogen-fixing symbiosis of guar (Cyamopsis tetragonoloba) with strains Bradyrhizobium retamae RCAM05275 and Ensifer aridi RCAM05276 in pot experiment. Agric Biol 57(3):555–565

Vyas S (2014) Impact and strategies for yield improvement of arid legumes under drought. Int J App Life Sci Eng 1:12–19

Wang L, Zhang L-M (2009) Viscoelastic characterization of a new guar gum derivative containing anionic carboxymethyl and cationic 2-hydroxy-3-(trimethylammonio)propyl substituents. Ind Crop Prod 29(2):524–529. https://doi.org/10.1016/j.indcrop.2008.10.003

Waraich EA, Ahmad R, Ashraf M (2011) Role of mineral nutrition in alleviation of drought stress in plants. Aust J Crop Sci 5(6):764–777

White JW, Hoogenboom G (2003) Gene-based approaches to crop simulation. Agron J 95(1):52–64. https://doi.org/10.2134/agronj2003.5200

Zambrano F, Despinoy P, Ormenese R, Faria E (2004) The use of guar and xanthan gums in the production of 'light' low fat cakes. Int J Food Sci Technol 39(9):959–966

Zamir MS, Khan MA, Hussain M, Haq I, Khan MK, Zaman Q, Afzal U, Islam N, Asim M, Ali I (2016) Quantitative behaviour of guar (Cyamopsis tetragonolobus L.) to various tillage systems and mulches and soil physical properties. Am J Plant Sci 7(7):1040–1045

The manufacturer's authorised representative in the EU is Springer Nature Customer Service Centre GmbH, Europaplatz 3, 69115 Heidelberg, Germany. If you have any concerns regarding our products, please contact ProductSafety@springernature.com

Printed and bound by CPI Group (UK) Ltd, Croydon, CR0 4YY
01/10/2026
02225864-0016